Fifth edition

CARL P. SWANSON
Ray Ethan Torrey Professor of Botany Emeritus
University of Massachusetts

PETER L. WEBSTER
Associate Professor of Botany
University of Massachusetts

Prentice-Hall, Inc., Englewood Cliffs, New Jersey 07632

Library of Congress Cataloging in Publication Data

Swanson, Carl P.
The cell.

Includes bibliographies and index.
1. Cytology. I. Webster, Peter L.,
1940- . II. Title.
QH581.2.S89 1985 574.87 84-17812
ISBN 0-13-121799-2
ISBN 0-13-121781-X (pbk.)

Cover Illustration. Nucleus from a living endosperm cell of the wild cucumber, *Echinocystis lobata*. The nucleus measures over 200 microns in diameter; two nucleoli and strands of chromatin are visible. (Courtesy of Professor Seymour Shapiro, Dept. of Botany, Univ. of Mass., Amherst, Mass.)

Editorial/production supervision and
interior design: Fay Ahuja
Cover design: Joe Curcio
Manufacturing buyer: John Hall

Printed in the United States of America

10 9 8 7 6 5 4 3 2 1

ISBN 0-13-121799-2 01
ISBN 0-13-121781-X {PBK.} 01

PRENTICE-HALL INTERNATIONAL, INC., *London*
PRENTICE-HALL OF AUSTRALIA PTY. LIMITED, *Sydney*
EDITORA PRENTICE-HALL DO BRASIL, LTDA., *Rio de Janeiro*
PRENTICE-HALL CANADA INC., *Toronto*
PRENTICE-HALL HISPANOAMERICANA, S.A., *Mexico*
PRENTICE-HALL OF INDIA PRIVATE LIMITED, *New Delhi*
PRENTICE-HALL OF JAPAN, INC., *Tokyo*
PRENTICE-HALL OF SOUTHEAST ASIA PTE. LTD., *Singapore*
WHITEHALL BOOKS LIMITED, *Wellington, New Zealand*

to D.N.S. and M.A.W.

CONTENTS

Preface

The success of the past four editions leads us to believe that we have been able to convey to our audience the dynamism as well as the substance of the field of cell biology; we trust that the changes made in this edition are no less successful. We recognize the difficulty of the challenge. New techniques and concepts leapfrog over each other today with almost bewildering rapidity, making any attempt to give an introductory overview a task not to be taken lightly. The seven years since the last edition have been especially fruitful ones in the cellular area. To meet this challenge we have brought every chapter and its list of appended references up to date, increased the size of the volume, strengthened the chemical aspects, and added pertinent illustrations when appropriate.

Attention is called to two aspects. Those familiar with cell biology as a discipline will recognize that there are three levels of references. The appended lists of textbooks and monographs of general cytological interest are for the general reader who might wish to consider other approaches to the subject, and thus to determine whether we are alone in our emphases and biases. The appended lists of journals and reviews are for those who desire, in a general way, to keep up with the broad trends; the bibliography at the end of each chapter is for those who wish to pursue a particular subject to greater depth or to gain some sense of historical perspective. Cell biology is not a narrow and isolated field. The information emerging, satisfying as it is in a basic sense, finds application in areas as diverse as medicine, agriculture, and the manufacture of industrial products, while the subject itself is constantly being broadened and deepened by information flowing in from biophysics and biochemistry as well as being reinforced by that arising out of the more traditional fields of biology.

The second aspect to which we call attention is that the final chapter dealing with the evolution of cells is pitched at a somewhat higher level than is the remainder of the volume. It is an area that is of interest to us, but it may be omitted without detriment to the subject as generally presented.

An introductory textbook, by its very nature, is historical; it deals with the past. Inevitably, therefore, we are mindful of our debt to those—unnamed for the most part, but somewhere in the background—on whose data and concepts we depend, and who through their efforts have shaped and molded the field. We hope that we have dealt fairly and accurately with their contributions. We express our thanks to our colleagues for their generosity in providing us with appropriate illustrations. The revised manuscript has been read by a number of reviewers. Their criticisms, gratefully received, have made the volume more accurate, pertinent, and readable. Any errors, however, reflect on no one other than ourselves.

Carl P. Swanson
Peter L. Webster

1

The Cellular Basis of Life

Every science has one or more periods of time when it seems to burst its bounds and exhibit an extraordinary pattern of growth. The stimulus for such change is usually a new discovery or a recently advanced theory which casts doubt on long-held beliefs, poses new problems, or allows old problems to be viewed in a totally different light. Since the early 1950s, this kind of change has characterized biology, and it would not be unreasonable to say that a veritable explosion of biological knowledge has occurred, with every facet of the science being affected. Further, it is not simply that the volume of biological knowledge has increased tremendously, although this is the case; the character of the information generated also has been different, and it in turn has altered the structure of the biological sciences. New instrumentation and techniques, more appropriate experimental organisms, and, more important, new ways of thinking and of asking critical and testable questions have all contributed to this burst of knowledge, and have influenced the direction, character, and promise of the biological sciences. Much of this recent activity falls within the broad areas of molecular and cellular biology, with new ideas and new techniques leading to the emergence of exciting and productive subdisciplines.

The significance of the information generated during the past quarter century is as great philosophically as it is scientifically. As human beings we have an interest in knowing who we are, where we exist in space and time, and how we came to be what and where we are. We recognize individual uniqueness, not only in ourselves and our fellow human beings, but in the plants and animals around us; the biological sciences can now supply a rational explanation for this uniqueness, not only in more readily understandable morphological and behavioral terms, but also in some instances in unambiguous, concrete, molecular terms. The purpose of this volume is to explore the cell, which we now know to be the organizational basis of this uniqueness.

It is through examination of the intimate details of nature that the unity, continuity, and diversity of the living world are made manifest. It is, however, one thing to recognize details and quite another to understand them. Understanding requires that we ask questions capable of being answered, devise critical experiments to prove or disprove some point of ambiguity or uncertainty, and construct testable hypotheses and theories that bind our isolated facts and observations into meaningful wholes. As a result of these observations, experiments, and speculations, we have come to recognize that the world around us, living and nonliving, is an ordered system governed by certain rules that we call natural laws. Science, in fact, makes sense only if it is assumed that nature is orderly.

The function of science, therefore, is to uncover that order and attempt to explain it. In the process the information acquired can often be exploited for social gain or ill—nuclear power is an example. It is frequently said that scientific information is amoral in character, and that the use or misuse of that information is a social and not a scientific judgment, but we need also to remember that science is a social operation carried out in a particular social environment, that scientists are members of their society, and that they should participate in, and contribute to the resolution of the problems raised by their discoveries.

Our contact with the world around us is through our five senses of touch, taste, hearing, smell, and sight. They are a part of the equipment which came into being as the result of evolutionary processes, and which enable us to come to terms with our environment, to apprehend and visualize the world of reality external to us; we have come to realize that this world is made up of matter and energy, which we detect, define, describe, and sometimes measure in specific ways. With our unaided senses and in unsubtle ways, we usually have no difficulty in distinguishing the sky and the land from the water, a gas and a solid from a liquid, the living from the nonliving. On a more refined level, we can distinguish degrees of roughness, intensity, and shade of color (if we are not color-blind), and an acid taste from one that is salty, sweet, or bitter. But human powers of sensory discrimination are limited. We all know, for example, that water, steam, and ice are made up of molecular H_2O, and that they have different characteristics, but our ordinary senses cannot tell us *why* this is so except to indicate that the differences are correlated with temperature. We hear only within a certain range of sound waves, and see only that portion of the light spectrum called the visible region (Figure 1.1). When we try to go beyond these limits, we can no longer directly perceive the physical nature of things. We, of course, turn to instruments, experiments, and speculation to try to understand that which we perceive, but when we attempt to penetrate areas outside the sphere naturally circumscribed by our senses, we are restricted to instruments to record that which we cannot.

Try to imagine how much of your knowledge of yourself and of the universe around you has been gained *only* through use of your five senses, as compared to

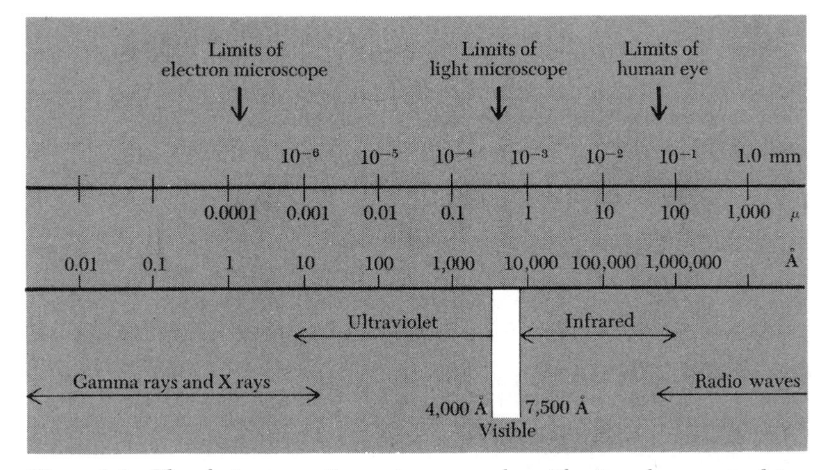

Figure 1.1 The electromagnetic spectrum on a logarithmic scale, measured in millimeters, microns, and angstrom units: 1 μm (micrometer) = 0.001 mm (millimeter) = 10,000 Å (Angstroms). The approximate lower limits of resolution of the human eye, the light microscope, and the electron microscope are given.

that derived through instruments and/or experiments. With knowledge only from your senses, you would find your world shrinking appreciably in size and diversity. The heavens, for example, would be a blue vault studded with stars at night, but at distances that could only be guessed at; light-years, galaxies, black holes, and neutron stars would be terms without meaning. The world of cells and microorganisms would disappear from your frame of reference, and the structure of all other organisms, including yourself, would be understood only in terms of larger, readily visible portions. We do, however, possess the means whereby we can extend the use of our senses beyond their normal capabilities. On a clear night, for example, the naked eye can discern stars that number only in the several thousands. But powerful telescopes show that the Milky Way, the galaxy of which our solar system is a part, contains billions of stars similar to our sun, while the universe as a whole contains billions of comparable galaxies. Light and electron microscopes, on the other hand, are concerned not with great distances but rather with the world of the minute, the microcosm of viruses, bacteria, cells, and large molecules, all of them ordinarily invisible because of their size. Similarly, photographic plates, more sensitive to certain portions of the electromagnetic spectrum (Figure 1.1) than are our eyes, extend the limits of our "visibility" so that we can detect the long infrared rays and radio waves on one side of the spectrum and the short ultraviolet rays, X rays, and gamma rays on the other.

Whatever means we use, we attempt to define the "things" we observe in terms of the units of which they are composed, and the more refined our knowledge, the more powerful and discriminating our instruments and techniques, the more precise become our definitions of these units—their limits, basic nature, and modes of aggregation into larger units. It would, indeed, be impossible for you to read these pages without understanding letters, the basic symbolic units of our language, the numbers that make up our decimal or metric systems, or the rules by which they are put together to express a thought or a quantitative value. One of the first goals of a science, therefore, whether it be physics, chemistry, or biology, is to determine the uniqueness of the units with which it is concerned, for unless such units are understood and accepted by everyone in a particular field, meaningful communication is difficult and scientific knowledge in that field cannot progress.

THE CELL DOCTRINE

The basic unit of life, the *cell*, is a physical entity; life, with its ability to reproduce, mutate, and respond to stimuli, does not exist in smaller units of matter. We can break cells apart and, by centrifugation, extract selected portions for study much as the physicist breaks up atoms. We find that these cellular fragments can carry on many of their activities for a time; they may consume oxygen, ferment sugars, and even form new molecules. But these activities individually do not constitute life, any more than the behavior of a subatomic particle is equivalent to the behavior of an intact atom. The disrupted cell is no longer capable of continuing life indefinitely; we therefore conclude that the cell is the most elementary unit that can sustain life, even though, as we shall see, the cell is highly complex. Viruses, on the other hand, are smaller and less complex than cells, but they cannot maintain life independently of the cells in which they exist (page 19).

Compared to the atom and the molecule, the cell is a unit of far greater size and complexity. It is a microcosm having a definite boundary, within which constant chemical activity and a flow of energy proceed. At ordinary temperatures, a chemically quiescent cell is dead. The cytologist, therefore, seeks to identify the kinds of cells that exist, to understand their organization and structure in terms of their activities and functions, and to visualize the cell not only as a total entity (as, for example, the unicellular bacterium), but also as an integral part of the elaborate organs and organ systems of multicellular plants and animals.

The now familiar idea that the cell is the basic unit of life is known as the *cell doctrine*. It is essentially a statement of fact, not a hypothesis of a debatable or controversial nature. It developed gradually through microscopical observations of the structure of many plants and animals, and eventually the presence of cells as a common structural feature of all biological organization was recognized. The Englishman Robert Hooke first saw the remains of dead cells in 1665 in a piece of cork as he was using his newly invented microscope (Figure 1.2), and he coined the word "cell" to describe the tiny structures, thinking that they resembled the unadorned cells occupied by monks. By 1800 good microscopes (Figure 1.3) were becoming available, as were techniques for the preserving and staining of cells, and there was general acceptance of the idea that organisms were cellular; however,

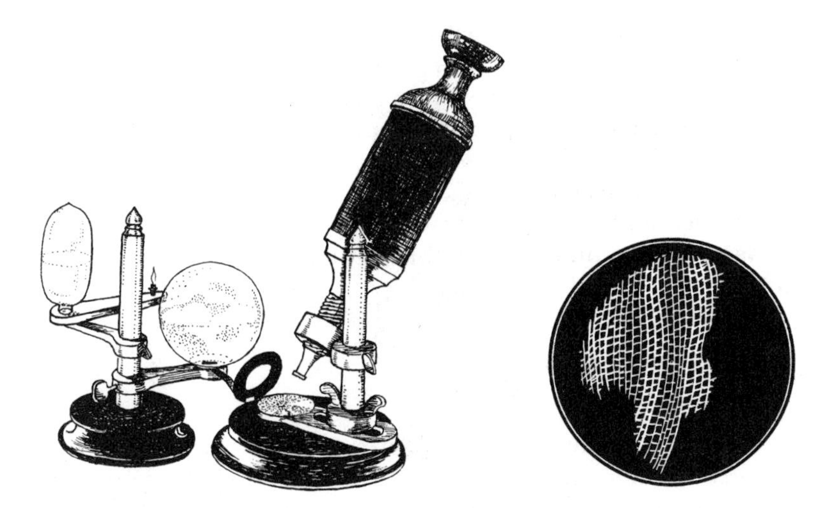

Figure 1.2 Robert Hooke's microscope with which he observed the microscopic structure of cork and his drawing of it (in circle). Here, in his own words, is a description of his experiment: "I took a good clear piece of Cork and with a Pen-knife sharpen'd as keen as a razor, I cut a piece of it off, and thereby left the surface of it exceeding smooth, then examining it very diligently with a Microscope, me thought I could perceive it to appear a little porous; but I could not so plainly distinguish them as to be sure that they were pores. . . . I with the same sharp pen-knife cut off from the former smooth surface an exceeding thin piece of it, and placing it on a black object Plate. . . . and casting the light on it with a deep plano-convex Glass, I could exceedingly plainly perceive it to be all perforated and porous, much like a Honeycomb, but that the pores of it were not regular . . . these pores, or cells, were not very deep, but consisted of a great many little Boxes, separated out of one continued long pore by certain Diaphragms . . . Nor is this kind of texture peculiar to Cork only; for upon examination with my Microscope, I have found that the pith of an Elder, or almost any other Tree, the inner pulp or pith of the Cany hollow stalks of several other Vegetables: as of Fennel, Carrets, Daucus, Bur-docks, Teasels, Fearn . . . & c. have much such a kind of Schematisme, as I have lately shewn that of Cork."

there was also a good deal of confusion over the definition of cells, the significance of their contents and cell walls, their mode of origin, and their role in organization. In 1838 and 1839, two German scientists, botanist M. J. Schleiden and zoologist Theodor Schwann, took the loose threads of ideas and observations available to them and wove them into a doctrine, which stated that cells with nuclei were the structural and functional basis of all living things. Many of their ideas concerning cellular structure, function, and origin have been proved erroneous, but by emphasizing the importance of the cell, they gave coherence to the biological thought of their time and focused attention on the one structure that had to be understood if biology was to advance beyond its purely descriptive stage. We now recognize that a study of the cells of such very different organisms as bacteria, orchids, and

Figure 1.3 A microscope in the collection of the museum of the History of Science, Oxford University, Oxford, England made about 1770 by George Adams for King George III of England, an active patron of the arts and sciences. Compare this with the microscopes of Robert Hooke and of today. An equally ancient microscope, made of bronze and once the gift of the French King Louis XV to his mistress Madame de Pompadour was sold at auction in 1976 for $74,000.

humans aids in the understanding of the structure and function of all organisms, an impossibility prior to the acceptance of the universality of the cell doctrine. Who would have thought, before this period, that a human being and an orchid had anything in common? It is now considered that the cell doctrine, however vague its beginnings and however long it took for formulation, ranks with Darwin's *theory of evolution through natural selection* and the *theory of the gene* as one of the foundation stones of modern biology.

Some 20 years after the announcements of Schleiden and Schwann, Rudolf Virchow, a great German physician, made another important generalization: *cells come only from preexisting cells.* When biologists further recognized that sperm and ova are also cells that unite with each other in the process of fertilization, it gradually became clear that life from one generation to another is an uninterrupted succession of cells. Growth, development, metabolism in all of its aspects, inheritance, evolution, disease, aging, and death are, therefore, but varied aspects of cellular behavior, even though each of these phenomena also can be viewed at higher and lower levels of biological organization.

Most generalizations have exceptions that cast doubt on their universal validity. This is true for the cell doctrine, but we shall not consider these exceptions until after we have examined cellular structure in some detail. Let us now see what the cell doctrine, as presently interpreted, embodies in the way of solid ideas. There are essentially three:

First, as we have already mentioned, the cell doctrine states that life exists only in cells; organisms are, therefore, made up of cells; the activity of an organism is dependent on the activities of cells, individually and collectively; and the cell is the basic unit through which matter and energy are acquired, converted, stored, and utilized, and in which biological information is stored, released, manipulated and expressed. Second, the cell doctrine has embodied within it the idea that the continuity of life has a cellular basis, which is another way of stating Virchow's generalization. Now, however, we can be more explicit, adding that genetic continuity in a very exact sense includes not only the cell as a whole but also some of its smaller components, such as genes and chromosomes, as well as the hereditary mechanism for transmitting its genetic substance to the next generation. The nature of viruses, as we shall see later, also reinforces the concept that the whole cell is the basic unit of heredity. A virus possesses genes and a chromosome, but it cannot reproduce itself without the aid of the cell it infects. Third is the idea that there is a relationship between structure and function. This has been called the principle of complementarity; it means, briefly, that orderly behavior and orderly structures are intimately related to each other, and that within the domain of cells the biochemical activities of cells occur within, and indeed are determined by, structures organized in a definite way. We shall encounter this idea again in our discussion of cellular components.

André Lwoff, the French microbiologist, has expressed the cell doctrine in yet another way:*

> When the living world is considered at the cellular level, one discovers unity. Unity of plan: each cell possesses a nucleus imbedded in protoplasm. Unity of function: the metabolism is essentially the same in each cell. Unity of composition: the main macromolecules of all living beings are composed of the same small molecules. For, in order to build the immense diversity of living systems, nature has made use of a strictly limited number of building blocks. The problem of diversity of structures and functions, the problem of heredity, and the problem of diversification of species have been solved by the elegant use of a small number of building blocks organized into specific macromolecules. . . . Each macromolecule is endowed with a specific function. The machine is built for doing precisely what it does. We may admire it, but we should not lose our heads. If the living system did not perform its task, it would not exist. We have simply to learn how it performs its task.

CELLS AND ENERGY

In the preceding statement Lwoff describes the cell as a "machine" that performs a task. More than likely, the word "machine" will evoke an image that is more a reflection of our technological and mechanical age than of the living world as we

* André Lwoff, *Biological Order* (Cambridge, Mass.: The MIT Press, 1962), pp. 11, 13.

know it. We touch a switch to flood a room with light, and we know, if we stop to think about it at all, that the switch is somehow connected to an outside source of electrical energy, with that energy being converted, by resistance in a filament, to incandescent heat, which we then see as radiant energy. A comparable thing happens when we start an automobile: a turn of a switch sets into motion a chain of events that converts the chemical energy of gasoline into heat energy, which, through expansion of gases, is translated into the mechanical energy for driving the gears and wheels of the machine. Switches, filaments, motors, and wheels are devices through which energy flows, or by which energy is controlled for a given purpose.

There is no switch that can turn life on or off; life is not that simple or that obviously mechanical. We may not think ourselves, or any other organism, to be as predictable or as automatic in response or performance as an electrical circuit or a gasoline-driven motor, but living systems, like machines, also manipulate energy in a controlled way. Like machines, living organisms are characterized by a high degree of order of structure and behavior, and it is this orderliness that either permits, or is necessary for, the controlled manipulation of energy. The maintenance of this necessary order in turn requires a continual input of energy; life, therefore, is basically an energy-capturing, energy-converting, and energy-consuming process, and it can continue only as long as a supply of appropriate kinds of energy is available (Figure 1.4).

Energy can be defined as the capacity to do work. Energy exists in a number of forms, and the work performed can be of many kinds: lifting, running, and an engine in action are obvious forms of work, while chemical processes often are less readily detectable. The latter are key factors in the maintenance of life; there is no cellular activity of any sort, however subtle it might be, that is not related to some chemical, energy-driven process.

The first law of thermodynamics states that energy cannot be created or destroyed. It can, however, be changed from one form to another, with each change leaving the energy in a form less able to do work. Thus the chemical energy bound up in the molecules of muscles can be converted into the mechanical energy required for the contraction needed for lifting, with a good part of the energy released to the outside as heat energy. The latter cannot be recovered to perform work, and hence is lost. The second law of thermodynamics tells us that all physical and chemical processes tend toward equilibrium, that is, they proceed from an ordered to a random, or disordered, state. Once a system reaches a state of equilibrium, the energy within it is said to be maximally randomized. A measure of the degree of randomization of energy is described as the *entropy* of the system; an ordered system has a low entropy value, a randomized one a high value. When in the latter state, the energy is less available to do work. Thus we can consider our universe to be "running down" in the sense that the energy that can be used to perform work is becoming less and less available. That component of the total energy that is available to perform work—that is, the useful energy—is known as *free energy*. For example, the energy of a well-defined system undergoing chemical

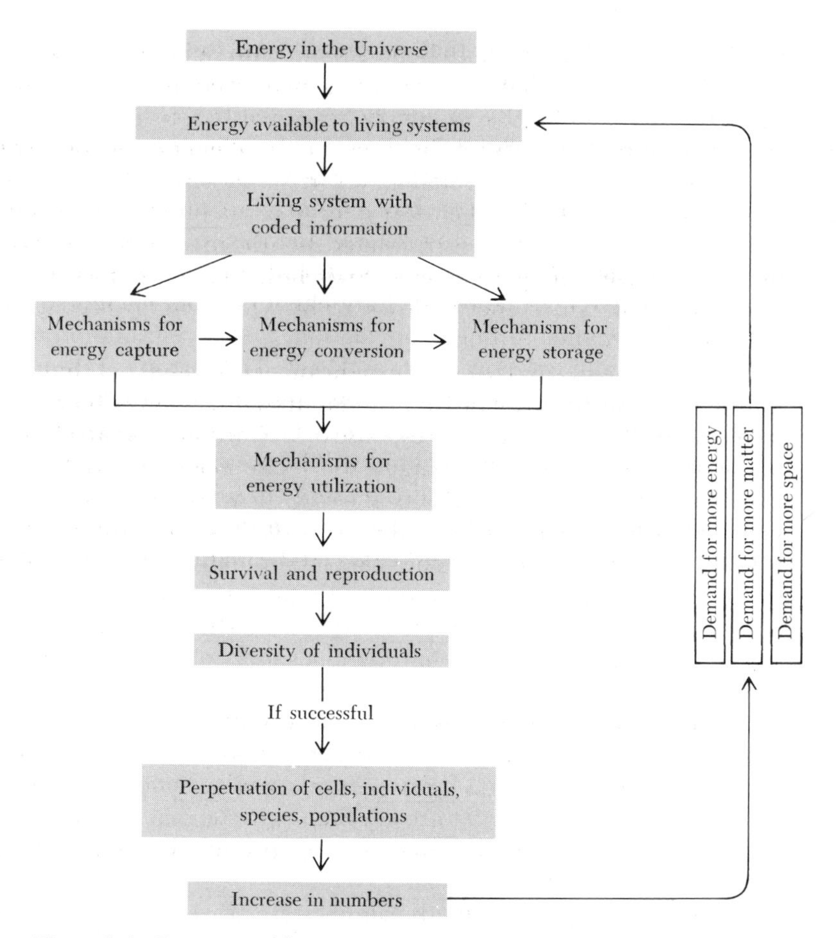

Figure 1.4 Dynamics of living systems viewed as an energy problem, with the coded information within the cell governing the movement and manipulation of energy through ordered channels. It should be recognized that since structure and behavior are intimately related, the mechanisms governing the flow of energy are as much a reflection of the coded information of the cell as is the flow of energy itself. It should also be realized that the diagram reflects hierarchies of explanatory convenience more so than physical realities, although an attempt has been made to depict the flow of energy as sequentially accurate as possible.

change can be measured at the start and at the end of the process. No process is totally efficient, so some of the energy will be lost to the surrounding environment, while additional energy is involved in changing the system from one state or condition to another. The difference in energy content of the system before and after the completion of the process is free or released energy; it is that which was required to do work, and thus to bring about the change. Once used, the energy is less available, with the result that processes tend toward a state of minimum free energy and a state of maximum entropy (Figure 1.5).

The following forms of energy are listed from highest to lowest in terms of free energy (or from lowest to highest entropy): gravitation, nuclear reactions, sunlight (radiant energy), chemical reactions, and microwave radiation. In a closed system, the direction of change from one form to another can *only* be toward increasing entropic values; the reverse direction of change is thermodynamically impossible. The primary energy of the universe, from which all other kinds of energy are derived, is that of gravitational attraction, leading to the contraction of massive objects and, through contraction, to the degradation of this energy into chemical reactions, motion, light, and heat. The entropy of the universe as a whole is increased by this degradative process, and the universe is, therefore, inexorably moving toward a state of inevitable death in which all energy will be maximally degraded, and no work of any kind will be possible. Life is an incidental if only a temporary beneficiary since it is the energy of sunlight that enabled living systems to come into being, and thereafter to continue to build order out of disorder through the process of *photosynthesis* and the manipulation of chemical energy. As

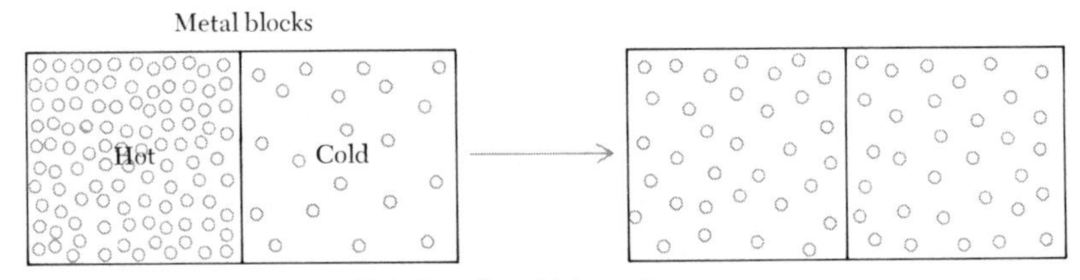

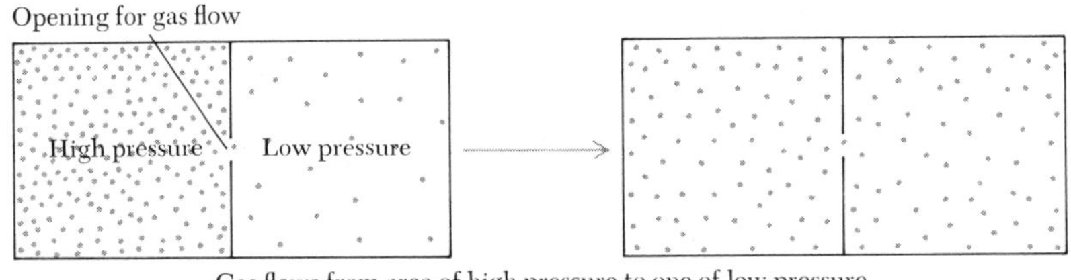

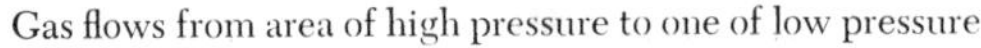

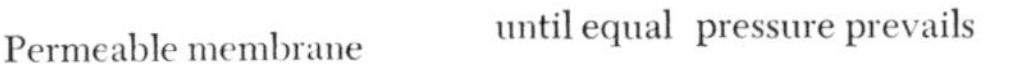

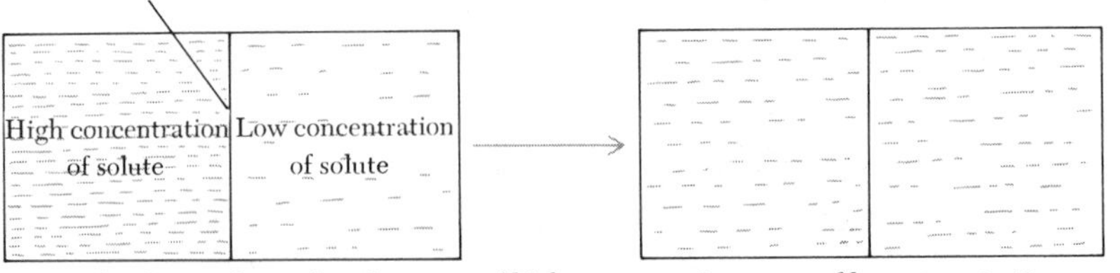

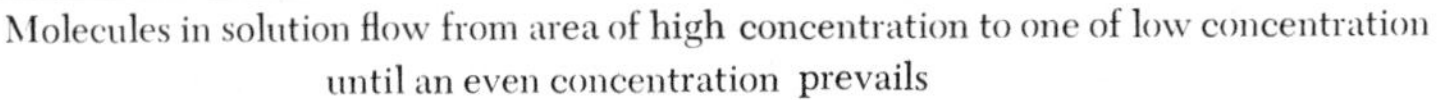

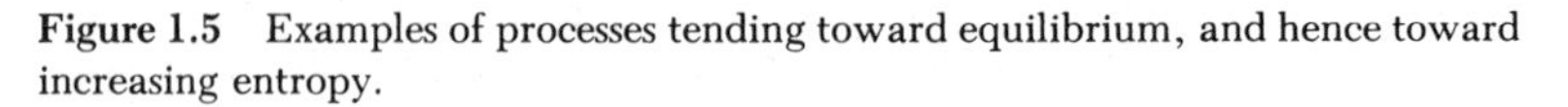

Figure 1.5 Examples of processes tending toward equilibrium, and hence toward increasing entropy.

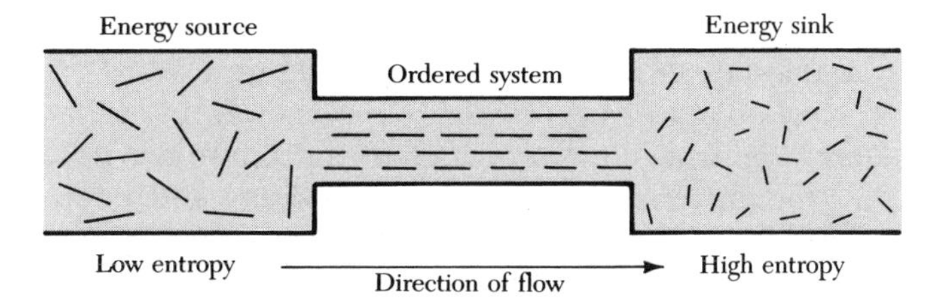

Figure 1.6 Diagram of the flow of energy in an open system from a source whose entropy is low to a sink whose entropy is higher. Not only is work performed during this process of change, that is, order is introduced by the flow of energy, but the system through which the flow takes place facilitates the ease and direction of flow. Life, therefore, can be considered an ordered system in which energy flows from a source (the radiant energy in sunlight) toward an energy sink (heat, or microwave energy) (after Morowitz, 1975).

Figure 1.6 indicates, life is an open, not a closed, system. The green plant, for example, maintains a high free energy content by trapping the external radiant energy of sunlight to build chemical bonds, which in turn lead to the formation of complex but orderly molecules and structures; animals maintain a similarly high free energy status, but they do so by utilizing a lower form of energy, that bound up as chemical energy in the organic molecules of their diet. They consequently are not only less efficient as energy utilizers when compared to green plants, but they also contribute more significantly to an increase in the entropy of the environment (Figure 1.7). Man is a prime example of this fact, particularly in the present human technological state with its high rate and inefficient use of energy. Life in general, however, delays an increase in entropy for a small fraction of the substance of this planet, and as long as sufficient sunlight is available it should be able to do so indefinitely. This may be a minor feature of no consequence in the total energy picture of the universe, but it is obviously of supreme importance to us as part of the living world.

Physical and chemical processes that proceed toward equilibrium are thought of as closed systems, although a completely closed system of change in which no energy is gained or lost is an ideal, not an actual one (with the possible exception of the universe itself). By comparison, life is an open system. It does not tend toward equilibrium for it is constantly energy-consuming. Life achieves a steady state—a dynamic situation far from equilibrium in that there is a continual interaction with the environment, with energy and matter being taken in to be converted into living substance, and energy and matter being given off, generally in a different form (Figure 1.6).

The remarkable feature of life, therefore, is not that it seems to function in a qualitatively different manner from the remainder of the universe, but that living systems are programmed to be order-creating both as to their structure and, intimately related to this, their management of energy. Life increases the free energy state of its own substance and lowers its entropy, a seeming contradiction of the

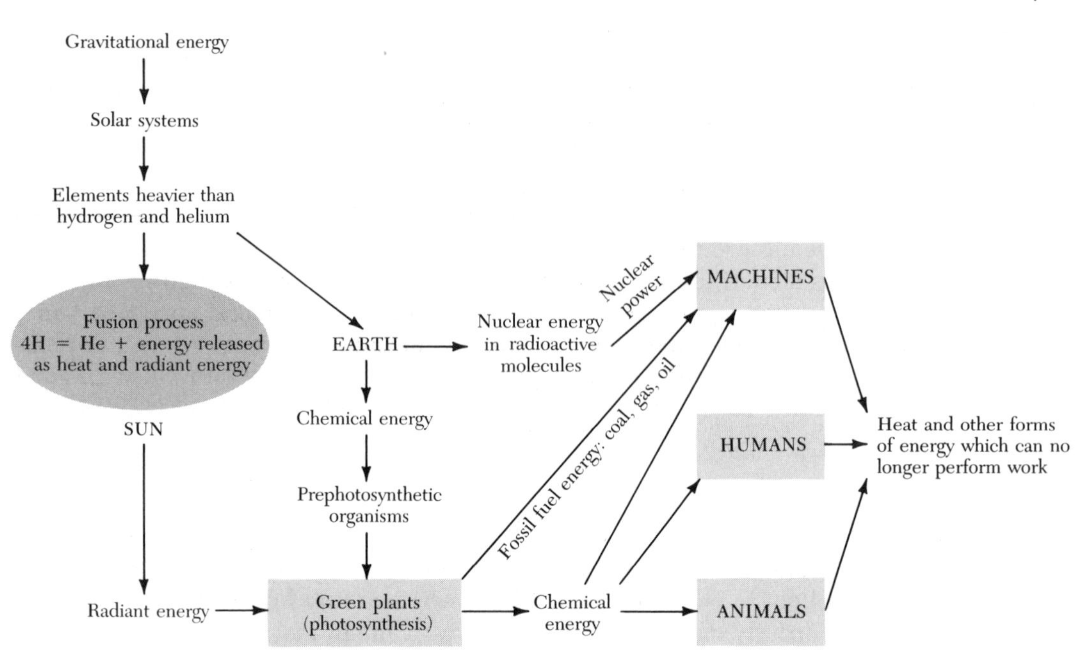

Figure 1.7 A diagram to illustrate the place of the green plants, animals, and humans (with their machines of civilization) in the energy picture of this planet. Gravitational energy, which produced the sun and governs its activities, has the lowest level of entropy; entropy increases as one proceeds toward the right of the diagram, until the point is reached where the character of the energy precludes the further performance of work.

laws of thermodynamics. Functioning as they do, however, living systems do not violate any physical law, since the programmed order represented by the maintenance and reproduction of life is "paid for" by energy commanded from the environment, and at the cost of a similar or greater increase in the entropy of the environment. Life, therefore, must be viewed as a dynamic, or steady-state system, not one that is at equilibrium with its environment.

What is true for life itself is true for the cell, since it is the basic unit of living organization. In a very general way, Figure 1.4 diagrams the channels of energy flow within and between cells, and between cells and their environment. Each step is an orderly process associated with particular and highly ordered intracellular structures, and we shall find that their organization and behavior are intimately related to the kind and manner of energy being managed. The diversity of organisms in the living world, with most possessing many kinds of cells, suggests that there are likely to be variations among cells as to the patterns of energy flow and the structures associated with this flow. Variations have been found, mainly in structures, but the basic energy patterns are surprisingly similar in all forms of life.

The last feature indicated in Figure 1.4 is that the energy management of living systems is not a haphazard one; rather it is directed by information coded within a particular portion of the cell. Life, with all of its varied attributes, was made possible in the long-distant past when matter somehow acquired a particular pattern of organization within which energy from the environment could be acquired and manipulated. Once life arose, therefore, it acquired a measure of control over its environment, and in a sense took command of certain external energy sources and directed their use to its own advantage. Evolution, which through the course of time has produced the diversity of cells and organisms now alive, can be thought of as a continuing experiment in the creation of new structures for, and new ways of, energy management, with each new experimental approach being determined by changes that take place in the informational code. We will be dealing with all of these aspects in subsequent chapters, but we might well, with good reason, subscribe to the words of the poet William Blake—"Energy is Eternal Delight."

CELLS IN GENERAL

Our purpose then, in the chapters that follow, is to examine cells by whatever techniques or instrumentation are needed to uncover and to understand their structure and function. There is no single ideal cell that serves our purpose, so whatever cell most appropriately illustrates the point under discussion will be used. This can be done without fear of bias or distortion because of the unity of plan, function, and composition of which Lwoff speaks: what is learned from one kind of cell can be applied, sometimes directly, sometimes with modification, to other kinds of cells. We do, on the other hand, recognize two general classes of cells: the *prokaryotic cell* typical of bacteria and blue-green algae, and the *eukaryotic cell* found in all other organisms, plant and animal.

The prokaryotic cell is the simplest kind known and, from what the fossil record can tell us, probably the first to come into existence perhaps 3 to $3\frac{1}{2}$ billion years ago. As a rule, these cells are small in dimensions: from 0.1 to 0.25 μm (micrometers) among the mycoplasmas, the smallest cells known (Figure 1.8); a few micrometers in length and somewhat less in width among the bacteria (Figure 1.9); and a bit larger in the Cyanobacteria, or blue-green algae (Figure 1.10). The living portion of the cells of bacteria and blue-greens is limited externally by a *plasma membrane*, outside which a more or less rigid cell wall and a jellylike, mucilaginous capsule or sheath are present. The composition of the wall, which determines the infectivity of some pathogenic groups, varies with the particular prokaryotic species; the bacterial wall, for example, contains lipids, carbohydrates, and complexes of mucopeptides derived from amino acids and amino sugars; the same is true for the wall of the blue-greens, although structurally it is more complex than those of bacterial cells.

The cellular contents consist of a less electron dense nuclear area and a very dense cytoplasm. Thin, tangled fibers, about 3 to 5 nm (nm = nanometer, 1/1000

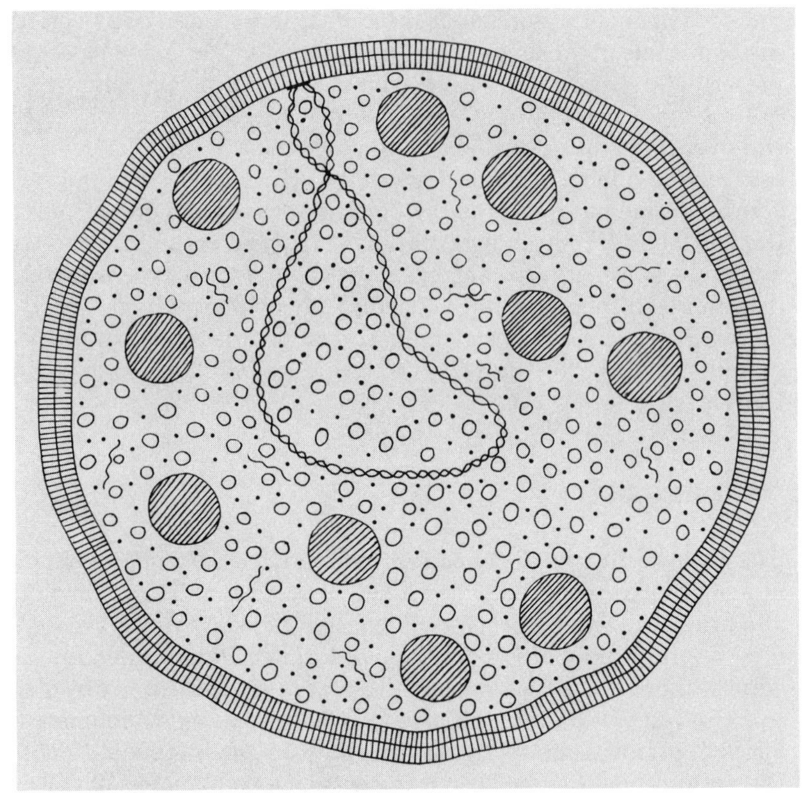

Figure 1.8 Schematic drawing of the morphology of a cell of mycoplasma, one of the smallest cells known. It consists of an outer plasma membrane, a circular double helix of genetic material attached to a structure in the plasma membrane called the mesosome, ribosomes (larger spheres), and other dissolved substances (smaller spheres and threads). The diameter of the cell is about 0.1 μm.

of a micron) in diameter, traverse the clear nuclear area. They are readily extractable, at least from bacteria, and it is known that they are strands of deoxyribonucleic acid (DNA), the hereditary material of the cell. The blue-green algae and some bacteria possess layered membranes that are involved in photosynthesis and that appear to be derived from infoldings of the plasma membrane. The photosynthetic pigment in bacteria is bacteriochlorophyll; the blue-green algae possess chlorophyll *a* (rarely chlorophyll *b*) and phycocyanin, the latter a blue pigment. *Ribosomes*, small, rounded structures about 15 to 20 nm in diameter, are also found in the cytoplasm, but no other internal structure is characteristic of the prokaryotic cell. Some species possess flagella, but their internal structure is quite different from that of a eukaryotic flagellum, lacking, as they do, the 9 + 2 fibrillar structure (see Figures 12.5 and 12.8, Chapter 12). The lack of visible structure in

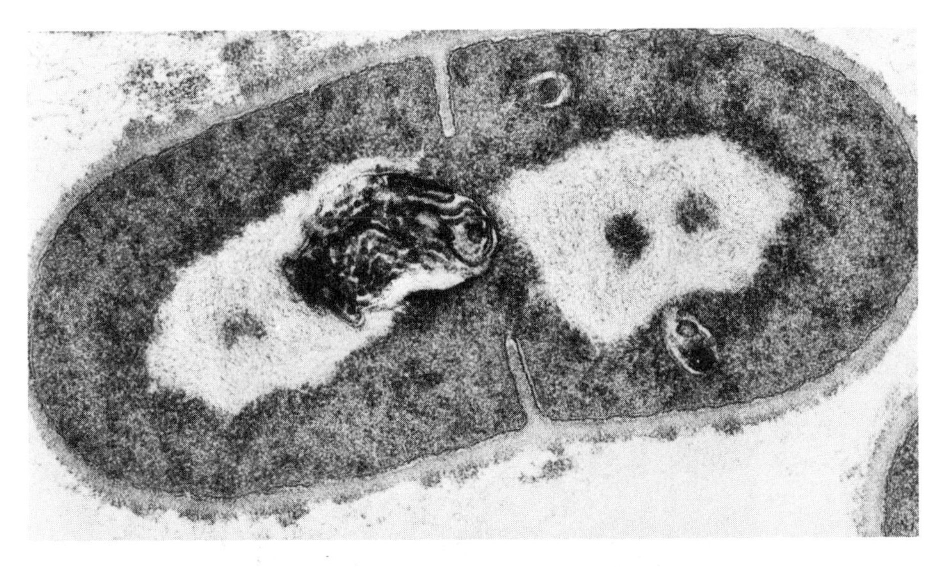

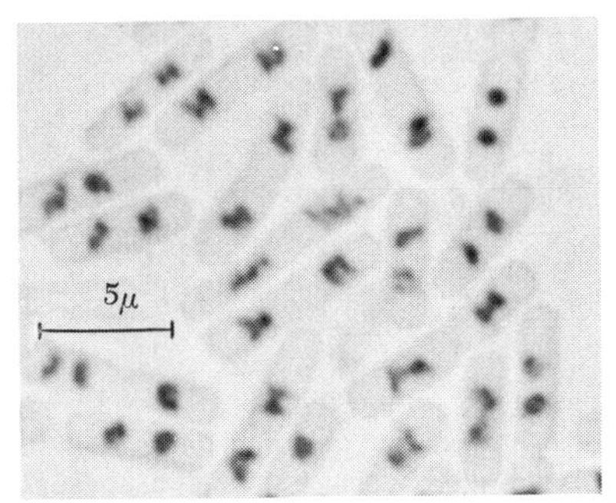

Figure 1.9 Cells of bacteria. Left: cells of *Bacillus cereus*, stained to reveal the nuclear areas under the light microscope. The cells will divide by fission, a process of pinching the cell in two between the nuclear areas (see also Figure 12.6). Above: electron micrograph of a dividing cell of *Bacillus megatherium* showing the rather heavy cell wall, the plasma membrane that lies just beneath the wall, the light nuclear areas containing thin, twisted strands of DNA, the relatively undifferentiated cytoplasm, which is, however, very rich in ribosomes, and the mesosomes (the large membrane-bound bodies), which are formed by in-foldings of the plasma membrane. (Courtesy of Dr. Stanley C. Holt.)

the cytoplasm may be somewhat misleading, however, for the prokaryotes as a group are versatile organisms, deceptively complex as to their biochemical activities and fully capable of carrying on an independent existence in the right environment.

The bacteria divide by simple fission (Figure 1.9). The DNA of the nuclear area appears to be attached to a *mesosome*, a structural feature of the plasma membrane. When the cell is about to divide, this is initiated by division of the mesosome; the two halves then separate and DNA, the hereditary material, is pulled apart into two masses, which will form the nuclear areas of the two

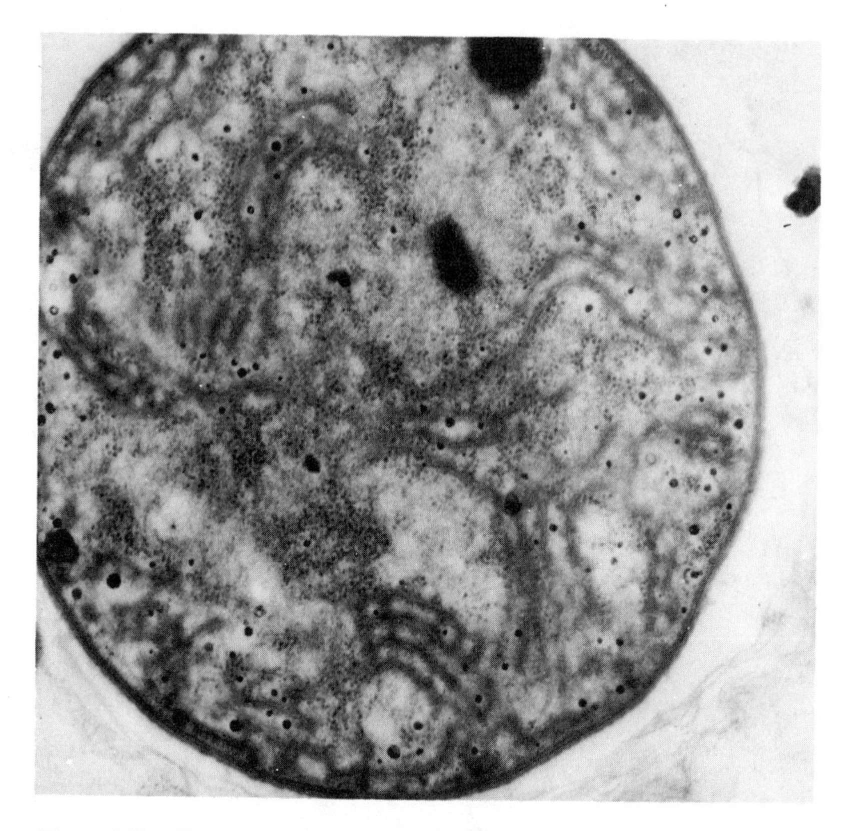

Figure 1.10 Electron micrograph of a cell of the blue-green alga, *Nostoc*, showing the photosynthetic membranes and free ribosomes. The nuclear area is not clearly defined in this illustration.

daughter cells. Whether this is also the manner of division of all the prokaryotes is not known.

Bacteria and blue-greens show an additional similarity in their ability to form resting spores under adverse conditions. A spore coat, at least in bacteria, forms around the nuclear area and a small amount of cytoplasm, and the cell can remain in a state of almost suspended animation until favorable conditions cause the spore to germinate. As a spore, the bacterium is most resistant to adverse conditions, even surviving temperatures of up to 120°C for a brief period. The blue-greens, on the other hand, may actually thrive in hot springs with temperatures ranging up to 70°C; they are responsible for the colored formations seen at such springs, the colors coming from varied accessory pigments.

The eukaryotic cell is a far more elaborately structured and partitioned unit than is the prokaryotic cell from which it is presumably derived (Figure 1.11). An internal division of labor has taken place, accomplished by the use of membranes. The exterior of the cell is bounded by a plasma membrane, to which, in the case of

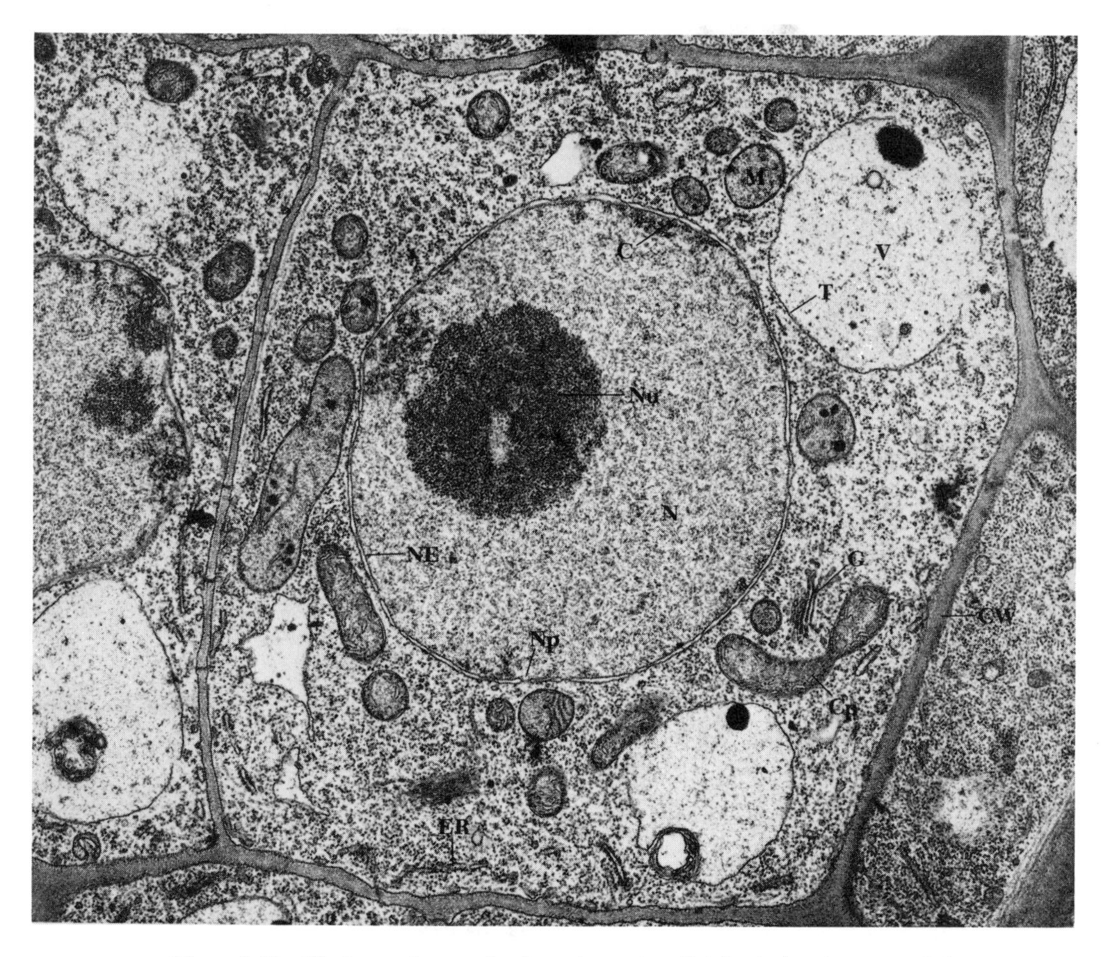

Figure 1.11 Electron micrograph of a eukaryotic cell (plant) showing most of the membranous structures of the cell. N: nucleus; Nu: nucleolus; NE: nuclear envelope; Np: pore in the nuclear envelope; C: chromatin concentrated at the nuclear envelope (the granular aspect of the nucleus is also due to the presence of chromatin); M: mitochondrion; Cp: chloroplast; G: Golgi bodies; ER: endoplasmic reticulum; CW: cell wall; V: vacuole; T: tonoplast, the membrane surrounding the vacuole. The granular aspect of the cytoplasm is due to the presence of many ribosomes.

plant cells, an outer wall of cellulose and other materials has been added; the hereditary material is enclosed in a *membrane-bound nucleus*, and is segmented into complex nucleoprotein bodies, or *chromosomes*, the number of which is characteristic for each species; *mitochondria* convert the chemical energy of carbohydrates and fats into energy forms which the rest of the cell can use when needed; the complex cytoplasmic membrane systems of the *endoplasmic reticulum* and *Golgi apparatus* are concerned with the synthesis and packaging of the macromolecules needed for cellular structure and function; sunlight is trapped by green plants in membrane-bound and internally layered *chloroplasts*, and converted into, and sometimes stored as, the chemical energy of carbohydrates; and *vacuoles*, *lysosomes*, and *peroxisomes*, all membrane-enclosed, play additional roles in the life of eukaryotic cells. Only the ribosomes, chromosomes, *microtubules*, and *microfibrils* are not membranous in nature. Each of these structures mentioned will be examined in detail in subsequent chapters, but while it is obvious that membranes and membrane-bound structures are concerned with energy manipulations in a direct way, so too are all other structures with the exception of the chromosomes. They are bearers of cellular information, but even they are concerned with energy, although as somewhat remote directors rather than manipulators. The expression of chromosomal information takes many forms, so we should anticipate that cellular structures must be flexible as to use and diverse in character to participate in the many reactions that take place in the eukaryotic

TABLE 1.1 FEATURES OF PROKARYOTIC AND EUKARYOTIC CELLS.

Feature	Prokaryote	Eukaryote
Plasma membrane	Present	Present
Nuclear membrane	Absent	Present
Mitochondria	Absent	Present
Endoplasmic reticulum	Absent	Present
Golgi apparatus	Absent	Present
Ribosomes	Present	Present
Cell wall	Present, composed of amino sugars and muramic acid	Lacking in animals; present in plants, with cellulose a major component
Capsule	When present, of mucopolysaccharides	Absent
Vacuoles	Absent	Present (particularly in plants)
Lysosomes	Absent	Present
Chromosomes	Single naked structure composed only of DNA	More than one present, and composed of DNA and proteins
Photosynthetic apparatus	Membranes with chlorophyll *a* and rarely chlorophyll *b*, and phycocyanin in blue-greens, bacterio-chlorophyll in bacteria	Chloroplasts with chlorophyll *a* and *b* in stacked grana
Flagella	Present in some species, but lacking 9 + 2 fibrillar structure	Present in some species, but possessing 9 + 2 fibrillar structure
Division	Simple fission	Spindle of microtubules

cell. We need to remember, however, that the simplicity of the prokaryotic cell's structure is not a reflection of simplicity of biochemical activity.

Table 1.1 summarizes the existing differences between prokaryotic and eukaryotic cells. There is, as far as we know, no graded series of cells that extends from the prokaryotic mycoplasmas to the complex eukaryotic cells of higher plants and animals. It is probable that the latter were derived evolutionarily from the former, or, more likely, that those forms existing today derived from a common ancestor, but if so the intermediate stages are missing or unknown. We will see, however, in subsequent chapters that both chloroplasts and mitochondria have their own DNA and ribosomes, and hence a degree of hereditary independence from the remainder of the eukaryotic cell. It has been suggested that both chloroplasts and mitochondria, on structural and functional grounds, were cellular invaders that took up residence in other cells, and that during the course of evolution have become adapted to a symbiotic existence. We will examine this question more fully in the last chapter.

VIRUSES, PLASMIDS, AND VIROIDS

Viruses are not cells in a structural sense, lacking cytoplasm as well as the organelles of other cells. However, so much information about how organisms function in an hereditary way has been derived from viruses that they warrant consideration in any discussion of cells.

Viruses exist in a wide variety of shapes, sizes, and complexity, and they infect or inhabit, in highly specific patterns, a wide range of host cells. Some of them, the bacteriophages (literally, bacteria-eaters), infect bacteria (Figure 1.12). Some restrict themselves to plant species, such as TMV, the tobacco mosaic virus, or the tobacco necrosis virus. The mosaic virus exists as long rods; the necrosis sometimes forms a crystalline complex of many individual viruses when precipitated from solution (Figure 1.13). Still others infect only animal cells. Smallpox, chicken pox, measles, mumps, and poliomyelitis are among the diseases of humans caused by animal viruses, while still others appear to be responsible for the transformation of normal into malignant cells.

As a group, the viruses exhibit certain characteristics. Some characteristics set them apart from cells, others enable an investigator to treat them experimentally as if they were cells or organisms. First, they are extremely small, able to pass through a filter that would strain out even the tiniest prokaryotic cells. Their presence in a filtrate that was infective led to their discovery long before they were pictured in an electron microscope. Thousands may be formed within a single bacterium, although in being so prolific they may lyse (kill) the cell at the time of their release. Others, such as ribonucleic acid (RNA) tumor viruses, are small enough to be released without disrupting the cell. Size, therefore, cannot be a criterion for distinguishing viruses from other elements within the cell, for there are, in fact, many molecules of biological origin that are larger than the very small viruses.

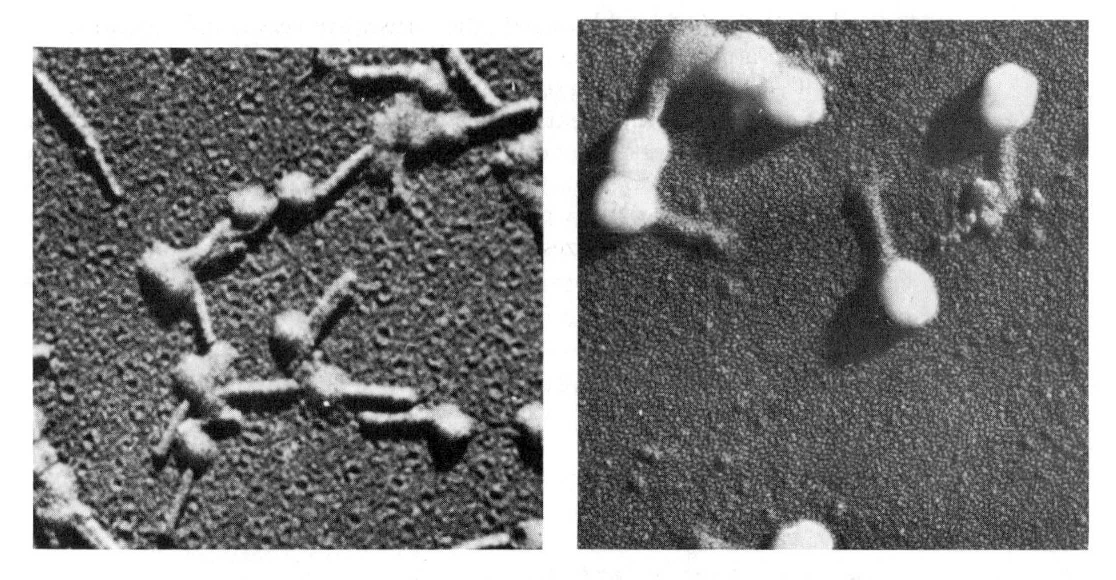

Figure 1.12 Electron micrographs of bacterial viruses. Both of these DNA-containing bacteriophages, P2 on the left and T6 on the right, attack the colon bacterium, *Escherichia coli*. Their detailed structure is indicated in Figure 1.13.

Figure 1.13 Electron micrographs of two RNA-containing viruses, which infect plants. Left: tobacco necrosis virus. The individual viral particle is spherical and about 250 Å in diameter, but when precipitated with ammonium sulfate they characteristically form a crystalline structure. Right: tobacco mosaic virus. Each virus is rod-shaped, with the protein on the outside and RNA on the inside (see Figure 1.15 for detailed structure).

Second, as their size would suggest, the viruses are very simply constructed compared to even a mycoplasma cell. Basically, they consist of a central core of nucleic acid (either DNA or RNA, a related ribonucleic acid) and an outer covering of protein. The nucleic acid represents their hereditary potential, while the protein serves two purposes: a protective covering for the nucleic acid when outside of the host cell, and a means of specificity for recognizing the kind of cell it can infect. The different ways of packing the two kinds of macromolecules provide the viruses with their distinctive shapes and sizes (Figures 1.14 and 1.15).

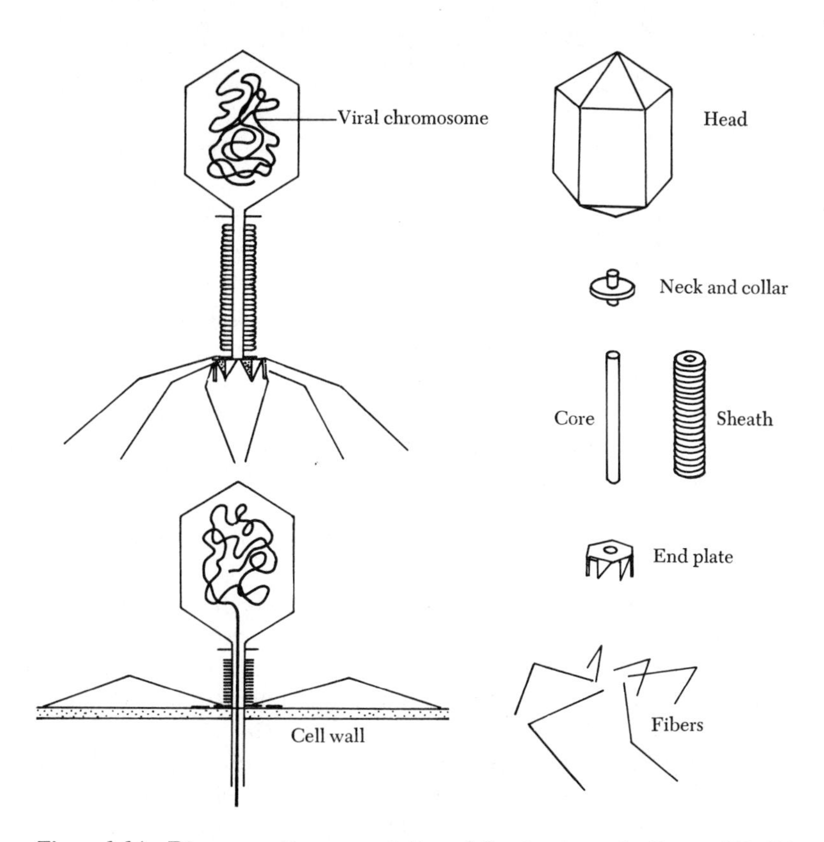

Figure 1.14 Diagrammatic representation of the structure of a T-even (T2, T4, T6) bacteriophage, with the intact virus, its component parts, and as it appears attached to the bacterial wall. The DNA is contained in the protein head. After attachment to the bacterial wall by the action of the end plate and fibers, the protein sheath contracts, driving the core through the wall; the DNA then passes from the head, through the hollow core, and into the interior of the cell where it can either multiply or become attached to the bacterial chromosome. If it multiplies, the bacterial contents will be used up in the process, and the wall will burst to release the virus particles into the medium where they can infect other cells. If the viral DNA attaches itself to the chromosome, it becomes an integral part, replicating along with the remainder of the chromosome, but without interfering with the behavior or integrity of the bacterial cell.

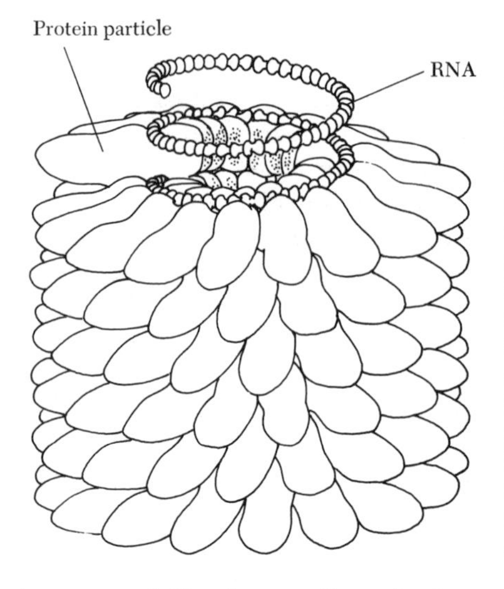

Figure 1.15 Diagrammatic representation of the structure of the tobacco mosaic virus. The protein particles, each having a molecular weight of 17,000, are assembled helically around the RNA molecule, eventually forming a hollow rod about 3,000 Å long, and a diameter of about 170 Å. It is possible to dissociate the protein and RNA. The RNA of the virus is infective by itself, the protein is not; the RNA, therefore, is the crucial hereditary molecule, the protein is more or less protective. If the RNA and the protein are added separately in solution, they will aggregate spontaneously to form typical TMV rods.

Third, viruses differ from cells, whether prokaryotic or eukaryotic, in that they are unable to lead an independent existence. They must infect a cell before reproduction can occur; otherwise, they are simply inert chemicals packaged in a special way. Their dependent existence as obligate cellular parasites suggests that to reproduce they must somehow commandeer the machinery of the invaded cell and make use of it for their own purposes. The host cell and the invading virus are, therefore, intimately related to each other, although how this relationship was achieved is not known. Some viruses, in fact, are so lacking in hereditary capabilities that they possess only enough information to code for their own protein coat.

Plasmids differ from viruses in that they lack a protein coat, but are similar in that they consist of small, double-stranded circles of DNA existing in the cytoplasm of certain bacterial species. Unlike viruses, however, they are not pathogenic, and as many as 1000 may be found in a single bacterium without detrimental consequences to the host cell. They enter and leave the bacterial cell through *pili*, slender projecting strands of the cell membrane and cytoplasm. One of them, the F, or sex, factor, possesses genes governing the formation of the pili, and thus facilitates the transfer not only of itself from cell to cell, but also any genetic material attached to it. Some plasmids possess genes for antibiotic resistance, or for forming the protein colicin, which may be lethal to other bacterial strains. Instead of being pathogenic, therefore, they provide the cell with genes not normally found in the host cell, but because they are dispensable, generally lead an independent existence in the cytoplasm, and do not replicate synchronously with the host chromosome, they are not considered to be a second chromosome.

The transfer of genes by plasmids from one cell to another probably occurs frequently in nature, but the discovery and use of this property has been taken advantage of by geneticists, who have devised ways of inserting genes from a wide variety of sources into a plasmid, and thus of introducing them into a bacterium, where they can continue to function. Genes coding for insulin, for example, have been handled in this manner, with the bacterium acting essentially as a miniature chemical factory. These aspects of genetic engineering will be dealt with at length in a later chapter (pages 120–124), but it is clear that the potential significance of the technique in medicine and industry is great.

Viroids are the most recently discovered nucleic acid particles found as cytoplasmic inclusions, in this instance in the cells of higher plants. They are very much smaller than the smallest viruses (Table 1.2), and are naked like plasmids, but unlike the latter they consist of circular or linear molecules of single-stranded RNA rather than DNA (ribose nucleic acid differs slightly from DNA in its nucleotide composition). Like viruses, however, they are pathogenic, causing such diseases as spindle-tuber disease of potatoes (PSTV), exocortis, an infectious disease of citrus trees, and chlorotic mottling of chrysanthemums. Their presence in animals has been suspected, but not yet proven, and it may well be that scrapie, a well-known disease of sheep and long thought to be caused by an elusive virus or viroid, is due to a *prion*, a proteinaceous particle isolated from the brains of infected sheep. The origin of this protein, and its mode of replication, are not yet understood, and it may be that this protein, like other cellular proteins, will eventually be traced back to a nucleic acid source which codes for it.

Viruses, plasmids, and viroids are not cells. In many respects, however, they

TABLE 1.2 COMPARISON OF THE TYPE AND AMOUNT OF NUCLEIC ACID FOUND IN REPRESENTATIVE CELLS, VIRUSES, PLASMIDS, AND A VIROID.

	Type of nucleic acid*	Amount of nucleic acid in nucleotides (n.) or nucleotide pairs (n.p.)
Cells		
Escherichia coli	d-s DNA	3.235×10^6 n.p.
Haemophilus influenzae	d-s DNA	2.47×10^6 n.p.
Viruses		
Bacteriophage T7	d-s DNA	3.25×10^5 n.p.
Adenovirus 2	d-s DNA	3.5×10^4 n.p.
Poliovirus	s-s RNA	ca. 8,000 n.
Tobacco mosaic virus	s-s RNA	ca. 6,100 n.
ϕX174	s-s DNA	5,386 n.
Bacteriophage R17	s-s RNA	ca. 3,000 n.
Plasmids		
F (sex factor)	d-s DNA	ca. 55,000 n.p.
ColE1	d-s DNA	ca. 3,500 n.p.
Viroid		
PSTV	s-s RNA	359 n.

* d-s = double-stranded; s-s = single-stranded (see Chapter 4 for the meaning of strandedness, and of nucleotides).

can be handled experimentally as if they were cells or unicellular organisms. Their relation to disease in both plants and animals makes them of much importance in human affairs, but for problems of heredity, mutability, immunity, biochemistry, and molecular biology in general, they are remarkably useful experimental objects. Their heredity is bound up in nucleic acid molecules, so that an understanding of their inheritance patterns is possibly transferable to higher forms; they mutate like any cell or organism and their genes are similar even if their patterns of hereditary transmission may be unique; their very simplicity of structure and their limited hereditary potential are advantages in trying to correlate structure and function; and some of them, such as the bacteriophages and plasmids, possess prodigious powers of reproduction, with thousands of individual particles being produced within a very short period of time after infection.

The origin of these cytoplasmic inclusions in the evolutionary scheme of life is uncertain. Their dependence upon living cells as a culture medium for growth and reproduction would indicate that they could not have preceded cellular organisms in origin. This fact, plus their lack of usual cellular machinery, would suggest that they might be very degenerate cells or fragments of cells, having somehow rid themselves of all unnecessary features except their hereditary apparatus in the form of nucleic acids and their protective and infective apparatus in the form of protein. In any event, they are intriguingly packaged units of genetic information, and much of what will be described in Chapter 4 was first apprehended and understood in these simple units.

TOOLS AND TECHNIQUES

Any science is mature to the extent that it has acquired a sound and acceptable theoretical base. The theory of evolution through natural selection, the cell doctrine, and the theory of the gene provide the fundamental structure of biology, but progress in theory often plays leapfrog with new instrumentation, new techniques, and new questions to be answered. This situation has been especially true for cytology. Some cells may be large enough to see with the unaided eye. But to identify their internal organization, we must magnify them greatly and often use specific dyes, radioactive atoms, or fluorescent molecules to highlight selected parts of the cell (Figure 1.16).

In attempting to study in detail the objects they observe, adequate magnification is as much of a problem for the cytologist, who has to overcome very small sizes, as it is for the astronomer, who has to overcome great distances. For our purposes, the problem of magnification can be best considered in terms of resolving power, which is the ability of an optical system to reveal details of structure, or more specifically, to show as discrete entities two small points or bodies that lie close together. In observing a double star (for example, one found in the handle of the Big Dipper, seen in the Northern Hemisphere), some of you will be able to discern but a single star; others, with better resolving power, will see two separate

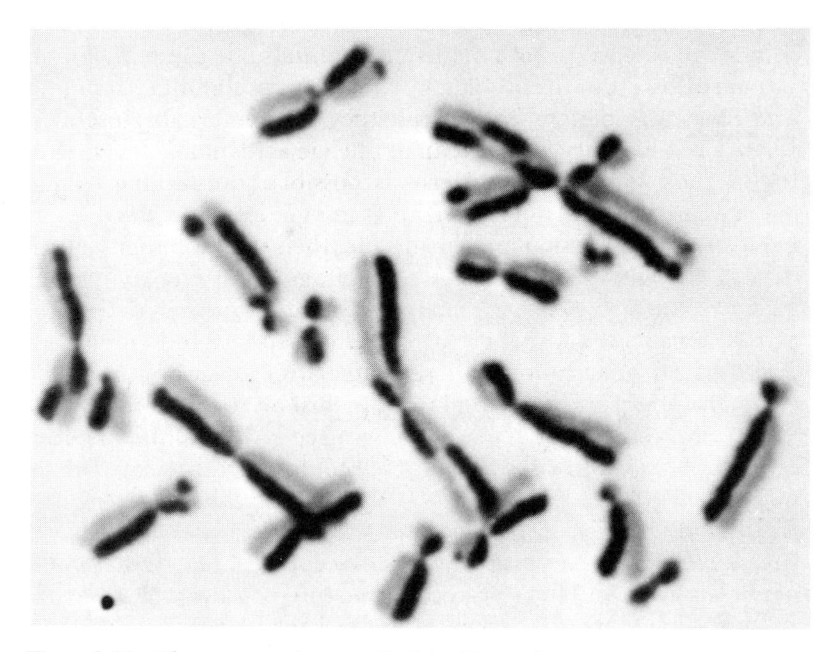

Figure 1.16 Chromosomes from a cell of the Chinese hamster that had replicated in the presence of bromodeoxyuridine, and then been stained with a combination of a fluorescent dye and Giemsa stain. The two chromatids of each chromosome stain differentially, and it can be seen that the sister chromatids of each chromosome occasionally exchange pieces of chromatin. (Wolff, S., and P. Perry, 1975, *Exp. Cell Research* 93, 23–30; courtesy of Dr. Sheldon Wolff.)

stars. In a compound microscope, the resolution of the first lens is the critical factor. As Figure 1.17 indicates, the lens nearest the specimen being examined, called the objective lens, is the key element of a compound microscope, because the projector lens, the ocular, can magnify only what the objective has resolved.

The unaided human eye has a resolving power of about 0.1 to 0.2 mm (millimeters). Lines closer together than this will be seen as a single line, and objects that have a diameter smaller than this range will be invisible or seen only as blurred images. The human eye, however, has no power of magnification; each of us, therefore, must calculate sizes mentally, and experience is probably the largest factor in our ability to judge accurately when we are using our unaided visual sense. Microscopes, of course, both resolve and magnify, but with the lenses obtainable today the ability to resolve is limited by the kind of illumination used. Objects that are smaller or closer to each other than about one-half the wavelength of the illuminating light may be seen, but cannot be clearly distinguished in a light microscope. Thus, even with the most perfectly ground lenses, and with white light having an average wavelength of 550 nm (nanometers), an oil immersion objective

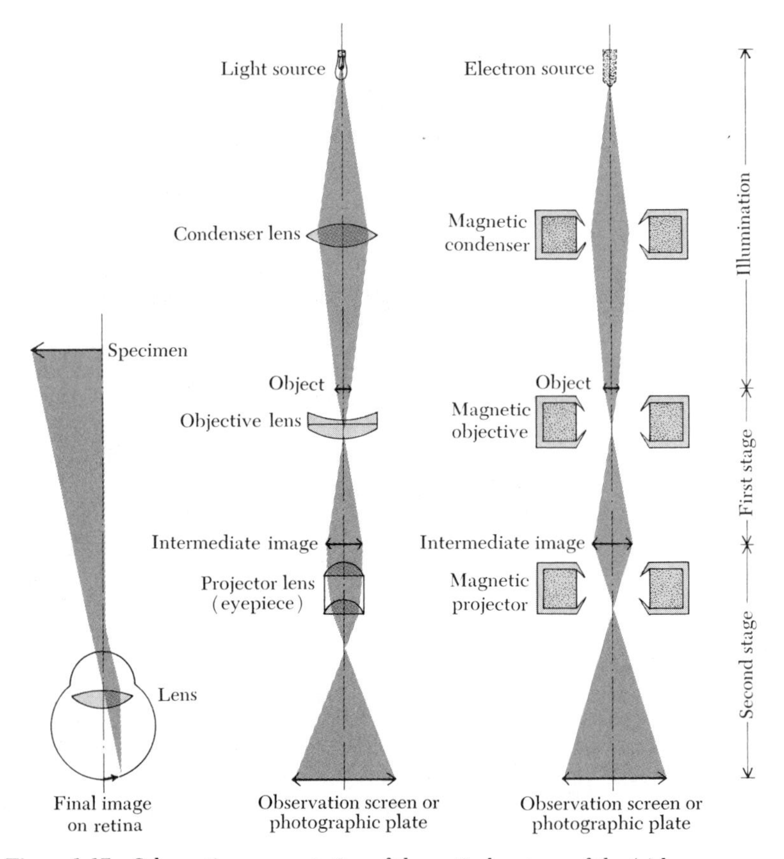

Figure 1.17 Schematic representation of the optical systems of the (*a*) human eye, (*b*) the light microscope, and (*c*) the electron microscope.

cannot resolve two discrete points separated from each other by a distance less than 270 nm, or 0.27 μm (micrometer). Since many parts of cells have smaller dimensions, their presence and structure were undetected until a means of greater resolution was found.

The electron microscope provides this increased resolving power by making use of "illumination" of a different sort. High-speed electrons are employed instead of light waves. As the electrons pass through the specimens being viewed, parts of the cell absorb, permit the passage of, or scatter electrons differentially because of differences in density. An image of the specimen is thus formed on an electron-sensitive photographic plate or fluorescent screen. The human eye is not stimulated by electrons, hence the need for plates or screens. The "optical" system is similar to that in the light microscope (Figure 1.17), except that the illumination is focused by magnetic lenses instead of conventional glass lenses.

Although electrons are particles and hence possess mass, they behave like radiant energy and can be characterized by their wavelength, which varies with voltage. When accelerated through the microscope by a potential difference of 50,000 volts, they have a wavelength of about 0.005 nm. This is 1×10^{-5} that of average white light. An electron microscope can thus theoretically resolve objects that are separated by a distance of about 0.001 nm. This dimension is far less than the diameter of an atom (the hydrogen atom has a diameter of 0.106 nm), but owing to limitations in the way that magnetic lenses function, the actual resolving power of the best modern instrument is about 0.5 to 1.0 nm. At this level, individual atoms cannot be distinguished, but large molecules of biological importance are readily visible (Figures 1.18 and 1.19). In approximate figures, then, the human eye can resolve down to 100 μm, the light microscope to 0.2 μm, and the electron microscope to 0.001 μm. Or, to put it another way, if the normal human eye has a resolving power of 1, that of the light microscope is 500, and of the electron microscope is 100,000. The electron microscope has thus opened a whole new domain to the cytologist by making visible a number of cellular structures that would have otherwise remained undetected.

The significance of magnification in microscopy, as distinguished from that of resolution, is that the smallest objects resolvable by the light microscope need to be magnified to about 2000 to 4000 diameters to be readily discernible visually. In the electron microscope, objects 1 nm in diameter require magnifications of 200,000 to 300,000 times.

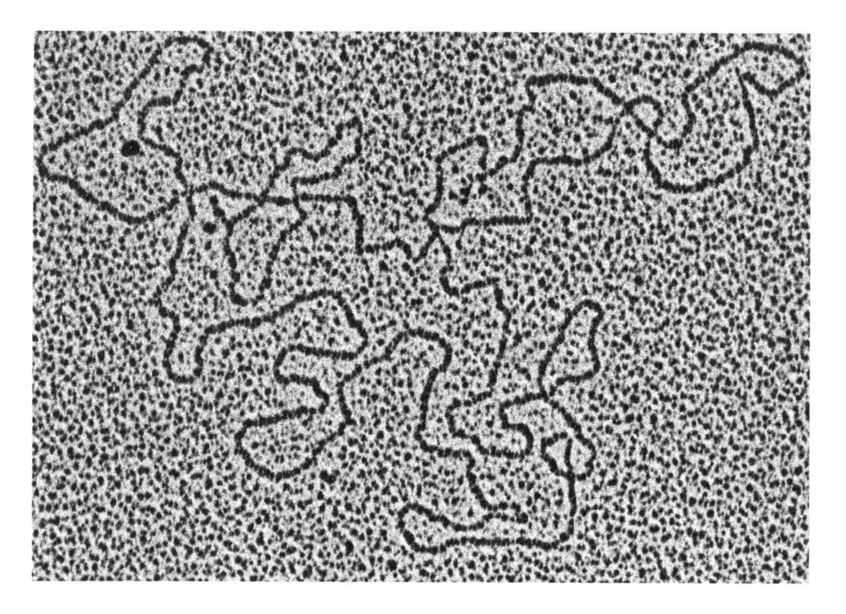

Figure 1.18 Electron micrograph of the circular chromosome of the P22 bacteriophage, which infects the bacterium, *Salmonella typhimurium*. This is basically a naked DNA molecule, released from a disrupted viral particle, spread on a protein film layered on water, and then picked up on a grid before being photographed. The background is the surface of the grid.

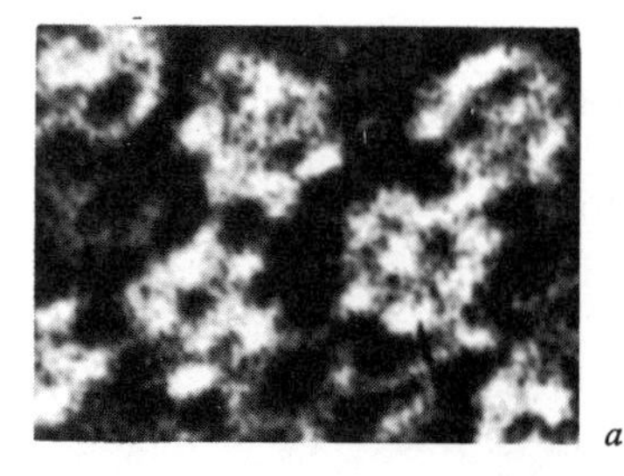

Figure 1.19 Electron micrographs of protein molecules: (*a*) aggregates of the carbon dioxide-fixing enzymes extracted from spinach chloroplasts; and (*b*) collagen, an animal protein found prominently in cartilage, showing its characteristic banded structure.

The increasing degree of resolution made possible by advancements in microscopy is indicated in Figure 1.20 as well as by the other electron micrographs in this book. How much more refined we can become in our visualization of cellular structure and organization is open to question. An electron microscope that would resolve structure at the level of 0.1 nm would enormously expand our field of vision, for it would enable us to visualize molecular organization directly, as well as molecular aggregates such as those in Figures 1.18 and 1.19. Reliable preservation of the fine structure of the cell seems to have been achieved through freeze-drying techniques (techniques that freeze an object very rapidly, and then evaporate all moisture without altering structure), as well as through the more conventional means of fixation, sectioning, and staining, so this aspect of microscopy does not seem to be the bottleneck for progress. Neither is voltage a problem, for present voltages seem more than adequate. Lens construction, however, presents inherent difficulties, and it well may be that as in light microscopy, a limitation has been reached, but with lens design rather than wavelength of radiation the limiting factor.

A word needs to be said about the differences in preparation of cells for light and electron microscopy, differences dictated by the character of the radiation

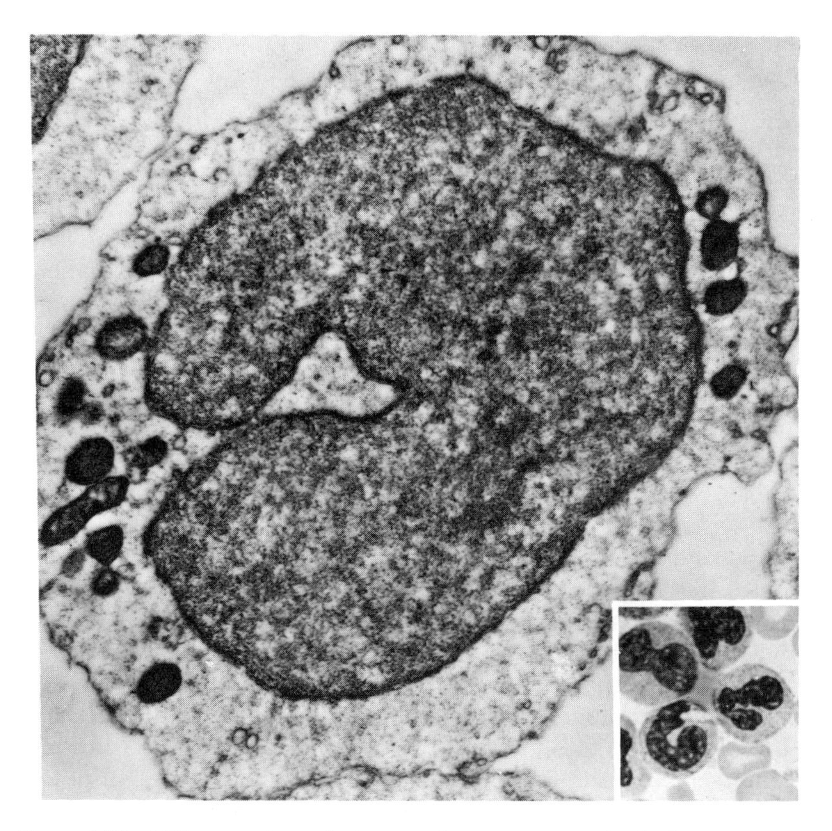

Figure 1.20 Differences in resolution: an electron micrograph of a human lymphocyte; and (inset) a group of comparable cells photographed in the light microscope.

used. Visible light readily passes through whole cells, or those that have been sliced in a microtome at thicknesses of 5 to 10 μm. Electrons of the voltage customarily used would be totally absorbed in such thick masses, so cells must be sliced at about 10 nm thickness to be usable. For electron microscopy, an ultramicrotome, using diamond or glass knives, is needed for precision cutting, and the cells must be embedded in a plastic rather than the customary paraffin used by the light microscopist.

Regardless of the instrument used or the method of preparation of cells for examination, the parts of the cell being investigated must be clearly distinguishable from their immediate surroundings.

In the electron microscope, this contrast is possible because some structures are or can be made more electron "dense" than others, and photosensitive film is darkened to the degree that it is struck by the electrons passing through a specimen. Since the degree to which electrons are scattered is a function of the mass of the atom, and since the light atoms of organic materials—hydrogen, carbon, nitrogen, and oxygen—have little scattering power, the parts of the cell to be examined may be selectively "stained" with heavy metals to show contrast. Heavy metals such as osmium, bismuth, uranium, and manganese are customarily used. Contrast is equally difficult to achieve with the light microscope because most parts of the cell are transparent to light. To overcome this problem, the cytologist uses the proper killing agent (fixative) and stain to color selectively the parts to be examined. Literally hundreds of fixing and staining procedures are known; they are the cytologist's recipes, continually improved upon in the search for better ways to study cells. Since many molecules, because of their chemical makeup, will selectively absorb or interact with particular dyes, some of them fluorescent, the staining procedures are used not only to reveal cellular structure but also to assist in the identification and distribution of molecules that could not be detected in any other way.

One of the limitations of both light and electron microscopy is that transmitted illumination, whether visible light or electrons, reveals only a two-dimensional image of the specimen being examined. A three-dimensional idea of the cell or its parts can be constructed only from serial sections, a difficult, if not impossible, task for the electron microscopist when one realizes that it would require 1000 to 3000 sections from an ordinary-sized cell of 30 μm in diameter. Furthermore, it is impossible in the transmission electron microscope to visualize whole cells. The development of the scanning electron microscope (SEM) overcomes these difficulties in part, and within a certain range of magnification.

The scanning electron microscope enables the development of three-dimensional images for several reasons. It does not record the electrons passing through the specimen, but collects rather than focuses all of the secondary electrons released from the specimen by the impinging electrons. Since the impinging electrons constitute needlelike probes, providing thereby a constant depth of focus regardless of magnification, the hills and valleys of the specimen are clearly revealed instead of simply a single plane of focus (Figure 1.21); compare with Figure 1.24). In addition, the scanning electron microscope possesses a continuous range of magnifications, and can yield clean images from a magnification of about 15 diameters, also possible from good hand lenses, to 20,000 diameters (Figure 1.22). The scanning electron microscope, therefore, spans the resolution and magnification ranges of both the light and transmission electron microscopes, but in a complementary way, thus adding greatly to the ability of the cytologist to study cells.

A word of caution should be entered here. Many of the techniques for the examination of cells in both light and electron microscopes are drastic, and capable of significantly altering cellular structure and arrangement. Artifacts are easily introduced, and the microscopist must constantly be aware of their possible existence.

Parts of cells also can be selectively studied through the use of molecules containing radioactive atoms, particularly phosphorus 32 (^{32}P), carbon 14 (^{14}C), and tritium (^{3}H), an isotope of hydrogen. When radioactive molecules are taken up and react chemically with particular parts of the cell, their presence and location can be detected through autoradiography. That is, a thin layer of photographic emul-

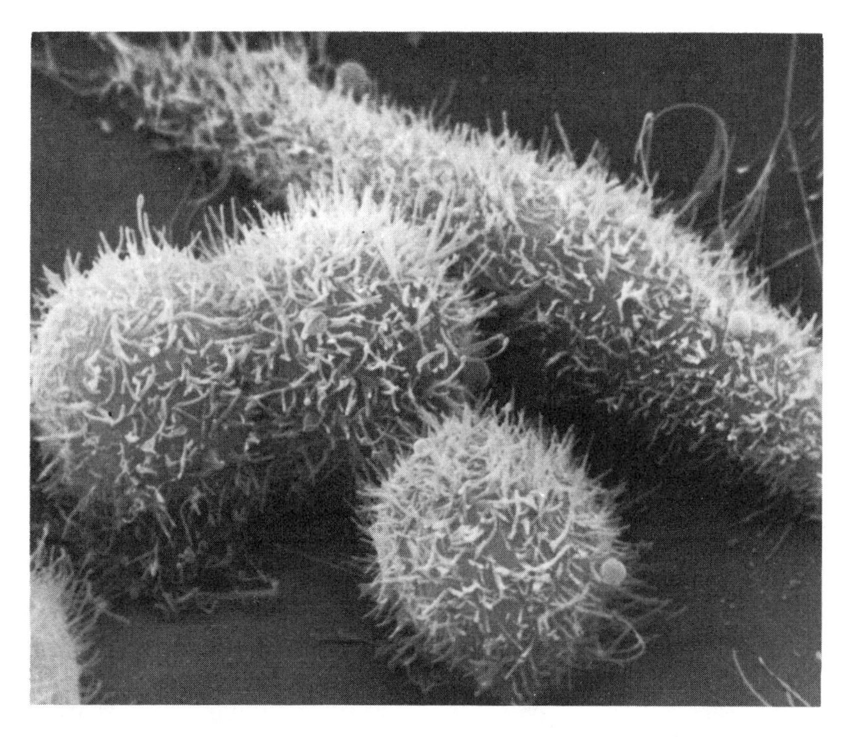

Figure 1.21 A scanning electron micrograph (SEM) of a living, malignant human (Hela) cell, cultured in the laboratory and growing on the surface of the medium. The rounded cells are about 30 μm in diameter. Since only the surfaces of objects can be viewed in the SEM, the thickness of the specimen is not of critical importance. Compare this figure with that in Figure 1.24, which is a similar cell taken in a phase microscope.

Figure 1.22 Images obtained with the scanning electron microscope. Left, pollen grain of the lotus, *Nelumbo lutea* ($\times$ 1900), and right, a higher magnification of the sculpturing of the pollen wall ($\times$ 10,000).

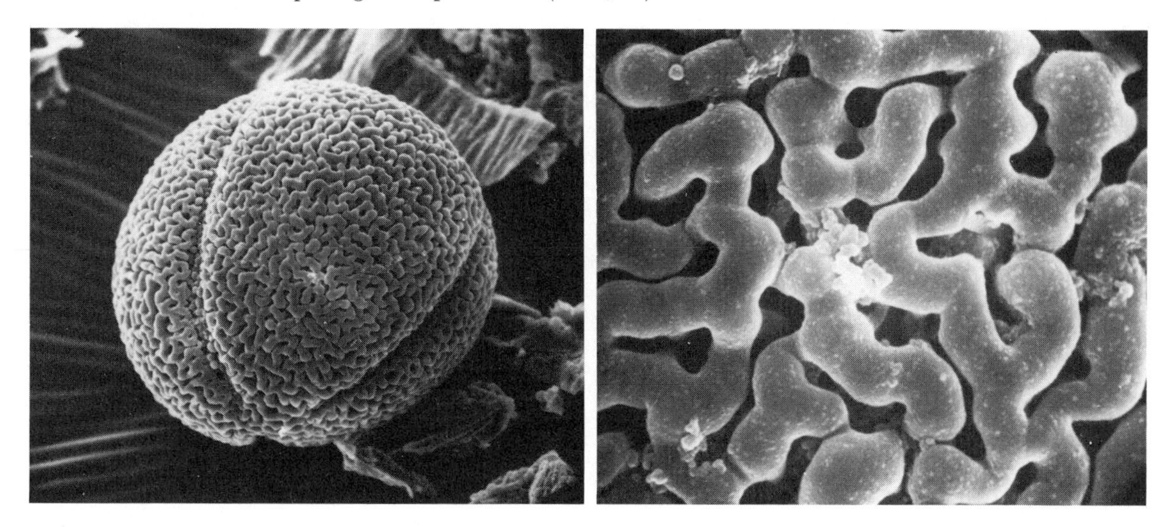

sion is spread over the flattened cells, and as the radioactive atoms disintegrate, the rays or particles released from these atoms pass into the emulsion and cause a darkening of the emulsion much in the manner that exposure to light causes a photographic film to darken. When such cells are viewed under the microscope they may appear as in Figure 1.23. Molecules possessing radioactive atoms are, therefore, like dyes or fluorescent molecules having a high degree of specificity of attachment: they are cellular "probes" or "labels" that enable the microscopist to distinguish one kind of molecule from another, or to follow the reaction pathway of particular molecules or their parts.

A living cell, however, is always more fascinating than a dead one. To watch cells divide is to witness one of the most dramatic of biological phenomena. This cannot be done at the level of the electron microscope because of the vacuum in which the specimen is placed, but a special light microscope, called a phase-contrast microscope, permits the examination of living material. It does so by taking advantage of the fact that the higher the refractive index of a medium (or cell part) the more is light retarded in velocity as it passes through. Transmitted light from one part of a cell is, therefore, out of phase with that from another part, and

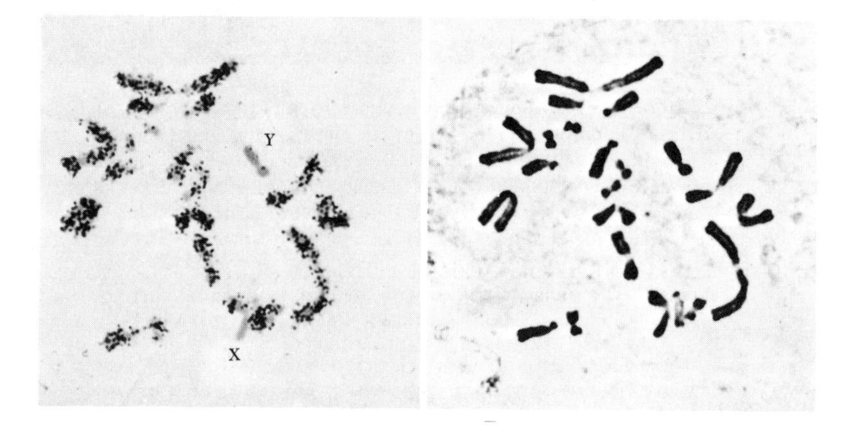

Figure 1.23 A metaphase cell of the male Chinese hamster that had gone through part of its life cycle in the presence of radioactive tritium (^3H). The tritium replaces hydrogen in thymidine, a molecule that becomes incorporated selectively in the DNA of the chromosomes. Right: the cell was stained with Feulgen and then photographed. The coverglass was then removed, and an emulsion was spread over the preparation and allowed to remain for some time before being rephotographed (left). When the tritium atoms disintegrate, they darken the emulsion as they pass through, much as a photographic film is darkened when struck by light. Notice that the X and Y chromosomes have fewer disintegrations over them than the other chromosomes; this is because they undergo replication at a different time than do the remainder of the chromosomes, and at a time when the tritiated thymidine was not available.

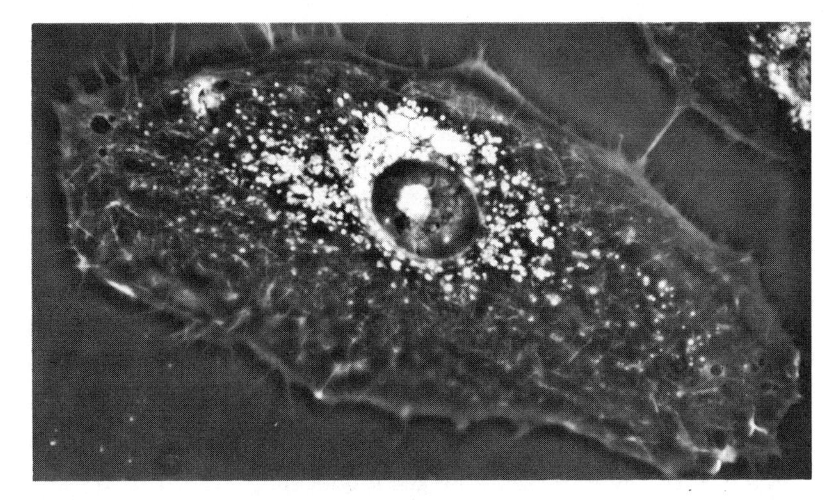

Figure 1.24 A Hela cell grown in tissue culture and photographed through a phase-contrast microscope. The nucleus with its nucleolus is visible in the center; the white mass is liquid refractile material taken in from the culture medium; the slender rods inside the cell are mitochondria; and the outer fine projection are microfibrils frequently formed by mammalian cells grown in tissue culture. Compare this cell with that in Figure 1.20.

the phase-contrast microscope converts these differences in phase into differences in brightness. Figure 1.24 is a photograph of a living human cancer cell taken through a phase-contrast microscope. Under the conventional light microscope, such a cell would appear almost structureless.

Another way to study living cells is to "grind" them up and examine their parts. This is done with a special mortar and pestle to burst the cells and release their contents into a solution. When this solution is centrifuged at carefully regulated speeds, the different parts of the cell, suspended in solution, settle out and sediment according to their density; that is, the denser portions settle out at lower speeds, the less dense ones at higher speeds (Figure 1.25). Even large molecules of different molecular weights can be separated from each other, but this is done by density gradient centrifugation rather than by the differential centrifugation depicted in Figure 1.25. In density gradient centrifugation, a concentrated solution of cesium chloride or sucrose is centrifuged at about 100,000 times gravity for about 20 hours. At this force the molecules of cesium chloride, for example, move toward the base of the centrifuge tube, forming a gradient of density from top to bottom. Any molecules suspended in this solution prior to centrifugation, say molecules of DNA or protein, will also move toward the bottom of the tube, and will come to rest at that place in the density gradient equal to the density of the molecules. The centrifuge tube then can be punctured at the bottom, and the solution can be collected in fractions and analyzed for content or tested for

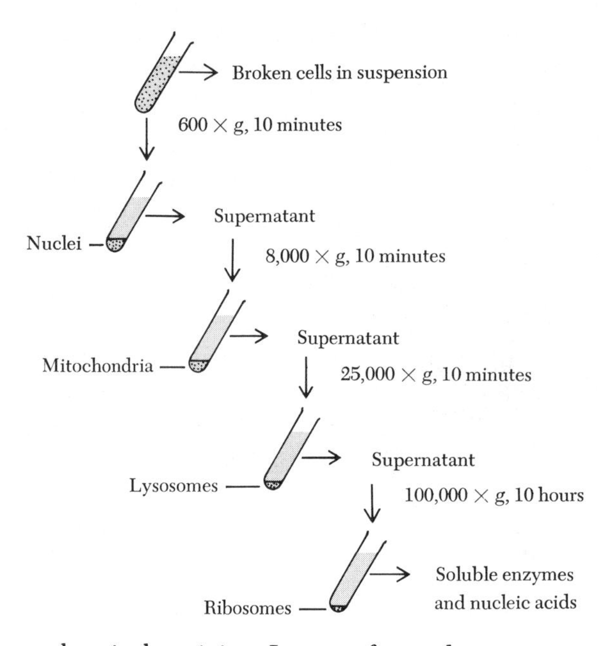

Figure 1.25 A flow diagram showing the fractionation of the cell into its component parts by means of differential centrifugation. If one were using plant cells, chloroplasts capable of carrying out photosynthesis can be obtained by centrifuging at 2,000 to 3,000 × g for 1.0 minute. The soluble enzymes, DNA, and the several RNAs could be separated still further by either chromatographic or electrophoretic techniques.

chemical activity. In a preformed sucrose gradient, molecules are separated by mass rather than by density.

Mixtures of molecules of biological importance also can be separated and individual molecular species often purified by a variety of techniques. In chromatographic techniques—there are many variations, each of particular usefulness—solutions of molecules are poured through a vertical column of specially prepared materials and collected in fractions of equal amounts at the bottom of the column. The molecules move down the column according to size or degree of attraction to the materials in the column. In electrophoretic procedures, the mixture in question is placed in an electrical field, and the net electrical charge of the molecules causes them to move along a moistened paper or a gel. The greater the charge on the molecule the faster and farther it will move. For molecules with similar charges per unit mass, such as DNA or RNA, smaller molecules will migrate faster, providing an additional powerful means of molecular separation. Individual molecular species will appear as spots or bands when properly stained, and then they can be isolated and examined for purity, chemical structure, and/or biological activity.

Centrifugation, chromatography, and electrophoresis are, of course, aids in studying cell function and chemical activity, but it should be realized that an isolated particle or molecule may or may not behave in a test tube as it does in a complex cell. Techniques for studying the intact cell are limited, particularly at the level of chemical activity and interaction, and other approaches such as those described are useful.

Living cells of plants and animals, including those of humans, can also be cultured in much the same manner as bacteria on artificial media, and can be subjected to a wide variety of manipulative procedures, providing new approaches to the solution of old problems that have been difficult to solve as well as opening up wholly new avenues of investigation. The fact that cells in culture can be induced to fuse with each other also has greatly increased the range of problems that can be attacked. In particular, problems relating to the genetic control of development, tumor formation, and antibody production are beginning to seem solvable, it is possible to locate genes on particular chromosomes without the aid of breeding techniques (always a barrier to the understanding of human inheritance), and genes can be isolated and transferred to other individuals and species in the absence of any form of sexual reproduction. Some of these problems will be addressed in later chapters. The cell biologist, therefore, has a large and varied arsenal of instruments, and an ever-expanding battery of procedures for making the cell give up its secrets. Any one tool or technique, however, is usually not enough, and several methods are often employed before an answer can be found to the particular problem being investigated. Such knowledge as we have already garnered from the cell has strengthened our belief that it is the basis of life, and at the same time has made us acutely aware of how little we know of its many complexities.

BIBLIOGRAPHY

ANDERSON, W. F., and DIACUMAKOS, E. G. 1981. Genetic engineering in mammalian cells. *Scientific American 245:* 106–121.

AVERS, C. J. 1978. *Basic Cell Biology.* Van Nostrand Reinhold Company, Inc., New York.

CLAUDE, A. 1975. The coming of age of the cell. *Science 189:* 433–435.

DIENER, T. O. 1981. Viroids. *Scientific American 244:* 66–73.

DUPRAW, E. J. 1973. *The Biosciences: Cell and Molecular Biology.* Stanford Cell and Molecular Council, Stanford, Calif.

DYSON, R. D. 1978. *Essentials of Cell Biology,* 2nd ed. Allyn and Bacon, Inc., Boston.

FAWCETT, D. W. 1966. *The Cell: Its Organelles and Inclusions.* W. B. Saunders Company, Philadelphia.

HSU, T. C. 1974. Longitudinal differentiation of chromosomes. *Annu. Rev. Genet. 7:* 153–176.

HUGHES, A. 1959. *A History of Cytology.* Abelard-Schuman Ltd., London.

LEDBETTER, M. C., and PORTER, K. R. 1970. *Introduction to the Fine Structure of Plant Cells.* Springer-Verlag, New York.

LEWIN, B. 1977. *Gene Expression: Plasmids and Phages,* Vol. 3. John Wiley & Sons, Inc., New York.

LURIA, S. E. 1970. Molecular biology: past, present and future. *BioScience 24:* 1289.

MCELROY, W. D. 1971. *Cell Physiology and Biochemistry.* Prentice-Hall, Inc., Englewood Cliffs, N.J.

PORTER, K. R., and BONNEVILLE, M. A. 1968. *Fine Structure of Cells and Tissues.* Lea & Febiger, Philadelphia.

PRUSINER, S. B. 1982. Novel proteinaceous infectious particles cause scrapie. *Science 216:* 136–144.

Scientific American. September 1970. The biosphere. The entire issue is devoted to energy and its relation to living systems.

Scientific American. September 1971. Energy and power. Articles by F. J. Dyson, D. M. Gates, C. M. Summers, M. Tribus, and E. C. McIrvine are pertinent to an understanding of energy systems and energy flow.

WOLFE, S. L. 1981. *Biology of the Cell*, 2nd ed. Wadsworth Publishing Company, Inc., Belmont, Calif.

2

Basic Chemical Concepts

The universe around us is enormously diverse in terms of form and action, and life in all of its dimensions contributes to that diversity. As was stressed earlier (Figures 1.4 and 1.6) life is a phenomenon that impresses a special kind of order on all of the matter that becomes a part of the living system, and that channels the energy needed to establish and maintain that order. Both the manipulation of energy and the organization of living substance are mediated through chemical reactions, and living things in no way violate the laws of physics and chemistry that govern chemical reactions. In fact, the contrary is true, and we have come to understand life in a deeper sense, and in all of its varied manifestations, to the extent that we understand the physics and chemistry of cellular existence. This chapter, then, serves the purpose of providing an elementary view of the properties of the atoms and molecules which constitute living substance and how these properties underlie biological structure and function; the material presented should also serve as a basis for the more critical examination of the meaning of structures and their relation to function to be explored in later chapters.

THE NATURE OF ATOMS

Matter is composed of a number of basic kinds of substances, or *elements*, which cannot be broken down by purely chemical means. The smallest particles of any element that still allow it to be identified as that element are *atoms;* although atoms can be considered as the fundamental units of matter, they themselves consist of three main types of subatomic particles, *protons*, *neutrons*, and *electrons.*

There are many other subatomic particles known, but other than being responsible for the nature of atoms, they play no direct role in living processes.

Protons and neutrons are found at the center of the atom, and together make up the *nucleus*, around which the electrons spin. The protons and neutrons are of equal mass, being about 1845 times heavier than electrons. Protons and electrons carry equal but opposite electrical charges, each proton being positive and each electron negative, while the neutrons, as their name would suggest, are electrically neutral. Protons and neutrons are held together by strong nuclear forces, while the electrons are held in the vicinity of the nucleus by electrostatic forces between them and the protons.

The smallest atom is hydrogen, consisting of one electron and one proton. It has, by definition, an atomic weight, or unit of mass, of one (1) (Figure 2.1), and since the electron is so small in comparison to the proton, atomic weights beyond the hydrogen atom are calculated on the basis of the number of protons and neutrons in the nucleus. Helium is the next largest atom, having two electrons and two protons, but an atomic weight of four (4) because the nucleus also contains two neutrons (Figure 2.1). The presence or absence of neutrons influences, therefore, the mass of an atom, but not its reactivity with other atoms; rather, it is the number of electrons that determines the chemical reactivity of an atom, and all chemical bonds between atoms depend on their ability to exchange or share electrons.

The positions that can be occupied by the electrons in an atom are characterized by discrete energy levels; in general, the higher the energy level, the farther from the nucleus is the electron likely to be found. These energy levels can be represented (for our purposes, at least) as a series of shells surrounding the nucleus. Seven such shells are recognized, from the innermost *K* shell, through the *L*, *M*, *N*, *O*, *P*, and to the outermost *Q* shell. Only a fixed maximum number of

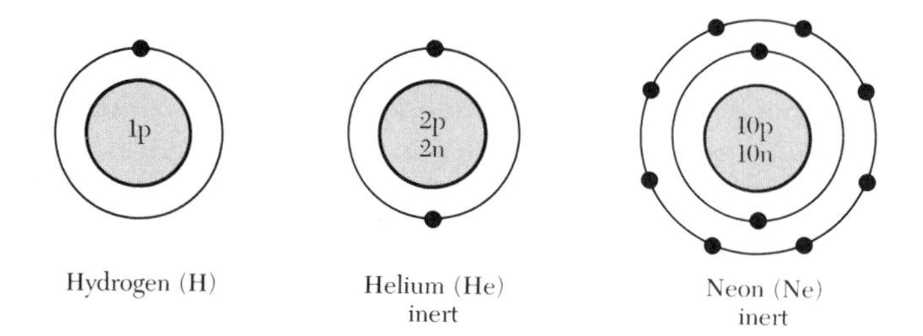

Figure 2.1 The structures of the atoms of hydrogen, helium, and neon. Hydrogen has a nucleus containing a single proton; one electron occupies the inner, or K shell. The helium nucleus has two protons and two neutrons, and has two electrons in the K shell. Neon, with ten protons and ten neutrons, has two filled electron shells, the K and the L. Hydrogen can react with other atoms since its K shell is not saturated, but helium and neon are inert gases, the former because its K shell is saturated, the latter because both its K and L shells are saturated.

TABLE 2.1 SOME ELEMENTS OF THE PERIODIC TABLE, WITH THEIR ATOMIC NUMBER, SYMBOL IN CHEMICAL FORMULAS, AND ENERGY LEVELS (OR THE NUMBER OF ELECTRONS IN THE VARIOUS SHELLS). THOSE ELEMENTS WITH AN ASTERISK (*) ARE VIEWED AS THOSE OF MAJOR BIOLOGICAL SIGNIFICANCE AND ABUNDANCE, THOSE WITH A PLUS SIGN (⁺) NECESSARY BUT IN TRACE AMOUNTS, THOSE WITH A SMALL CIRCLE (°) EQUALLY NECESSARY BUT IN EVEN SMALLER AMOUNTS, AND THOSE WITH A TRIANGLE (△) INERT GASES.

Atomic Number	Element	Symbol	Energy levels K	L	M	N	O	P	Q
1	Hydrogen*	H	1						
2	Helium△	He	2						
3	Lithium	Li	2	1					
4	Beryllium	Be	2	2					
5	Boron	B	2	3					
6	Carbon*	C	2	4					
7	Nitrogen*	N	2	5					
8	Oxygen*	O	2	6					
9	Fluorine	Fl	2	7					
10	Neon△	Ne	2	8					
11	Sodium⁺	Na	2	8	1				
12	Magnesium⁺	Mg	2	8	2				
13	Aluminum	Al	2	8	3				
14	Silicon	Si	2	8	4				
15	Phosphorus⁺	P	2	8	5				
16	Sulfur⁺	S	2	8	6				
17	Chlorine⁺	Cl	2	8	7				
18	Argon△	Ar	2	8	8				
19	Potassium⁺	K	2	8	8	1			
20	Calcium⁺	Ca	2	8	8	2			
21	Scandium	Sc	2	8	9	2			
22	Titanium	Ti	2	8	10	2			
23	Vanadium	V	2	8	11	2			
24	Chromium	Cr	2	8	13	1			
25	Manganese⁺	Mn	2	8	13	2			
26	Iron°	Fe	2	8	14	2			
27	Cobalt°	Co	2	8	15	2			
28	Nickel	Ni	2	8	16	2			
29	Copper°	Cu	2	8	18	1			
30	Zinc°	Zn	2	8	18	2			
36	Krypton△	Kr	2	8	18	8			
47	Silver	Ag	2	8	18	18	1		
53	Iodine°	I	2	8	18	18	7		
56	Barium	Ba	2	8	18	18	8	2	
79	Gold	Au	2	8	18	32	18	1	
92	Uranium	U	2	8	18	32	21	9	2

electrons can be accommodated at each level; at the lowest level, represented by the *K* shell, only two electrons are allowed. Atoms with more than two electrons must maintain the additional number at a higher level, with up to eight being allowed in the *L* shell, and so on (Table 2.1). Any atom with a filled outer shell tends to be extremely stable, for example, helium, with two electrons, has a filled *K* shell, while neon, with ten electrons, has a filled outer *L* shell (Figure 2.1). Atoms with numbers of electrons other than required to complete shells are chemically reactive, since they have a strong tendency toward achieving a complete outer shell. Chemical reactivity, therefore, depends on the number of electrons in the outer shell, and hence on how many electrons must be gained, lost, or shared to reach the stable condition.

CHEMICAL BONDS

Let us consider a relatively simple bond, that between two atoms of hydrogen to form a hydrogen molecule. Although each atom is electrically neutral, at very close range the negatively charged electron of each atom is attracted toward the positively charged proton of the other. However, the electrons are also still attracted to their own nuclei; this mutual attraction between the electrons and the two nuclei holds the atoms together in a more stable state. Since the electrons are equally shared, the *K* shell of each atom is filled (Figure 2.2). Each such shared pair of electrons is known as a *covalent* bond, and is shown conventionally as H–H.

Covalent bonds between carbon atoms and between carbon and other atoms, in particular hydrogen, oxygen, nitrogen, phosphorus, and sulfur, are of great importance in the molecules of life. The carbon atom, with four electrons in its outer shell, is able to form four separate covalent bonds; for example, the compound methane consists of a carbon atom attached to four separate hydrogen atoms (Table 2.2). Once again the shells of each atom are completed as a result of the sharing of electron pairs. If two pairs of electrons are shared, a double bond is formed; the difference between ethane and ethylene, for example, is determined

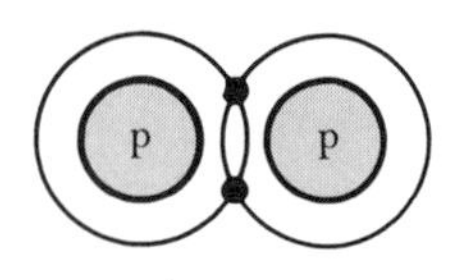

H : H

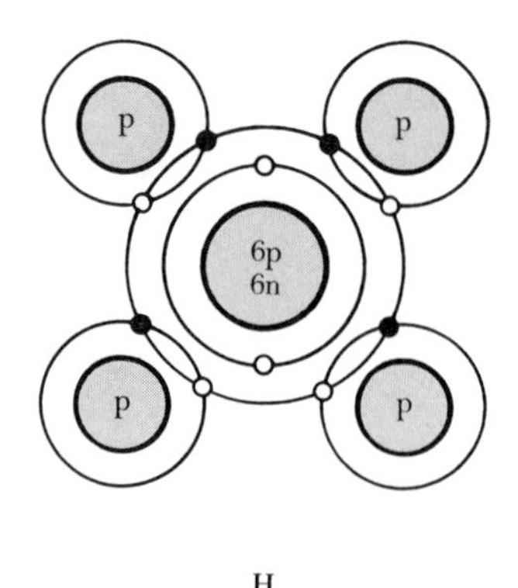

H
H : C : H
H

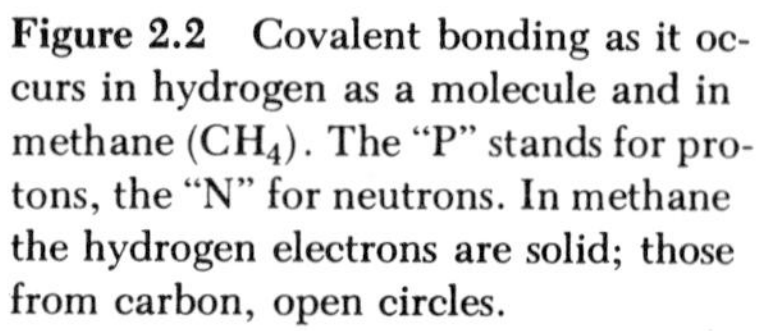

Figure 2.2 Covalent bonding as it occurs in hydrogen as a molecule and in methane (CH_4). The "P" stands for protons, the "N" for neutrons. In methane the hydrogen electrons are solid; those from carbon, open circles.

TABLE 2.2 COMMON, SMALL CARBON COMPOUNDS CONTAINING GROUPS OF BIOLOGICAL INTEREST. THE LETTER R STANDS FOR RADICAL. IN METHANE, R = H, IN ETHANE R = CH_3, AND IN PROPANE R = C_2H_5. ACETONE HAS TWO RADICALS, BOTH BEING CH_3.

Structural formula	Shorthand form	General name	Compound	Common name
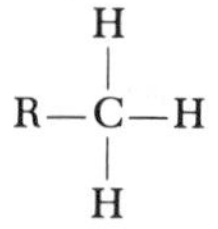 R—C(H)(H)—H	RCH_3	Alkane	CH_4	Methane
			C_2H_6	Ethane
			C_3H_8	Propane
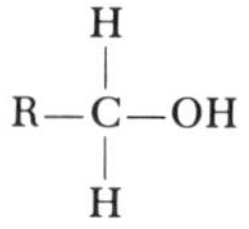 R—C(H)(H)—OH	RCH_2OH	Alcohol	CH_3OH	Methanol
			C_2H_5OH	Ethanol
			C_3H_7OH	Propanol
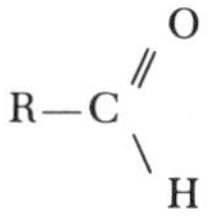 R—C(=O)—H	RCHO	Aldehyde	HCHO	Formaldehyde
			CH_3CHO	Acetaldehyde
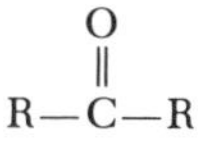 R—C(=O)—R	RCOR	Ketone	CH_3COCH_3	Acetone
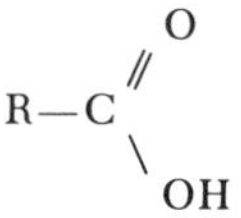 R—C(=O)—OH	RCOOH	Carboxylic acid	HCOOH	Formic acid
			CH_3COOH	Acetic acid
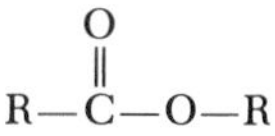 R—C(=O)—O—R	RCOOR	Ester	$CH_3COOC_2H_5$	Ethyl acetate

by such a double bond (Figure 2.3). Double bonds are sites of increased reactivity, since there is a tendency for the second bond between the same two atoms to break and form single bonds with other atoms. The versatility of carbon allows it to participate in the formation of complex "skeletons" of the large, biologically important molecules.

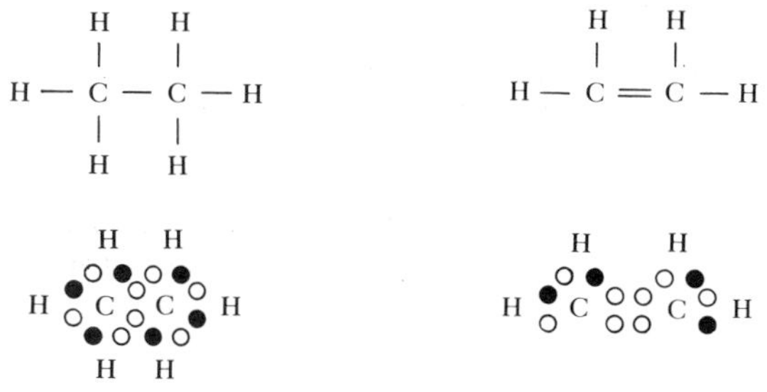

H—C(H)(H)—C(H)(H)—H   H—C(H)=C(H)—H

Figure 2.3 The molecules of ethane (left) with only single covalent bonds, and ethylene (right) with a double covalent bond between the two carbon atoms.

Polar Covalent Bonds

In the covalent bond joining the two H atoms of the H_2 molecule, the electron pair is shared equally by the two nuclei, and the bond is symmetrical. The C–H single bond of the type found in methane is slightly different, in that the larger carbon nucleus attracts the electrons more strongly than does the H nucleus. However, because the arrangement of atoms and bonds in the methane molecule is symmetrical, no one part of the molecule is any more or less charged than another; that is, the molecule is *nonpolar*. Other bonds that form between different atoms result in molecules that are *polar;* that is, that have a slightly positive and a slightly negative end (Figure 2.4). Water, the medium in which most of the reactions of life take place, is an extremely important polar substance. The water molecule consists of two hydrogens covalently bonded to an oxygen (Figure 2.5). However, the orientation of the hydrogens relative to the oxygen is such that they both lie to one side of the oxygen (Figure 2.5). Furthermore, the oxygen atom is highly *electronegative* compared to the hydrogen; that is, it exerts a much stronger pull on the bonding electrons than does the hydrogen. Thus, the oxygen end of the molecule is relatively negative, while the end with the hydrogens is relatively positive (Figure

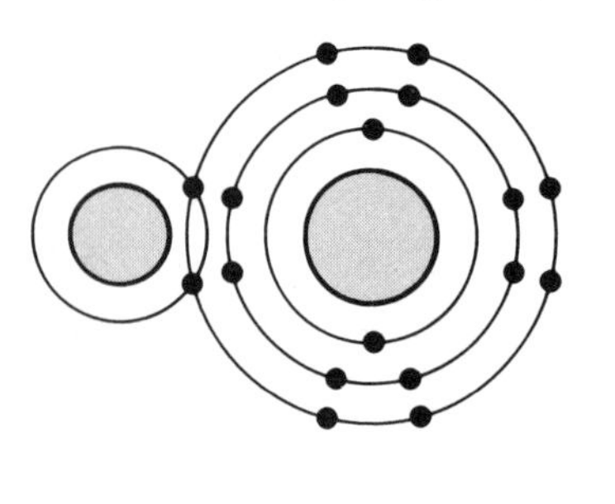

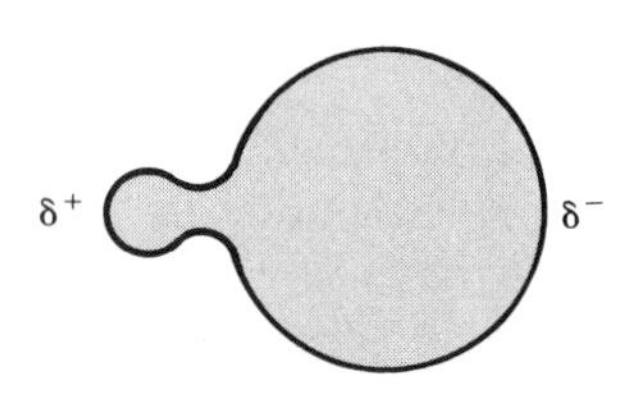

Figure 2.4 The formation of a molecule of hydrochloric acid (HCl) depends on the formation of a polar covalent bond. The shared pair of electrons are pulled more toward the Cl atom; thus the H end of the molecule is partially positive (δ+) and the Cl end partially negative (δ−).

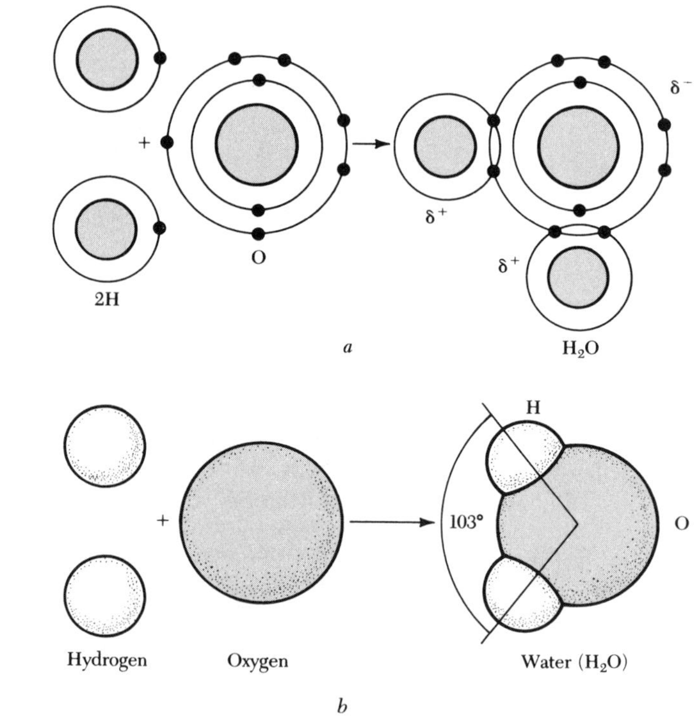

Figure 2.5 The structure of water in terms of its shared electrons (*a*), and the disposition of the two hydrogens to give a positive and a negative end to the molecule as a whole (*b*).

2.5). As we shall see, it is the polar nature of the water molecule that determines many of its unique properties.

Ionic Bonds

The extreme case of a polar bond is the ionic bond, in which the unequal pull of one atom on the electrons is so strong that an electron is actually transferred from one atom to the other; the electron acceptor therefore becomes negatively charged and the electron donor becomes positively charged. For example, in the formation of common table salt, sodium chloride, the tendencies of sodium (Na) to reject its outer electron (Figure 2.6) and of chlorine (Cl) to attract electrons are so great that the chlorine takes over completely the shared electron pair, with the result that an electron is transferred from Na to Cl. Both the *L* shell of sodium and the *M* shell of chlorine are then saturated; however, the sodium now carries a positive charge (Na^+) and the chlorine a negative charge (Cl^-). These charged atoms, or *ions*, are held very tightly together by *ionic bonds*, the strong electrostatic interactions between the oppositely charged ions. However, as we shall see, the behavior of ionic substances in water is very different, and the attractions between oppositely charged atoms or molecules very much weaker.

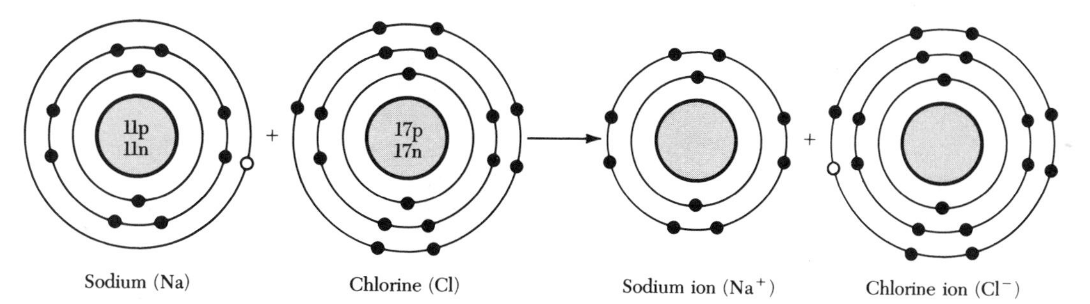

Figure 2.6 When sodium chloride (NaCl), or ordinary table salt, is formed, an electron from the L shell of sodium (open small circle) is transferred to the L shell of chlorine. Both atoms are therefore stabilized by having their outermost shell completed.

Weak Interactions

As we have implied, the nonpolar covalent bond, such as that found between two H atoms, and the ionic bond are extremes of the types of electron interactions found; most bonds in biologically important molecules lie in between, with varying degrees of polarity. Polar bonds, with their partially positive and partially negative ends, allow molecules to interact in various ways with other molecules, with each other, or with different regions of the same molecule. As we shall see, such secondary interactions, although relatively weak, play a vital role in the chemistry of life, since they contribute to both the shape and activity of many of the structures of cells.

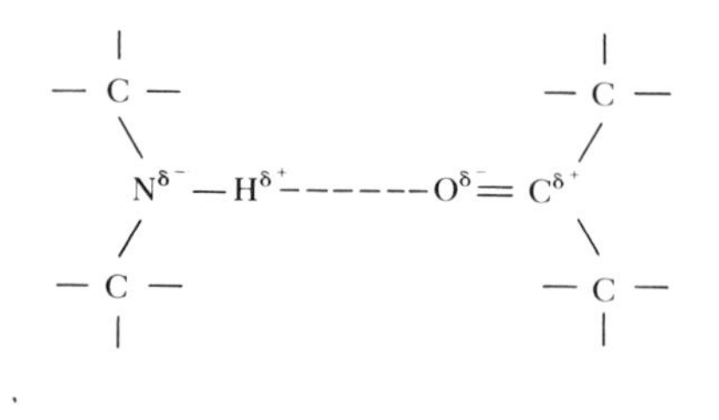

a

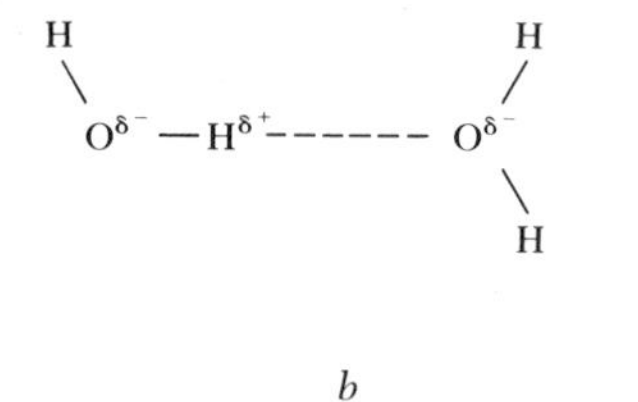

b

Figure 2.7 Hydrogen bonding (*a*) between polar atoms of two different groups, and (*b*) between two water molecules. The hydrogen, which is shared between two different atoms, generally remains closer to one of the atoms than to the other.

One type of weak interaction which is very important in the chemistry of life is the *hydrogen bond.* In our discussion of polar covalent bonds we pointed out that when hydrogen is covalently bonded to a relatively electronegative atom it tends to carry a partially positive charge. This hydrogen can therefore be attracted to some partially negative charge carried by an electronegative atom participating in a different polar covalent bond. Figure 2.7a shows an example of such an electrostatic interaction; the covalent bond between the nitrogen and hydrogen of one molecule is polar, and the H is partially positive. Similarly, the oxygen covalently bonded to the carbon of the second molecule has a partial negative charge, and can attract the electropositive hydrogen if the two atoms are brought sufficiently close together. Although individual H bonds represent relatively weak attractions, large numbers of them can play a critical role in determining molecular interactions in the cell.

One extremely important example of hydrogen bonding is that which occurs between water molecules (Figure 2.7b), which, as we have already stressed, are strongly polar. We shall now see how the polar nature of water and its capacity to form such H bonds are crucial in determining many of the physical and chemical properties of water, including other types of weak interactions that occur in an aqueous environment.

WATER

No understanding of the chemistry of life is possible without considering the unique features of water, the medium in which the reactions of life processes take place, and how these features determine the behavior of other substances in water.

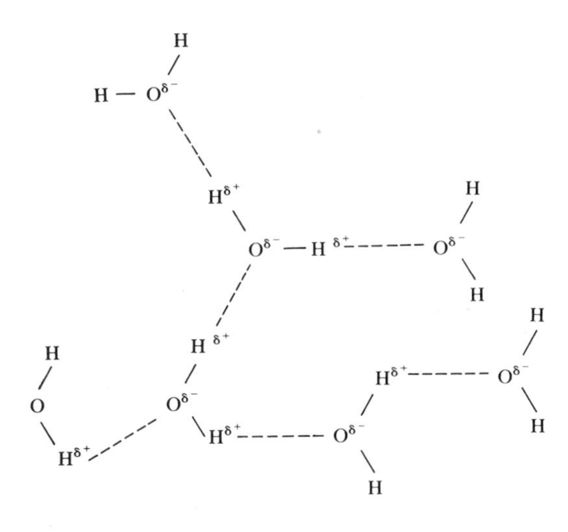

Figure 2.8 Hydrogen bonding between adjacent water molecules (dashed lines) is responsible for giving water its organized structure, both as a liquid and as a solid (ice).

As we have already pointed out, the water molecule is extremely polar, and this polarity enables it to interact with a variety of other molecules, including other water molecules (Figure 2.8). As a polar molecule water can interact with other polar substances, and allow them to dissolve. For example, when sodium chloride is added to water, the Na^+ and Cl^- ions separate, and become surrounded by water molecules in such a way that they are prevented from combining (Figure 2.9); that is, the salt is *ionized*, or *dissociated* into ions. Thus the opposite charges of ions in solution are partially neutralized, with the consequence that ionic interactions in the aqueous environment of the cell are greatly weakened. Water is also able to act as a solvent for covalently bonded polar compounds by forming hydrogen bonds with them; as we shall see shortly, the attraction of the charges between water and other polar molecules may also be sufficiently great for ionization to occur.

Nonpolar substances, on the other hand, do not attract the polar water molecules; indeed, large nonpolar molecules, or nonpolar regions of large

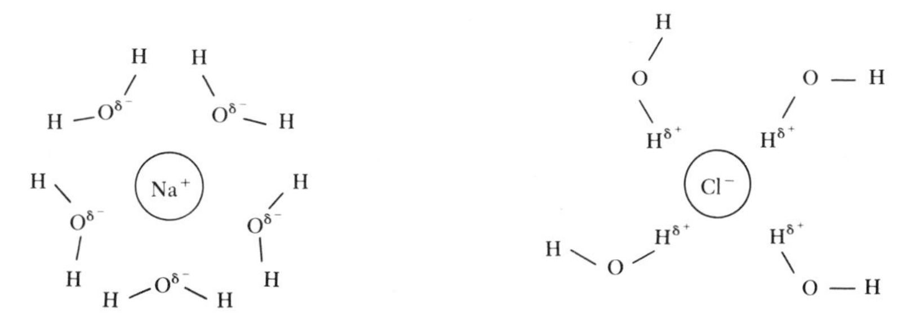

Figure 2.9 The polar nature of water allows it to orient itself with respect to ions in such a way that the ions are surrounded by water molecules. The strong charges on the ions are therefore shielded from each other; in water the compound, in this case NaCl, dissociates.

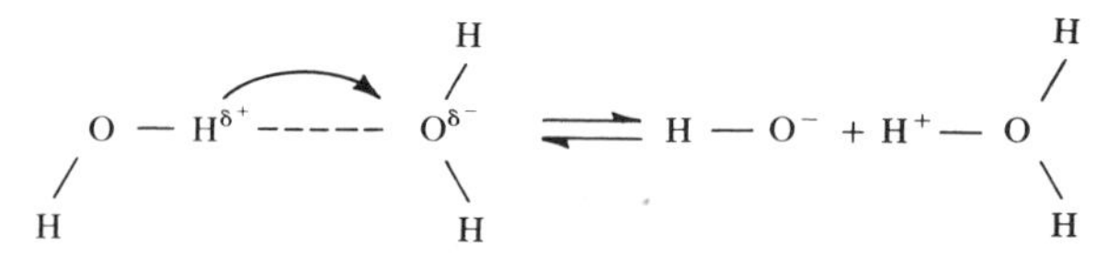

Figure 2.10 When water dissociates, this is usually indicated as the formation of H^+ and OH^- ions, but in actual practice a proton (H^+) leaves one molecule of water and attaches itself to another, producing an OH^- and an H_3O^+, or hydronium, ion.

molecules, tend to associate with each other to exclude water. Such *hydrophobic* interactions among nonpolar substances are important in the maintenance of cellular membranes, which can act as barriers between different aqueous components of the cell, restricting movement of polar molecules or ions between these compartments.

Another important feature of water is its capacity to itself dissociate to a degree into ions (Figure 2.10), thus promoting still more its reactivity. Although the water molecule is held together by strong covalent bonds, a small fraction of molecules (about 1 in 550 million) are able to interact in such a way that a proton from one molecule is transferred to another (Figure 2.10), thereby producing two ions, H_3O^+ and OH^- (although the positively charged *hydronium* ion is really a water molecule plus a proton, it is generally symbolized as H^+). This capacity of water to self-ionize enables it to accept or donate protons from or to other compounds, a capacity that, as we shall now see, can influence the activity of other molecules in the cell.

ACIDS, BASES, AND pH

An acid is any chemical substance that can donate protons, or hydrogen ions, and a base (or alkali) is any chemical group that can accept protons. For example, if hydrochloric acid (HCl) is added to water, it undergoes a high degree of ionization:

$$HCl + H_2O \longrightarrow H_3O^+ + Cl^-$$

thereby increasing the concentration of H^+ ions in solution. Thus HCl is a strong acid. On the other hand, when ammonia (NH_3), a base, is dissolved in water, it attracts the already dissociated H^+ ions and itself becomes ionized:

$$NH_3 + H_3O^+ \longrightarrow NH_4^+ + H_2O$$

thereby reducing the level of H^+ ions. Addition of a hydroxide such as NaOH produces the same effect as ammonia, but by increasing the concentration of OH^- ions in solution as a result of its strong tendency to dissociate into Na^+ and OH^-.

The degree of acidity or alkalinity of a solution is expressed in units related to the H^+ concentration. Since the range of H^+ concentrations that can be encountered is great, the scale used is the pH scale, where

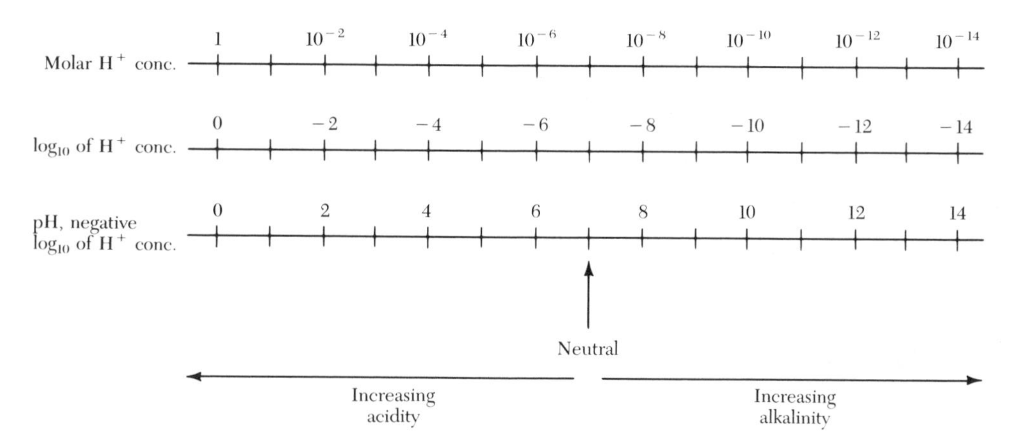

Figure 2.11 Three ways of representing degree of acidity or alkalinity: by the H^+ ion concentration, H^+ concentration to the $\log_{10}$, or as a pH value.

$$pH = \log_{10}[H^+]$$

(square brackets are used to denote concentration.) In pure water the H^+ concentration is 10^{-7} molar; hence, in the case of water:

$$\begin{aligned} pH &= \log_{10}(10^{-7}) \\ &= -(-7) \end{aligned}$$

Since in pure water the concentrations of H^+ and OH^- are equal, pH 7.0 is considered to be neutral. Solutions with pH values of less than 7.0 (that is, H^+ concentrations of greater than 10^{-7}) are acidic, whereas those with pH values higher than 7.0 are alkaline (Figure 2.11).

Buffers

Many biologically important molecules carry ionizable groups, or particular aggregates of bonded atoms, and many of the properties of molecules depend on whether these groups are charged or not. Hence dramatic changes in cellular pH are likely to affect cellular activities by changing the degree to which these groups are dissociated. Thus the cell must ensure that it is protected against changes in pH that would impair its functioning. This is done by substances found in the cell that can act as *buffers;* buffers are able to maintain pH values within a limited range by accepting H^+ ions when their concentrations in solution would otherwise increase, and releasing them into solution to balance any increase in the concentration of OH^- ions. The mode of action of buffers can be illustrated by considering the behavior of acetic acid, a weak acid that dissociates only partially:

$$\underset{\text{acetic acid}}{CH_3COOH} \rightleftharpoons \underset{\text{acetate}}{CH_3COO^-} + H^+$$

In water, the degree of dissociation is such that the pH is about 4.8. If a strong base, for example NaOH, is added, it will immediately dissociate into Na^+ and OH^-. The OH^- combines with the H^+ already in solution, thereby tending to raise the pH; however, as the H^+ is removed from the right-hand side of the equation in this way, more of the CH_3COOH will dissociate, releasing more H^+ and maintaining the pH at close to 4.8. Similarly, if an acid, say HCl, is added, the excess H^+ ions will be neutralized by combining with the acetate ions (acting as a base) to form acetic acid.

ORGANIC MOLECULES

Most of the molecules that make up living things are organic molecules; that is, they all contain carbon. The tremendous diversity of organic substances is based in large part on the versatility of the carbon atom, and its capacity to form chemical bonds with other carbon atoms as well as with such widely distributed and abundant atoms as oxygen, hydrogen, nitrogen, and sulfur. As we pointed out earlier, each carbon atom has four electrons in its outer shell, and therefore can participate in four separate bonds.

The carbon–carbon bonds are extremely important in determining the overall framework of organic molecules, since C "skeletons" of various sizes and shapes can be built up. Figure 2.12 depicts a series of molecules with increasing numbers of carbon linearly arranged, and with hydrogen atoms attached to them. In Figure 2.13 are ring structures, in one of which only carbon atoms are used in completing the ring, while in the other an oxygen atom is also involved. Ring compounds of this sort generally have either five or six atoms in the ring skeleton; compounds with fewer than five or more than six rarely form rings because the spatial considerations put too much strain on the bonds to allow them to remain stable.

As is indicated in Figure 2.13, the hydrogens attached to the carbons of the skeleton can be replaced by other atoms or groups of atoms which form *functional groups*. For example, if a *hydroxyl* group (—OH) replaces one of the hydrogens in ethane (Figure 2.14), the result is *ethyl alcohol*. An alcohol, in turn, can be converted to an aldehyde, characterized by a C=O or *carbonyl* group (Figure 2.14). If the carbon of the carbonyl group is also attached to two other carbons, rather than to one carbon and one hydrogen, a ketone is formed (Figure 2.15).

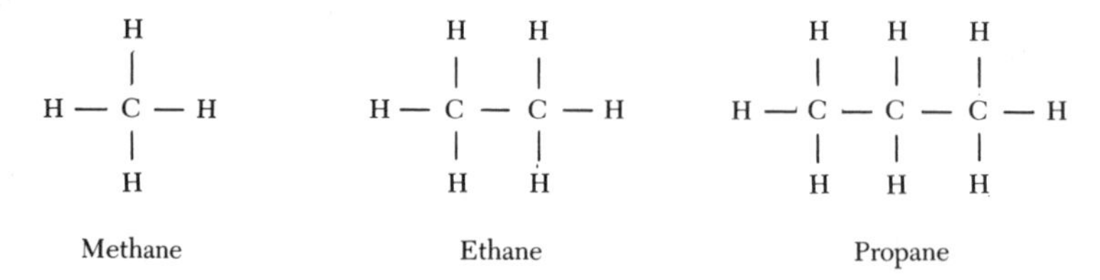

Figure 2.12 One-, two-, and three-carbon hydrocarbons, indicating the ability of carbon to act as the backbone of larger and larger molecules.

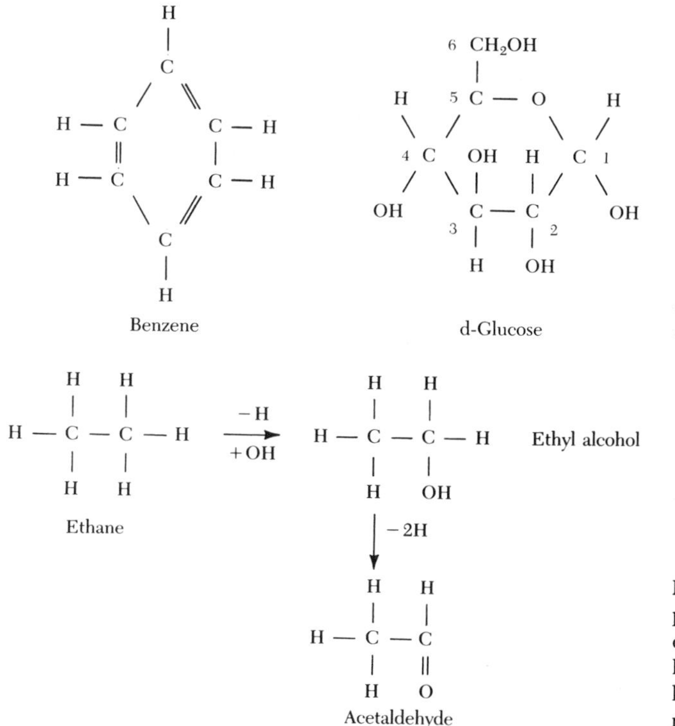

Figure 2.13 Ring structures of two organic molecules in which the ring skeleton contains only carbon atoms, as in benzene, or carbons plus an oxygen, as in glucose. The conventional numbering of C atoms is shown in the glucose molecule.

Figure 2.14 When a hydrogen is replaced by a hydroxyl group, ethane is converted to ethyl alcohol; when two hydrogens are removed from the alcohol, acetaldehyde results from the formation of a carbonyl (C=O) group.

The importance of such functional groups is that, unlike C–C or C–H bonds, they are polar, and hence more reactive. Thus the chemical reactions in which a particular organic molecule can participate are determined to a great extent by which functional groups that molecule carries. Two extremely important functional groups in organic molecules are the *carboxyl* group and the *amino* group. As is shown in Figure 2.16, these groups can accept or donate protons, and hence act as acids and bases, respectively.

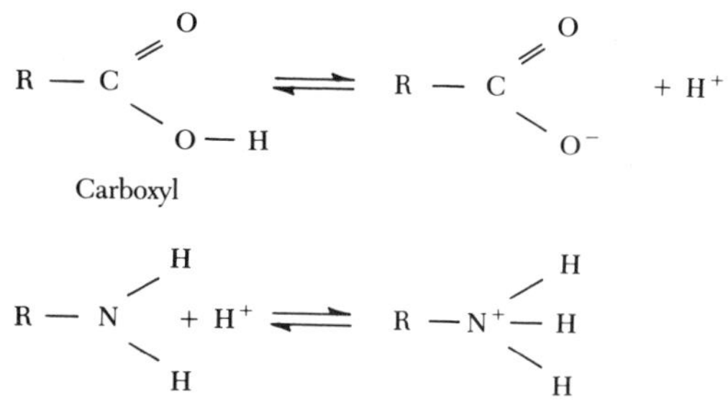

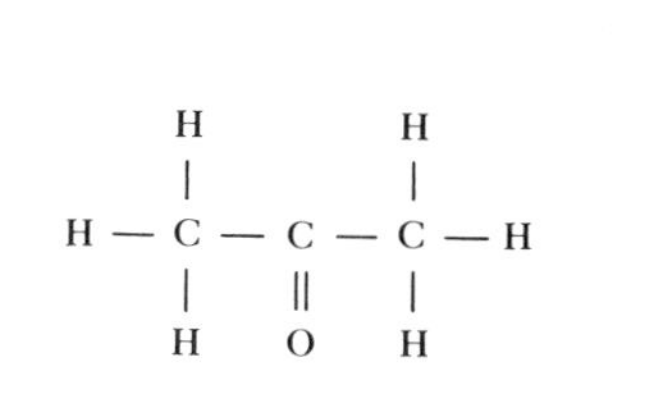

Figure 2.15 If two valences of the carbon of a carbonyl group are used to bind other carbons, a ketone results; in the above instance, acetone is the result.

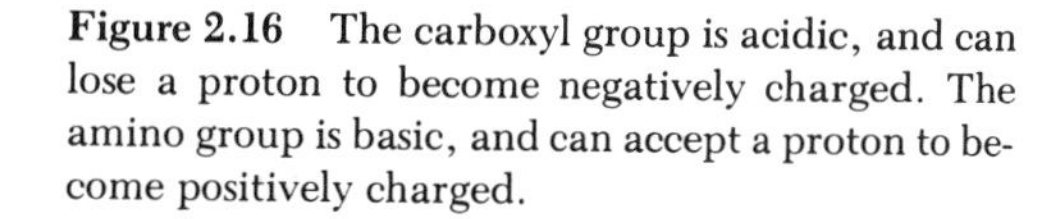

Figure 2.16 The carboxyl group is acidic, and can lose a proton to become negatively charged. The amino group is basic, and can accept a proton to become positively charged.

CLASSES OF CELLULAR MOLECULES

Any cell probably contains molecules that are peculiar to that cell type, but all cells and organisms, with all of the diversity of structure and function which they represent, make use, as Lwoff (page 7) has pointed out, of a limited number of small "building blocks" organized into specific macromolecules, with each playing a particular role in the life of the cell. The important classes of molecules are the carbohydrates, lipids, proteins, and nucleic acids. The structure of the first three will be considered here, as well as some molecules that are combinations of them. Because of the central importance of the nucleic acids as the source of biological information that conveys uniqueness on all living things, they will be the subject matter of a subsequent chapter.

Carbohydrates

The carbohydrates, with the sugars, starch, and cellulose being typical representatives, have carbon as the principal structural element to which other atoms are attached. The amount of hydrogen and oxygen combined with carbon is fixed so that the general formula, $(CH_2O)_n$, describes them all. This formula also explains the term *carbohydrate;* these molecules are *hydrates of carbon*, with the equivalent of a water molecule for each carbon.

The simplest of the sugars are the monosaccharides, with these in turn being further classified on the basis of the length of their carbon chains. That is, three-carbon sugars are *trioses*, four-carbon sugars *tetroses*, five-carbon sugars *pentoses*, and so on. Each of these sugars may be given other names as well (Figure 2.17), but note that all are characterized by the *-ose* ending.

The two most common sugars utilized by cells for energy and structural purposes are the hexoses, *glucose* and *fructose*. They are formed in green plants by the process of photosynthesis from carbon dioxide and water, and with visible light as

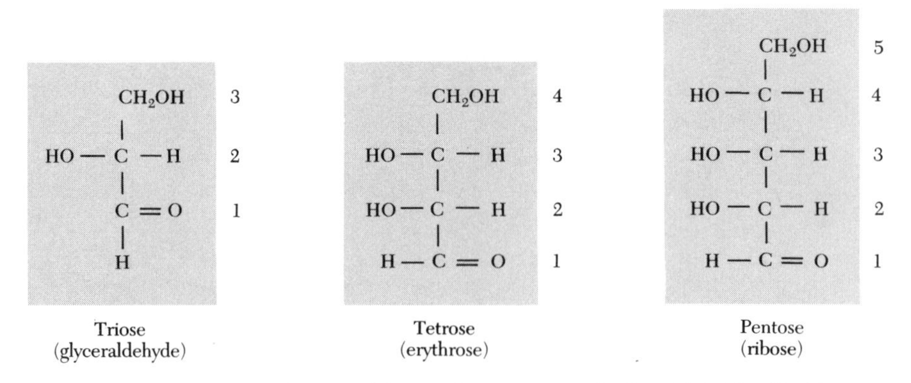

Figure 2.17 Three-, four-, and five-carbon sugars, each of which is identified by the number of carbons as well as by other names. Note that the carbon at the top, that is, the $-CH_2OH$, is given the highest number, even as it is in Figure 2.13.

$$6CO_2 + 6H_2O \xrightarrow{\text{Light}} C_6H_{12}O_6 + 6O_2$$

Figure 2.18 The general formula of the reaction taking place when CO_2 and H_2O are converted into sugars ($C_6H_{12}O_6$) and O_2 during the process of photosynthesis. The stored energy, derived from the absorption of visible light by chlorophyll, amounts to 672,000 calories per mole of sugar, a calorie being defined as that amount of energy required to raise one gram of water one degree centigrade. That energy can be released for use by the cell when the sugars are broken down during metabolism.

the energy source driving the reaction (Figure 2.18). Both can be represented by the formula $C_6H_{12}O_6$, but, as Figure 2.19 indicates, there are differences in the groups attached to the number 1 and 2 carbons. Other variations within the hexose formula are also known, and relate to the disposition of the H and OH groups at the number 4 and 5 carbons in the chain to give such sugars as *galactose* and *mannose* (Figure 2.20). These molecules have the same chemical composition, but different chemical properties; they are said to be *isomers* of each other.

The open-chain form of the hexoses, when in solution, exists in equilibrium with a ring form (Figures 2.21 and 2.22). The rings result from an interaction of the OH group at the number 4 or 5 carbon position with the carbonyl group in position 1 or 2. As a result, glucose forms a six-membered ring (Figure 2.21), fructose a five-membered ring (Figure 2.22). Since the structure of a molecule determines its reactivity, linear and ring variations of the same sugar are bound to behave differently.

The pentose sugars are of critical importance as structural elements in the formation of the nucleic acids. The two with which we will deal in the next chapter are *ribose* and *deoxyribose*, with the latter differing by having the OH group in the number 2 position replaced by a H (Figure 2.23).

The monosaccharides are among the "small building blocks" used to form

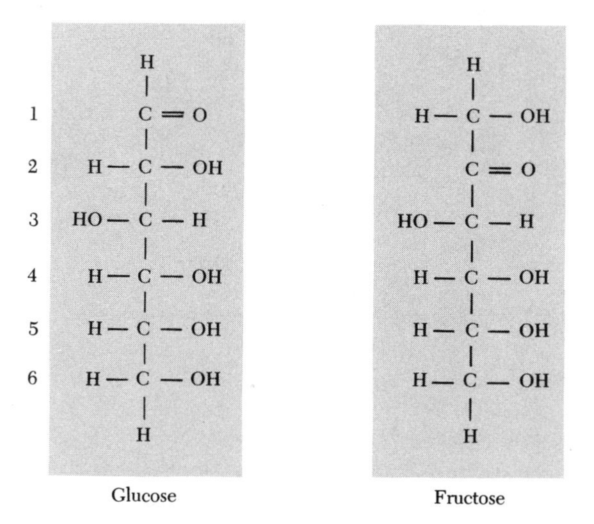

Figure 2.19 The linear structure of glucose and fructose. Note the differences at the one and two carbon positions. Both can be represented by the general formula $C_6H_{12}O_6$, so they are isomers of each other.

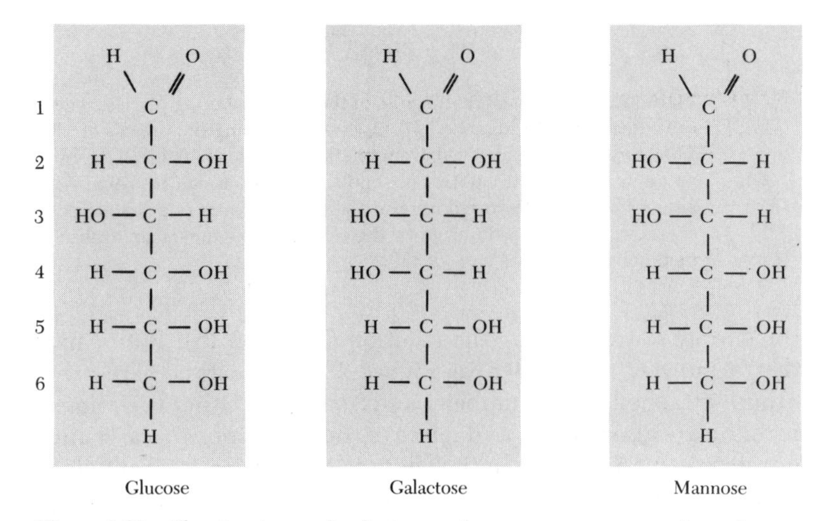

Figure 2.20 The structures of galactose and mannose as compared to glucose. Note that here, as compared to the difference between glucose and fructose, the alteration revolves around the disposition of H and OH groups at the two and four carbons.

larger molecules through linkage with each other. For example, two molecules of glucose unite to form the *disaccharide maltose* (Figure 2.24); glucose and fructose to form *sucrose*, the common table, or cane, sugar; glucose and galactose to form *lactose*, or milk sugar. A molecule of water is removed to bring about such union, but is required when the disaccharide is broken down to its monosaccharide units.

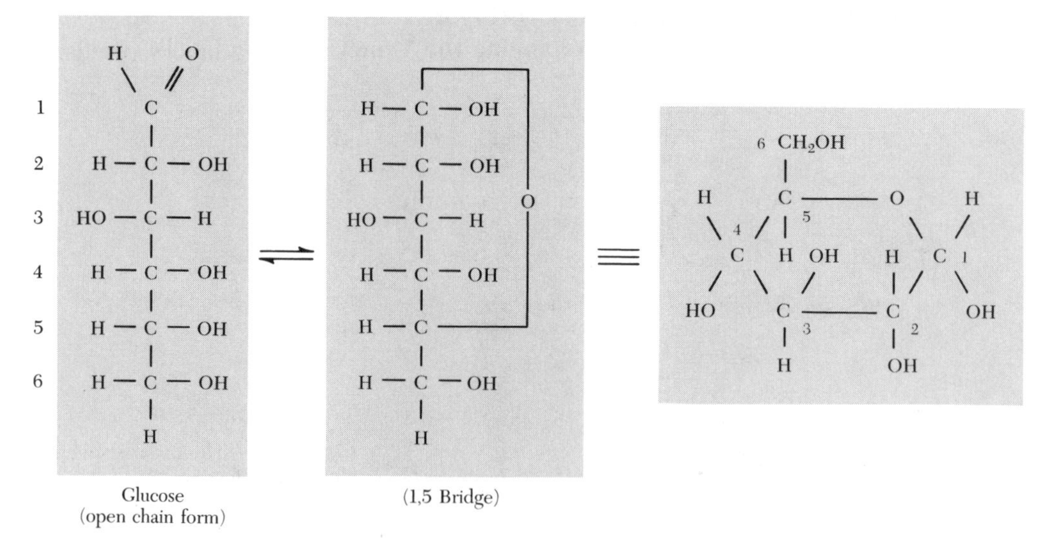

Figure 2.21 The chain and ring forms of glucose, with the six-membered ring the result of a linkage of the one and five carbons.

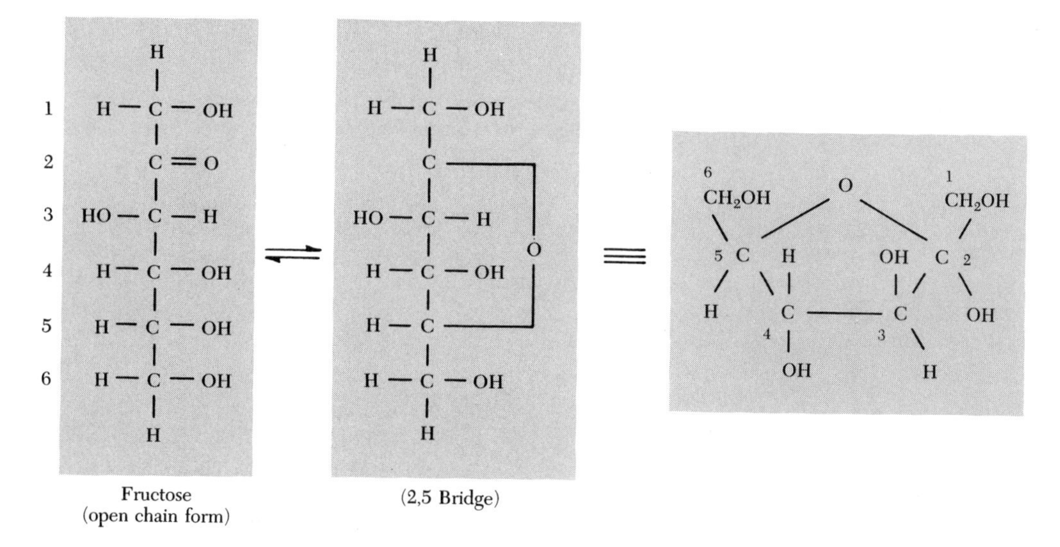

Figure 2.22 The chain and ring forms of fructose, with the five-membered ring resulting from a linkage of the two and five carbons.

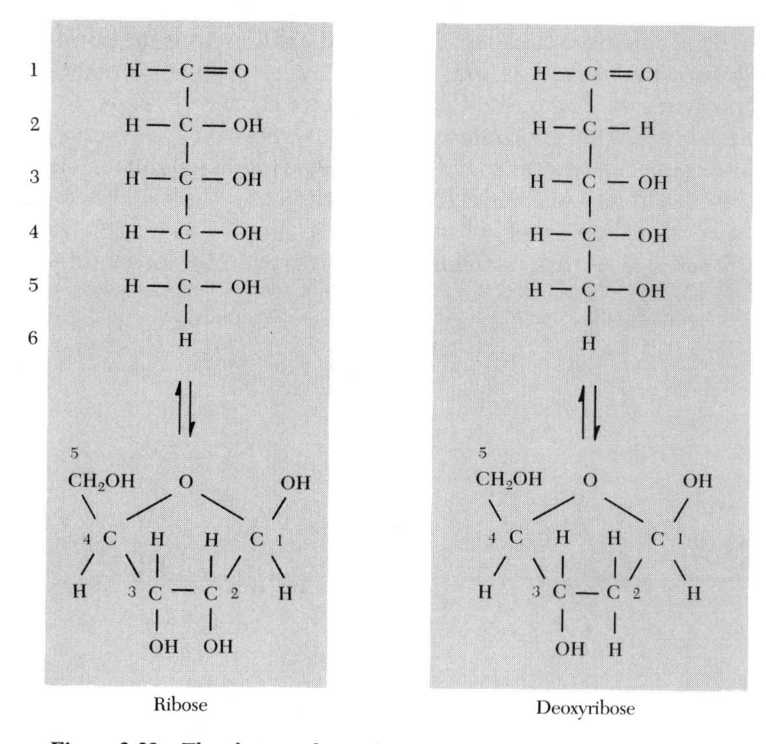

Figure 2.23 The chain and ring forms of ribose and deoxyribose sugars.

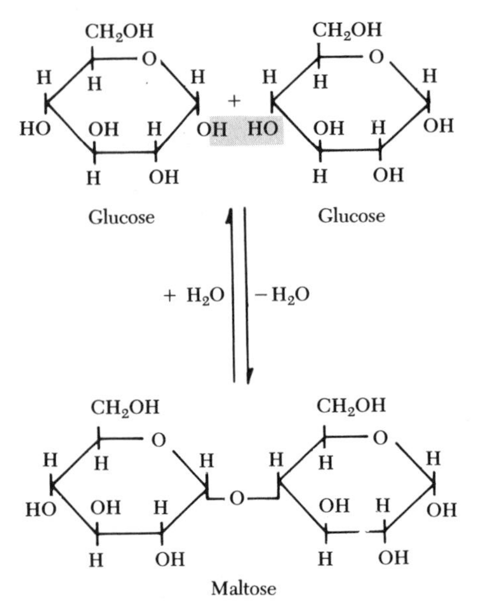

Figure 2.24 The union of two monosaccharides (glucose) to form a disaccharide (maltose) by linkage of the one and four carbons, and with loss of a molecule of water.

Oligosaccharides contain from two to ten monosaccharides joined by such linkages, while the *polysaccharides* are longer-chain molecules made up of many repeated monosaccharides. *Starch* and *cellulose* are the common plant examples, *glycogen* the typical animal example. As glucose is formed during the process of photosynthesis, the individual hexoses are rapidly shifted into an insoluble state as starch within the chloroplasts, where it is retained as a reserve carbohydrate. Each starch molecule contains a large number of glucose units in either a straight-chain form or a highly branched form. In both cases the backbone linkage between the glucose molecules is a 1–4 linkage as in maltose. The branch points in the branched form, however, are 1–6 linkages (Figure 2.25). In cellulose, which is also a

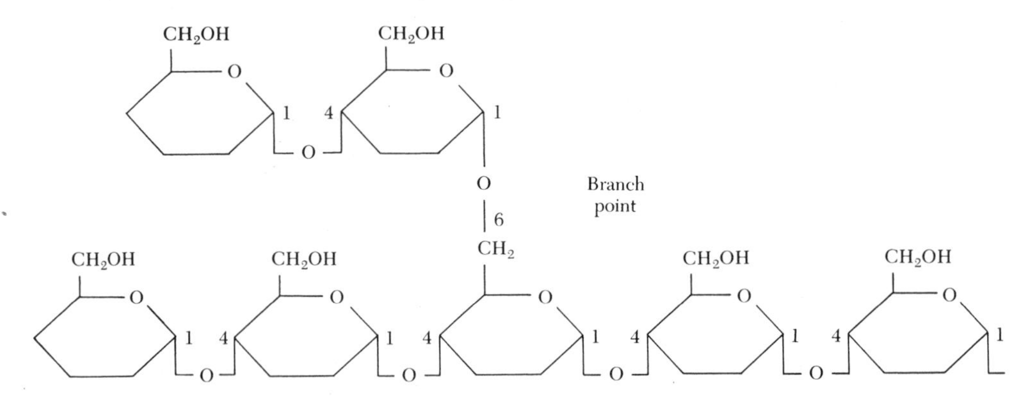

Figure 2.25 Two types of linkages found between glucose subunits of starch. The one–four linkage is characteristic of both the straight chain form and of the straight chain portions of the branched form. The branch points are one–six linkages.

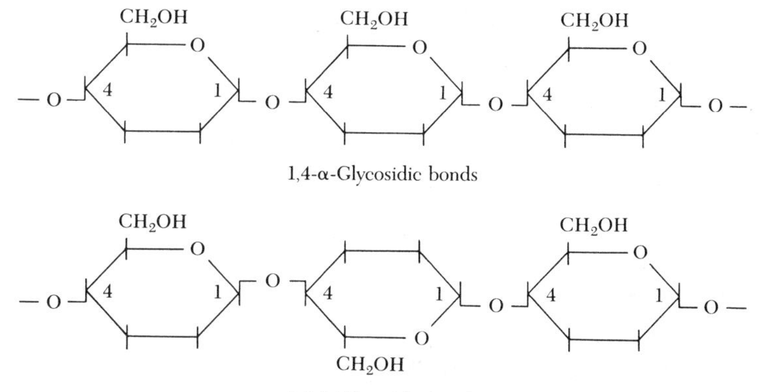

Figure 2.26 Two of the types of glucosidic linkages in polysaccharides. The top one is most commonly found in a variety of sugars, but the bottom one is characteristic of cellulose, and contributes to its great tensile strength.

straight-chain form, the repeated units of glucose are linked by a slightly different 1,4-β-glucosidic bond (Figure 2.26).

Cellulose is the most prevalent polysaccharide on the planet Earth, with at least 50 percent of all of the carbon in plants in this form. Its different form of linkage makes it insoluble in water and in most organic solvents, and it is digestible only by certain microorganisms. Termites, for example, use the cellulose in wood for food, but must rely on intestinal microorganisms to break the cellulose down into smaller, usable units.

The linear strands of cellulose are also linked to each other and aggregated into bundles by hydrogen bonding to give macromolecules of high tensile strength. We are familiar with cellulose in its purest state as cotton or linen, while that in wood and straw is intermingled with other plant products to give the many types of wood their varied characteristics of color and hardness.

Glycogen is a branched molecule which can contain as many as 30,000 glucose residues, connected to each other by α-glucosidic bonds (Figure 2.27). Stored in the cytoplasm of liver and muscle cells, glycogen, like starch, is a reserve carbohydrate which can be drawn upon when energy needs demand.

Lipids

The lipids are a diverse group of molecules having the common property of being generally insoluble in water, but fully soluble in nonpolar organic solvents such as ether, chloroform, or benzene. They include the fatty acids, neutral fats, oils, waxes, and steroids and sterols, as well as other complex molecules made up of fatty acids combined with glycerol, a three-carbon alcohol, or with nitrogen- and phosphorus-containing substances.

The *fatty acids* are unbranched hydrocarbon chains terminated at one end by

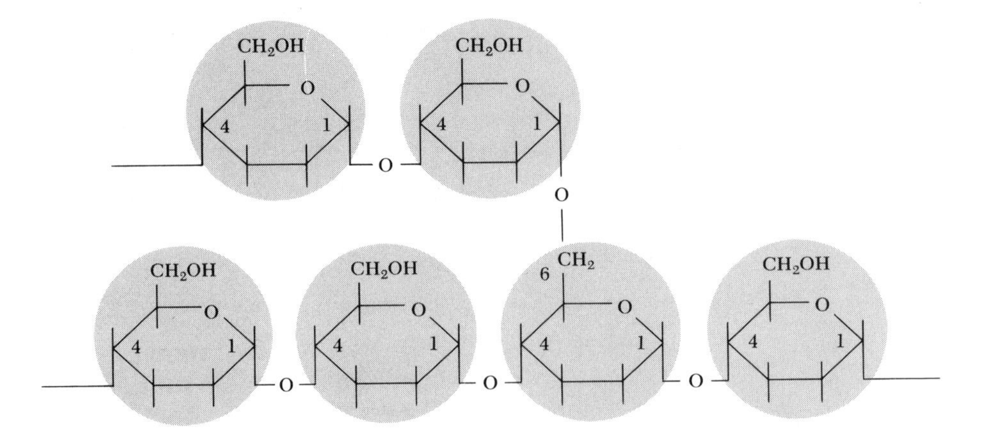

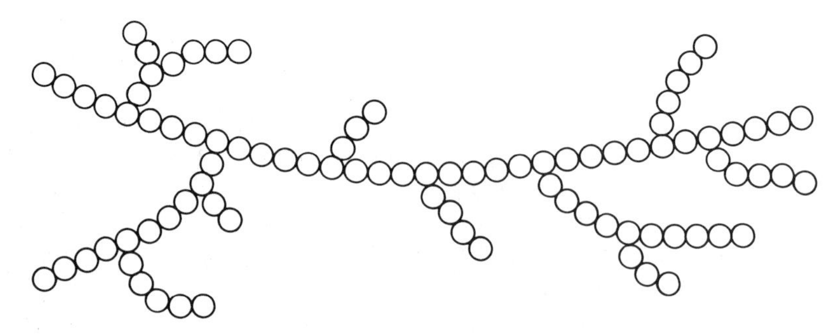

Figure 2.27 Two diagrams of the glycogen molecule, with the carbon skeletons of glucose residues (above), and with the glucose molecules shown as circles (below). The glucose molecules are held together by bonds between the one carbon of one glucose and the four carbon of its next neighbor. Branching occurs when a bond is established between the six carbon of one glucose and the one carbon of a side branch.

a carboxyl group. They include both saturated and unsaturated forms (Figure 2.28), the latter having one or more double bonds along their length. The double bonds of the unsaturated chains alter the configuration from a straight to a "kinked" geometry (Figure 2.29), as well as conferring additional reactivity on the molecule at the double-bonded carbons.

Although fatty acids are not found free in the cell to any great extent, they are important building blocks of lipids. *Triglycerides*, or *neutral fats*, which together with the carbohydrates represent stored forms of energy are formed by the condensation of three fatty acids to a molecule of the three-carbon alcohol, *glycerol* (Figure 2.30). At room temperature, animal fats tend to be solid or semisolid, while vegetable fats or oils are liquid. The difference is due to the degree of saturation of the fatty acid chains; the more unsaturated, the more readily liquefied. This is the

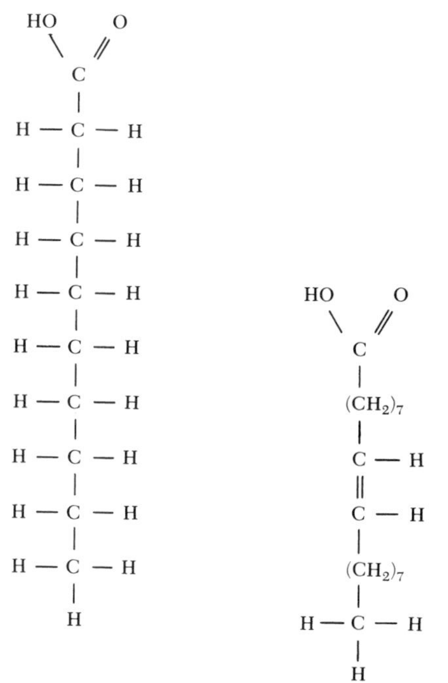

Figure 2.28 Two examples of fatty acids, each with a C backbone and a carboxyl group at one end. Caproic acid (saturated) on the left, oleic acid (unsaturated) on the right.

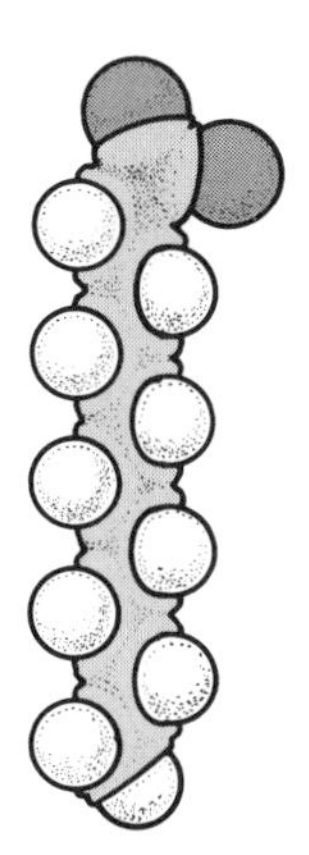

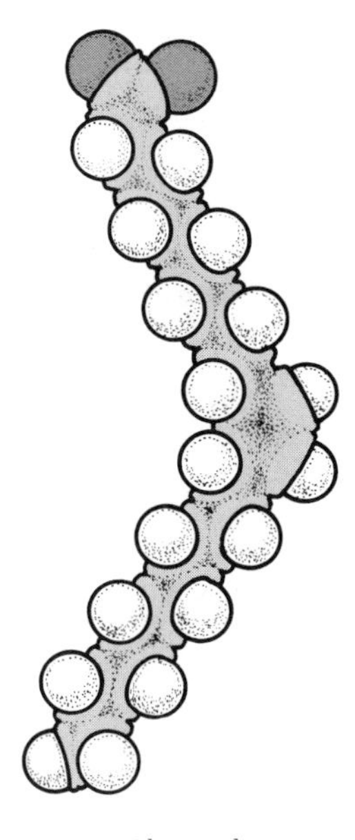

Figure 2.29 Space-filling three-dimensional models of (*a*) a saturated and (*b*) an unsaturated fatty acid, showing the effect of the double bond on molecular shape.

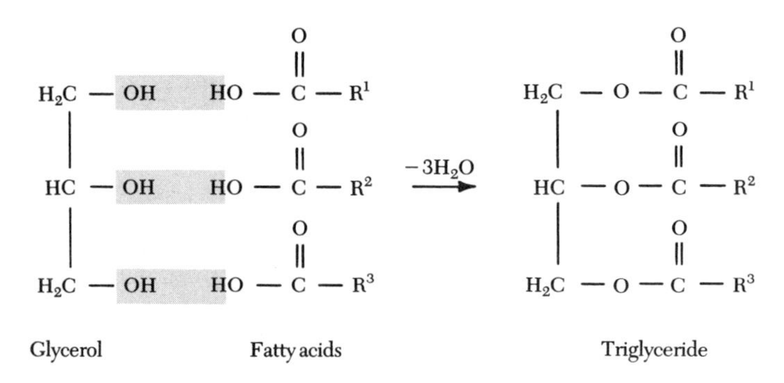

Figure 2.30 The synthesis of triglycerides from glycerol and fatty acids, with the removal of three molecules of water as a result of the reaction of the hydroxyl (OH) groups of the glycerol with the carboxyl (—C—OH) groups of the fatty acids. Such a synthesis requires energy which would be released with the breakdown of the triglycerides. The R groups can be identical or a mixture of fatty acids (see Figure 2.31). The water-repelling, or hydrophobic, portion of the triglyceride consists of the long-chain R groups of the fatty acid portion, while the water-seeking, or hydrophilic, portion is due to attached groups having OH, COOH, H_3PO_4, etc., in their makeup.

reason soap can be made more readily from animal fats than from vegetable fats. The latter, however, can be hydrogenated and made less unsaturated; this is the technique by which margarine is made.

One very important class of lipids is comprised of the *phosphoglycerides*, which play very central cellular roles. In phosphoglycerides, fatty acids are attached to only two of the carbon atoms of glycerol; the third carbon atom carries one of several phosphorus-containing compounds, as seen in Figure 2.31, which depicts the structure of one of the phosphoglycerides, *phosphatidyl choline*, or *lecithin*. Although the fatty acid portions of phosphoglycerides are nonpolar, and insoluble in water, that part of the molecule attached, via its phosphate group, to

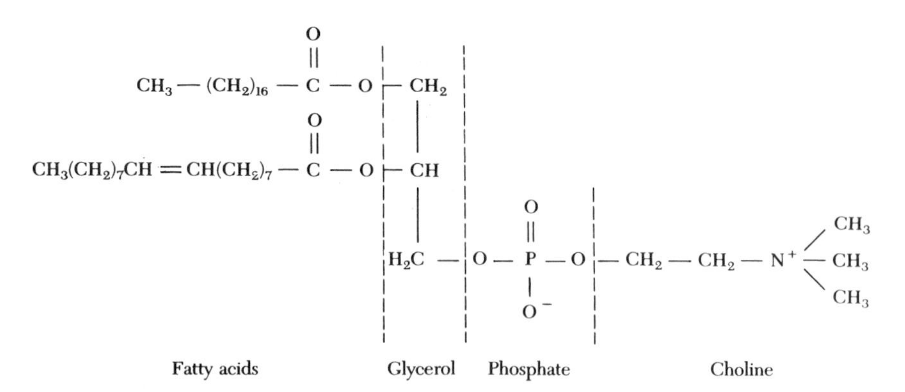

Figure 2.31 The structure of the phosphoglyceride, phosphatidyl choline.

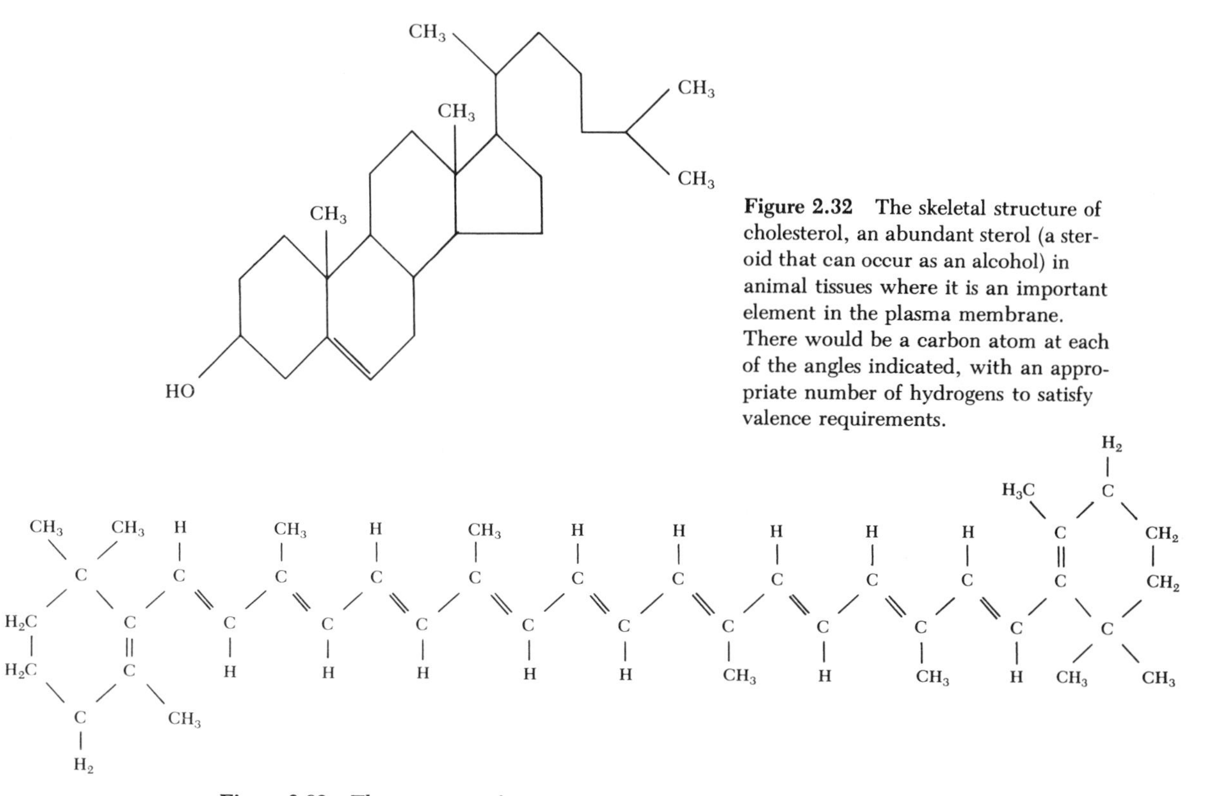

Figure 2.32 The skeletal structure of cholesterol, an abundant sterol (a steroid that can occur as an alcohol) in animal tissues where it is an important element in the plasma membrane. There would be a carbon atom at each of the angles indicated, with an appropriate number of hydrogens to satisfy valence requirements.

Figure 2.33 The structure of β-carotene, a molecule of central importance in the vision of animals. Animals cannot manufacture β-carotene themselves and must obtain it from plant materials. In plants it is involved in photosynthesis, and is also responsible for the orange color of carrots.

the third carbon is usually highly polar, and therefore soluble in water. As we shall see later (Chapter 5), this dual nature of phospholipids is a key feature in making them well suited for their structural and functional roles in cellular membranes.

Other lipids of importance include such *steroids* as the sex and adrenal cortical hormones, *cholesterol* (Figure 2.32), *carotene* (Figure 2.33), which is manufactured by plants but vital to animals because of its involvement in the chemistry of vision, and *vitamins* such as A, D, E, and K. Their structures are quite different from each other and from those previously discussed, but despite this and their varied roles, they conform to the definition of a lipid in being soluble in non-polar organic solvents.

Proteins

Proteins are the most predominant kind of molecules found in cells, and they perform a variety of roles of both a structural and regulatory nature. Indeed, proteins of one kind or another are associated with every structure in the cell and are involved in almost every cellular activity. Enormous numbers of different kinds of proteins are known, and they display great diversity in size and shape, each kind of protein being highly specialized in its function. They are extremely large molecules, with molecular weights of from 5000 up to 1 million. Some proteins are *fibrous* in nature, and form long rodlike structures, whereas others are folded upon themselves to assume a *globular* conformation. The fibrous proteins generally are structural elements, while the globular proteins include some of the hormones which regulate the internal environment of the animal body, the *antibodies* which are concerned with the destruction or immobilization of disease-producing agents, and most important, the *enzymes* which govern the chemical reactions occurring within cells.

When proteins are hydrolyzed in aqueous solution they yield a mixture of smaller molecules, all of which contain nitrogen and some of which contain sulfur as well. These are the *amino acids*, of which there are 20 in number. It is the varied number and sequential arrangement of the amino acids which counts for the great diversity among the proteins.

The amino acids have a common structure:

$$\begin{array}{c} R\text{—}C\text{—}COOH \\ | \\ NH_2 \end{array}$$

in which a carboxyl (COOH) group and an amino (NH_2) group are attached to the same carbon atom. Also attached to that C atom is the rest of the molecule, a side chain denoted by R. At pH levels close to those generally encountered in the cell, both the carboxyl and amino groups of the amino acid are charged, the former being negatively charged and the latter positively charged (Figure 2.34).

The differences between the 20 amino acids found in proteins reside in the R portion of the molecule. *Acidic* amino acids, such as aspartic acid, have an additional carboxyl group attached to their chain, while the *basic* amino acids are those with a second amino group, such as arginine. Such additional carboxyl and amino groups are also charged at normal intracellular pH values; *neutral* amino acids are those with uncharged R groups. While the acidic and basic amino acids are strongly hydrophilic, some of the neutral amino acids have relatively nonpolar R side chains and hence are hydrophobic. Such different properties of amino acids all greatly influence the structure and activity of the proteins in which they are present.

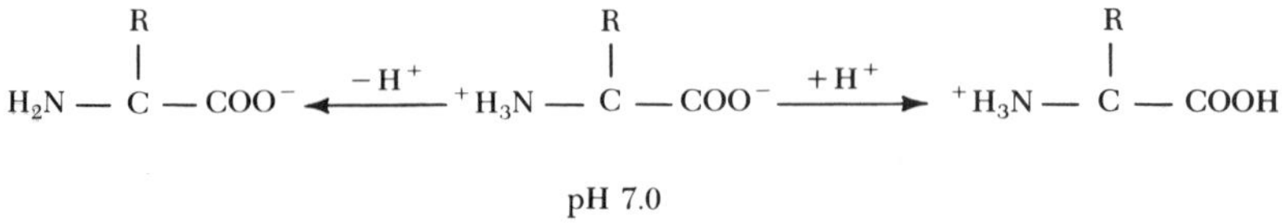

$$H_2N - \underset{}{\overset{R}{\overset{|}{C}}} - COO^- \xleftarrow{-H^+} {}^+H_3N - \overset{R}{\overset{|}{C}} - COO^- \xrightarrow{+H^+} {}^+H_3N - \overset{R}{\overset{|}{C}} - COOH$$

pH 7.0

Figure 2.34 At neutral pH, both the amino and the acid groups are charged. Increasing or decreasing the pH can alter the charge on one or another of the groups.

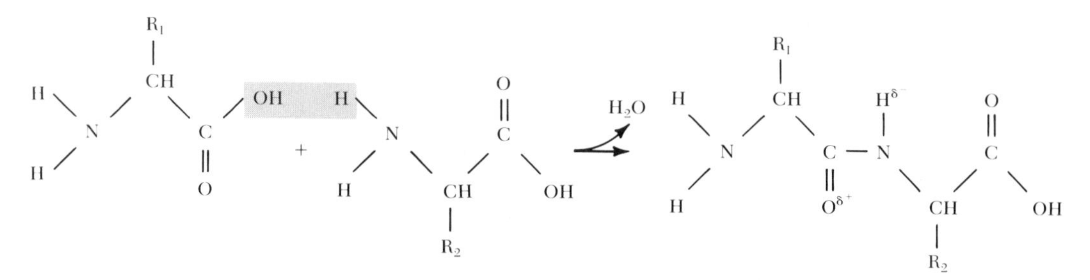

Figure 2.35 The formation of a peptide bond between two amino acids involves the removal of a molecule of water. The carboxyl and amide groups of the bond carry slight charges, and can participate in H-bonding with other polar groups.

The amino acids are linked together in a linear sequence in proteins by *peptide* bonds, in which the carboxyl of one amino acid joins with the amino group of another; in the process a molecule of water is removed (Figure 2.35). Another important feature of peptide bonds is that the oxygen of the carboxyl group has a slight negative charge and the hydrogen of the amide group a slight positive one, allowing them to participate in hydrogen bonding.

The *primary structure* of a protein is determined by the number, type, and linear order of the constituent amino acids in the *polypeptide* chain. In the next chapter we shall see that this sequential arrangement is determined by the information contained in the nucleic acids of the chromosomes. The linear sequences can themselves assume various conformations, depending on interactions between the amino acids of the primary sequence.

Secondary structure is the extent to which a particular three-dimensional orientation of amino acids relative to one another is maintained in the polypeptide chain. One such example is the α helix (Figure 2.36), which is the most stable or "comfortable" conformation for certain sequences of particular amino acids to assume. The α helix is stabilized by hydrogen bonding between nonadjacent amino acids, and the R groups of the amino acids project outward from the axis of the helix. The ability of a polypeptide sequence to form an α helix depends on the nature of the R groups, which in some cases can disrupt the H bonds and hence the α-helical coil. A different secondary structure is called the β sheet; in this configuration stretches of the polypeptide chain are extended rather than coiled, and different extended chains, or parts of chains are held parallel to one another by H bonds to form a sheet.

The *tertiary structure* of a protein results from folding of the polypeptide chain upon itself in a highly specific fashion. This tertiary, or globular three-dimensional structure is stabilized by various weak interactions between different parts of the molecule, such as H bonds and ionic interactions between charged amino acid side chains, as well as by occasional disulfide bonds which can form between two residues of the amino acid cysteine (Figure 2.37). The three-dimensional conformation assumed also depends on the types and positions of the different R groups, since the chain will tend to fold in such a way that hydrophobic

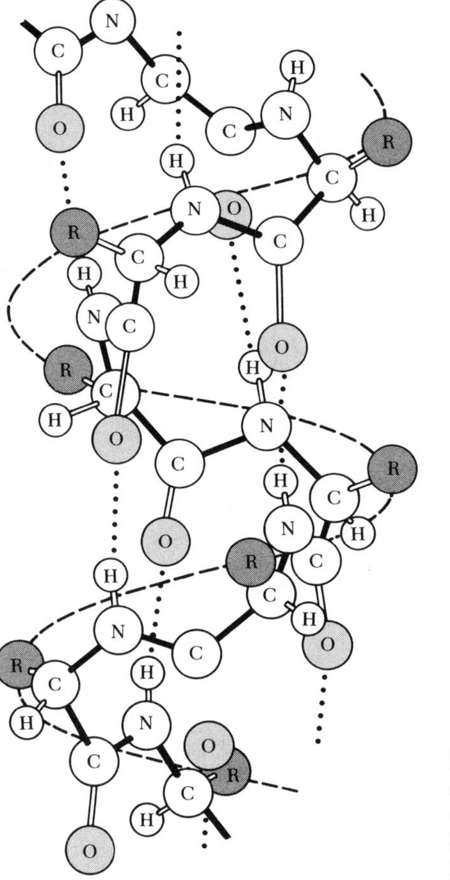

Figure 2.36 The α-helical arrangement of a protein, giving it a secondary structure stabilized by H-bonds (dotted lines) between the $>C{=}O$ and $>NH$ groups of different peptide bonds.

groups will be shielded from the surrounding water, whereas hydrophilic groups will tend to be exposed at the surface.

Quaternary structure results when two or more polypeptides unite to form a multimeric protein. For example, hemoglobin A, the prevalent hemoglobin in human adults, consists of two α-polypeptide chains and two β chains, grouped together around a heme center (Figure 2.38). Ferritin, with a molecular weight of 480,000, is an even more complex mammalian protein; it consists of 20 similar polypeptides of about 200 amino acids each, grouped into a large globular structure around atoms of iron (Fe). Once again weak interactions between the polypeptide chains are primarily responsible for the maintenance of quaternary structure.

Denaturation. It can be seen that the structure of a particular protein is not a haphazard one. The sequence of amino acids that give a protein its primary structure will automatically determine its three-dimensional conformation under specific conditions, depending on the various intramolecular forces between the amino acids, and the forces between the protein and its surrounding chemical en-

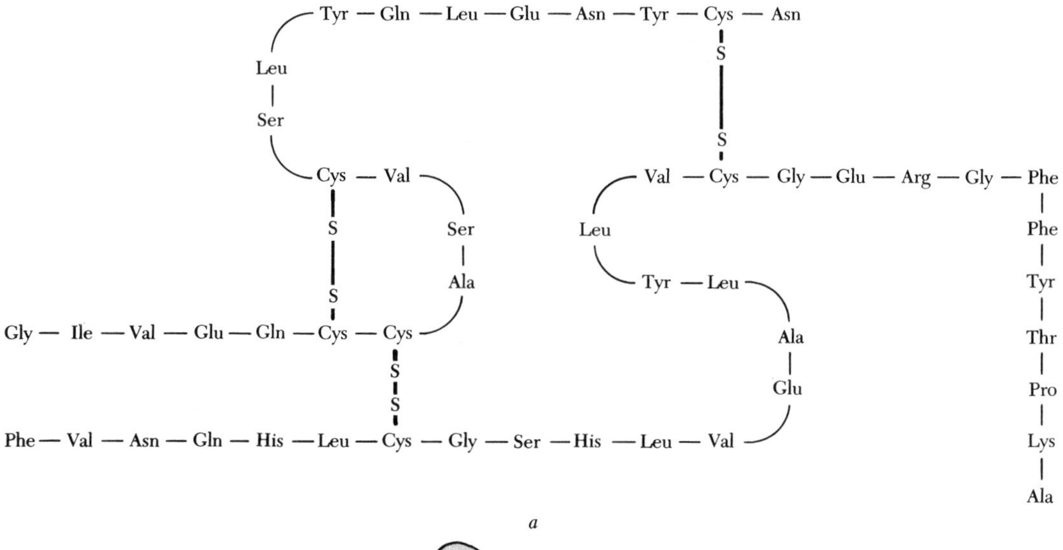

a

b

Figure 2.37 (*a*) Tertiary structure of a molecule of insulin showing the folding that results from the formation of disulfide bonds between cysteine residues. The molecule is flattened in this illustration; it would normally have a more globular shape. (*b*) Schematic representation of a globular protein, with the spiral polypeptide folded and stabilized in disulfide and other bonds.

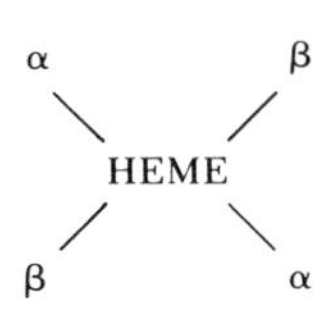

Figure 2.38 The molecule of hemoglobin A, made up of a central heme group to which two molecules of α-protein, and two of β-polypeptide, are attached to provide a globular, quartenary form to the functional unit.

vironment. Since the three-dimensional conformation is required for a protein to function properly and in a specific *manner*, any factors which disrupt the forces responsible for maintenance of this conformation will *denature* the protein, resulting in an altered structure and hence in malfunction. Two such factors are extremes of pH, which can affect the charges on different parts of the molecule, and of temperature, which can disrupt H bonds. Hence proteins can function properly only within limited ranges of temperature and pH.

ADDITIONAL READING

HELMPRECHT, H. L., and FRIEDMAN, L. T. 1977. *Basic Chemistry for the Life Sciences.* McGraw-Hill Book Company, New York.

LEHNINGER, A. L. 1982. *Principles of Biochemistry.* Worth Publishers, Inc., New York.

MCELROY, W. D. 1971. *Cell Physiology and Biochemistry*, 3rd ed. Prentice-Hall, Englewood Cliffs, N.J.

STRYER, L. 1981. *Biochemistry*, 2nd ed. W. H. Freeman and Company, Publishers, San Francisco.

3

Energy, Enzymes and Chemical Reactions

As we have seen in Chapter 1, life can be viewed as a process that organizes matter and energy into ordered states; as such it runs counter to the overall tendency toward an increase in entropy. If the levels of order and complexity that are characteristic of living substance are to be maintained and increased, energy must be acquired from the surroundings and funneled into biological processes. Furthermore, there must be chemical and structural features of cellular organization and activity that permit the energy flow to be regulated and directed in specific ways. In this chapter we introduce some important aspects of, and general concepts relating to, processes in which energy can be managed by the cell.

The ultimate source of energy for all life today is that part of solar radiation which we call sunlight; some of this radiant energy is captured by green plants in the process of *photosynthesis* and converted into chemical energy that can be conserved in the storage molecules, primarily carbohydrates produced during the process. As we shall see, this energy is locked into the structure of these storage molecules. Animals, of course, cannot carry out photosynthesis and must depend for their energy supply on the storage products made by plants; this dependence may be direct, as for plant-eating herbivores, or indirect, as for carnivores, which eat other animals, including the herbivores. The plant or animal cell can then draw on these fuel reserves by breaking down the molecules in such a way that the energy is made available in forms that can be utilized to carry out the work of the cell; for example, in the formation of other molecules and structures needed by the cell, or for the various kinds of mechanical, electrical, and osmotic work that must be done. The controlled series of chemical reactions in which the storage molecules are broken down and which result in an efficient recovery of the energy is the process known as *aerobic respiration*.

The way in which energy is captured in photosynthesis and made available in respiration for the work of the cell will be considered in detail later (Chapter 6), as will the cell organelles, the chloroplasts and mitochondria, respectively, in which these processes take place.

ENERGY AND CHEMICAL REACTIONS

Once radiant energy has been trapped and converted to chemical energy, all other events taking place within the cell are through the medium of chemical reactions; we must, then, consider briefly how energy is manipulated in chemical reactions occurring in the cell.

All compounds are characterized by certain energy contents; some of this energy is free energy (Chapter 1), that is, energy that is available and can be retrieved and used to perform work. The rest of the energy, which is no longer available to do work, is related to the degree of entropy. Since it is the useful energy that is important, we are concerned primarily with the free energy changes that occur during chemical reactions. The amount of free energy available from a molecule depends on which atoms it contains and how they are arranged. When chemical bonds are broken and re-formed during chemical reactions, energy is redistributed within the molecules. In some reactions the free energy level of the final state is lower than that of the initial state. Such *exergonic* reactions are said to be *spontaneous*, and the difference in free energy levels is the amount available to do work. In other reactions, including many that occur in the cell, the final free energy state is higher than the initial one; such *endergonic* reactions cannot take place without an input of energy, which must be supplied from elsewhere. In the cell the endergonic, or uphill, reactions must be coupled to the exergonic, or downhill, reactions in such a way that the free energy available from the latter can be used to perform the work involved in driving the former.

How, then, is the free energy yield from the exergonic reactions, primarily those of aerobic respiration, transferred to the endergonic reactions which represent chemical work? Part of the answer to this is that the cell uses a metabolic system involving molecules that we can think of as energy-carrying intermediates—molecules that can carry some of the free energy from downhill reactions and supply it to uphill reactions.

The ATP-ADP Cycle

The common intermediate used by the cell as an energy "shuttle" is the compound adenosine triphosphate, or ATP (Figure 3.1). ATP can be hydrolyzed to adenosine diphosphate, or ADP, by removal of the terminal phosphate (Figure 3.1). Since the free energy state of ATP is considerably higher than that of ADP and inorganic phosphate, a large amount of free energy is made available by the reaction. Conversely, when ATP is formed by the addition of a terminal phosphate, free energy

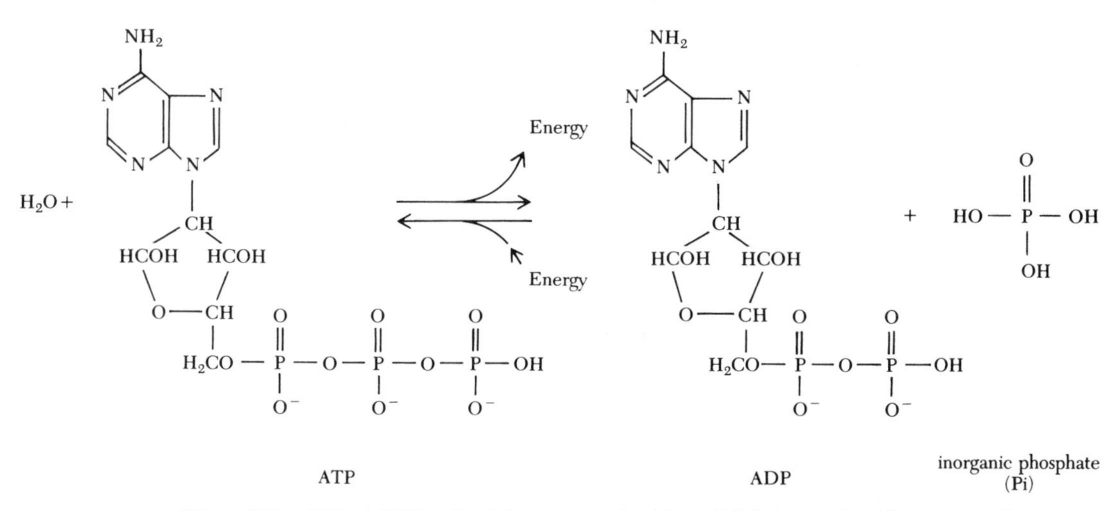

Figure 3.1 ATP ⇄ ADP + P_i. Adenosine triphosphate (ATP) is a nucleotide carrying three phosphate groups; adenosine diphosphate (ADP) has only two. The difference in free energy between the two compounds is approximately 7,000 kcal/mole. This free energy is used to do the work of the cell when ATP is hydrolyzed to ADP and inorganic phosphate.

must be supplied. The position of ATP as the intermediate between exergonic and endergonic reactions is illustrated in Figure 3.2. Some of the free energy available from the exergonic reaction A ⟶ B can be used to form ATP from ADP and phosphate. The subsequent hydrolysis of ATP, if coupled directly to the endergonic reaction C ⟶ D, can supply the necessary energy to this uphill reaction. Thus the energy-yielding reactions, particularly those of aerobic respiration, need not be linked directly to the energy-requiring reactions. Rather, ATP acts as the intermediate carrier in energy transfers; it is the molecular currency in the energy economy of the cell.

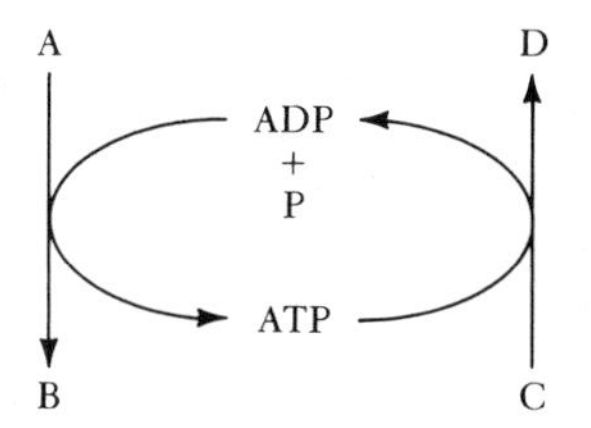

Figure 3.2 The transfer of energy from the exergonic reaction A→B to the endergonic reaction C→D is mediated by the intermediate energy carrier ATP.

Energy Transfer from ATP

In order that the free energy of hydrolysis of ATP be conserved to do the work involved in driving an endergonic reaction, for example, C ⟶ D (Figure 3.3a), the reactions must be directly coupled. The way in which this coupling is most commonly accomplished in the cell is by two sequential reactions (Figure 3.3b) in

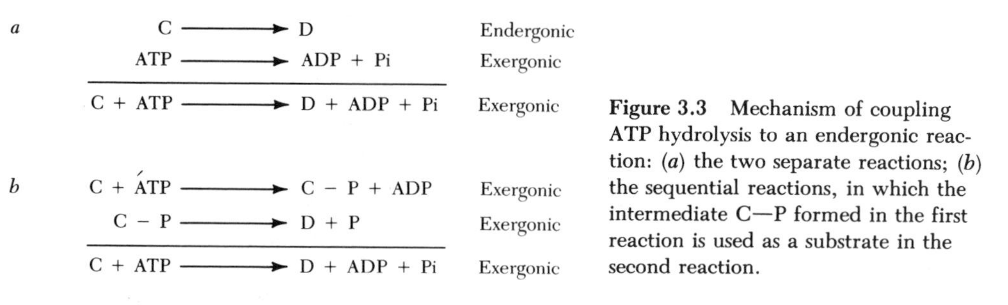

Figure 3.3 Mechanism of coupling ATP hydrolysis to an endergonic reaction: (*a*) the two separate reactions; (*b*) the sequential reactions, in which the intermediate C—P formed in the first reaction is used as a substrate in the second reaction.

which a product of the first is used up in the second. In the first reaction the terminal phosphate group of ATP is transferred to the substrate, C, in such a way that some of the free energy of hydrolysis of ATP is conserved and stabilized in the new association of the phosphate group with the substrate C. The free energy level of the phosphorylated species is now sufficiently high for its conversion to the final product, D, to be a thermodynamically favored, that is, downhill, reaction.

The transfer of the phosphate group from ATP to the substrate is accomplished with the aid of enzymes (see later); the intervention of the enzymes not only permits efficient transfer of energy, but can confer specificity; that is, the enzymes can channel the flow of energy into specific chemical reactions.

Oxidation-Reduction Reactions

The chemical reactions which yield the energy for ATP formation are predominantly those of aerobic respiration in which fuel molecules such as glucose are *oxidized* to CO_2 and oxygen is reduced to H_2O. While the details of respiration will be covered in Chapter 6, oxidation-reduction reactions are very important in energy manipulation by the cell, and should also be considered here.

Oxidation is defined as loss of electrons, while *reduction* refers to the gain of electrons. Since electrons can exist only as parts of atoms or molecules, each oxidation must be accompanied by a concomitant reduction, as electrons are transferred from the molecule being oxidized to the molecule being reduced. Also, in many, but not all, biological oxidation-reduction reactions the electrons being transferred are accompanied by protons or hydrogen ions (H^+); hence oxidation and reduction may also be thought of as removal and gain respectively of hydrogen atoms.

Let us consider a hypothetical example of an oxidation-reduction reaction, in which one compound, A, is oxidized and some other compound, B, is reduced. We can consider this reaction as the sum of two half-reactions, as shown:

$$AH_2 \xrightarrow{\text{oxidation}} A + 2e^- + 2H^+ \quad (1)$$

$$B + 2e^- + 2H^+ \xrightarrow{\text{reduction}} BH_2 \quad (2)$$

The overall combined reaction, in which electrons and protons have been transferred from A to B, is, therefore,

$$AH_2 + B \longrightarrow A + BH_2 \quad (3)$$

Since the half-reactions (1) and (2) involving loss or gain of electrons (and in this case protons) result in changes in the chemical bonds within the molecules, the energy levels of the molecules A and B are also changed. The capacity of a compound to give up its electrons (to act as a reducing compound) is a measure of its *redox potential*, which in turn is related to its free energy level. Thus, if, in the example above, A loses more energy in being oxidized than is gained by B in being reduced, the final state, $A + BH_2$, will be at a lower potential than the initial state $AH_2 + B$, and the difference in free energy will be available to do work. Such electron transfer reactions are commonly used by the cell in its energy exchanges and, as we shall see later, are the major source of the energy for the formation of ATP.

Electron Carriers

As we have pointed out earlier in this chapter, the breakdown of fuel molecules such as glucose occurs in a series of carefully controlled steps, some of which are exergonic oxidations involving loss of electrons and their ultimate transfer to oxygen. However, these electrons, with or without accompanying protons, are not transferred directly from oxidizable compounds to oxygen; rather, special carrier molecules are used to transfer the electrons to their site of utilization in much the same way that ATP can transfer phosphate group potential.

The electron carrier we shall consider here is nicotinamide adenine dinucleotide (NAD), the structure of which is shown in Figure 3.4. NAD can exist in two forms, one oxidized, the other reduced. In its oxidized form, NAD is positively charged and can be represented as NAD^+. NAD^+ can accept two electrons, together with one proton, and be reduced to NADH. The two half-reactions involved in the oxidation of a substrate by NAD^+ are shown on the following page.

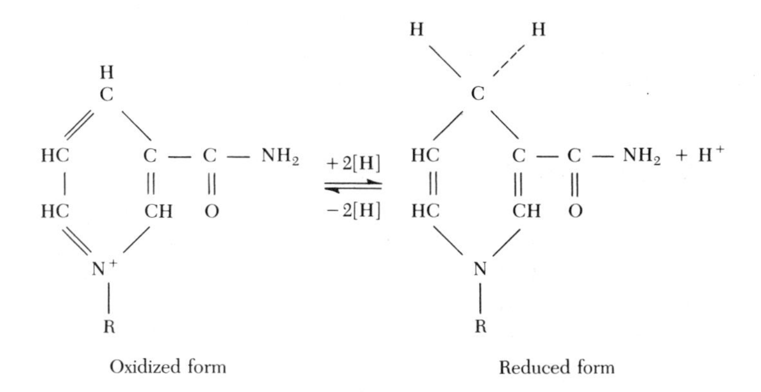

Figure 3.4 Structure of the electron-accepting portion of NAD. The ring of the nicotinamide moiety can accept two electrons and one proton and become reduced. Similarly, the reduced form can donate electrons and a proton in reducing some other substrate.

$$AH_2 \xrightarrow{\text{oxidation}} A + 2e^- + 2H^+ \quad (4)$$

$$NAD^+ + 2e^- + H^+ \xrightarrow{\text{reduction}} NADH \quad (5)$$

Again, the overall reaction is shown as

$$AH_2 + NAD^+ \longrightarrow A + NADH + H^+ \quad (6)$$

We can see that only one hydrogen ion is transferred to NAD; the other is released into the medium, thereby contributing to a lowering of pH.

Once the NAD has been reduced, it can transfer the electron potential to some other compound, thereby effecting a second oxidation-reduction reaction. The most important reduction it effects is that of oxygen, the ultimate electron acceptor in aerobic respiration. Although the transfer of electrons from NADH to oxygen is indirect and complex, involving many steps (see Chapter 6), we can represent it again by two half-reactions.

$$NADH \longrightarrow NAD^+ + 2e^- + H^+ \quad (7)$$

$$\tfrac{1}{2}O_2 + 2H^+ + 2e^- \longrightarrow H_2O \quad (8)$$

Overall,

$$NADH + \tfrac{1}{2}O_2 + H^+ \longrightarrow NAD^+ + H_2O \quad (9)$$

The free energy of oxidation of NADH is very large, and it is this free energy that is used to form most of the ATP of the cell. The complexity of electron transfer from NADP to oxygen is related to the necessity of harvesting the free energy in as efficient and useful a manner as possible. The connection between oxidation of the electron carrier NADH and phosphorylation of ADP to form ATP will be discussed in detail in a later chapter.

Electron carriers other than NADH are utilized by the cell in shuttling "reducing power" from place to place. In all cases, however, and as we pointed out for ATP utilization, the intervention of enzymes is required, thereby ensuring that the "reducing power" is channeled down specific pathways.

ENZYMES AND CHEMICAL REACTIONS

In order for a reaction or series of reactions to be useful to the cell, they must proceed at rates that can provide sufficient products to satisfy the requirements of the cell. Although particular chemical reactions can occur if the overall free energy change is negative, this does not mean that they will occur, or that they will occur at measurable rates. For example, the burning of a log of wood represents an overall reaction in which carbohydrates are completely oxidized, with a concomitant release of free energy as heat. Two important points can be made about this. First, although the reaction is clearly exergonic, a log of firewood will not combust unless it is first lit; that is, before the stored energy can be obtained, thermal energy must be supplied, suggesting that there is some initial energy barrier which must be overcome for the reaction to proceed at a measurable (and useful) rate. As we shall see, this is indeed the case for all chemical reactions. Second, while the end result of oxidation of the carbohydrates of the log is similar to that of the oxidation of carbohydrates in aerobic respiration, that is, free energy is made available, the pathways are very different. In aerobic respiration there is a very precise and complex series of reactions which take place to ensure that some of the energy is conserved as useful chemical energy, and not all is dissipated as heat; that is, in aerobic respiration the pathway is directed and controlled. These two points are related in that the precise pathways followed in the metabolism of the cell are regulated by *enzymes*, proteins which act by lowering energy barriers to specific reactions and thereby directing the flow of energy through specific channels.

Before describing how enzymes do this, we must look a little more closely at the nature of chemical reactions. Molecules consist of two or more atoms united to each other by chemical bonds, representing particular stable arrangements of electrons. In order for these molecules to react and arrive at a new stable state, the electronic structures must first be rearranged; the amount of energy required to excite the electrons to a more unstable state in which the molecules can react is called the *energy of activation* and represents the energy barrier referred to earlier (Figure 3.5). Also, if molecules are to react with each other, they must collide or be brought into close juxtaposition. Thus the rate at which a reaction will proceed will depend on the number of collisions involving molecules of energy levels above the level of the energy of activation. One way of increasing the rate of a reaction is by heating the system; this increases the internal energy of the molecules and also increases the likelihood of their colliding (Figure 3.6). Thus a higher proportion of the molecules will be able to react.

The cell, however, functions only within a limited range of temperatures and the reacting molecules are rarely in high concentration. Another means of ensuring that specific reactions can occur at appreciable rates must therefore be used; this is

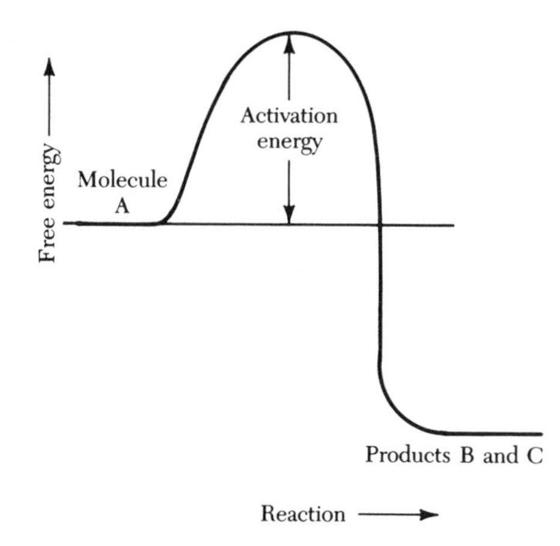

Figure 3.5 Although an exergonic reaction may yield free energy, an initial energy level must be achieved. The activation energy represents an energy barrier to the reaction.

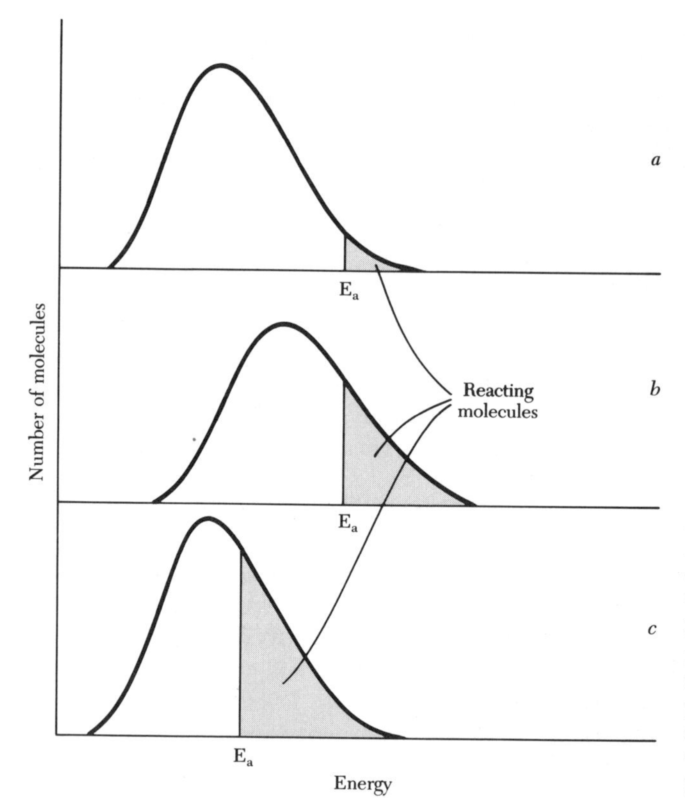

Figure 3.6 (*a*) In this population of molecules only a small proportion has energy greater than the energy of activation (E_a); hence few will react. (*b*) By raising the temperature a higher proportion of molecules has sufficient energy to react; hence the reaction will proceed more rapidly. (*c*) By lowering the energy of activation the reaction will proceed more rapidly even at the lower temperature.

to lower the energy of activation itself (Figures 3.6c and 3.7) to a level at which a higher proportion of molecules can participate in the reaction, thereby accelerating the rate of the reaction.

Agents which speed up the rates of reactions by lowering the energies of activation are called *catalysts.* In the cell the catalysis of specific reactions is accomplished by the intervention of enzymes. Enzymes are long-chain proteins made up of a sequential array of amino acids. The number, kind, and sequence of the amino acids, determined by the coded information of the cell, in turn determines the three-dimensional geometry of the enzyme, an important feature in its activity. Enzymes lower the energy of activation by providing surfaces with specific sites to which reactants, or *substrates*, can bind; such interaction between enzyme and substrate not only can facilitate the rearrangement of electrons necessary for the substrate to react but also allows substrate molecules to be oriented in appropriate configurations for interactions to occur.

Figure 3.8 is a diagrammatic and simplified representation of the formation of an enzyme–substrate complex and of catalysis. There are several points, however, which may not be obvious from the diagram and which should be remembered. First, the formation of the enzyme–substrate complex is itself a

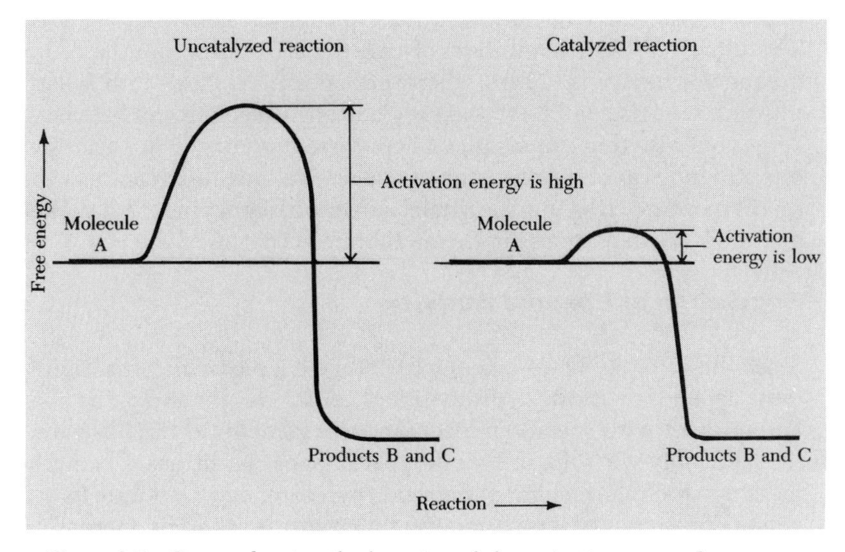

Figure 3.7 Curves showing the lowering of the activation energy by an enzymatic catalyst as contrasted with the larger amount of energy needed to accomplish the same thing in the absence of the enzyme. The enzyme makes the molecule more reactive without itself being consumed in the process.

chemical reaction which must have a sufficiently low energy of activation and high probability of contact between enzyme and substrate to be likely to occur. Enzymes as a rule do not float around freely in the cell, particularly in a eukaryotic cell, but tend to be strongly compartmentalized or localized; presumably this is related to the requirement for both spatial and temporal proximity of enzymes and substrates. Second, enzymes are not used up in the reaction; once the products are

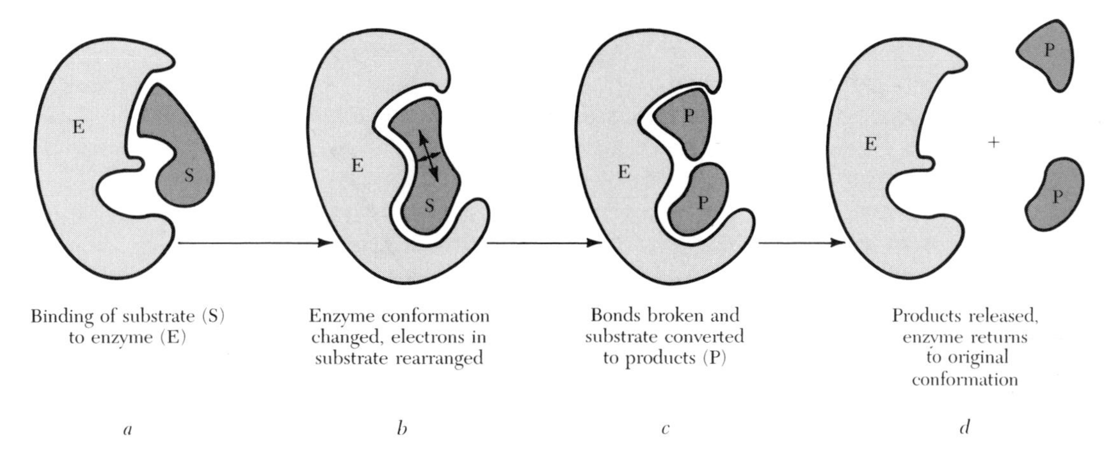

Figure 3.8 Representation of how an enzyme–substrate might form and how the conversion of substrate to products might be catalyzed.

released, the enzyme molecule can participate in another identical catalysis. Therefore only small numbers of enzyme molecules may be required for a large amount of catalysis. Third, there are specific regions into which the substrates must fit, ensuring not only that only certain molecules can be recognized by the enzyme, but also that any change in conformation of the substrates will be a specific one, thereby favoring only one of the several possible reactions that the substrate might undergo. It is this specificity of action of enzymes that makes their role such a crucial one in channeling the metabolic activities of the cell.

Regulation of Enzyme Activity

Since the capacity of an enzyme to catalyze a particular reaction (or type of reaction) depends on its three-dimensional configuration and on its ability to recognize and interact with specific molecular arrangements of its substrate, it can work effectively only within a fairly narrow range of conditions. Changes in the environment in which the enzyme functions, therefore, can modulate its activity in various ways. For example, increasing the temperature (thereby increasing the frequency of successful contacts) will tend to accelerate the rate of catalysis (Figure 3.9a). However, above a certain optimal temperature, bonds are broken and the enzyme is denatured, resulting in loss of activity (Figure 3.9a). Similarly, enzyme activity is pH dependent; above and below optimum pH values changes in the charges on the amino acids reduce the rate of the reaction (Figure 3.9b).

Interactions with other molecules in the cell are also important in regulating enzyme activity. *Competitive inhibitors* are molecules that are similar enough to the normal substrate that they can be recognized by, and bind to, the enzyme. However, they are dissimilar enough that the reaction cannot take place. Thus high concentrations of inhibitor molecules can tie up the enzyme molecules and prevent them from reacting with the proper substrate. *Noncompetitive* inhibitors,

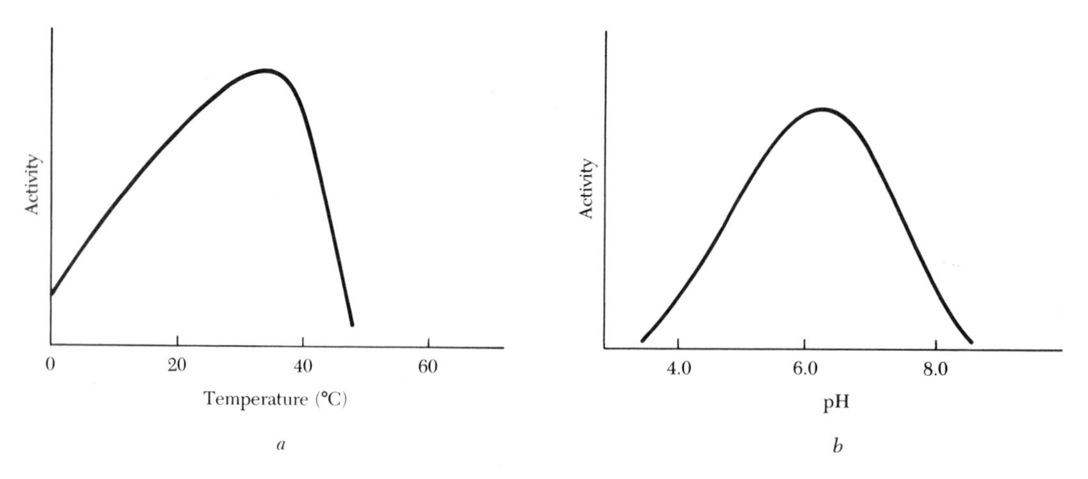

Figure 3.9 Effects of (*a*) temperature and (*b*) pH on enzyme activity.

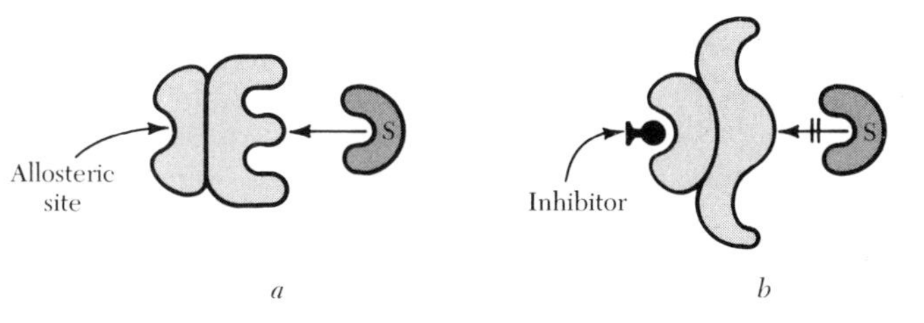

Figure 3.10 Allosteric inhibition of enzyme activity. (*a*) The enzyme in its active configuration is able to recognize and bind the substrate, S. (*b*) When the inhibitor binds to the allosteric site, the configuration at the active site is changed, and the substrate does not bind. Thus the enzyme is rendered inactive.

instead of competing for the active site, bind elsewhere to the enzyme and in some way prevent access of the substrate to the active site. Unlike competitive inhibition, noncompetitive inhibition cannot be reversed by an increase in the concentration of substrate molecules.

Allosteric control of enzyme activity is an important phenomenon in regulating the rates at which chemical reactions occur in the cell. Allosteric enzymes generally consist of two or more subunits that can interact in such a way that the enzyme can assume two different conformations, one active and the other inactive. As well as the catalytic site, such enzymes have another site that can bind specific regulatory molecules. In the case of allosteric inhibition, binding of the regulator causes a change from the active to the inactive form, and catalysis is prevented (Figure 3.10). Stimulation of activity, however, is also known; in such cases the regulator is an activator molecule that produces an allosteric change converting the enzyme to the active form. Allosteric effects are particularly important in *end-product inhibition*, where the end product of a series of reactions in a biochemical pathway can inhibit the activity of the enzyme catalyzing an early step in the pathway (Figure 3.11); in this way unnecessary production of the end product is suppressed.

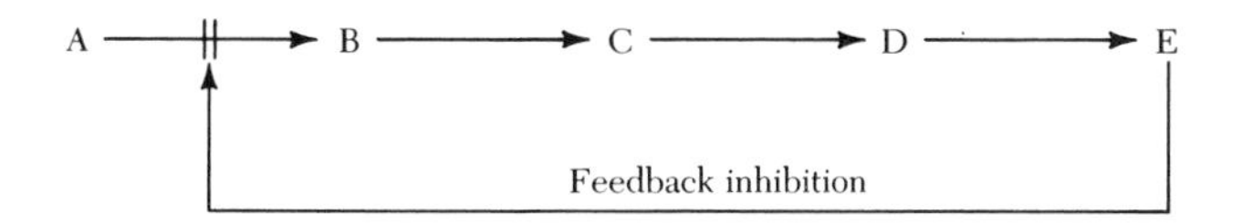

Figure 3.11 In the biochemical pathway leading to the end-product E, the enzyme that catalyzes the first reaction in the pathway is subject to allosteric inhibition by compound E. Thus the rate of production of the end product is regulated by its levels in the cell.

ADDITIONAL READING

BAKER, J. J. W., and ALLEN, G. E. 1974. *Matter, Energy, and Life*, 3rd ed. Addison-Wesley Publishing Co., Inc., Reading, Mass.

BECKER, W. M. 1977. *Energy and the Living Cell.* J. B. Lippincott Company, Philadelphia.

DEWITT, W. 1977. *Biology of the Cell: An Evolutionary Approach.* W. B. Saunders Company, Philadelphia.

LEHNINGER, A. L. 1971. *Bioenergetics*, 2nd ed. W. A. Benjamin, Inc., Menlo Park, Calif.

LEHNINGER, A. L. 1982. *Principles of Biochemistry.* Worth Publishers, Inc., New York.

MCELROY, W. D. 1971. *Cell Physiology and Biochemistry*, 3rd ed. Prentice-Hall, Englewood Cliffs, N.J.

STRYER, L. 1981. *Biochemistry*, 2nd ed. W. H. Freeman and Company, Publishers, San Francisco.

4

Cellular Information

The diagram in Figure 1.4, Chapter 1, indicates that the flow of energy, and substances as well, through a living system is not haphazard or random, but is directed by an informational source within the cell. Since structure and function are intimately related, structure also must be determined by this same source of information. Hence the cell as a whole, structurally as well as functionally, is an expression of its own information; it directs and regulates the flow of substances and energy in and out of the cell, and provides the mechanisms for doing so.

That this is so should be expected. Each individual, plant or animal, that owes its existence to some form of sexual reproduction, began life as a single cell, the fertilized egg. Each such cell is different in its informational content. The cell from an oak tree will, through growth and differentiation, produce another oak tree; a human egg eventually will produce a human being. Not only will the eggs of different species be different, so will the eggs within a species, although to a different and less obvious degree. We can witness this most dramatically if we examine closely the characteristics of identical twins. Such twins result when a mass of cells, produced from a single fertilized egg, splits in two at a very early stage of development. Each segment, consisting at that time of a few hundred undifferentiated cells, will produce an entire individual made up of many billions of cells, yet the twins will show such remarkable physical and behavioral similarities, even those of a rather trivial nature, that they are often told apart only with difficulty.

The striking similarity of identical twins forces us to recognize that their very existence, as well as that of any other individual, is an example of biological engineering of great precision. Our mass production assembly lines are scarcely more exact than these processes of life. A number of questions, therefore, arise.

Where is the "blueprint" of the individual to be found? In what form is the information stored? How is the information released so that it can be acted upon? Or, to put our questions in terms of Figure 1.4, how does the information govern the flow of matter and energy in a living system so that growth, reproduction, and individual expression are achieved?

Our answers must be sought at the cellular level, for the cell is the basic unit of organization. We now know that this information is found largely within the nuclei of eukaryotic cells and the nuclear area of prokaryotes, and it is coded in the structure of deoxyribonucleic acid (DNA), one of the nucleic acids of the cell. Such a brief statement, however, compresses the results of over a century of investigation before a full understanding of hereditary information was gained. Let us now consider in more detail the nature of cellular information, and how it is expressed.

THE NUCLEUS

The membrane-bound nucleus is the most prominent feature of the eukaryotic cell (see Figures 1.20 and 1.24, Chapter 1). Schleiden and Schwann, when setting forth the cell doctrine in the 1830s, considered that it had a central role in growth and development. Their belief has been fully supported even though they had only vague notions as to what that role might be, and how the role was to be expressed in some form of cellular action. The membraneless nuclear area of the prokaryotic cell, with its tangle of fine threads, is now known to play a similar role.

Some cells, like the sieve tubes of vascular plants and the red blood cells of mammals, do not possess nuclei during the greater part of their existence, although they had nuclei when in a less differentiated state. Such cells can no longer divide, and their life span is limited. Other cells are regularly multinucleate. Some, like the cells of striated muscles or the latex vessels of higher plants, become so through cell fusion. Some, like the unicellular protozoan Paramecium, are normally binucleate, one of the nuclei serving as a source of hereditary information for the next generation, the other governing the day-to-day metabolic activities of the cell. Still other organisms, such as some fungi, are multinucleate because cross walls, dividing the mycelium into specific cells, are absent or irregularly present. The uninucleate situation, however, is typical for the vast majority of cells, and it would appear that this is the most efficient and most economical manner of partitioning living substance into manageable units. This point of view is given credence not only by the prevalence of uninucleate cells, but because for each kind of cell there is a ratio maintained between the volume of the nucleus and that of the cytoplasm. If we think of the nucleus as the control center of the cell, this would suggest that for a given kind of cell performing a given kind of work, one nucleus can "take care of" a specific volume of cytoplasm and keep it in functioning order. In terms of materials and energy (see Figure 1.4), this must mean providing the kind of information needed to keep the flow of materials and energy moving at the correct rate and in the proper channels. With the multitude of enzymes in the cell, materials and energy can of course be channeled in a multitude of ways; it is the function of some informational molecules to make some channels of use more preferred than others at any given time. How this regulatory control is exercised is not entirely clear.

The nucleus is generally a rounded body. In plant cells, however, where the center of the cell is often occupied by a large vacuole, the nucleus may be pushed against the cell wall, causing it to assume a lens shape. In some white blood cells, such as the polymorphonucleated leucocytes, and in the cells of the spinning gland of some insects and spiders, the nucleus is very much lobed (Figure 4.1). The reason for this is not clear, but it may relate to the fact that for a given volume of nucleus, a lobate form provides a much greater surface area for nuclear-cytoplasmic exchanges, possibly affecting both the rate and the amount of metabolic reactions.

The nucleus, whatever its shape, is segregated from the cytoplasm by a double membrane, the *nuclear envelope*, with the two membranes separated from each other by a *perinuclear space* of varying width. The envelope is absent only during the time of cell division, and then just for a brief period. As Figure 7.1, Chapter 7, shows, the outer membrane is often continuous with the membranes of the endoplasmic reticulum, a possible retention of an earlier relationship, since the envelope, at least in part, is formed at the end of cell division by coalescing fragments of the endoplasmic reticulum. The cytoplasmic side of the nucleus is frequently coated with ribosomes, another fact that stresses the similarity and relation of the nuclear envelope to the endoplasmic reticulum (see Chapter 7). The inner membrane seems to possess a crystalline layer where it abuts the nucleoplasm, but its function remains to be determined.

Everything that passes between the cytoplasm and the nucleus in the eukaryotic cell must traverse the nuclear envelope. This includes some fairly large molecules as well as bodies such as ribosomes, which measure about 25 nm in diameter. Some passageway is, therefore, obviously necessary since there is no indication of dissolution of the nuclear envelope in order to make such movement possible. The *nuclear pores* (Figure 4.2) appear to be reasonable candidates for such passageways. In plant cells these are irregularly and rather sparsely distributed over the surface of the nucleus, but in the amphibian oocyte, for example, the pores are numerous, regularly arranged, and octagonal and are formed by the fusion of the outer and inner membrane.

The diameter of these pores is about 100 nm, ample enough for the movement of relatively large structures or molecules. The ringlike structure of the pore is called an *annulus*, with the octagon shape resulting from the presence of eight regularly spaced, fibrous particles which extend from the outer through to the inner membrane (Figure 4.3). The openings of the pores seem often to be plugged with a fibrous material distinct from that found in the surrounding particles, suggesting that the pores are not simply openings in the membrane, allowing free passage of materials. Differences in pH (hydrogen ion concentration) between

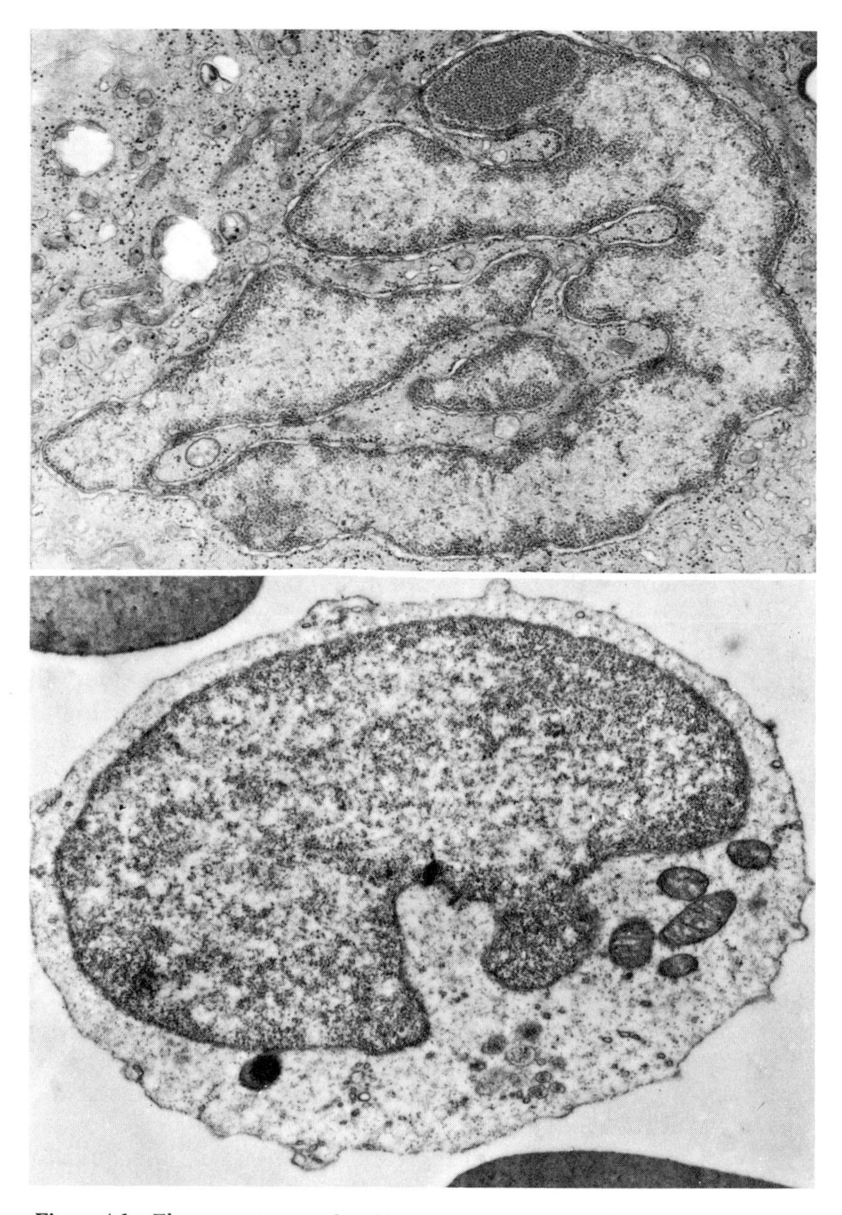

Figure 4.1 Electron micrographs of human white blood cells. Top: highly lobed nucleus of a polymorphonucleated leucocyte; concentrations of chromatin can be seen adjacent to the nuclear membrane and filling the lobe at the top of the nucleus. Bottom: small leucocyte with its nucleus lobed to a lesser extent. Both of these cells have characteristically large nuclei relative to the amount of cytoplasm.

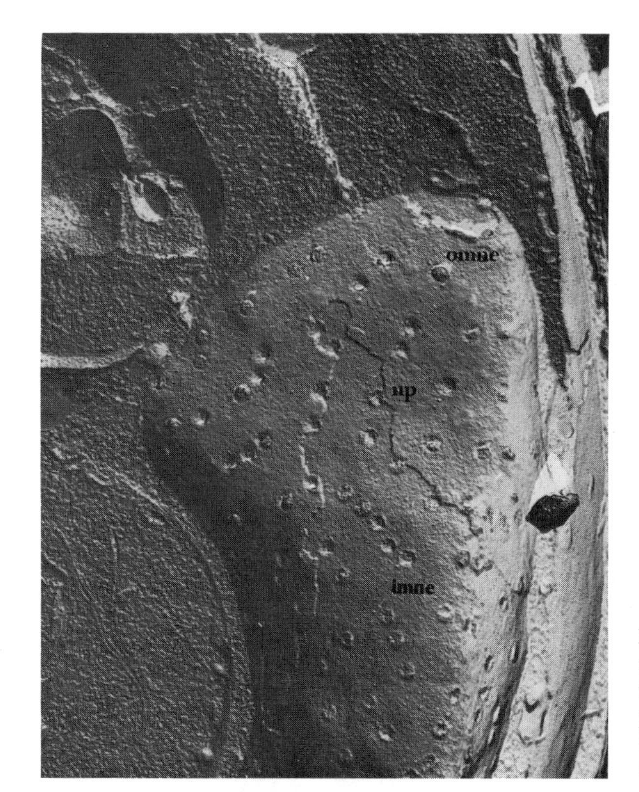

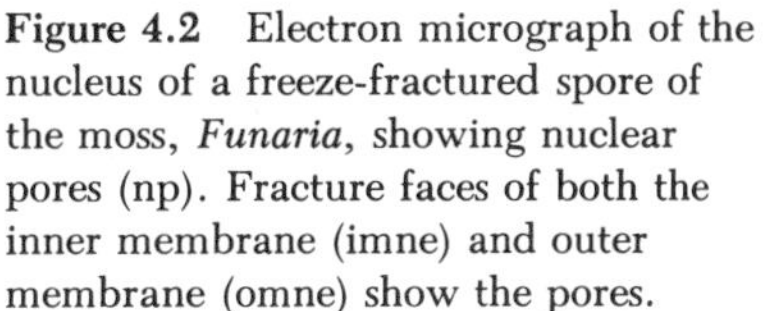

Figure 4.2 Electron micrograph of the nucleus of a freeze-fractured spore of the moss, *Funaria*, showing nuclear pores (np). Fracture faces of both the inner membrane (imne) and outer membrane (omne) show the pores.

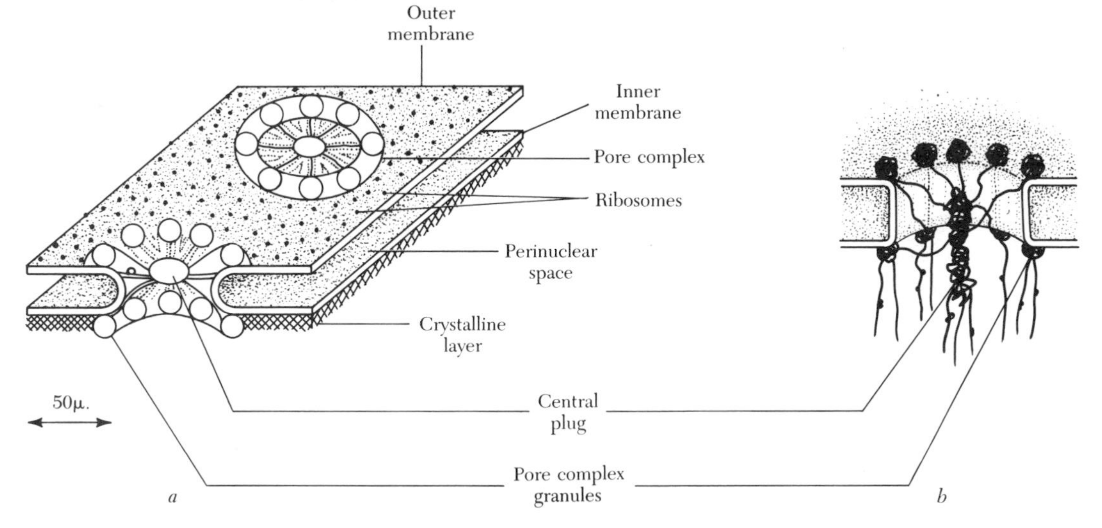

Figure 4.3 The structure of the pores in the nuclear envelope. The octagonal shape of the pore derives from the regularly spaced fibrous bundles situated around the outer edge, while the pore opening can be plugged by a somewhat similar bundle extending from the nucleoplasm into the cytoplasm.

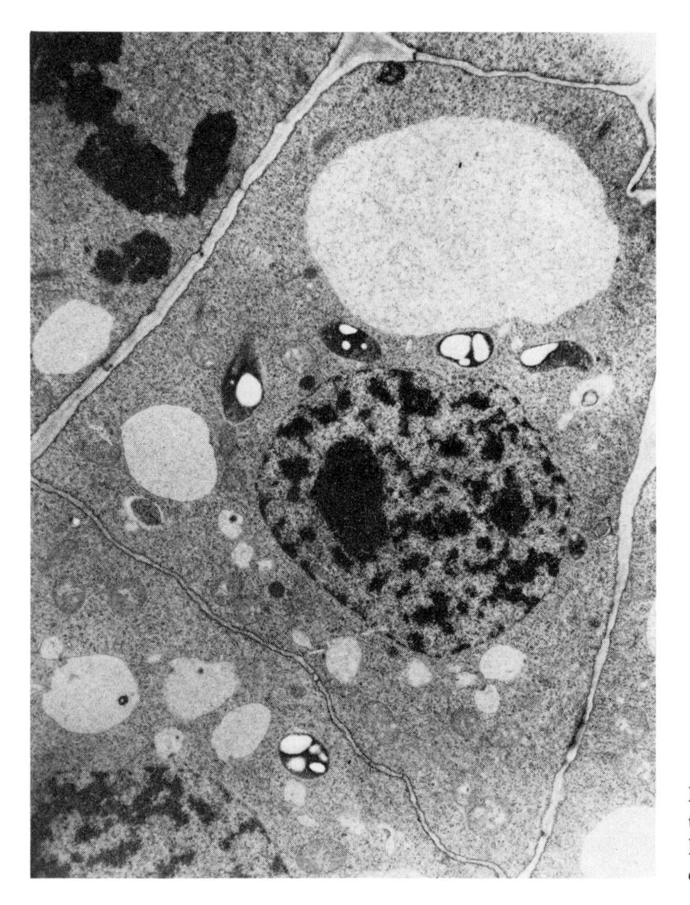

Figure 4.4 An interphase cell of a root tip of the pea, *Pisum sativum*, showing how the chromatin is often appressed closely to the nuclear envelope.

nucleus and cytoplasm also suggest that the passage of materials is not as free as the membrane structure might suppose. The structure, significance, and origin of the pores remain uncertain.

The nucleus stains readily with basic dyes after acid fixation to reveal a network of fine threads, among which coarser masses of stained material are distributed, much of it adjacent to the inner nuclear membrane (Figure 4.4). The electron microscope, with its greater powers of resolution, reveals much the same network of fine threads (Figure 4.5), with no additional details of striking significance except that the threads are more numerous and of smaller diameter (Figure 4.6). This stainable material is the *chromatin*, with the fine threads being *euchromatin*, the coarser masses *heterochromatin*. Chromatin, through condensation and contraction during cell division, will transform itself into chromosomes (Figure 4.7). It is also the chromatin, and particularly the DNA contained within it, that is stained in a highly specific manner by the Feulgen reaction; by measuring spectroscopically the amount of Feulgen dye per nucleus, it is possible to show that the amount of DNA per nucleus is constant for each species (Table 4.1).

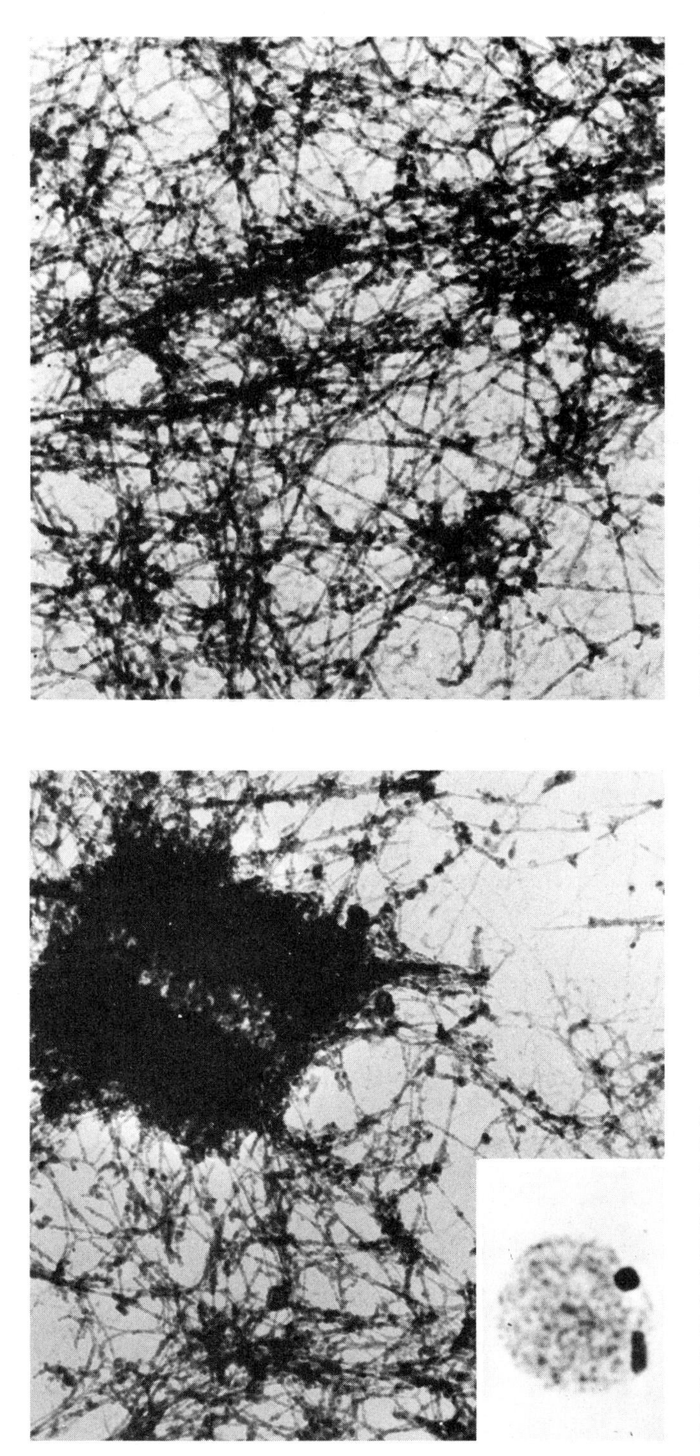

Figure 4.5 Electron micrograph of a portion of a nucleus, prepared by floating it on water, at a very early stage in the division of a cell from the testis of the milkweed bug, *Oncopeltus fasciatus*. The strands of chromatin are about 10 nm in diameter and reveal no distinguishing features. (Courtesy of Dr. S. Wolfe.)

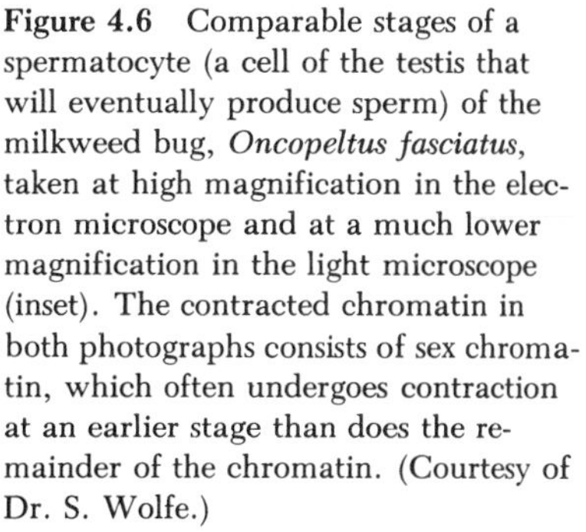

Figure 4.6 Comparable stages of a spermatocyte (a cell of the testis that will eventually produce sperm) of the milkweed bug, *Oncopeltus fasciatus*, taken at high magnification in the electron microscope and at a much lower magnification in the light microscope (inset). The contracted chromatin in both photographs consists of sex chromatin, which often undergoes contraction at an earlier stage than does the remainder of the chromatin. (Courtesy of Dr. S. Wolfe.)

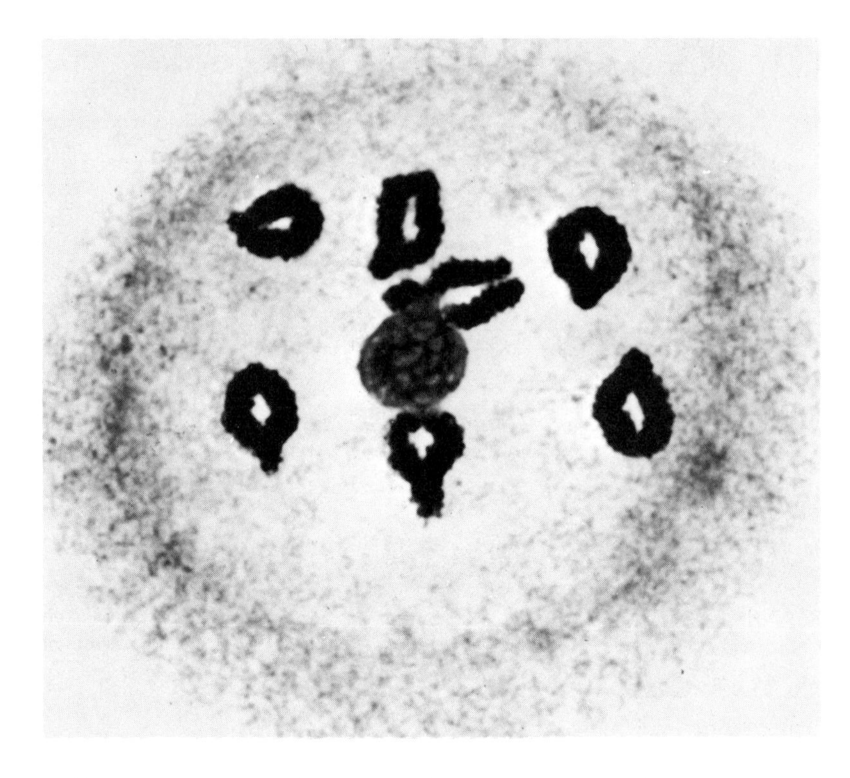

Figure 4.7 Dividing cell in rye, showing the attachment of the nucleolus to particular chromosomes (the centrally located pair).

In addition to the chromatin, the nucleus contains one or more dense bodies, or *nucleoli* (singular, *nucleolus*). These are attached to, and formed by, special regions of particular chromosomes called *nucleolar organizer regions* (Figure 4.7; see also Figures 10.10 and 10.15, Chapter 10), which not only synthesize the nucleic acid portion of the nucleolar material but also organize it into a dense body. The character of the nucleolus varies with the type of cell and its metabolic state; it is larger and more dense in active cells, rapidly growing embryonic cells, and those engaged in protein synthesis. It disappears and reappears during the course of cell division (see Chapter 7).

Electron microscopy reveals an internal differentiation of the nucleolus in the form of a loose network of strands and granular material (Figure 4.8). As will be discussed later in this chapter, it is now known that the nucleolus, with its associated chromatin, is concerned with the formation of ribosomes, which eventually will accumulate in the cytoplasm to become organized as part of the protein-synthesizing machinery of the cell. This is demonstrated by the fact that radioactive uridine, a nucleoside that becomes incorporated as a base into the RNA of the ribosomes, is first concentrated in the nucleolus, after which it passes to the cytoplasm. The fact also that the nucleolus does not stain with Feulgen dye is in-

TABLE 4.1 DNA PER NUCLEUS, IN PICOGRAMS (1 PICOGRAM EQUALS 10^{-9} MILLIGRAM AND IS EQUIVALENT TO ABOUT 9.5×10^8 NUCLEOTIDE PAIRS).*

Mammals		*Amphibians*	
Human	3.65	*Amphiuma*	86.0
Beef	2.82	*Necturus*	24.2
Horse	3.18	Frog	7.5
Dog	2.93	Toad	4.5
Mouse	3.26	*Xenopus*	3.15
Marsupial		*Fish*	
Bandicoot	4.62	Carp	1.64
Kangaroo	3.13	Shad	0.91
Birds		Lungfish	50.0
Chicken	1.44	*Miscellaneous*	
Duck	1.30	Maize	8.4
Goose	1.46	*Drosophila*	0.085
Pigeon	1.72	Jellyfish	0.33
Reptiles		Squid	4.50
Snapping turtle	2.56	Sea urchin	0.67
Alligator	2.94	*Neurospora*	0.20
Water snake	2.51	*E. coli*	0.004
Black snake	1.48		

*From C. P. Swanson, T. Merz, and W. J. Young, *Cytogenetics*, 2nd ed. (Englewood Cliffs, N.J.: Prentice-Hall, Inc., 1981), p. 63.

Figure 4.8 The nucleus of a mouse fibrocyte, showing the fibrous nature of a portion of the nucleolus. The dark strands are the nucleolonema; although not evident in this illustration, there is often a distinct clear area in the nucleolus called the pars amorpha. The granular material consists of ribosomes or of other RNA-containing particles. The nucleolus is subject to many changes in the cell and can assume many forms; this is often true in malignant or diseased cells.

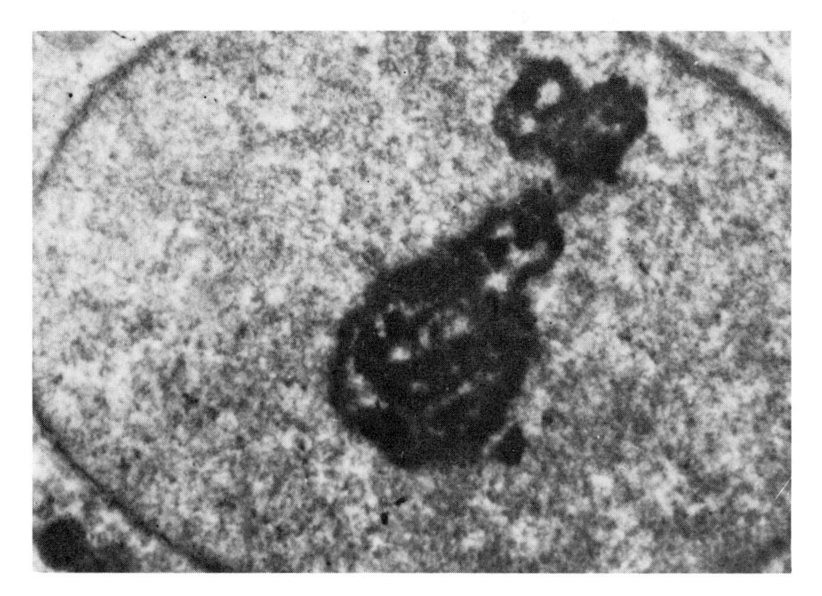

dication that the nucleolus and the chromatin differ in the kind of nucleic acid present in each structure.

Chromatin can be extracted quite readily from cells and, by chemical analysis, shown to be a complex substance. It consists of four species of molecules: DNA; RNA; basic proteins of low molecular weight called *histones*, rich in the amino acids lysine and arginine, and existing in five distinct forms; and acidic, nonhistone proteins of variable amount and molecular weight. During isolation procedures, the DNA and histones are associated as a nucleoprotein complex, with the weight ratio of DNA to histone very close to 1:1, even when the amount of DNA per nucleus varies with the particular species. The RNA and acidic proteins vary in amount quite widely from one kind of cell to another even within an individual, being greater in the more actively metabolizing cells.

Nucleus as a Control Center

A number of observations support the idea of the nucleus as not only an integral part of the eukaryotic cell, but also as its center of control. The whole science of genetics, in fact, is based on this assumption, with the chromosomes the vehicles of hereditary transmission, and the DNA within the chromosomes the molecular basis of heredity.

The mammalian red blood cell lacks a nucleus as it passes from its origin in the bone marrow and enters the bloodstream. It is a cell with a limited, albeit crucial, respiratory role—that of moving oxygen from the lungs to tissues, and carbon dioxide from the tissues to the lungs. It is also limited in length of life (about 120 days), and it is incapable of further division and of continuous repairs. From these observations, coupled with the almost universal occurrence of a single nucleus per cell, the equal contribution of the sperm to the zygote even though it is little more than a nucleus, and the fact that in cell division it is principally the nucleus that goes through an exact partitioning of its chromatin content, we can point to the nucleus as a control center of the cell. Three other experiments confirm the validity of this concept. If the nucleus of a fertilized frog egg is removed, a not particularly difficult operation, and the nucleus of a cell of a frog of a different species is substituted, a tadpole and eventually a frog will result, and it will be of the character of the frog that donated the substitute nucleus. The cytoplasm of the first frog exerts little influence after the very early stages of embryogenesis. It is also possible to remove the nucleus of an amoeba by pinching off a piece of the cytoplasm containing the nucleus. The remaining cytoplasm is not damaged, and the fragment lacking a nucleus can metabolize for days or even weeks afterward. The cell will eventually run down, however, and will be unable to rejuvenate itself unless another nucleus is put back into the cytoplasm. A cell without a nucleus is, therefore, a cell with a limited future and no posterity, and the nucleus is the necessary organelle that provides information and/or parts to keep the cytoplasm functioning properly for an indefinite period.

The influence of the nucleus in controlling the morphology of an organism is

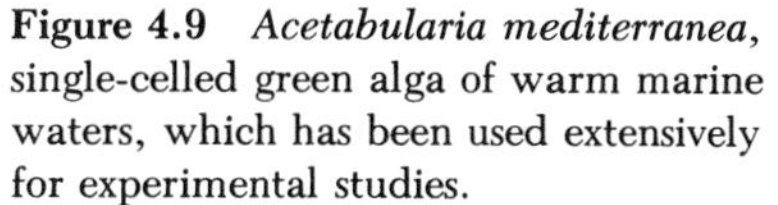

Figure 4.9 *Acetabularia mediterranea,* single-celled green alga of warm marine waters, which has been used extensively for experimental studies.

made strikingly evident by the work done on the green alga *Acetabularia*, a single-celled organism of large dimensions—5 to 10 cm (centimeters) in height—found in warm marine waters, and fastened to a substrate of sand or rocks by a rhizoid (Figure 4.9). *A. mediterranea* and *A. crenulata* are two species that differ as to the morphology of their caps. If a cap is cut off, a new one will be regenerated (Figure

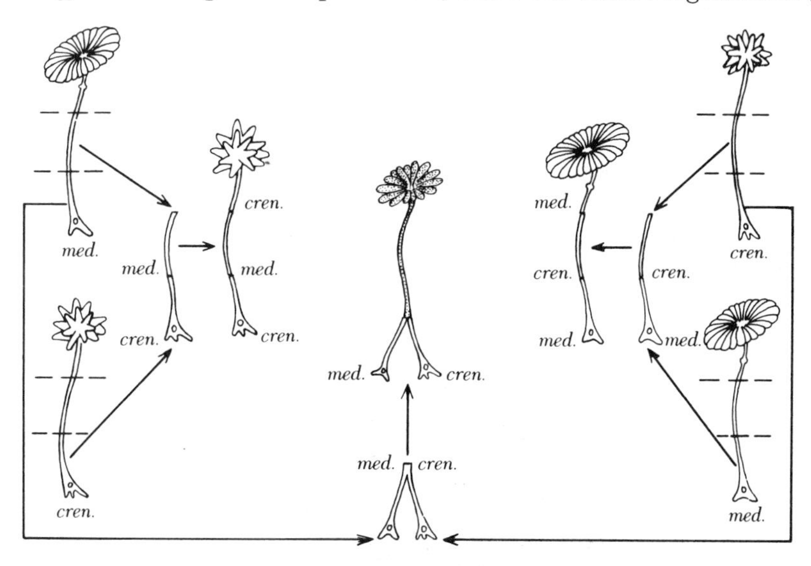

Figure 4.10 Influence of the nucleus on development in Acetabularia. Stalk segments of *A. mediterranea* grafted onto nucleus-containing rhizoids of *A. crenulata*, and vice versa produce caps characteristic of the species contributing the nucleus. When two nucleus-containing rhizoids are grafted together, the cap consists of loose rays, as in *A. crenulata*, but their points are more rounded, as in *A. mediterranea*.

4.10). It is also possible to graft a piece of stalk of one species onto the decapitated, nucleus-containing rhizoid of the other species. When the cap forms, it will be somewhat comparable to the species that donated the stalk, but if it is removed and another allowed to regenerate it will be characteristic of the species that contributed the nucleus; the stalk cytoplasm will have exhausted its influence and that of the nucleus will become evident. The nucleus, therefore, causes the cell to do as it dictates; the next problem is to determine how this jurisdiction is exercised.

THE GENETIC MATERIAL

Cellular control and activity must be exercised through chemical means; there are no other means. Furthermore, since a human cell does what a human cell is supposed to do, and the cell of an orchid does what it is supposed to do, the information which is contained within the nucleus must be capable of being decoded or "read," and must be acted upon in such a manner that the instructions from the nucleus have been carried out.

Attention was early directed at the chromosomes in the nucleus as the site of coded information. Chromosomes are constant in number, shape, and size within an individual and, generally, within a species (Figures 4.7 and 4.11). The number of chromosomes characteristic of a species is referred to as the *diploid*, or *2n*, number; as we shall discuss later, the cells involved in fertilization, that is, the sperm and egg, each contain half of this number, and are said to be *haploid*, or *n*. When chromosomes are changed, spontaneously or experimentally, the change can often bring about a change in the individual (Figure 4.12). Thus a human being who has an extra chromosome 21 in all the cells of the body invariably exhibits Down's syndrome, and is generally retarded to some degree. When genes that control the expression of specific characters in an individual are followed from one generation to the next, they follow the same pathways and show the same inheritance patterns as do the chromosomes. We can, therefore, trace the source of

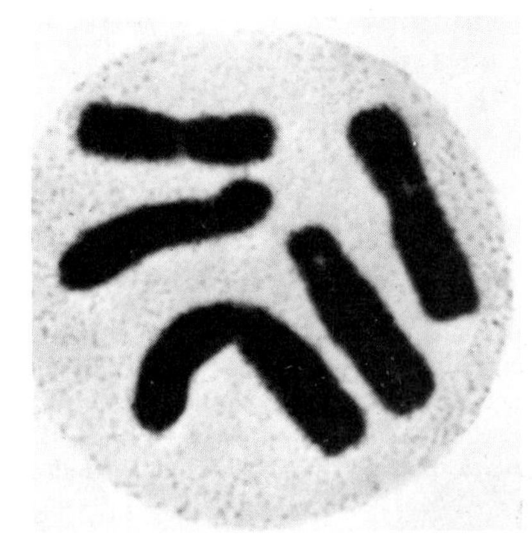

Figure 4.11 The haploid set of chromosomes of the wakerobin, *Trillium erectum*. The length of each chromosome and the position of the centromeres permits the identification of individual chromosomes.

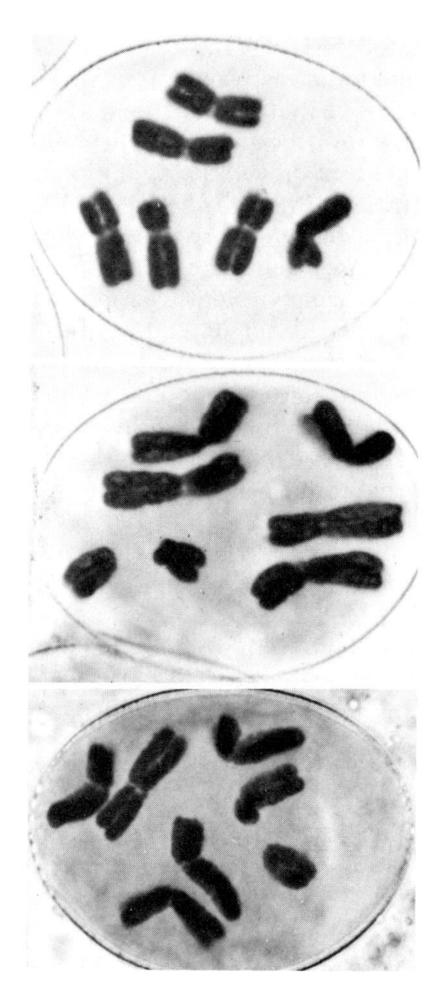

Figure 4.12 Chromosomes in the microspores of *Tradescantia paludosa*, the spiderwort. Those at the top are normal; one chromosome in each of the other two cells has been broken by x rays. The two fragments without centromeres will be unable to move properly during cell division and will be lost. The large size of these fragments would, in all probability, be lethal because of the amount of chromatin (genetic material) involved, but lesser losses can act as mutations to bring about phenotypic changes in the individual possessing such aberrant chromosomes.

cellular control, and of individual uniqueness, to the nucleus, to the chromosomes, and then to the genes. The next step in that search was to continue on to the molecular levels, and eventually it became clear that the only molecule meeting all informational criteria was the DNA in the chromosomes.

A number of experiments have confirmed this conclusion and have demonstrated that the coded and controlling information in the cell is bound up in the molecules of DNA. It took a long time to prove this beyond any reasonable doubt. It was, for example, long thought that the DNA molecules were of monotonous regularity, being made up of units of the four bases (adenine, guanine, cytosine, and thymine), regularly repeated. By this time hundreds and hundreds of genes, from many organisms and governing the expression of every imaginable kind of character, had been discovered, tested in inheritance tests, and located on particular chromosomes; it was then argued that only a protein consisting of many amino acids of 20 different kinds, and linked together in every possible combination and permutation, could provide the necessary diversity demanded.

A series of experiments with bacteria, which until this time had been little used for genetic studies, pointed research in the right direction and eventually settled the issue. Frederick Griffiths, a British bacteriologist, was carrying out a series of studies on *Pneumococcus*, a bacterium responsible for respiratory illness in various mammals, including mice, guinea pigs, and people. One strain of the bacterium, called Type I, was highly virulent, causing death when injected into mice. Type I bacteria, when plated out on culture dishes, also formed shiny, regularly spherical colonies. Another strain, Type II, was avirulent; it did not produce the disease when injected into mice, and it differed further in producing dull, irregularly shaped colonies when cultured. Both the degree of virulence and the shape and character of the colonies depended upon the presence or absence of a polysaccharide capsule external to the cell membrane; Type I possessed the capsule, Type II lacked it.

In one experiment, Griffiths injected heat-killed Type I bacteria into a group of animals. This, by itself, would not induce symptoms of the disease, but he then injected a dose of avirulent (Type II) bacteria into the same animals. One would normally assume that the second injection would be equally ineffective in causing death, but the doubly injected animals died. The bacteria isolated from these dead animals were Type I, virulent and capsulated (Figure 4.13).

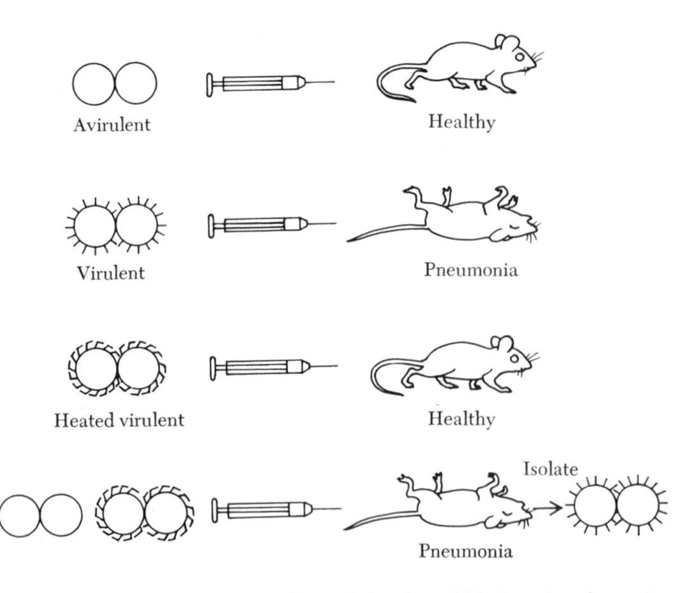

Figure 4.13 Griffiths' experiment showed that heat-killed virulent bacteria could contribute "something" to avirulent strains and cause them to be changed to a virulent form. It was eventually showed that this "something" was DNA.

Two hypotheses can be advanced to explain these observations. Either the heat-killed Type I cells somehow had been reactivated by the presence of the avirulent Type II cells, or the Type II cells had received something from the dead Type I cells that converted them into virulent cells. The conversion was permanent, not simply a temporary phenomenon. The latter explanation proved to be correct, because the contents of burst Type I cells added to a Type II culture in a test tube brought about a similar transformation of some Type II cells. The rate of transformation was low but easily detectable. This experiment also eliminated the idea that the mice or guinea pigs might have had something to do with the experimental results. *Bacterial transformation*, as the phenomenon came to be known, was a one-way transfer of heritable characteristics from one cell to another.

Griffiths carried out his experiments in 1928, but it was not until 1944 that the contents of these bacteria had been fractionated and tested for transforming ability, and DNA was identified as the transforming principle. It was further shown that not only virulence but also any other heritable character could be transferred in a similar way; therefore, if genes are responsible for heritable characters, then genes must be made of DNA. The central problem then became one of determining the structure of the DNA molecule to understand how heritable information could be stored in, and released from, a chemical structure, and subsequently be given expression by way of the chemical reactions taking place within the cell.

The One Gene–One Enzyme Hypothesis

A heritable change in an organism is most commonly detected by an observed variation in a given trait. Eye color in humans exhibits a wide range in shades of brown, blue, and green; albinism of the skin, eyes, and hair are obviously different from normal coloration; and we are all acquainted with the fact that we exhibit various blood types. Since these are heritable it is usually said that a gene determines the trait. A trait, however, particularly one expressed at a morphological level, is far removed from the gene, and comes to be expressed only through the processes of growth and development, and by means of all of the chemical reactions taking place during that period of maturation. That being so, the variation in a trait should be reflected in an equivalent variation in the gene, or genes, responsible for the trait, so the logical place to detect such changes is at the level of chemical reactions. A variation in a chemical reaction is as legitimate a trait as, say, color blindness, and is obviously closer to the source of initial change in the gene.

In the preceding chapter the statement was made that specific chemical reactions in the cell are controlled by specific enzymes. It is then logical, and as it turned out, correct as well, to assume that there is a relation between genes and enzymes, and that variation in a gene should be reflected in some kind of variation in an enzyme. Two possibilities offer themselves: either the enzyme governing a reaction is missing if the reaction is faulty or absent, or the enzyme, if present, fails to function properly. In either instance, an alteration in the chemical processes of the cell would occur, to be detected, possibly, by alteration of a visible trait. To extend the argument further, it is also logical to postulate a relation between the substance DNA, of which the genes are composed, and proteins, since enzymes are protein in nature.

The cellular synthesis of arginine, one of the essential amino acids, provides a means by which the problem can be explored. An organism appropriate for such exploration is the pink bread mold, *Neurospora crassa*, although the results obtained are applicable to virtually all organisms. Ordinarily, *Neurospora* can synthesize its own arginine from a minimal culture medium containing nitrate, a vitamin (biotin), and glucose, the latter providing both energy and carbon. By exposing the spores of *Neurospora* to X rays, ultraviolet light, or certain chemicals, it is possible to alter DNA, and hence cause the genes to mutate.

These spores are tested for mutations by being grown first on a complete medium (containing all necessary substances, and then transferring a bit of the mycelium to the minimal medium. If a mutation has occurred, affecting, say, the synthesis of arginine, the mycelium transferred will not grow unless arginine is added to the growth medium.

From other studies it was known that the synthesis of arginine from glucose required a number of intermediate steps. Here we concern ourselves with only the last three (Figure 4.14). If our argument is correct that each chemical reaction in cells is governed by a specific enzyme, and that the relation between genes and enzymes is a direct one, then the three synthetic steps depicted in Figure 4.14 should

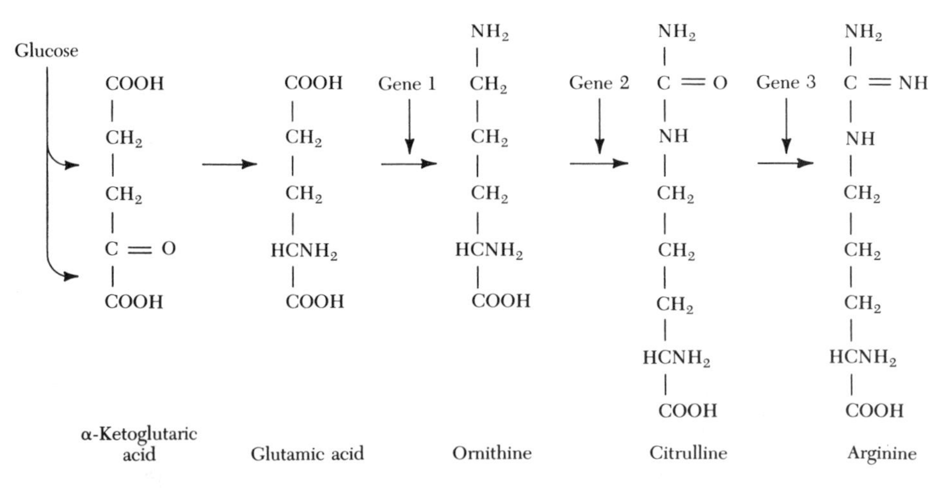

Figure 4.14 The biochemical reactions required for the formation of arginine from glucose. The reactions in the conversion of ornithine to arginine are unique to arginine formation, while the earlier steps are common to the formation of other substances. The enzymes governing the reactions indicated are themselves under the control of certain genes, which have been identified. If the genes mutate and produce defective enzymes, the steps leading to the next reaction are blocked.

be controlled by three different enzymes, and hence by three different genes. Thus, in a collection of mutants all of which are unable to synthesize arginine, three distinct types should be distinguishable: all would grow if arginine were added to the minimal medium, but those in one group would include these whose mutation occurred at step 1, and which could grow if ornithine, citruline, or arginine were added, but not if only glutamic acid were added. A second group, with step 2 affected, would grow only if citrulline or arginine were added, and a third group, with step 3 affected, would grow only if arginine were available. The specific enzymes governing these three steps would be either absent or modified to the extent that they were no longer functional.

As a result of studies such as these, Beadle and Tatum, in 1941, advanced the one gene–one enzyme hypothesis, which states that there is a 1:1 relationship between genes and enzymes. It is now known that the hypothesis is oversimplified. There are, for example, some dimeric enzymes consisting of two different polypeptides linked together to form a functional enzyme, while others, such as lactic dehydrogenase, consist of more than two polypeptides. Also, some proteins serve a structural rather than an enzymatic function, as for example, those found in hair or in collagen, so the hypothesis has been modified to state that one gene is responsible for the formation of one polypeptide.

The Molecular Nature of DNA

Chemical analysis had revealed that DNA was a polymer of high molecular weight, that is, a macromolecule of repeated units. Each of these units proved to be composed of three smaller molecules linked together to form a *nucleotide*: these included a five-carbon sugar (deoxyribose), phosphoric acid, and a nitrogen-containing moiety generally referred to as a base (Figure 4.15). These bases are of four kinds: two are pyrimidines (thymine and cytosine) and two are purines (adenine and guanine). At one time it was believed that the DNAs from all organisms were similar, with the four bases existing in a ratio of 1:1:1:1, but clearly

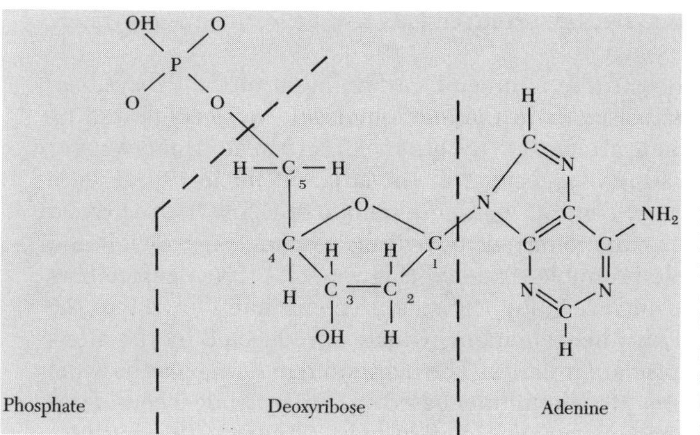

Figure 4.15 The linkage of the three kinds of molecules to form a nucleotide, in this instance, deoxyadenylate. Linkage to form a chain of nucleotides will be through an oxygen in the phosphate portion and the three carbon in the sugar portion. Linkage to form a double helix will be by hydrogen bonds as indicated in Figure 4.17.

TABLE 4.2 BASE RATIOS OF SEVERAL WELL-KNOWN ORGANISMS.*

	Adenine	Thymine	Guanine	Cytosine	Ratio of $\frac{A + T}{C + G}$
Man	29.2	29.4	21.0	20.4	1.53
Sheep	28.0	28.6	22.3	21.1	1.38
Calf	28.0	27.8	20.9	21.4	1.36
Salmon	29.7	29.1	20.8	20.4	1.43
Wheat	27.3	27.1	22.7	22.8	1.19
Yeast (fungus)	31.3	32.9	18.7	18.1	1.19
Virus, vaccinia	29.5	29.9	20.6	20.0	1.46
Bacteria					
Staphylococcus	31.0	33.9	17.5	17.6	1.85
Pseudomonas	16.2	16.4	33.7	33.7	0.48
Colon bacterium	25.6	25.5	25.0	24.9	1.00
Clostridium bacterium	36.9	36.3	14.0	12.8	2.70
Pneumococcus	29.8	31.6	20.5	18.0	1.88

* Values are arbitrary but accurate as ratios.

this notion had to be revised if the genes, of which there were endless kinds, were composed of DNA. More critical analyses of DNAs from a variety of organisms showed that the base ratios of different species were indeed different (Table 4.2), and that the differences could not be blamed on faulty isolation procedures. In addition, it was also demonstrated that in DNAs from most sources the purines equalled the pyrimidines in molar amounts, that the amount of adenine (A) was equal to that of thymine (T), and that cytosine (C) was equal to that of guanine (G). Thus, A + G = T + C, and A = T and C = G, but A or T is not necessarily equal to C or G. There was no indication that the bases were arranged in any particular order, suggesting that ATGTTCACG might very well have a different genetic meaning from ACACTTGTG, even though the number and kinds of bases are similar. In a sense this would be no different from the words *team*, *mate*, and *meat;* the letters are the same but the arrangements and hence the meanings are not.

Each base was attached to the number 1 carbon atom of the deoxyribose sugar (Figure 4.15); successive sugars in the macromolecule were connected by linkages between the phosphate attached to the number 5 carbon atom of one sugar and the number 3 carbon atom of the sugar of the adjacent nucleotide (Figure 4.16). Remembering that A = T and C = G in a sample of DNA, it was argued that the bases are arranged in pairs connected by hydrogen bonds (Figure 4.17) and that the molecule as a whole is double-stranded (Figure 4.18). Even before base pairing had been proposed, however, physical measurements had shown that the DNA molecule consisted of two helical chains, which were formed by the alternating linked sugar and phosphate residues. The question remaining was how the bases fitted into this structure. Model building based on stereochemical considerations not only supported the concept of a double helix (Figure 4.19), but also

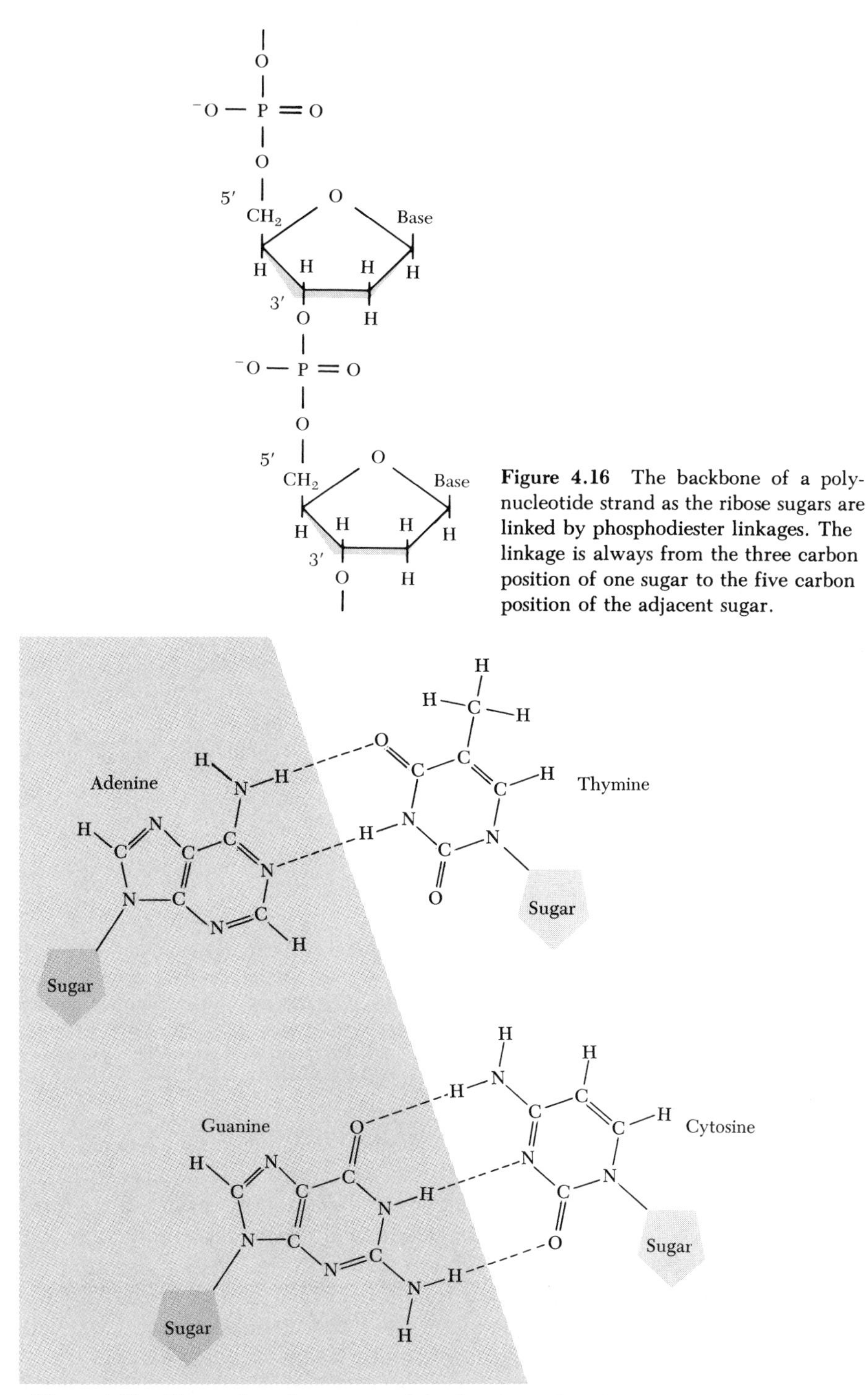

Figure 4.16 The backbone of a polynucleotide strand as the ribose sugars are linked by phosphodiester linkages. The linkage is always from the three carbon position of one sugar to the five carbon position of the adjacent sugar.

Figure 4.17 Chemical configurations of the four bases found in the DNA molecule, arranged as base pairs. Thymine and cytosine are pyrimidines; adenine and guanine are purines.

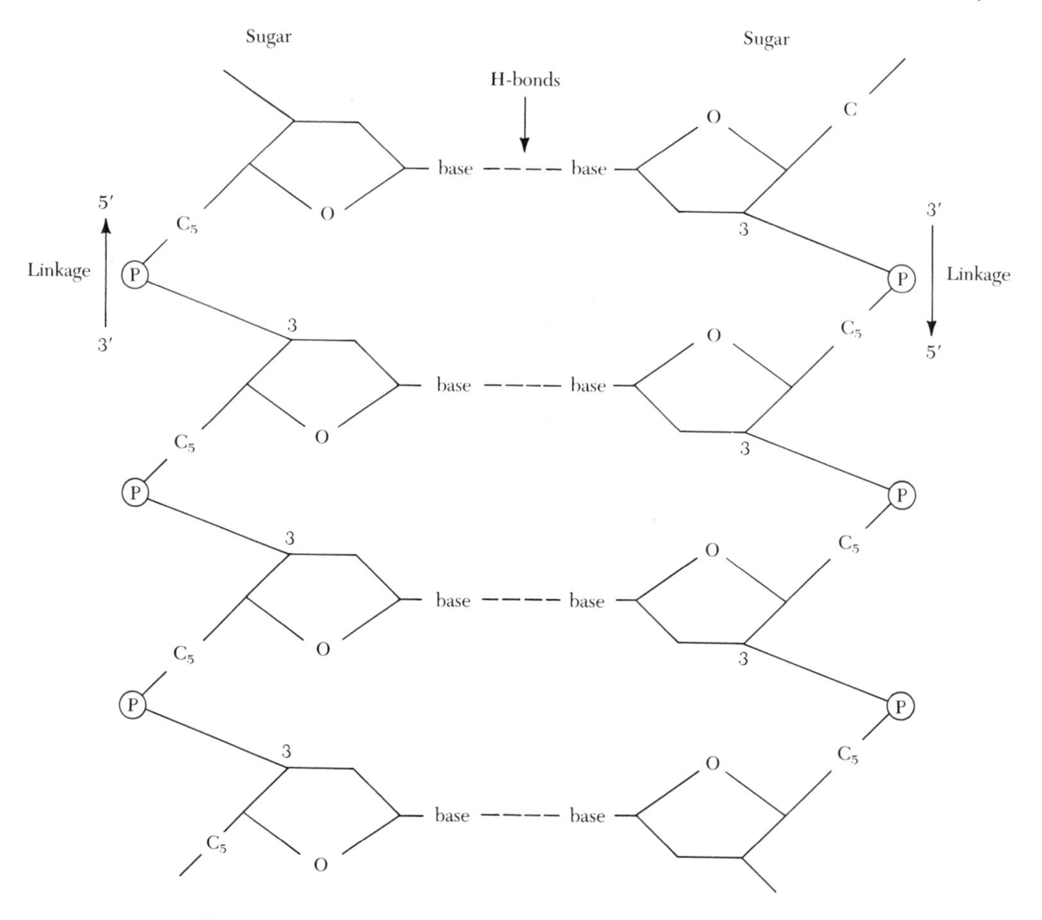

Figure 4.18 Schematic and flattened arrangement of phosphates, sugars, and bases to form the DNA molecule, with the right- and left-hand portions of the molecule held together by hydrogen bonds. The sugar-phosphate backbones of the two strains are anti-parallel. This arrangement, twisted into a helix by molecular forces, gives the configuration seen in Figure 4.19.

demonstrated how the stability of the DNA molecule was achieved and explained the significance of the base ratios. An enormous amount of critical experimental evidence has confirmed this structure beyond any reasonable doubt. The molecule of DNA consists, therefore, of two helically entwined molecular strands, each having a sugar–phosphate backbone and with the bases projecting inward from the sugar moiety. The bases are joined between the two strands by hydrogen bonds, thus giving stability to the molecule. The pairing, however, is strictly complementary in that A is always paired with T and C with G. The two strands of the molecule are, as a consequence, complementary with each other throughout their entire length. We find, as a result, that the only base pairs that are possible are

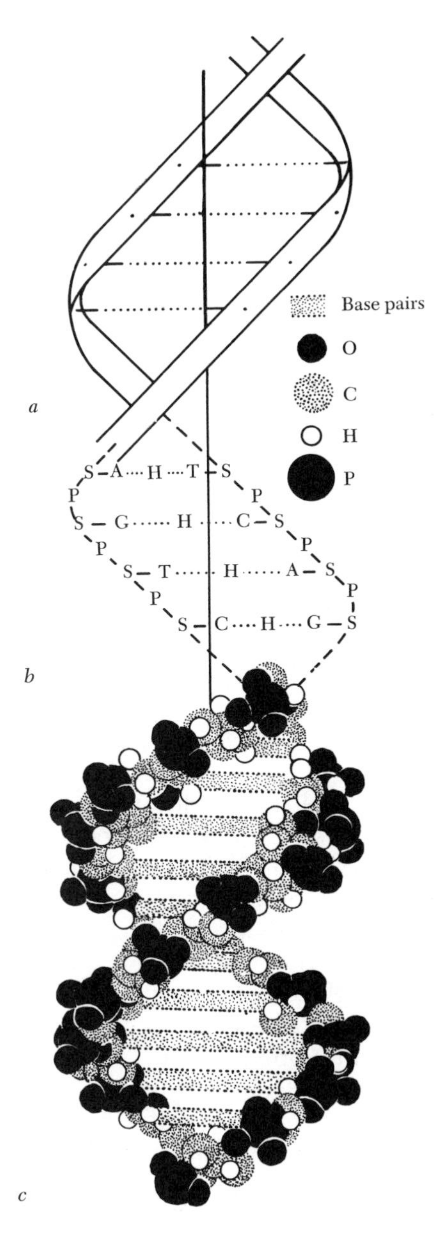

Figure 4.19 The DNA helix, with three different ways of representing the molecular arrangement. (*a*) General picture of the double helix, with the phosphate-sugar combinations making up the outside spirals and the base pairs the cross bars. (*b*) A somewhat more detailed representation: phosphate (P), sugar (S), adenine (A), thymine (T), guanine (G), cytosine (C), and hydrogen (H). (*c*) Detailed structure, showing how the space is filled with atoms: carbon (C), oxygen (O), hydrogen (H), phosphorus (P), and the base pairs.

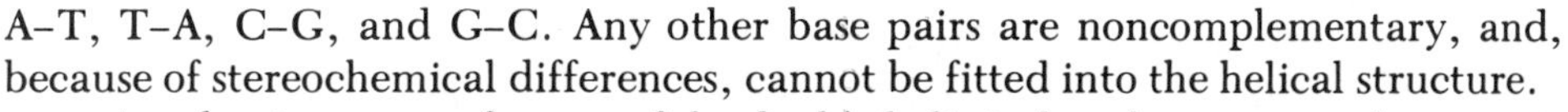

A–T, T–A, C–G, and G–C. Any other base pairs are noncomplementary, and, because of stereochemical differences, cannot be fitted into the helical structure.

Another important feature of the double helix is that the two strands are antiparallel, that is, the sugar–phosphate–sugar linkages run in the $3' \rightarrow 5'$ direction in one strand from one end of the helix to the other, while in the other strand, starting

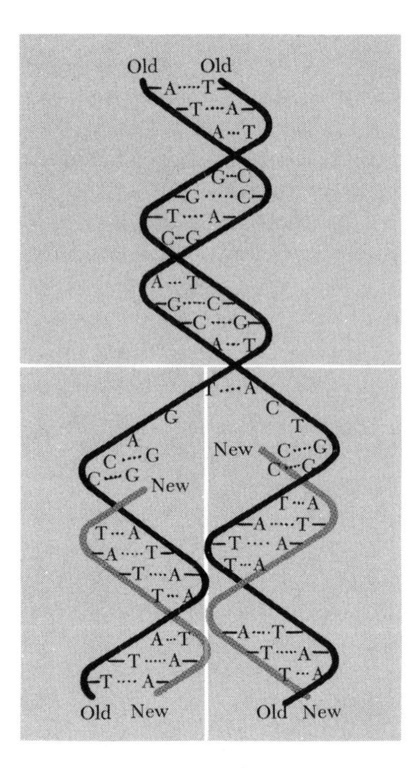

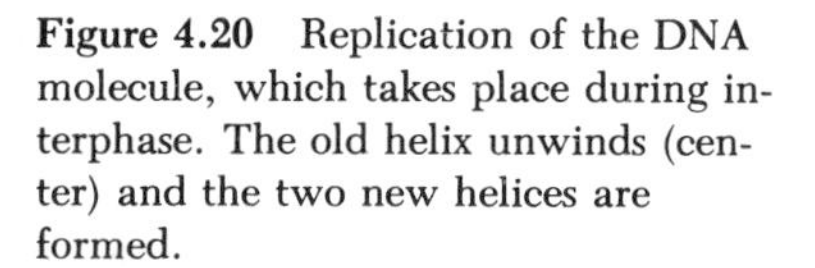
Figure 4.20 Replication of the DNA molecule, which takes place during interphase. The old helix unwinds (center) and the two new helices are formed.

from the same end of the double helix, they run in the $5' \rightarrow 3'$ direction (Figure 4.18). Thus the two strands are said to be of opposite polarity.

In addition to accounting for the base ratios set forth in Table 4.2, the nature of the double helix explains other phenomena. It reveals how the hereditary substance of the cell can be so stable in its metabolic characteristics and its amount per cell; it permits us to understand how the DNAs of a wide variety of organisms can be so similar physically and chemically, and yet so different in a hereditary sense; it offers an explanation of how a sequence of base pairs can be the source of coded information in the cell; and it provides a mechanism for the exact distribution of genetic information from one cell to another.

It was known, for example, that when a cell divides, the two daughter cells are genetically identical. The mother cell must, therefore, have replicated, in a highly precise manner, its hereditary and controlling substance prior to cell division so that during division each daughter cell would be, qualitatively and quantitatively, similar in genetic content. Figure 4.20 indicates how this is accomplished. The double helix separates into its two constituent polynucleotide strands by breaking the hydrogen bonds that hold the bases together in pairs. As each strand unwinds, the metabolic machinery of the cell forms a complementary copy alongside, so that when the process of replication is complete two identical helices are formed. Confirmation of this process of replication has been proven through the use of tagged molecules, and is explained in Chapter 9. By this mechanism, the hereditary content of all of the organism's cells remains constant, and human cells form only human cells, elephant cells form only elephant cells.

EXPRESSION OF INFORMATION

Recognizing that the DNA in each cell is the critical molecule of inheritance, and that the structure of DNA, as depicted in Figures 4.18 and 4.19, has been fully verified, we may deduce that the information needed by the cell to carry out its functions, and to build its necessary structures in an orderly and predetermined manner, must reside in coded form in the sequence of nucleotides in each molecule. In other words, the information in a nucleotide sequence of DNA must somehow be translated into, and expressed by, the proteins of the cell, the specificity of which resides in the sequence of their constituent amino acids. It is no accident, therefore, that both DNA and proteins are macromolecules made up of repeated units, and that it is the uniqueness of these sequences and their action that determine the uniqueness of every cell, individual and species. In fact, the sequences of amino acids in proteins are determined by sequences of nucleotide bases in the DNA. We shall now consider the nature of this relationship and look at the mechanisms by which the informational content of DNA is expressed.

The first thing we must recognize is that the assembly of amino acids into proteins takes place at some distance from the DNA itself. In eukaryotic cells, for example, the DNA is in a membrane-bound nucleus, while proteins are made in the cytoplasm, on structures called *ribosomes*. Somehow, therefore, a message must be sent from the DNA to the ribosomes, where it can specify the sequence of amino acids in the protein to be made. The only logical means by which this can be done is a chemical one. A series of brilliant experiments has shown that indeed the DNA does not act directly, but rather through the medium of several kinds of *RNA* (*ribonucleic acid*).

RNA is similar to DNA in that it is composed of a linear sequence of nucleotides; however, it is different in some very important respects. As its name suggests, the sugar moiety of ribonucleotides is *ribose* instead of deoxyribose; three of the nucleotide bases, adenine, guanine, and cytosine, are the same as those found in DNA, but the pyrimidine uracil (U) replaces thymine (Figure 4.21); finally, RNA, for the most part, is single stranded. RNA is synthesized in close association with DNA; in fact, it is copied from, and therefore complementary to, one of the strands of DNA, the latter acting as a template to determine the

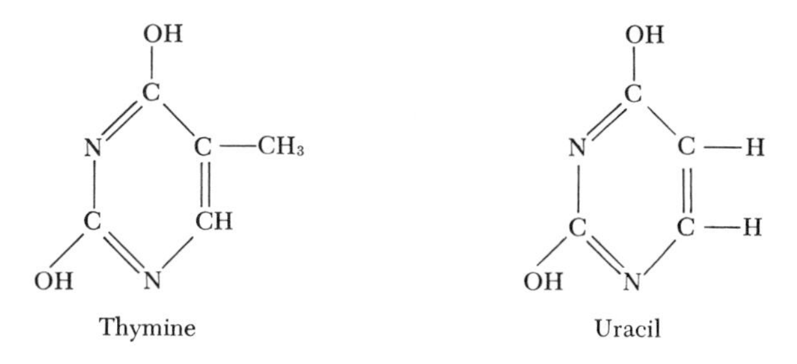

Figure 4.21 The molecular configurations of thymine, found in DNA, and uracil, which replaces it in RNA.

nucleotide sequence of the former. This *transcription* of DNA into RNA is accomplished by one of several enzymes, known as RNA polymerases. The RNA polymerase causes the DNA double helix to unwind and at the same time joins ribonucleotides together to form an RNA chain that grows in length in the $5' \rightarrow 3'$ direction as the polymerase moves along the DNA. Thus, if the portion to be transcribed is as follows:

5′ T—A—G—C—C—T—T—G—A—C—T—G—C—A—C 3′

3′ A—T—C—G—G—A—A—C—T—G—A—C—G—T—G 5′

and only the bottom strand of the helix is to be transcribed, the RNA sequence will be

5′ U—A—G—C—C—U—U—G—A—C—U—G—C—A—G 3′

The completed RNA molecule can then leave the DNA, which returns to its original configuration, and move to its destination in the cell.

Several different kinds of RNA, all formed in this way but from specific RNA polymerases, are involved in the synthesis of proteins; these include *ribosomal RNAs* (*rRNA*), which form part of the ribosome; *transfer RNAs* (*tRNA*), which recognize and bind to specific amino acids and bring them to the ribosome; and *messenger RNA* (*mRNA*), the RNA which codes for proteins. Messenger RNA, as its name suggests, is the important intermediary in information flow between DNA and proteins; it is the mRNA that carries the coded information of DNA, now transcribed into a nucleotide sequence in mRNA, to the ribosome. These nucleotide sequences can then be translated, at the ribosome, into the amino acid sequences of proteins.

The Genetic Code

Figure 4.22 illustrates the amino acid sequence of an enzyme, bovine chymotrypsinogen, concerned with digestion in the small intestine of cattle. It will be immediately evident that there are far more kinds of amino acids (there are 20 different ones) that enter into the structure of proteins than there are nucleotide bases (4). Quite obviously, then, if DNA, in any kind of direct manner, determines the sequence of amino acids in the protein, one nucleotide in DNA cannot specify the kind and position of any single amino acid in a protein; such a singlet code would take care of only four amino acids. A doublet code is similarly inadequate, since adjacent nucleotides, taken in pairs and making use of all possible combinations, could specify only 16 (4×4) amino acids (Table 4.3). A triplet code, however, is more than adequate, since 64 ($4 \times 4 \times 4$) triplets are possible, but the triplet code has been shown to be correct. By means of artificially constructed nucleic acids of known composition, which can function in *in vitro* circumstances, the triplet codes have been determined in the laboratory and have been found universally applicable to all tested species. If all of the triplet codes are functional, it would mean

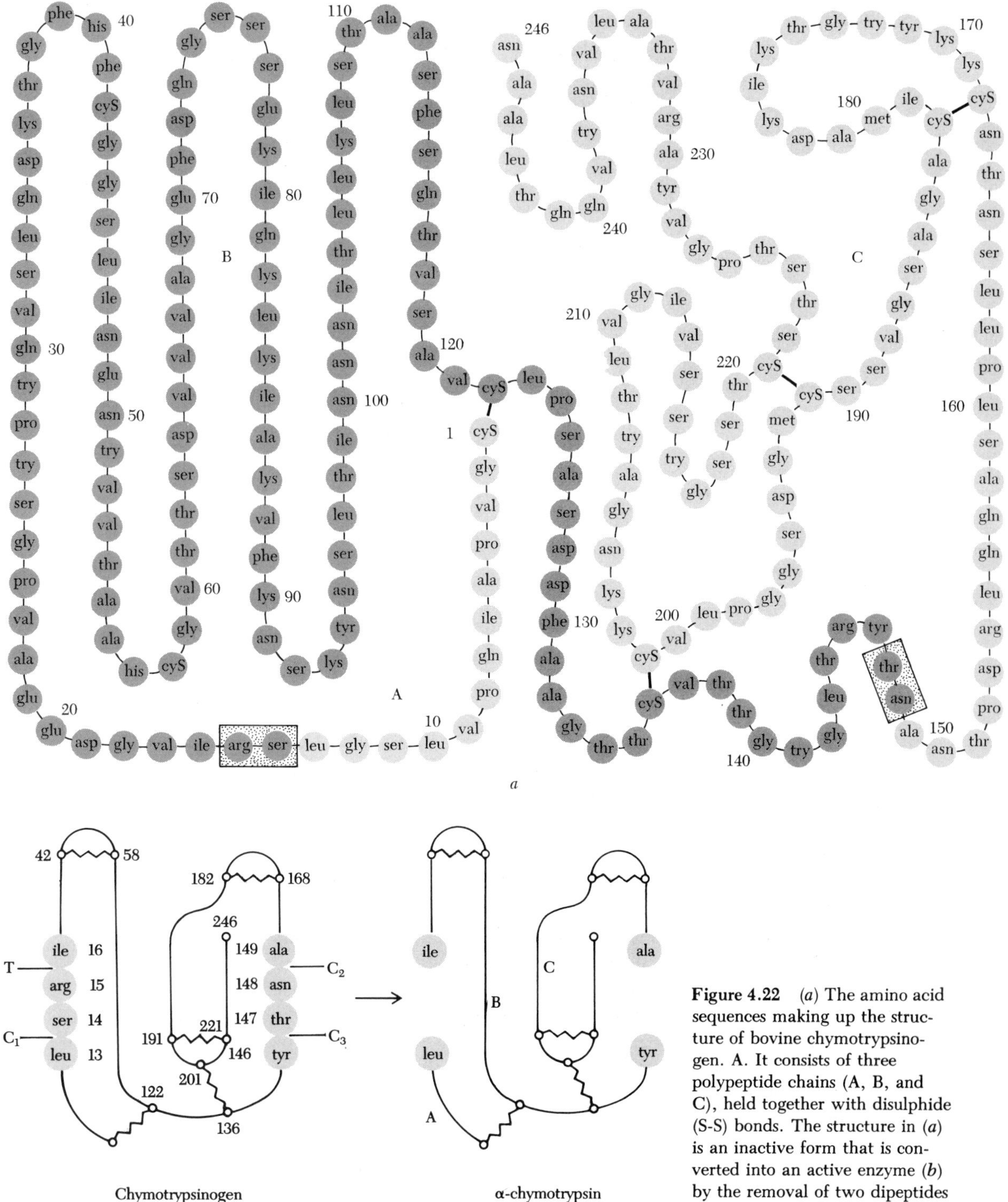

Figure 4.22 (*a*) The amino acid sequences making up the structure of bovine chymotrypsinogen. A. It consists of three polypeptide chains (A, B, and C), held together with disulphide (S-S) bonds. The structure in (*a*) is an inactive form that is converted into an active enzyme (*b*) by the removal of two dipeptides (in the shaded boxes).

TABLE 4.3 THE POSSIBLE NUMBERS OF CODE WORDS BASED ON SINGLET, DOUBLET, AND TRIPLET CODES ARE SHOWN HERE. WE NOW KNOW THAT THE TRIPLET CODE IS CORRECT.

Singlet code (4 words)	Doublet code (16 words)				Triplet code (64 words)			
					AAA	AAG	AAC	AAU
					AGA	AGG	AGC	AGU
					ACA	ACG	ACC	ACU
					AUA	AUG	AUC	AUU
					GAA	GAG	GAC	GAU
					GGA	GGG	GGC	GGU
A	AA	AG	AC	AU	GCA	GCG	GCC	GCU
G	GA	GG	GC	GU	GUA	GUG	GUC	GUU
C	CA	CG	CC	CU	CAA	CAG	CAC	CAU
U	UA	UG	UC	UU	CGA	CGG	CGC	CGU
					CCA	CCG	CCC	CCU
					CUA	CUG	CUC	CUU
					UAA	UAG	UAC	UAU
					UGA	UGG	UGC	UGU
					UCA	UCG	UCC	UCU
					UUA	UUG	UUC	UUU

that most amino acids can be specified by more than one code word. Table 4.4 indicates that this is indeed the case.

Since the code is a triplet one, the mRNA sequence, copied from a DNA sequence, represents a series of code words, or *codons*, each consisting of three nucleotides specifying a particular amino acid (Table 4.4). The recognition between the triplet codons of the mRNA and the amino acid is mediated by molecules of tRNA. Transfer RNAs, of which there are many different ones, can be represented in two dimensions as cloverleaf-shaped molecules (Figure 4.23), although the native three-dimensional conformation is different. The shape of the molecules results from the fact that internal complementary base pairing, mostly C⋮G pairs, causes the molecule to loop back on itself in several places. At one end of each type of tRNA is the sequence -CCA; this -CCA tip is the place where amino acids become covalently attached to the tRNA (Figure 4.24). Enzymes called aminoacyl-tRNA synthetases catalyze this reaction, and, by recognizing both the amino acid and corresponding tRNA, ensure that each type of tRNA can accept only one type of amino acid. This high degree of specificity is very important,

TABLE 4.4 THE GENETIC CODE THAT HAS BEEN FOUND TO EXIST IN ALL ORGANISMS. UAA (OCHRE), UAG (AMBER), AND UGA (UMBER), ARE CHAIN-TERMINATING CODONS; THE TRANSLATING MECHANISM IS UNABLE TO READ THESE CODONS, AND THE PROTEIN-FORMING ACTION CEASES. AUG PLAYS TWO ROLES, DEPENDING UPON WHERE IT IS LOCATED IN A MESSAGE: IT IS A CHAIN-INITIATING CODON FOR THE FORMATION OF RNAs BY TRANSCRIPTION; IF FOUND IN THE MIDDLE OF A MESSAGE IT CODES FOR METHIONINE.

First letter	Second letter: U		C		A		G		Third letter
U	UUU	Phe	UCU	Ser	UAU	Tyr	UGU	Cys	U
U	UUC	Phe	UCC	Ser	UAC	Tyr	UGC	Cys	C
U	UUA	Leu	UCA	Ser	UAA	OCHRE	UGA	UMBER	A
U	UUG	Leu	UCG	Ser	UAG	AMBER	UGG	Tryp	G
C	CUU	Leu	CCU	Pro	CAU	His	CGU	Arg	U
C	CUC	Leu	CCC	Pro	CAC	His	CGC	Arg	C
C	CUA	Leu	CCA	Pro	CAA	GluN	CGA	Arg	A
C	CUG	Leu	CCG	Pro	CAG	GluN	CGG	Arg	G
A	AUU	Ileu	ACU	Thr	AAU	AspN	AGU	Ser	U
A	AUC	Ileu	ACC	Thr	AAC	AspN	AGC	Ser	C
A	AUA	Ileu	ACA	Thr	AAA	Lys	AGA	Arg	A
A	AUG	Met	ACG	Thr	AAG	Lys	AGG	Arg	G
G	GUU	Val	GCU	Ala	GAU	Asp	GGU	Gly	U
G	GUC	Val	GCC	Ala	GAC	Asp	GGC	Gly	C
G	GUA	Val	GCA	Ala	GAA	Glu	GGA	Gly	A
G	GUG	Val	GCG	Ala	GAG	Glu	GGG	Gly	G

since, as we shall see, it is the tRNA that is responsible for ensuring that the amino acid is inserted into its proper position in the polypeptide chain; that is, the tRNAs are the readers of the message encoded in the mRNA.

Each type of tRNA molecule, attached to a specific type of amino acid, also includes a particular sequence of three nucleotides at one of the loops. These *anticodon* sequences can recognize, by base pairing, complementary codon sequences in the mRNA. As successive codons in the mRNA are recognized by the anticodons in the tRNA molecules, the correct amino acids are added sequentially to a growing polypeptide chain on the ribosome. In this way the protein structure is specified by the mRNA, the nucleotide sequence of which is in turn determined by that of the DNA from which it is transcribed. The basic features of this pathway by which the information is decoded and expressed as protein structure are shown in Figure 4.25; we shall now look in more detail at the processes involved and see that the elegant simplicity of these relationships masks a tremendously complex and integrated series of events.

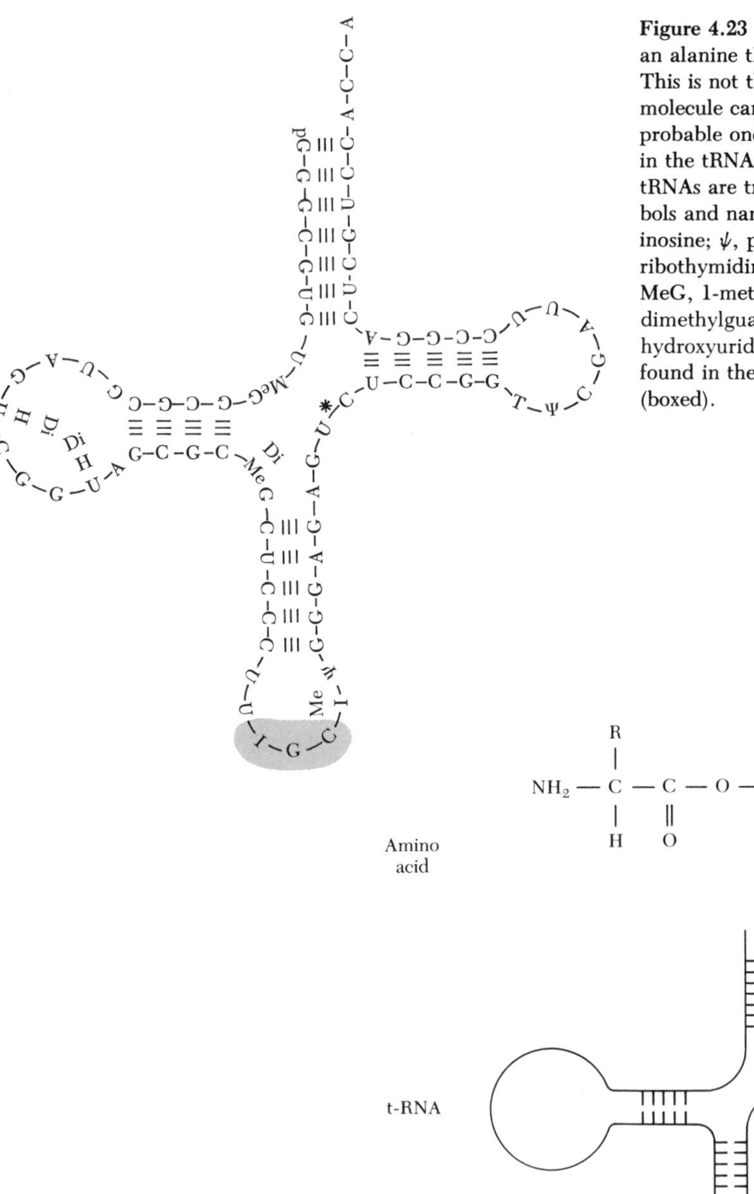

Figure 4.23 The cloverleaf model of an alanine tRNA molecule from yeast. This is not the only secondary form the molecule can take, but is the most probable one. The unusual nucleotides in the tRNAs are formed after the tRNAs are transcribed, and their symbols and names are as follows: I, inosine; ψ, pseudouridine; T, ribothymidine; MeI, 1-methylinosine; MeG, 1-methylguanosine; DiMeG, N-2-dimethylguanosine; and DiHU, 5,6-dihydroxyuridine. The anticodon, I-G-C, is found in the base of the middle loop (boxed).

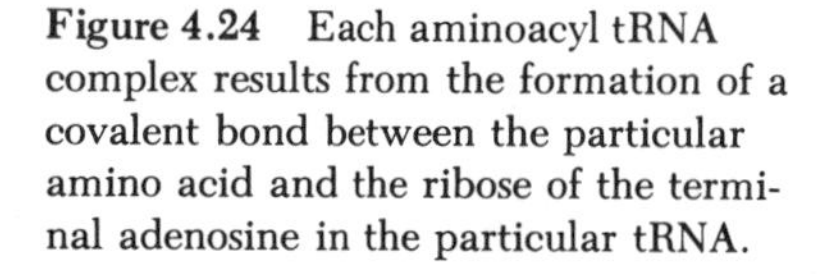

Figure 4.24 Each aminoacyl tRNA complex results from the formation of a covalent bond between the particular amino acid and the ribose of the terminal adenosine in the particular tRNA.

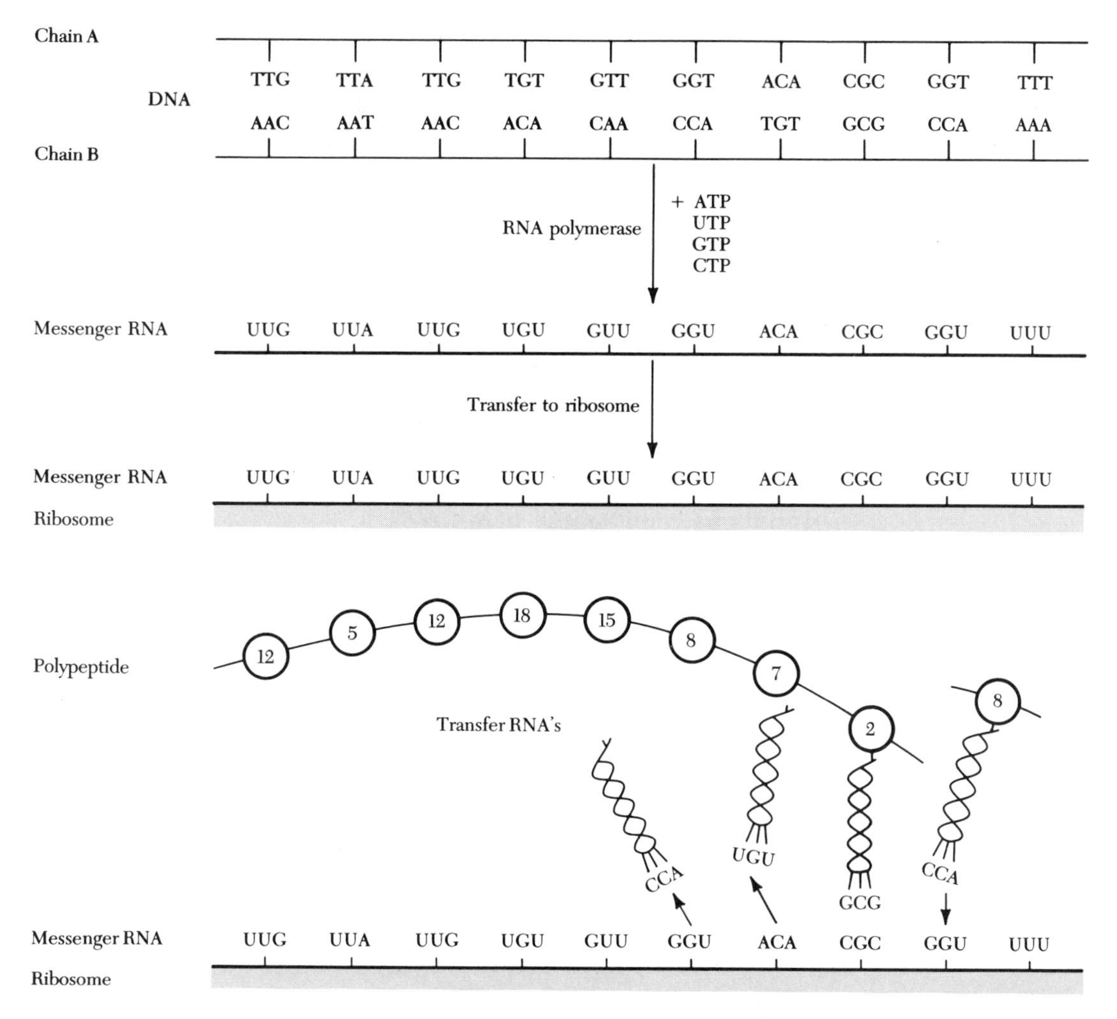

Figure 4.25 The sequence of events whereby the information coded in the DNA molecule is transcribed into messenger RNA (mRNA) and then, by translation, into the formation of a particular protein. It is assumed here that it is the Chain B of the DNA that is being "read." The mRNA contains the code to be read, while the transfer RNAs (tRNAs) possess an anti-code, which matches the code by base-pairing. The tRNA at the right carries an activated amino acid that has not yet been inserted into the growing polypeptide.

RNA Synthesis and Processing

Initiation of transcription. RNAs are transcribed from double-stranded DNA by an RNA polymerase. Studies of RNA synthesis in the prokaryote, *E. coli*, have shown that in order for transcription to be initiated at the correct position in the DNA, the RNA polymerase must first be able to recognize particular promoter sequences of nucleotides close to the initiation point. However, before the polymerase can recognize these sequences, it must first be coupled with an additional protein called *sigma*. Without sigma the polymerase is unable to make the correct selection of which of the two polynucleotide strands of the helix is to be transcribed; instead, it begins to transcribe at any point along the molecule, resulting in a jumble of meaningless and generally incomplete molecular sentences. Termination of the RNA chain must be similarly precise, and this, as in initiation, depends on a specific sequence of nucleotides together with a termination protein, *rho*. These two proteins, therefore, provide prokaryotes with a high degree of specificity in beginning and ending the process of transcription.

Transcriptional promotor sequences have also been identified in eukaryotic DNAs, again located some fixed distance from the initiation sites. Also, initiation factors other than the polymerase and the promotor sequences are required in order for recognition of the sequences by the polymerase, although these proteins are less well characterized than the sigma initiation factors of *E. coli*.

Post-transcriptional modification of mRNA. Although the polymerase synthesizes a primary transcript which is an exact complement to the DNA strand, this transcript is extensively modified, at least in eukaryotic cells, before it can function as mRNA. The RNA molecules that are present in the nucleus and that include the eventual mRNA sequences are known as heterogeneous nuclear RNA (hnRNA). The first modification after transcription begins is the addition of a "cap" at the start (5′) end of the molecule (Figure 4.26); this cap is an unusual nucleoside, 7-methyl guanosine, and appears to be responsible for attachment of the mRNA to the ribosome. At the 3′ end of the transcript, a poly-A "tail" is added (Figure 4.26); this consists of a sequence of about 200 adenine-containing nucleotides, the role of which is not fully understood.

Another feature of hnRNA molecules is that they are very much longer than the mRNA molecules to which they give rise. This is because they include noncoding sequences, or *introns*, which intervene between coding sequences and which are removed before the mRNA leaves the nucleus. The introns are excised from the molecule, and the ends of the coding regions (or exons) are spliced together (Figure 4.26). Small ribonucleoprotein particles in the nucleus appear to be involved in some way in the excision and splicing process. Other noncoding sequences are left in the mRNA, but these are present on either side of the final coding sequence and do not disturb the colinear relationship between the coding sequence of nucleotides and the amino acid sequence of the protein. Finally, the mRNA becomes associated with proteins before being transported to the ribosomes.

Ribosomes and ribosomal RNA. As we have already mentioned, the ribosomes are the organelles in the cell on which the directed assembly of amino acids into the polypeptides of proteins takes place. Individual ribosomes consist of rRNA and protein, and their sizes are generally expressed in S values. These numerical values are measures of how fast the particles sediment when centri-

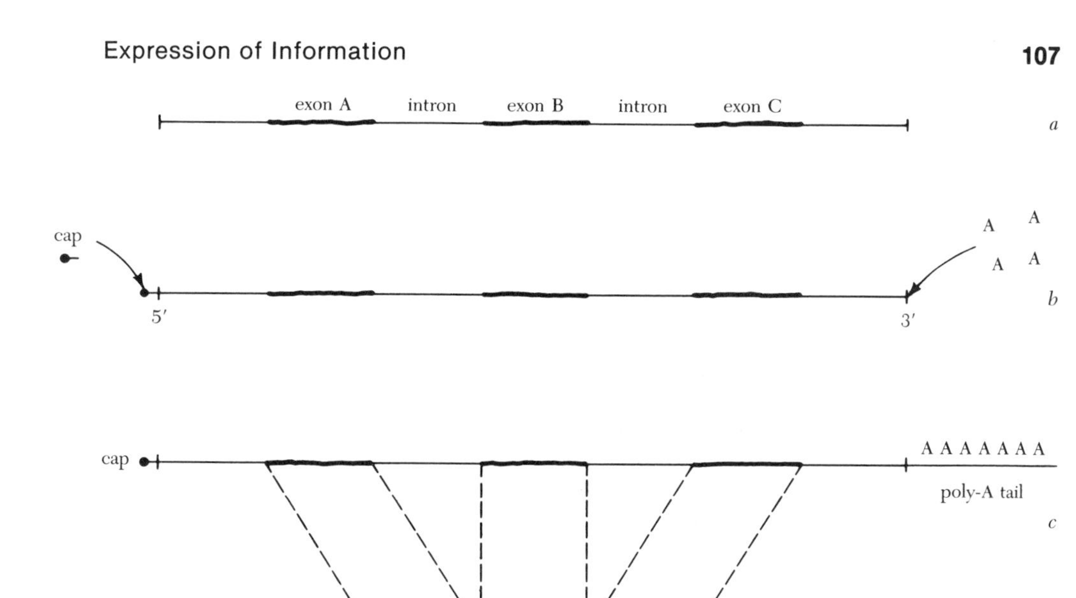

Figure 4.26 Processing of eukaryotic mRNA. (*a*) The DNA template includes coding sequences (exons—heavy lines) separated by intervening non-coding sequences (introns). Non-coding sequences are also present between the initiation site and the first exon, and between the last exon and the termination site. (*b*) The primary RNA transcript is capped at the 5′ end, and a poly-A tail added at the 3′ end. (*c*) The introns are excised, and the exon sequences spliced together before the functional mRNA reaches the cytoplasm, where its coding sequence ABC can be translated into a polypeptide.

fuged, and are related to the size of the particles. Although some variation in ribosome size is encountered, we shall discuss size in terms of mammalian ribosomes.

Each 80S ribosome is composed of a large, 60S subunit and a small, 40S subunit. (S values are not additive.) The large subunit contains one molecule each of 28S rRNA, 5.8S rRNA, and 5.0S rRNA, and about 50 different proteins. The smaller subunit consists of about 30 proteins, associated with one molecule of 18S rRNA (Figure 4.27). The molecular complexity of ribosomes is presumably related to the complexity of protein synthesis, and must provide the configurations and activities required.

Assembly of ribosomes occurs in the nucleolus, and all of the rRNAs, with the exception of 5S rRNA, are synthesized from the DNA of the nucleolar organizer regions of the chromosomes (Figure 4.28a). Ribosomal RNA transcripts, like mRNA, are extensively processed before leaving the nucleolus. The primary RNA transcript from the rDNA is a 45S molecule which includes the 28S, 18S, and 5.8S sequences. This 45S molecule is then cleaved and degraded in a series of steps into the three functional rRNA molecules (Figure 4.28b). This mechanism clearly en-

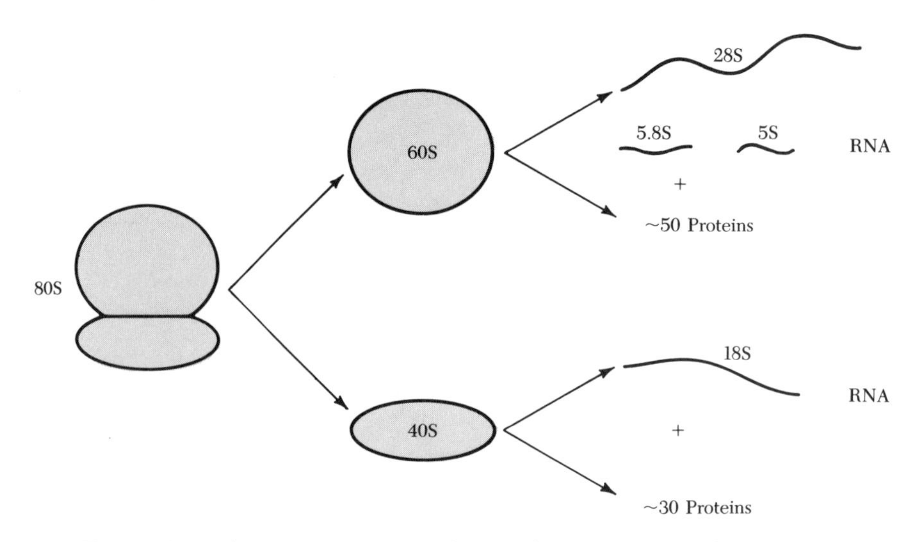

Figure 4.27 Ribosome organization. The 80S ribosome is composed of a 60S subunit and a 40S subunit. The former contains 28S, 5,8S, and 5S RNA molecules plus about 50 proteins, the latter an 18S RNA and about 30 proteins.

sures that equivalent amounts of the 28S, 18S, and 5.8S molecules will be synthesized by the cell; however, since the 5S rRNA is made from DNA elsewhere in the nucleus, expression of the different rDNAs must be coordinated in some other fashion.

Protein Synthesis

Now that we have discussed the general nature of the flow of information from DNA to protein, and some of the components involved, we can consider the processes that take place at the ribosomes and that result in the formation of functional proteins, that is, translation of the mRNA. Once the mRNA reaches the ribosome, it must become attached to it and be read, from its 5′ end, three nucleotides at a time, starting and ending at specific codons if the correct amino acid sequence of the protein is to be assembled. The initial attachment of mRNA is to the small subunit, and is thought to depend on interactions between the capped end of the mRNA and a complementary region of the rRNA.

When the ribosome is associated with the mRNA, the first code word in the message becomes aligned in a specific way with reference to a site on the ribosome known as the P site (Figure 4.29). The P site is occupied by an initiator molecule of tRNA carrying the amino acid methionine; the first codon in mRNA is always AUG, and therefore the first amino acid in the sequence formed is always methionine. Next, another amino acyl-tRNA, that with an anticodon complementary to the second codon in the mRNA, binds to a second site on the ribosome, the A site (Figure 4.29). A peptide bond is then formed between the methionine and the

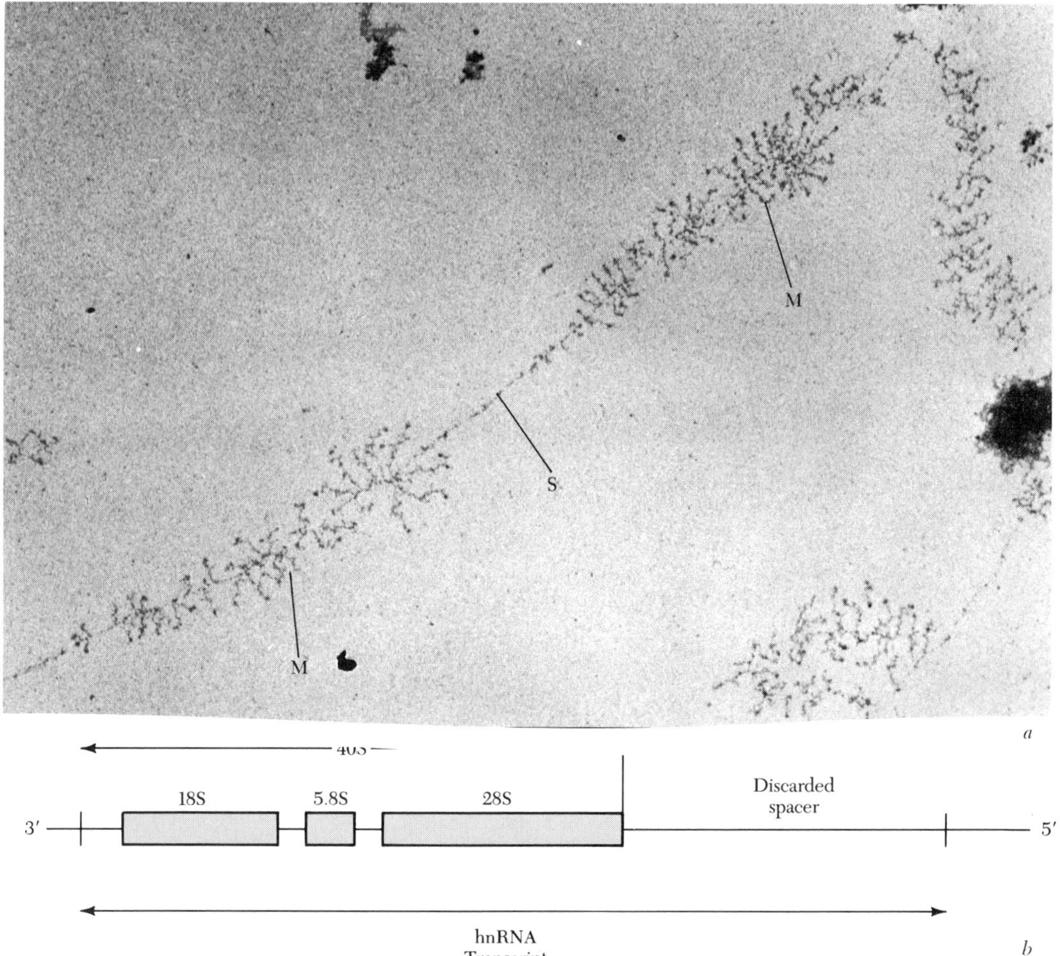

Figure 4.28 (*a*) Genes in the nucleolar region of the oocyte of the newt, *Triturus viridescens*, transcribing the ribosomal precursor RNA (probably the 45S unit), which will give rise to the 28S and 18S and 5.8S units, which will enter into the structure of the mature ribosome. The long axis is DNA, possibly existing as a DNA-histone complex. Each 45S coding sequence, many of which are arranged in series, is 2.5 μm long, and successive ones are separated from each other by a spacer DNA (S). Each matrix consists of a series of RNA fibrils, which increase in length as one goes from one end of the matrix to the other; these are released when they reach the proper length and are then processed in the nucleolus for incorporation into the ribosomes. Each matrix is, therefore, engaged in the simultaneous transcription of many rRNA molecules; the shortest have just begun to be formed, the longest are about ready to be released. (Courtesy of Dr. C. Woodcock.) (*b*) The structure of a ribosomal RNA gene in *Xenopus*, the African clawed toad. An hnRNA transcript of 8.8×10^6 daltons (a dalton is equivalent to one hydrogen atom) is first produced, the 5 end spacer is discarded, and the remainder is processed in the nucleus to form the 18S, 5.8S, and 28S rRNAs. The 5S rRNAs are produced elsewhere in the genome. The rRNA genes may number in the hundreds and are concentrated in the centric heterochromatin.

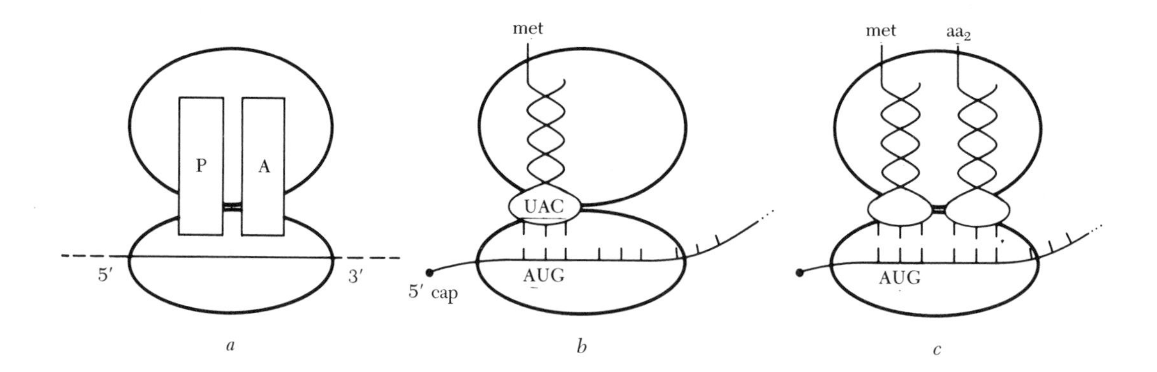

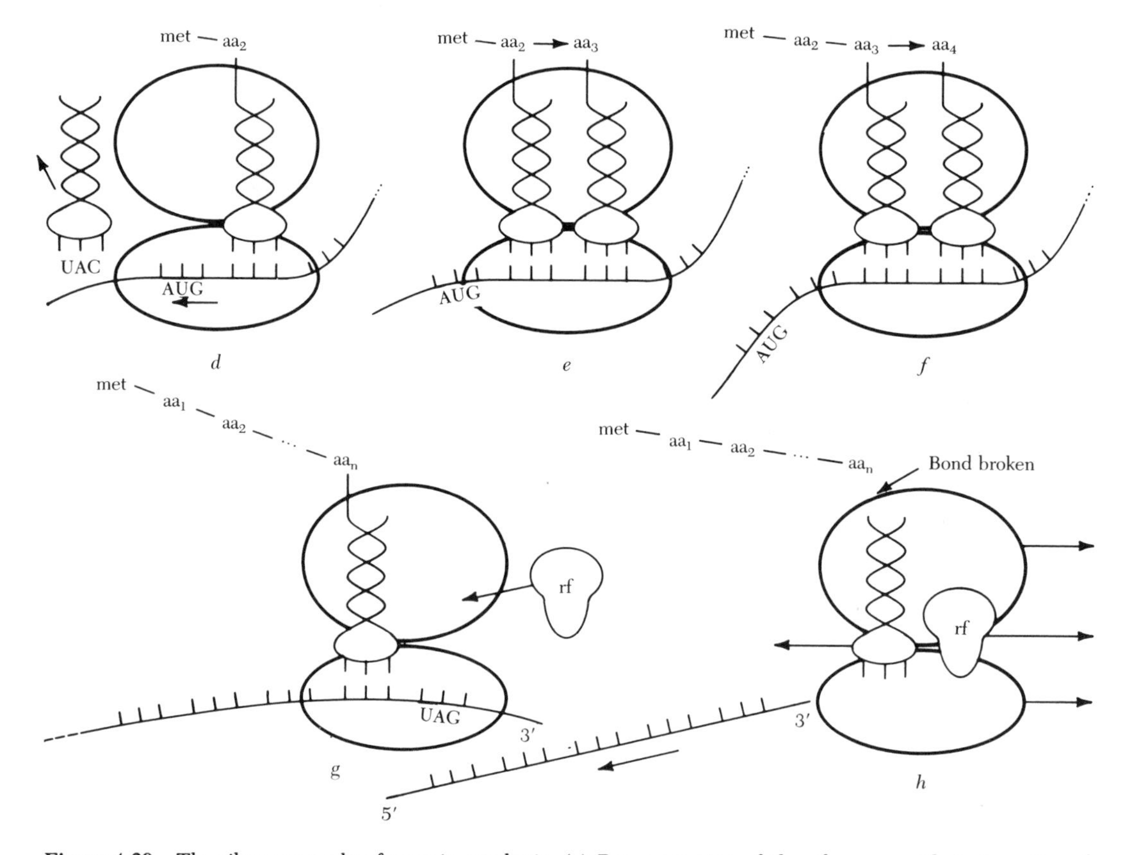

Figure 4.29 The ribosome cycle of protein synthesis. (*a*) Representation of the ribosome with its two sites of activity and the relation of mRNA to the two sites. The P, or peptidyl site, is occupied by a tRNA to which the growing polypeptide chain is attached, while the A site is occupied by the tRNA carrying the next amino acid to be added to the growing chain. (*b*) An initiator tRNA carrying methionine recognizes the initiator codon in the mRNA. (*c*) The next aminoacyl tRNA base pairs with the next mRNA codon at the A site, the methionine is transferred to amino acid 2 by peptide bond formation and the first tRNA is released from the P site. (*d*) The second tRNA, with its attached dipeptide, moves to the P site, the mRNA moves three nucleotides over so that the next triplet can be recognized by an incoming aminoacyl tRNA at the A site (*e*). The process is repeated (*f*) until the termination codon is reached (*g*). A release factor (rf) binds to the ribosome and causes the completed polypeptide chain to be released (*h*).

incoming amino acid to form a dipeptidyl-tRNA at the A site, and the methionine tRNA is released from the ribosome. The ribosome then transfers the tRNA with the attached dipeptide to the P site, leaving the A site open. At the same time it moves along the mRNA in such a way that the next codon can specify the aminoacyl tRNA to occupy the A site. This process is repeated until the end of the message is reached and all the amino acids have been joined together.

The completed polypeptide is, of course, still attached to the tRNA which brought the last amino acid to be specified to the ribosome. As Table 4.4 indicates, termination of the translational process occurs when one or other of the three termination codons—UAA, UAG, or UGA—is encountered in the reading of the message. No normal tRNA can bind to the vacant A site and interact in complementary fashion with these triplets; instead, a termination factor binds to the ribosome and causes the polypeptide to be released from the tRNA at the P site into the cytoplasm (Figure 4.29).

Polyribosomes. Although we have described the events in translation as they occur on individual ribosomes, each molecule of mRNA can be read many times. In the cell protein synthesis occurs on polyribosomes, or polysomes, aggregates of ribosomes all reading the same message simultaneously (Figure 4.30). As the ribosomes move along the message, others can initiate translation and start to form a second molecule of the protein. This sequential and repeated translation

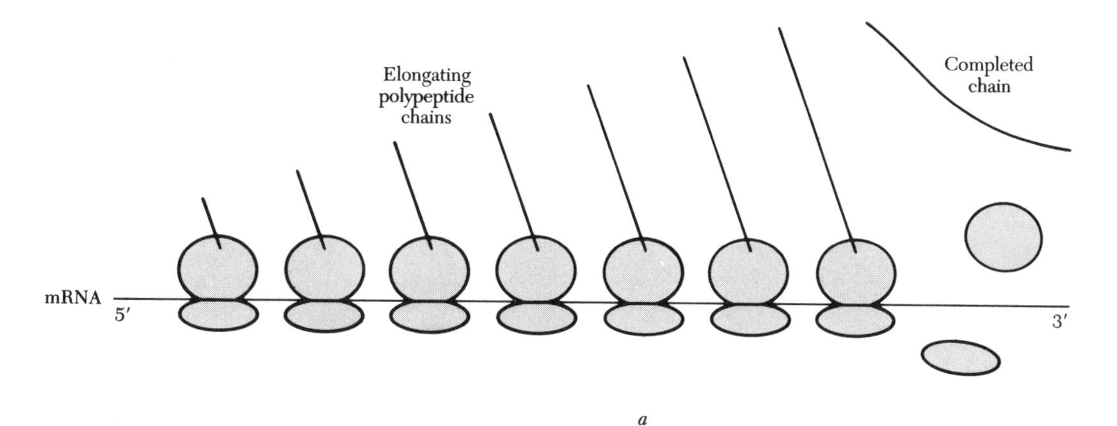

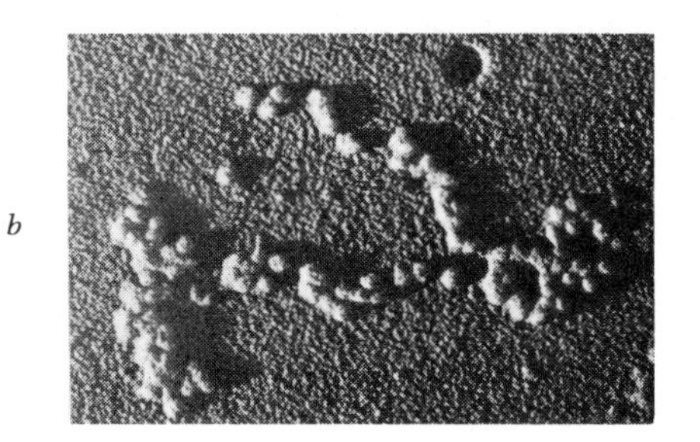

Figure 4.30 (*a*) A single mRNA molecule being read simultaneously in a polyribosome complex. As each ribosome moves along the message, it is followed by others. Thus several polypeptides of various lengths are in the process of being formed. (*b*) Electron micrograph of polyribosome clusters.

of mRNAs allows for many copies of the polypeptides to be produced in a much shorter time than would otherwise be the case.

Post-translational modifications. Once a polypeptide has been completed, it can be modified by the cell in several ways. For example, although methionine is always the first amino acid in the sequence, it may be enzymatically clipped off before the protein becomes functional. As we shall see later (Chapter 7), other amino sequences are often removed from the end of the coded polypeptide; these are sequences which may be important in allowing proteins to enter membrane-enclosed compartments, such as the endoplasmic reticulum, chloroplasts, and mitochondria, but which are not required for the ultimate activity of the protein. Other polypeptides contain sequences that must be removed before enzyme activity is acquired (see Figure 4.22b)—the hormone insulin is another such protein (see Figure 7.10).

THE ORGANIZATION OF CHROMOSOMES

Although we now have a good understanding of the nature of DNA, the primary informational substance of a cell, how it is replicated, and how the information is expressed, we know much less about how it is organized within the nucleus, or how that organization is related either to its activities or to how those activities are regulated.

Prokaryotes and eukaryotes differ in the number, size, and organization of their chromosomes. Prokaryotes have only a single circular chromosome, although in some cells multiple copies may be present. The haploid genome of eukaryotes may have as few as two (*Haplopappus gracilis*, a plant in the daisy family), or as many as several hundred chromosomes (*Ophioglossum reticulatum*, the adder's tongue fern). No normal eukaryotic chromosome is ever circular. Differences are even greater when one considers the amount of DNA per genome. A micrometer length of duplex DNA has a molecular weight of approximately 2.0×10^6. The 1300 μm of *E. coli* chromosome would, therefore, have a molecular weight of 2.6×10^9. By way of contrast, the DNA in the longest chromosome in *Drosophila melanogaster* ($n = 4$) has a molecular weight of $4.13 - 4.3 \times 10^{10}$, and a length of about 20,000 μm. The *Drosophila* genome is a relatively small eukaryotic one when contrasted with the nearly 10 meters of DNA in the newt, *Triturus cristatus*, so it is quite obvious that vast differences in the amount of DNA per genome separate these two groups of organisms (Table 4.5).

Although eukaryotic chromosomes contain very much more DNA than those of prokaryotes, the DNA of each normal eukaryotic chromosome is present as one long uninterrupted double-helical molecule. By carefully extracting DNA from nuclei and measuring the size of the molecules, it has been possible to demonstrate the existence of individual double helices that are sufficiently long to account for all of the DNA present in a chromosome. Clearly, there can be only one double

TABLE 4.5 GENOME SIZE IN PICOGRAMS (g X 10^{-12}), PERCENTAGE OF UNIQUE, SINGLE-COPY DNA IN EACH GENOME, AND SINGLE-COPY COMPLEXITY AS MEASURED IN NUCLEOTIDE PAIRS X 10^{8}. THE REMAINDER OF THE GENOME WOULD CONSIST OF MIDDLE-REPETITIVE AND SATELLITE SEQUENCES. (COMPILED FROM A VARIETY OF SOURCES, BUT SEE SWANSON ET AL., 1981, *CYTOGENETICS: THE CHROMOSOME IN DIVISION, INHERITANCE AND EVOLUTION.* PRENTICE-HALL, INC., ENGLEWOOD CLIFFS, N.J., P. 134.

Species	Common name	Genome size (pg)	Percent of single-copy DNA	Single-copy complexity (n.p. × 10^8)
Animals				
D. melanogaster	Fruit fly	0.12	75	0.82
Crassostrea virginica	Oyster	0.69	60	3.8
Aurelia aurita	Jellyfish	0.73	70	4.7
Strongylocentrotus purpuratus	Sea urchin	0.89	75	6.1
Spisula solidissima	Surf clam	1.2	75	8.2
Cerebratulus lacteus	Worm	1.4	60	7.7
Aplysia californía	Sea cucumber	1.8	40	10.7
Xenopus laevis	Clawed toad	2.7	75	18.5
Limulus polyphemus	Horseshoe crab	2.8	70	17.9
Rattus norvegicus	Norway rat	3.2	75	22.3
Plants				
Vigna radiata	Mung bean	0.48	70	4.0
Spinacia oleracea	Spinach	0.76	26	2.07
Gossypium hirsutum	Cotton	0.77	68	5.0
Petroselinum sativum	Parsley	1.83	30	6.0
Nicotiana tabacum	Tobacco	1.93	45	9.0
Avena sativa	Oats	4.15	25	11.0
Pisum sativum	Garden pea	4.63	30	14.0
Triticum aestivum	Wheat	5.4	25	14.0
Hordeum vulgare	Barley	5.5	30	16.0
Secale cereale	Rye	8.3	25	22.0
Vicia faba	Broad bean	12.5	20	26.0

helix per chromosome in such cases. (Some multistranded chromosomes do exist, however, and we shall discuss these later.)

The great variation in the amount of DNA among organisms suggests that such differences cannot be due to differences in the number of genes in each genome. It is a reasonable assumption that more complex organisms require more genes coding for specific proteins (also called structural genes) to attain their complexity than do simpler ones, but such relationships are by no means evident when DNA values per genome are considered. Much of the DNA, therefore, must be of a nonstructural, or noncoding kind, but how much is regulatory, in that it governs the expression of structural genes, and how much performs no known function is difficult to determine. Even many of the transcribed sequences are noncoding, such as the introns between coding sequences, and the sequences whose transcripts are excised from pre-rRNA and pre-tRNA molecules.

Another feature of eukaryotic DNA is the presence of sequences that are represented by varying numbers of copies. The following classes of DNA sequences have been characterized by fragmentation of the DNA molecules and the use of the DNA/DNA hybridization technique (Figure 4.31).

About 20 to 70 percent or more of the eukaryotic genome consists of unique, or single-copy, stretches of DNA, each piece differing from the others in nucleotide sequence (Table 4.5). These units are large enough to code for proteins of substantial length, but it is also clear that much of even the unique DNA is never transcribed, or if transcribed, never translated. The structural, coding genes are, of course, included among the pieces of unique DNA, but are interspersed among regions of noncoding DNA.

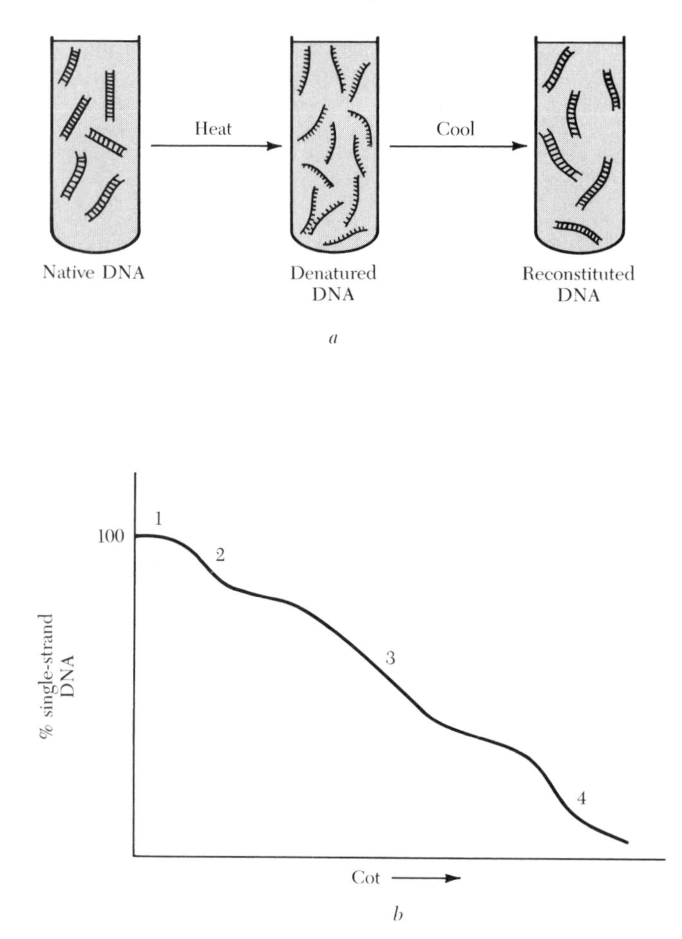

Figure 4.31 (*a*) Denaturation and reannealing of DNA. If a solution containing DNA double helixes is heated above a certain temperature (the melting temperature), the H-bonds between complementary bases are broken and the strands separate. Below the melting temperature complementary sequences are able to reanneal and duplex helixes are re-formed. The rate at which single-stranded DNA reverts to the double-stranded state is a function of the concentration of complementary sequences. (*b*) Reannealing of eukaryotic DNA showing three classes with respect to frequencies of repeated sequences. Double-stranded DNA is first sheared to give fragments of about 400 nucleotide pairs in length, then denatured. The rate of reannealing can be followed as a function of time. In practice the degree of reassociation is plotted against a Cot value, the log of the product of the initial concentration of DNA (Co) and time (t). Initially all of the DNA is single-stranded (1). Rapid reassociating DNA (2) represents highly repeated sequences present in higher concentration than other sequences. A second class, of middle-repetitive sequences (3), reassociates next, followed by (4) a class of single-copy sequences present in low concentration, and therefore reannealing slowly.

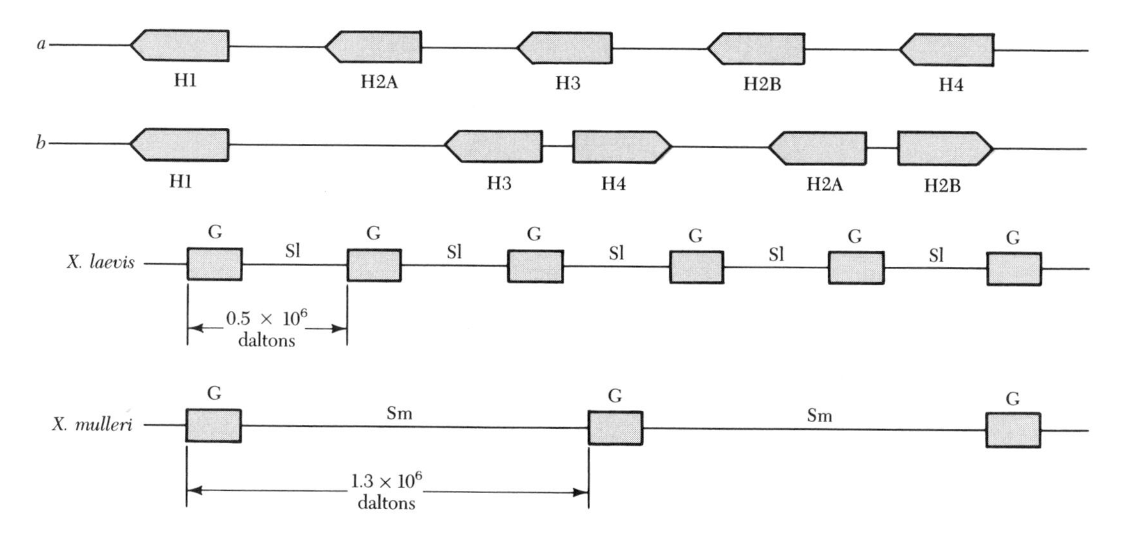

Figure 4.32 Above, the cluster of histone genes in *D. melanogaster* (*a*) and in the sea urchin (*b*). In *D. melanogaster*, at least, the histones are located in one region of the chromosomes, and are duplicated many times. Below, the clustering of 5S rRNA genes in *Xenopus laevis* and *X. mulleri*; G = gene, Sl = spacer DNA in *X. laevis*, Sm = spacer DNA in *X. mulleri*. These genes constitute 0.7% of the total genome in each species, but *X. laevis* has 24,000 copies of the same gene, *X. mulleri* only 9000.

A second class of DNA found in eukaryotes, known as middle-repetitive DNA, consists of sequences 300 to 500 nucleotides long (the *Drosophila* pattern), or 2000 to 3000 nucleotides long (the *Xenopus* pattern), scattered throughout the genome. There are families of these repetitive units, which would include the families of the histone (Figure 4.32), rRNA (Figure 4.28), and tRNA genes, with each family being represented by a few hundred to 100,000 or more copies. They constitute a variable portion of the genome, and their role in the life of a cell is not yet fully understood, although their regulatory nature is suspected.

A last class is that known as satellite DNA, so called because they represent a distinct class when sedimented to equilibrium in a CsCl density gradient. Largely concentrated in the heterochromatin adjacent to the centromeres of eukaryotic chromosomes, satellite DNA consists of sequences that are very short, with each being represented several million times in a genome. In *Drosophila virilis*, for instance, there are three different satellites, differing only slightly from each other in density and nucleotide composition (Figure 4.33); they make up 41 percent of the entire genome, and each is represented by several million copies. *D. melanogaster*, on the other hand, possesses six different satellites of short but varying length, each represented by over 1 million copies; together they make up about 20 to 25 percent of the genome.

Another major difference between the DNA of prokaryotic and eukaryotic chromosomes is in the manner in which it is associated with other molecules. The circular DNA of *E. coli* is highly compacted within the cell into a "nucleoid" about

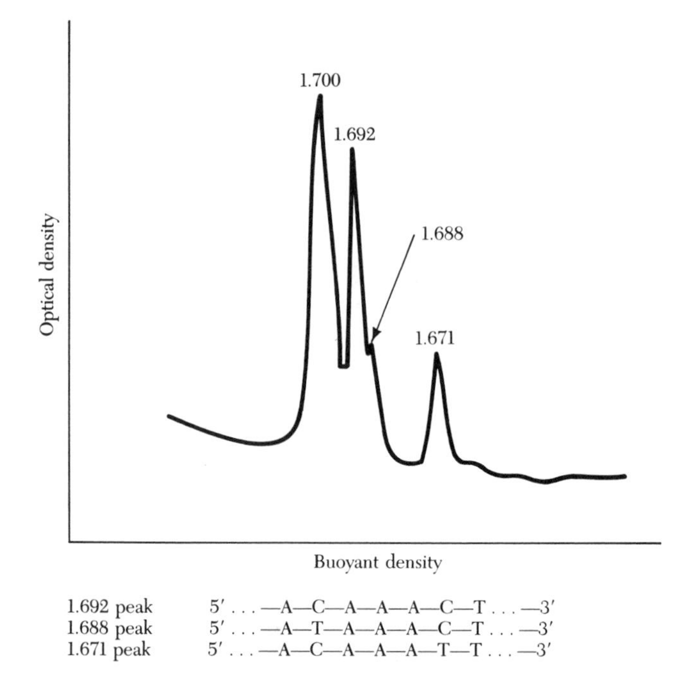

Figure 4.33 The buoyant density patterns obtained when DNA from the brains and imaginal discs of *D. virilis* was centrifuged to equilibrium in neutral cesium chloride; the larger the number, the heavier the DNA. The main peak at 1.700 consists largely of unique, or single-copy, DNA, while the satellite peaks of 1.692, 1.688, and 1.671 consist of highly repetitive, small pieces of DNA concentrated in the heterochromatin localized around the centromeres of the chromosomes. Their differences in molecular weight depend upon the amount of cytosine and thymine contained in each one. The nucleotide content of each satellite is indicated below.

1 μm in diameter. The maintenance of this highly folded and supercoiled structure probably depends on associated proteins, although neither the proteins responsible nor the nature of the association is well understood. In eukaryotic cells, however, the fundamental organization of the chromatin fiber, that is, the DNA molecule plus the associated proteins, is much clearer.

Chromatin Structure

Each chromosome in a eukaryotic nucleus consists of a single, long, double-helical molecule of DNA associated with histone proteins, of which five types are present, plus various nonhistone proteins. When chromatin is extracted from nuclei under appropriate conditions, it appears in the electron microscope as a fiber of about 10 nm in diameter. The fiber has a beaded appearance (Figure 4.34), and each repeating particulate structure in the continuous fiber is called a *nucleosome.* Mild digestion of chromatin with DNAase enzymes preferentially breaks the DNA where it connects adjacent nucleosomes, while leaving that within the nucleosome intact. Chemical analysis of such digested chromatin has shown that each nucleosome contains a length of about 200 nucleotide pairs of DNA associated with an *octamer* of eight histone molecules, two of each of histones H2A, H2B, H3, and H4, plus one molecule of a fifth histone, H1 (Figure 4.35). The DNA is wrapped around the octamer core, and appears to be stabilized in the form of two complete twists by the H1 (Figure 4.35). Some nonhistone proteins may also be associated

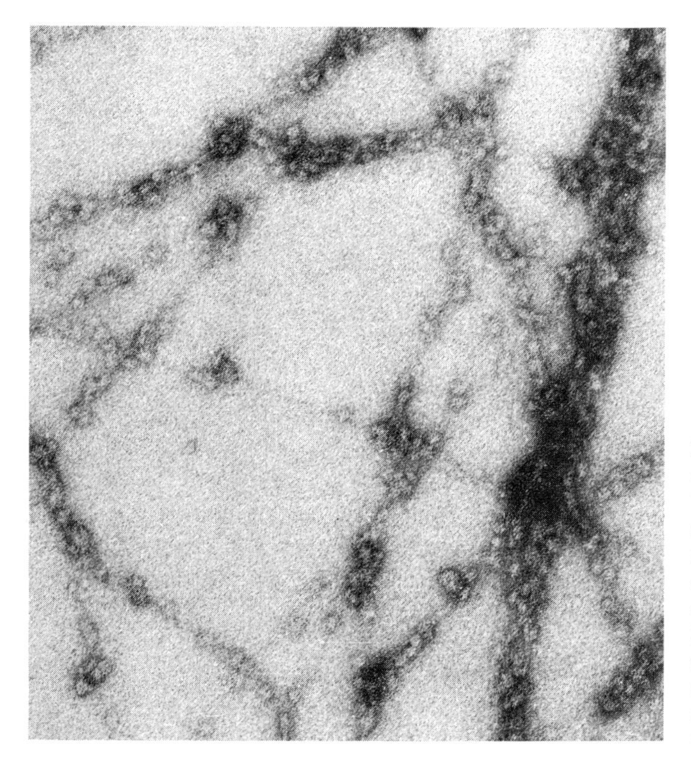

Figure 4.34 Isolated chromatin from chicken erythrocytes showing the beaded nature of the strands. Each strand has a diameter of 10 nm. Four of the five known histones are associated with the nucleosomes, and it is assumed but not proved that the fifth histone is coupled with the DNA of the interbead regions. Some investigators consider the interbead areas to be artifacts of preparation. (Courtesy of Dr. C. Woodcock.)

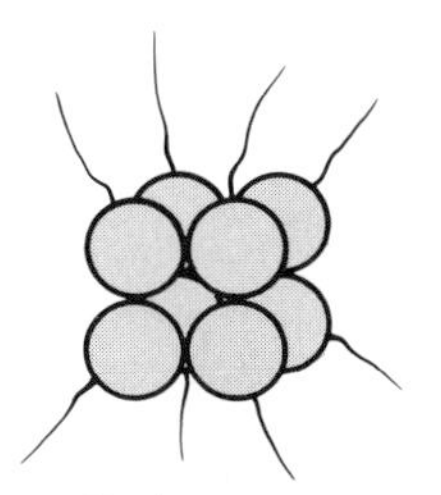

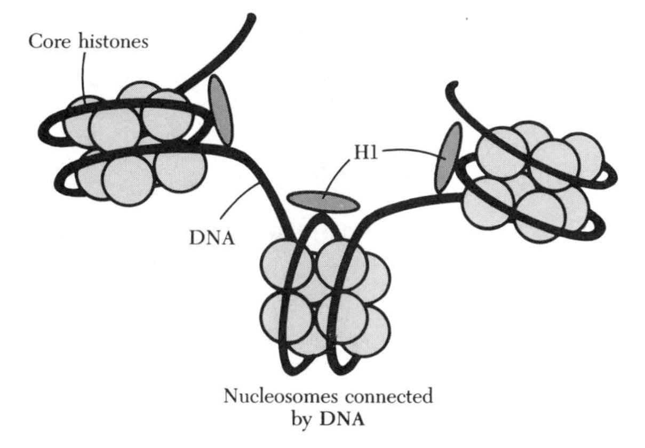

Figure 4.35 The structure of nucleosomes. Each consists of a core of eight histone molecules, two each of H2A, H2B, H3, and H4, around which the DNA is wrapped. About 160 nucleotide pairs of DNA form two complete twists around the core, and are stabilized by a molecule of histone H1. The remaining 40 nucleotide pairs form the links. The histone H1 may also be involved in further compaction of the chromatin fiber.

with the nucleosomes, although it is not known if they are regular components or how they might participate in the structure.

The basic 10-nm fiber is, in turn, coiled, six to seven nucleosomes per turn, into a fiber about 3.0 to 4.0 nm in diameter. Further compaction occurs prior to cell division, when the chromatin becomes resolved into the visibly distinct and highly contracted mitotic chromosomes (Figure 4.36). Even in interphase, however, the chromatin fibers appear to be folded in a regular fashion, with specific points of attachment to a proteinaceous framework in the nucleus. Such a *nuclear matrix* would be expected not only to impose spatial order on the nucleus, but also to provide a structural basis for replication, segregation, and expression of the DNA.

Eukaryotic chromosomes also exhibit a great deal of linear differentiation, depending on the kind of chromosome being investigated and the method of preparation (Figures 1.16 and 4.37). All have distinct ends, or *telomeres;* a specific *centromere* region, which engages the spindle during cell division and allows duplicated chromosomes to segregate; and blocks of heterochromatin of various sizes and placement. Some also have a *nucleolar organizer region* (NOR) that is specialized for the formation of nucleoli.

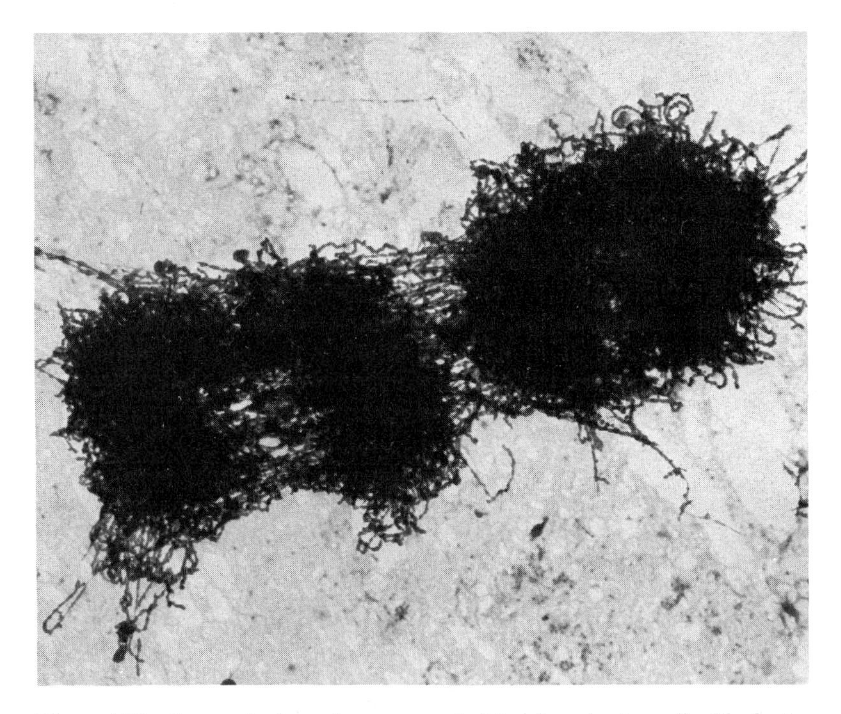

Figure 4.36 Two metaphase chromosomes isolated from bovine cells. The basic chromatin strands are very tightly compacted in the condensed chromosomes. (Courtesy of Dr. S. Wolfe.)

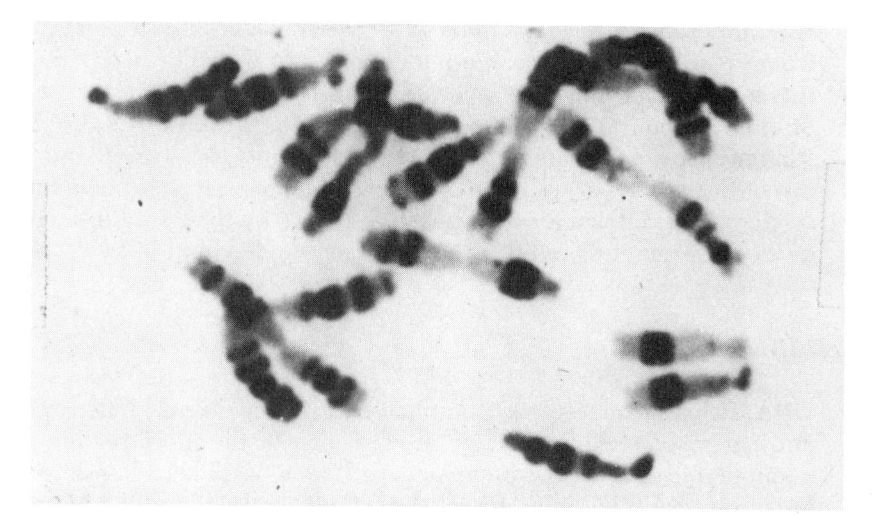

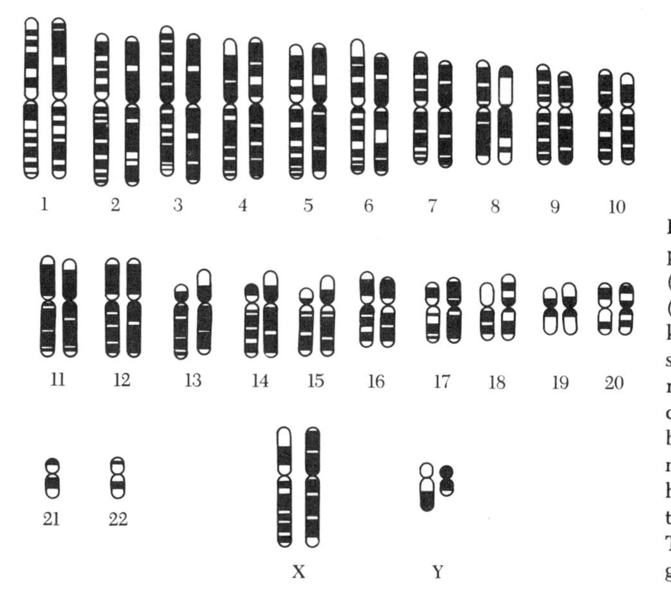

Figure 4.37 The banding patterns of plant chromosomes, *Anemone blanda* (above) and of human and macaque (*Macaca mulatta*, an Old World monkey) chromosomes. The plant banding, stained with giemsa, shows few but rather massive bands, while the animal chromosomes show far more, with each being smaller. The human (left) and macaque (right) chromosomes show how similar these are to each other, although differences are also obvious. The chromosomes of humans and the great apes are far more alike.

The linear differentiation existing among eukaryotic chromosomes can also be demonstrated by means of selective staining with fluorescent or Giemsa dyes. Plant and animal chromosomes, for unknown reasons, respond differently to these dyes, but each chromosome in a genome can be identified by its characteristic bands (Figure 4.37). The banding patterns vary with the dye employed and the treatment of the cell preceding its use. Comparative studies of individuals within a species have revealed that a good deal of chromosome polymorphism may exist, as it does in human populations, while comparisons between related species reveal how similar or dissimilar their genomes might have become through time. Figure

4.37, for example, suggests that the great behavioral differences that separate man from his primate relatives are not reflected in their chromosomes. Each species, therefore, has its own pattern of coding and noncoding sequences of DNA, with the differences coming to light through the techniques of DNA/DNA hybridization and staining. How each came into existence, and how the several kinds of DNA fit, or do not fit, into the overall expression of the genome, is not known for certain in any single species, but some of the questions raised by these findings will be dealt with in the subsequent chapters on development and evolution.

RECOMBINANT DNA

DNA is clearly a large and complicated macromolecule, but recently discovered and perfected techniques have made it one of the more readily sequenced and manipulable of cellular components. It is now possible to transfer from one cell or organism to another selected and functional pieces of eukaryotic DNA, with these recombinant genes being capable of transcription and translation in what is essentially a foreign environment. The transfer can be made within a species, between species, and even between genera as widely different as bacteria and human beings. Recombinant DNA, or gene-splicing, techniques have therefore created a huge gene pool to which all organisms can contribute. Among the steps involved are procedures for the isolation of particular pieces of DNA, the means for testing for their presence and functional integrity, and the means for transporting these pieces of DNA from one cell to another. The steps involved are elegant and sophisticated, but so precise are they that some of the manipulations can be carried out on a commercial scale.

The isolation of a piece of DNA for recombinant purposes can be achieved by one of several routes. To isolate a piece from a larger mass of DNA, use is made of *restriction enzymes*. These are found in, and isolated from, bacterial cells where their function is to destroy any foreign DNA that might, by chance, have entered the cell. Their usefulness resides in the fact that they cleave DNA in highly specific ways, and within *restriction sequences* of four to six nucleotides which have a particular nucleotide symmetry (Table 4.6). When single nucleotide strands are cleaved in a staggered manner (Figure 4.38), cohesive single-stranded ends are formed, and these, under appropriate conditions, can reanneal with other pieces of carrier DNA similarly cleaved by the same restriction enzyme (Figure 4.39). The open ends can be closed by another enzyme called a *DNA ligase*.

In bacterial systems, the recombinant DNA is customarily inserted into *plasmids*, small circular pieces of double-stranded DNA, about 1/1000 the size of the *E. coli* chromosome. These can readily enter or leave a bacterial cell. Although dispensable, plasmids serve as a second chromosome since they often contain genes for resistance to a number of antibiotics (Table 4.7). They can also replicate independently of the bacterial genome, and hence any piece of recombinant DNA inserted into them can be similarly replicated, or cloned. If the recombinant DNA

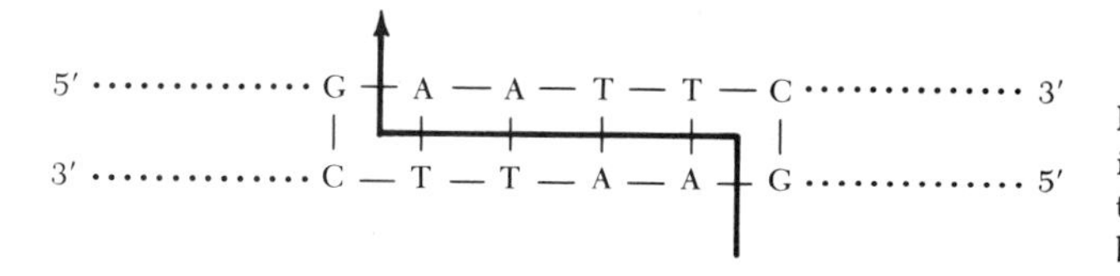

Figure 4.38 The staggered cut made in double-stranded DNA by the restriction enzyme, EcoR1. As the hydrogen bonds are broken the ends fall apart.

leads to the formation of a protein, the bacterial cell becomes a miniature protein factory. Proteins such as insulin are commercially produced in this manner.

How does the investigator know when the desired piece of recombinant DNA has been isolated, inserted into a transporting vehicle, and cloned? Obviously, there must be some kind of selective process which allows one to determine whether the DNA is not only present, but functioning as well. The product (protein) formed must not interfere with the functioning of the bacterial cell, and it must be identifiable before it can be isolated from the other proteins also present. These techniques of identification have been worked out to such an extent that it is now possible to construct a catalogue of all of the DNA in the human genome, and thus to be able to investigate the action of any given segment of DNA.

The cloning of certain genes can be accomplished more readily by the isolation first of the mRNA which they produce. If a particular cell produces a given kind of protein in large amounts—for example, the globins which make up the hemoglobin molecule—this cell must therefore first transcribe the mRNA, which is then translated into the polypeptide. The mRNAs can be isolated and through the

TABLE 4.6 SOME RESTRICTION ENZYMES USED IN RECOMBINANT DNA STUDIES.

Enzyme	Bacterial source	Recognition sequences	Results of cleavage
Eco R1	*E. coli*	G↓AATTC	Cohesive termini
Hae III	*Haemophilus aegyptius*	GG↓CC	Flush ends*
Hha I	*Haemophilus haemolyticus*	GCG↓C	3'-Dinucleotide extension
Hpa I	*Haemophilus parainfluenzae*	GTT↓AAC	Flush ends
Hpa II	*Haemophilus parainfluenzae*	C↓CGG	5'-Dinucleotide extension
Mbo I	*Moraxella bovis*	↓GATC	5'-Tetranucleotide extension
Bam I	*Bacillus amyloliquefaciens*	G↓GATCC	5'-Tetranucleotide extension
Hind III	*Haemophilus influenzae*	A↓AGCTT	Cohesive termini

* When flush ends are produced, single-stranded tails must be added for cohesiveness. For example, when Hae III cuts

···G-G ↓ C-C···
···C-C   G-C···

to leave flush ends, it is necessary, by means of appropriate transferases, to add to the cut pieces as follows:

···G-G-A-A-A-A and C-C···
···C-C and T-T-T-T-G-G···

The A and T tails, being complementary, can act as the cohesive ends.

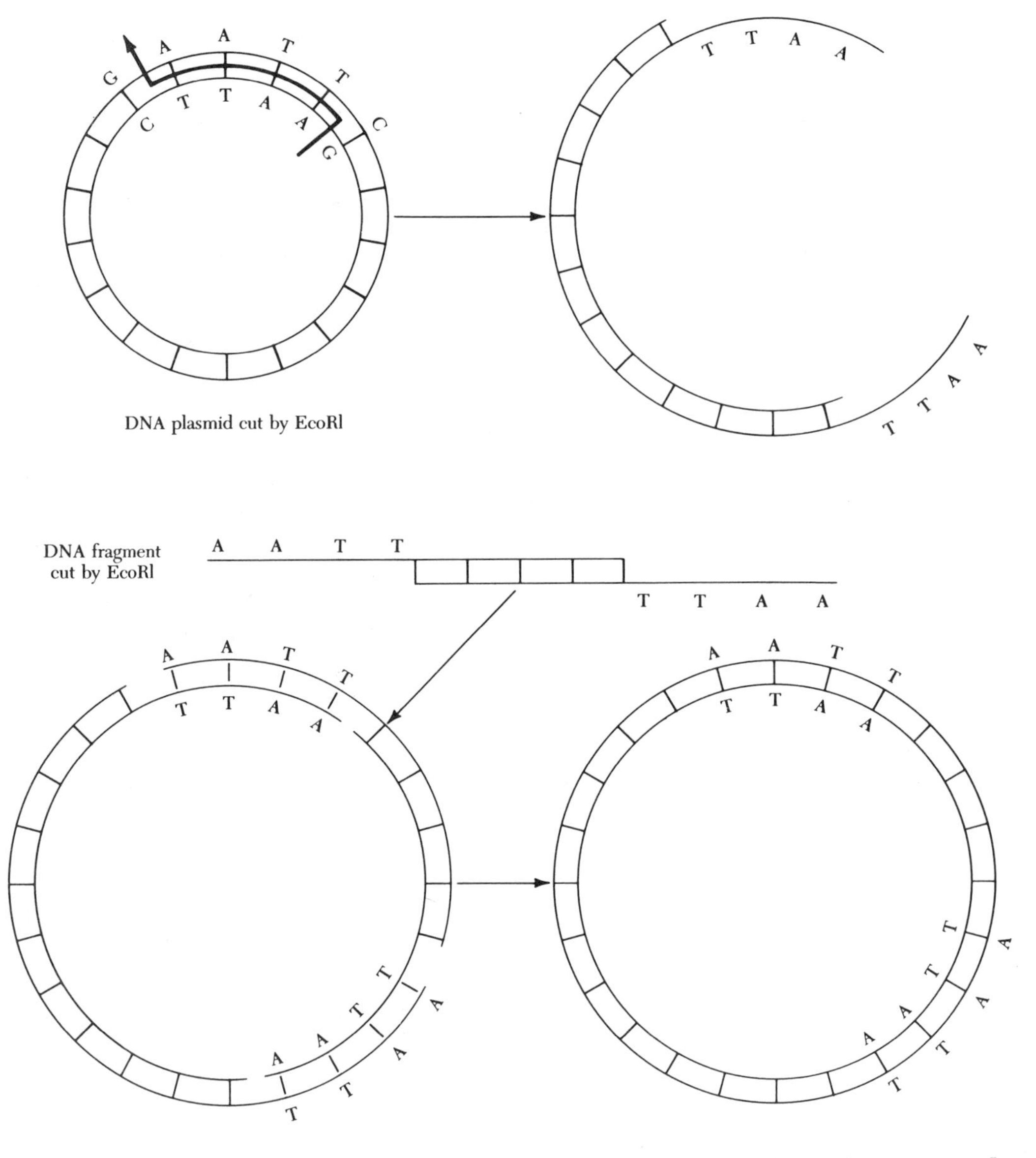

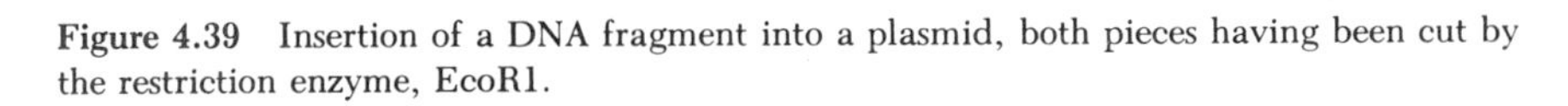

Figure 4.39 Insertion of a DNA fragment into a plasmid, both pieces having been cut by the restriction enzyme, EcoR1.

TABLE 4.7 COMMONLY USED PLASMIDS AND THEIR PHYSICAL AND CHEMICAL CHARACTERISTICS.

Plasmid	Molecular weight (10^6 daltons)	Number of copies per cell	Selective factor*
ColE1	4.2	10–15	E1 imm.
pCR1	8.7	10–15	Kn^R or E1 imm.
pTK16	2.8	10–15	Tc^R or Kn^R
pMB9	3.5	10–15	Tc^R or E1 imm.
pBR322	2.6	10–15	Ap^R or Tc^R
R6K	25.0	13–38	Ap^R or Sm^R
R1	65.0	1–3	Ap^R, Cm^R, Sn^R, Sm^R, Kn^R
R6	65.0	1–3	Tc^R, Cm^R, Sn^R, Sm^R, Kn^R

* Abbreviations of selective factors are as follows: El imm., immunity to colicin E1; Kn, kanamycin; superscript R, resistance; Tc, tetracycline; Ap, ampicillin; Sm, streptomycin; Sn, sulfonamide; Cm, chloramphenicol. Those plasmids beginning with p, and indented under ColE1, are derivatives of ColE1, and together with ColE1, they can, in the presence of chloramphenicol, increase the number of plasmids per cell to 1000–2000.

use of a viral-associated enzyme (a *reverse transcriptase*), the mRNAs can produce *cDNA*, or *complementary DNA*, which corresponds to the gene which made the mRNA in the first place (Figure 4.40). The cDNA can then be cloned by means of a plasmid system. It is also possible to start with a given protein, even one made in very minute amounts, and work backward to the gene which produced it; the techniques are far more complicated, however.

If it is desired to locate the source of a piece of recombinant DNA in a genome, the cloned DNA can be made radioactive, and then reannealed by DNA/DNA hybridization to chromosomes which had been previously fixed and spread. Autoradiography will reveal the precise position of the reannealed DNA.

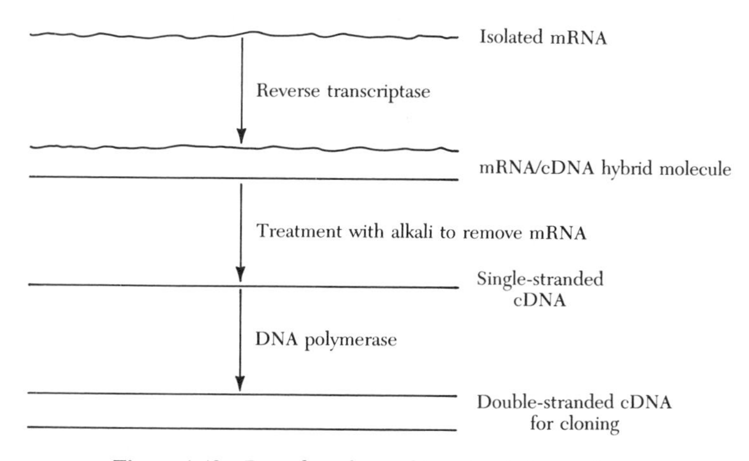

Figure 4.40 Procedure for making cDNA from mRNA.

The use of recombinant techniques with eukaryotic cells presents additional problems because of the nuclear envelope, the complex nature of chromatin, a paucity of transporting vehicles, the hydrolytic enzymes of the cytoplasm, and in higher plants, cellulose walls. Several plasmids from plant-associated bacteria offer promise of success for introducing, for example, genes for nitrogen fixation into species lacking them. The prospects for correcting genetic defects in humans by these techniques have arisen, but such procedures involve social implications, and cannot be solved solely by scientific means.

ADDITIONAL READING

ABELSON, P. H. (ed.). 1983. Biotechnology. *Science 219:* 609–746.

BRADBURY, E. M., MACLEAN, N., and MATTHEWS, H. R. 1981. *DNA, Chromatin, and Chromosomes.* John Wiley & Sons, Inc., New York.

CALLAN, H. G., and KLUG, A. (eds.). 1978. Structure of eukaryotic chromosomes and chromatin. *Phil. Trans. R. Soc., Lond. 8 283:* 231–416.

CHAMBON, P. 1981. Split genes. *Sci. Am. 244*(5): 60–71.

CHILTON, M. D. 1983. A vector for introducing new genes into plants. *Sci. Am. 248*(6): 50–59.

FRANKE, W. W., SCHEER, U., KROHNE, G., and GARASCH, E. D. 1981. The nuclear envelope and the architecture of the nuclear periphery. *J. Cell Biol. 91:* 39s–50s.

GALL, J. G. 1981. Chromosome structure and the *c*-value paradox. *J. Cell Biol. 91:* 3s–14s.

HOAGLAND, M. 1981. *Discovery: The Search for DNA's Secrets.* Houghton Mifflin Company, Boston.

KORNBERG, R. D., and KLUG, A. 1981. The nucleosome. *Sci. Am. 244*(2): 52–64.

LEWIN, B. 1980. *Gene Expression*, Vol. 2: *Eukaryotic Chromosomes*, 2nd ed. John Wiley & Sons, Inc., New York.

MILLER, O. L. 1981. The nucleolus, chromosomes, and visualization of genetic activity. *J. Cell Biol. 91:* 15s–27s.

PERRY, R. P. 1981. RNA processing comes of age. *J. Cell Biol. 91:* 28s–38s.

SIEKEWITZ, P., and ZAMECNIK, P. C. 1981. Ribosomes and protein synthesis. *J. Cell Biol. 91:* 53s–65s.

SUZUKI, D. T., GRIFFITHS, A. J. F., and LEWONTIN, R. C. 1981. *An Introduction to Genetic Analysis*, 2nd ed. W. H. Freeman and Company, Publishers, San Francisco.

SWANSON, C. P., MERZ, T., and YOUNG, W. J. 1981. *Cytogenetics: The Chromosome in Division, Inheritance and Evolution.* Prentice-Hall, Inc., N.J.

WATSON, J. D. 1976. *The Molecular Biology of the Gene,* 3rd ed. W. A. Benjamin, Inc., Menlo Park, Calif.

WATSON, J. D., and CRICK, F. H. C. 1953. Molecular structure of nucleic acids: a structure for deoxyribonucleic acid. *Nature 171:* 737–738.

5

The Cell Surface

In order for cells to maintain themselves as autonomous units and to continue to function, it is necessary that some control over the exchange of materials between the cell and its surroundings be exercised. The differences between the internal chemical composition of a cell and that of its external environment represent a degree of order, or nonrandomness, that can be maintained only by a barrier to free movement of material into and out of cells. Furthermore, since cells must also assimilate matter from their environment to function, to grow, and to reproduce, they must be able selectively to allow certain molecules or ions to enter across this barrier, often against concentration gradients, while restricting or excluding others.

All exchange of material between the cell and its environment must occur across the outer boundary of the cell. Energy must be expended both to maintain the structural organization of this cell surface and to surmount the barrier that it represents. It is also at the cell surface that information is first received, both from the immediate environment or from other cells, the latter type of communication being of fundamental importance to the integration of function characteristic of multicellular organisms. The capacity for recognition, then, is also an important feature of the surface of the cell.

THE PLASMA MEMBRANE

The existence at the surface of cells of a specialized, selectively permeable film, the plasma membrane, was inferred long before it could be demonstrated directly by the powerful resolution of the electron microscope. This inference was based on observations that when cell surfaces are torn by a very fine needle point the contents spill out, that certain compounds can enter and spread throughout the interior of the cell only after the surface has been punctured, and that certain molecules can penetrate intact cells more readily than others. Indeed, it was from physiological studies of the *permeability* of membranes (that is, their capacity to permit substances to cross them) that the first indications of the chemical composition and structure of the plasma membrane were derived. It is generally true (although there are important exceptions) that ionic, or electrically charged, compounds penetrate more slowly than nonionic compounds, that small molecules penetrate more rapidly than large molecules, and that the more soluble a compound is in lipid the more readily it can enter the cell. These generalizations led to the idea of a membrane carrying an electrical charge that would impede passage of ions, containing pores through which small molecules can readily pass, and consisting at least in part of a lipid, or fatty film into and out of which fat-soluble compounds can move.

The presence of proteins in the plasma membrane was also inferred, first from measurements of the surface tension, elasticity, and electrical resistance of the cell surface, the values obtained being closer to those expected of protein than of lipid. As we shall see, it is in the protein moiety of the membrane that much of the specificity of membrane function and behavior resides, although other components play an important role in certain types of cells.

We now know that lipids and proteins together constitute the major components of membranes. Some of the proteins and lipids, however, may have carbohydrates covalently attached to them, generally as short oligosaccharides. The sugar-containing sequences of these *glycoproteins* and *glycolipids* also play a role in determining some of the characteristics of the cell surface.

MOLECULAR ARCHITECTURE OF MEMBRANES

Membranes from various cells and organisms can differ widely both in the relative amounts and in the types of their constituent proteins and lipids, this variation providing a basis for the wide range of physiological activities displayed by different membranes. The various structural and functional properties of membranes, influencing such diverse activities as control of transport, cell-to-cell recognition, immunological response, and even cell movement, must themselves depend on which enzyme systems comprise the protein complement of the membrane and on how these proteins are arranged in relation to each other and to the various lipids also present. Furthermore, as pointed out earlier (Chapter 1), eukaryotic cells contain membrane systems other than the plasma membrane. Like the plasma membrane, the internal membranes form closed compartments, and appear to have structural features in common with each other and with the plasma membrane. We may, therefore, consider the structural organism of membranes in general, although we must recognize that it is unlikely that any one model can account for the properties of all membranes.

The Lipid Bilayer

It is now generally accepted that the way in which the lipids are arranged in membranes is one of the most important features of their structural and functional organization. Early analyses of red blood cell "ghosts" (the plasma membranes freed of the inner contents of the cells) indicated that these membranes contained approximately twice the amount of lipid necessary to form a monomolecular film equivalent to the surface area of the original cell, suggesting that the lipids were arranged in a bimolecular layer within the membrane (Figure 5.1). This lipid bilayer depends on the nature of the *phospholipids* which are the major lipid components of most membranes. These molecules are *amphipathic;* that is, one end of the molecule carries an electrical charge and is hydrophilic, while the other end is neutral and hydrophobic (Figure 5.2). Although lipids other than phospholipids are present in, and in some cases can be the major constituents of, membranes, all are characterized by their amphipathic nature. Thus the most probable arrange-

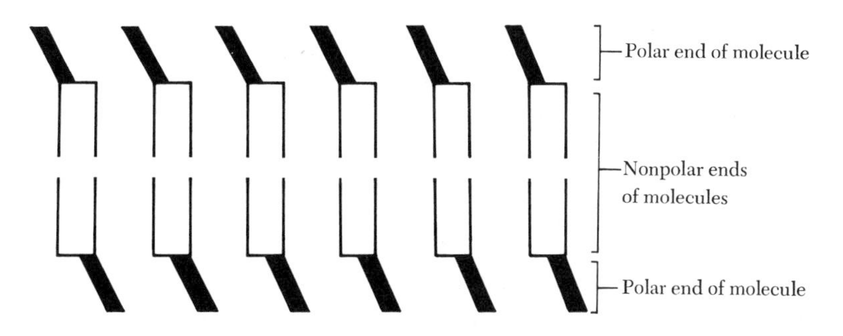

Figure 5.1 Diagrammatic representation of the arrangement of lipid molecules in a bilayer. The polar ends of the molecules are oriented toward the outer faces.

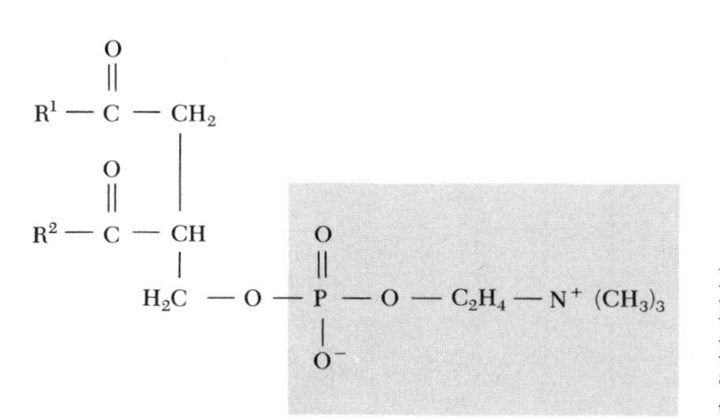

Figure 5.2 Structure of the phospholipid lecithin, showing the two hydrophobic fatty-acid chains (R^1 and R^2) and the hydrophilic phosphorus-containing group (in box).

ment of lipids in the membrane is one in which the polar ends of the molecules lie toward the aqueous environments at both the inner and outer surfaces of the membrane, while the apolar ends are oriented toward the interior of the membrane (Figure 5.1).

More recent estimates of the chemical composition of different membranes indicate that while the bilayer can account for all the lipids in the membrane, the fraction of the total surface area of the membrane occupied by the bilayer is rarely as high as 1.0. This appears to be a consequence of the membrane proteins being embedded in the lipid bilayer in such a way that they, too, contribute to the area of the membrane. For example, while a lipid bilayer could constitute the total surface area of the myelin sheath membranes of nerve cells, membranes which contain very little protein, it is generally true that the greater the amount of protein in a membrane, the less of its surface area can be accounted for by a lipid bilayer.

Membrane Proteins

Analysis of red blood cell membranes has shown that two classes of membrane proteins can be considered: the *peripheral*, or *extrinsic* proteins, which are easily removed from the membrane by mild extraction procedures and which are loosely associated with the cytoplasmic surface, and the *integral*, or *intrinsic* proteins, which are rather tightly bound within the membrane and which can be extracted only by treatments which disrupt the membrane structure. The intrinsic membrane proteins are amphipathic, with some regions, or domains, that are nonpolar and are associated with the hydrophobic interior of the lipid bilayer. The regions containing the polar amino acids are exposed at the aqueous surface (Figure 5.3).

By using chemical probes that can react with and label those regions of membrane proteins that are exposed at the surfaces, it has been possible to show that some intrinsic proteins extend all the way through the membrane, while others

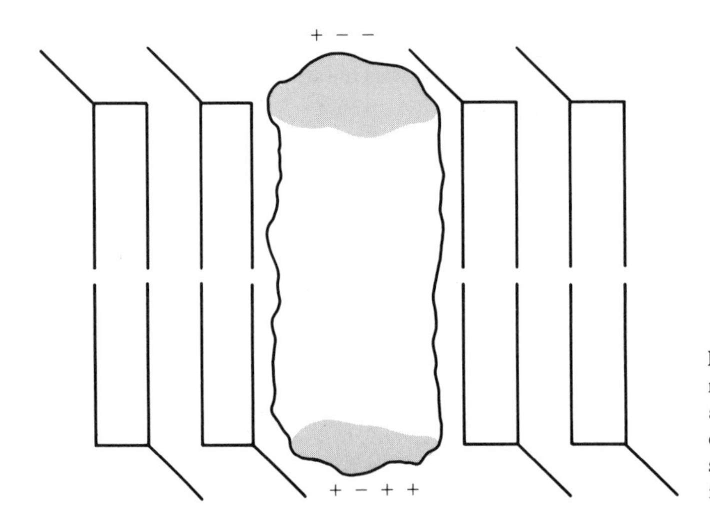

Figure 5.3 Amphipathic nature of membrane proteins. The hydrophobic amino acid sequences are in the interior of the membrane, while the hydrophilic sequences are oriented toward the surfaces.

may be embedded in the lipid bilayer in such a way that regions are exposed to only one or the other of the surfaces. The distribution of the intrinsic proteins is also known to be asymmetric; not only are some restricted to either the inner or the outer surface, but also those proteins that extend across the width of the membrane are oriented in a fixed manner. Such asymmetry is not surprising, since the specific interactions occurring between the cytoplasm and the membrane must be different from those that occur between the membrane and the outside environment of the cell.

Glycoproteins and Glycolipids

Some plasma membrane proteins and lipids, particularly in animal cells, are bound to oligosaccharides, and are known as *glycoproteins* and *glycolipids*. These surface compounds, which show in the electron microscope as a fuzzy layer at the cell surface (see Figure 5.24), appear to be involved in the recognition of, and the response of the cell to, factors in the environment. For example, some of the antigens that determine blood type are membrane glycoproteins and glycolipids, and the specific clumping reactions of blood cells when exposed to blood sera of different kinds depend on which antigens are present. Cells from the same tissue also exhibit the ability to "recognize" one another when in liquid suspension. For example, if heart and kidney cells from a chick embryo are dissociated to form a mixed single-cell suspension, and then allowed to remain undisturbed in culture, heart cells seek out and aggregate with heart cells, kidney cells with kidney cells. Such tissue-specific recognition between cells is a surface phenomenon that also may be mediated by the glycoproteins or glycolipids of the membrane. Glycoproteins also have been shown to be the receptors at the cell surface for *mitogens*, compounds that trigger cell proliferation in certain lymphocytes; the binding of the mitogen to the surface leads to changes in the permeability and transport properties of the membrane, which can in turn affect metabolic processes inside the cell. Both plant and animal cells are known to respond in specific ways to a range of hormones produced elsewhere in the body. The ability to recognize such chemical signals resides at the cell surface, and the initial response involves some change at the membrane itself.

ELECTRON MICROSCOPY OF MEMBRANES

Since a bimolecular lipid layer with its associated proteins is well below the limits of resolution of the light microscope, direct visualization of the plasma membrane was not possible until the development of the electron microscope and the various associated techniques of preparing material for examination. It is clear from electron micrographs that the cell surface is organized in a specific manner, usually appearing in thinly sectioned fixed cells as a tripartite structure consisting of two parallel dark lines separated by a central clear area (Figure 5.4). Similar tripartite

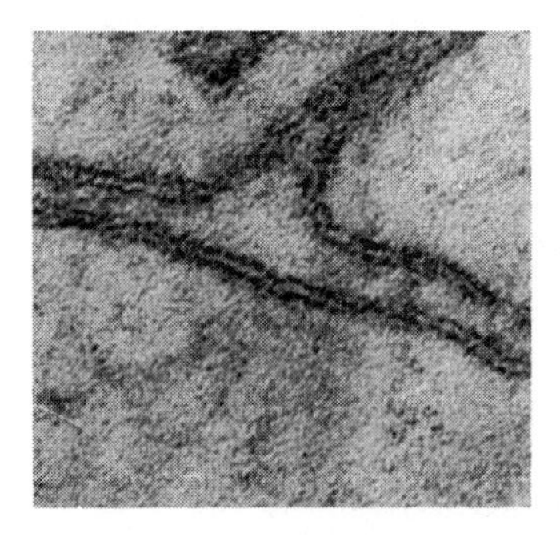

Figure 5.4 Electron micrograph of plasma membranes at the junction of three cells. Each membrane shows the typical trilaminar appearance.

organization also can be seen in sections of artificial membranes formed by mixing phospholipids or phospholipids and proteins in water. Although the dimensions of the three layers can vary from membrane to membrane, they are remarkably close to those expected of a bimolecular lipid layer. Thus the central clear area is thought to represent the hydrophobic ends of the two lipid layers, and the dark lines to represent the polar ends in association with proteins, at both the interior and exterior faces of the membrane. As well as revealing a specific organization of the cell surface, electron microscopy demonstrated the existence of the internal membranes of the cell; the appearance of these is similar to that of the plasma membrane, suggesting that a similar fundamental type of organization, based on a lipid bilayer, is common to all membranes.

Many membranes, including plasma membranes, can display a particulate appearance in the electron microscope. This is especially striking following preparation of membranes by the techniques of freeze-fracturing and freeze-etching. Both of these techniques involve freezing the specimen rapidly and then sectioning it. In freeze-fracturing, the frozen membranes are cleaved along the plane of weakness represented by the hydrophobic groups at the center of the lipid bilayer (Figure 5.5). Thus the internal organization of the membrane can be examined. In freeze-etching, the ice crystals formed are allowed to sublimate, and the true membrane surfaces can be exposed (Figure 5.5). Particles of various sizes are observed on both the interior and exterior faces (Figures 5.6 and 5.7), and these

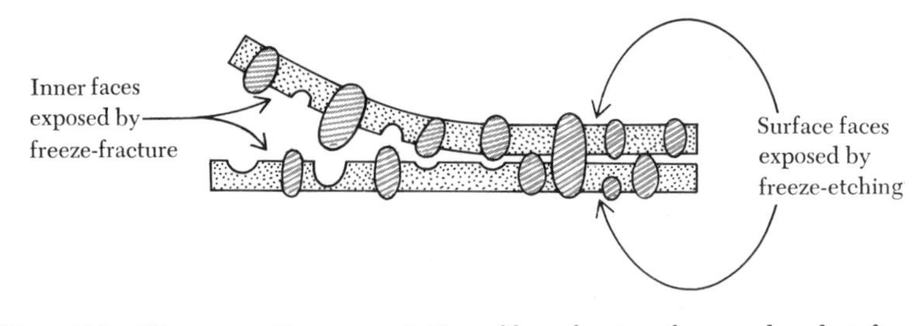

Figure 5.5 Diagrammatic representation of how fracture faces and surface faces of membranes are exposed by freeze-fracturing and freeze-etching.

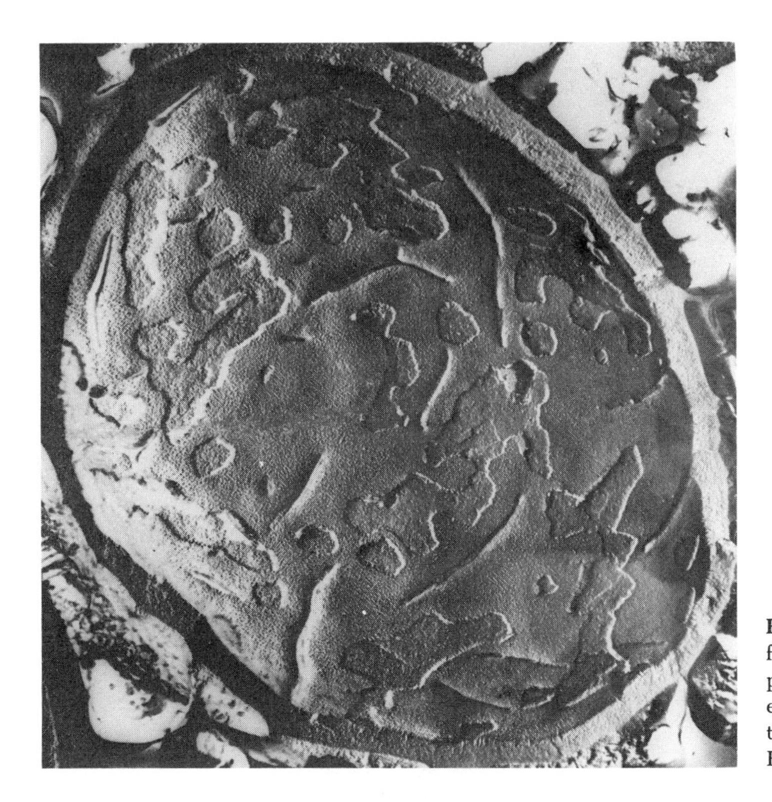

Figure 5.6 Electron micrograph of faces of yeast plasma membrane exposed by freeze-fracture and freeze-etching. Note the arrays of particles in the membrane. (Courtesy of Dr. S. C. Holt.)

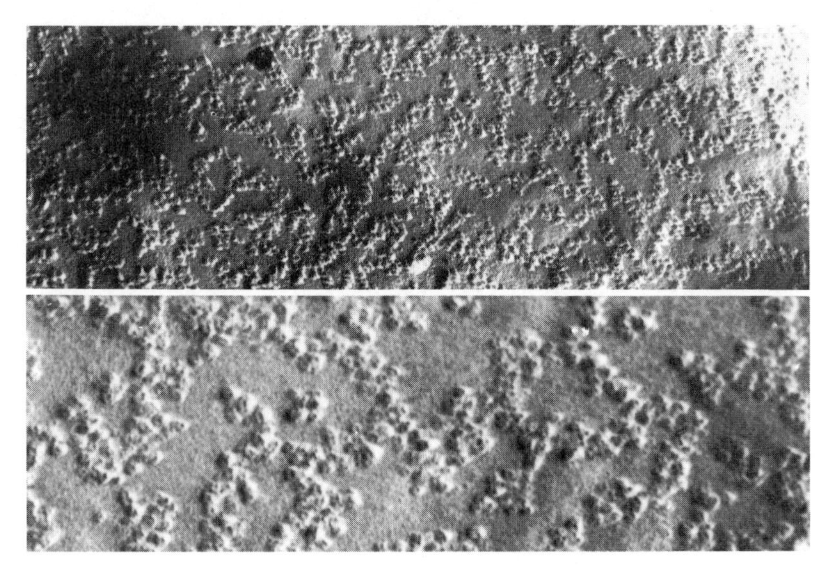

Figure 5.7 Fracture face of red blood cell plasma membranes, shown at two different magnifications. (Courtesy of Dr. S. C. Holt.)

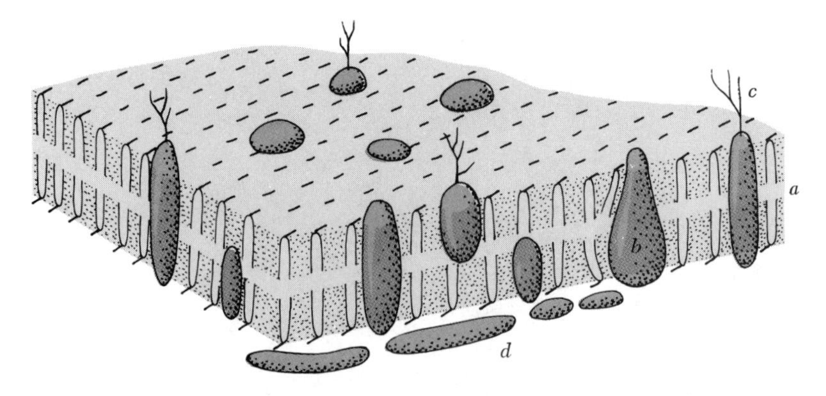

Figure 5.8 Model of membrane structure showing proteins (*b*) embedded in phospholipid bilayer (*a*). Carbohydrates associated with membrane proteins are on the outer surface of the membrane (*c*), and less tightly associated proteins (*d*) are at the inner surface.

are thought to represent the globular intrinsic proteins, or aggregates of proteins, embedded within the lipid matrix of the membrane (Figure 5.8). Different membranes display ranges of both particle size and particle distribution, again indicating a considerable amount of variation in structure and composition.

MEMBRANE FLUIDITY

An extremely important characteristic of membranes is that they are by no means static, and that there can be considerable freedom of movement within the membrane of its constituent molecules. The ability of membranes to reform rapidly over exposed regions of cytoplasm, to fuse with one another, and to expand and contract during cell movement or changes in cell shape suggests the fluid and dynamic nature of such membranes. Both the lipids and the proteins of plasma membranes can be shown to be mobile to varying degrees within the membrane. This has been beautifully demonstrated by the use of cell-fusion heterokaryons, which are formed by the fusion in culture of two cells of different species. Species-specific antigens, compounds located in the plasma membrane, can be identified by labeling the cell surface with colored fluorescent antibodies to the antigens. Following fusion of mouse and human cells with different-colored labels attached, the antigens could be shown to be segregated initially. However, within less than an hour after fusion, the mouse and human antigens completely intermingled on the surface of the heterokaryon, indicating that lateral diffusion of these membrane components had occurred (Figure 5.9).

The fluid mosaic concept of membrane structure, developed by Singer and Nicolson, considers the lipid bilayer to be a liquid state through which the embedded proteins can "float." Lateral diffusion of molecules must be limited and controlled in some way, however, if the functional integrity of the membrane is to

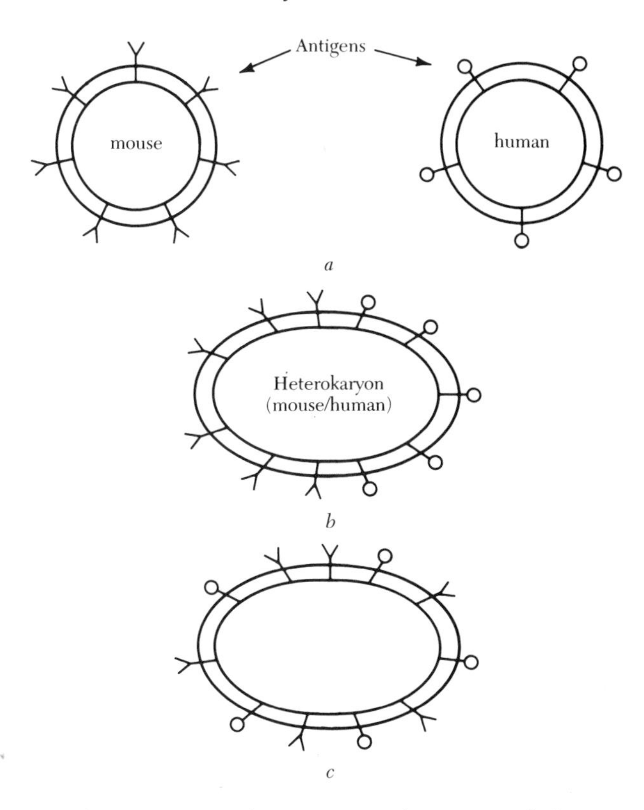

Figure 5.9 Diagram of experiment illustrating lateral mobility of membrane antigens: (*a*) mouse and human cells with specific antigens before fusion; (*b*) distribution of antigens immediately after fusion; (*c*) distribution of antigens one hour after fusion-intermixing has occurred.

be maintained. For example, some of the membrane proteins may be organized into multienzyme complexes, and would be expected to maintain fixed spatial relationships to one another. In such cases, and in membranes with high protein/lipid ratios, protein–protein interactions may be important, not only for functional reasons but also in contributing to the structural integrity of the membranes. Such proteinaceous membranes, or regions of membranes, would be much less fluid than one in which free movement in a lipid bilayer is possible.

Interactions between the inner face of the membrane and the cytoplasm immediately adjacent to it are believed to play a role in determining how the membrane components move and how they are distributed. For example, *spectrin*, one of the extrinsic proteins associated with the cytoplasmic surface of the red blood cell membrane, forms a network just beneath the membrane, to which it binds via another extrinsic protein, *ankyrin;* when this network is partially dissociated, the rate at which one of the major intrinsic glycoproteins of the membrane is able to diffuse laterally within the membrane is increased. Spectrinlike proteins have been identified in other types of cell and may represent a ubiquitous component of cytoplasm/membrane interactions. Other components of the membrane-associated cytoplasm, such as microtubules and microfilaments (see Chapter 8) may also be involved; treatment of cells with drugs that disrupt microtubules and microfilaments can block lateral mobility of membrane components.

MOVEMENT ACROSS MEMBRANES

As discussed at the beginning of this chapter, the main role of the plasma membrane is to regulate exchange of materials between the cell and its environment. We must now consider some of the mechanisms believed to be involved in transport of matter across the plasma membrane, as well as the forces responsible for such movement.

The first broad class of transport mechanisms to be discussed involves the movement of molecules or ions through the membrane itself and can be referred to as *transmembrane transport.* Such transport may be passive, in that no expenditure of energy by the cell is required, or *active*, in which case the transport is "uphill" in terms of free energy and, therefore, requires an input of energy from metabolic activities of the cell. The second general method, which we can refer to as *bulk transport*, includes endocytosis and exocytosis. *Endocytosis* is accomplished by the engulfing of material by part of the membrane and the subsequent deposition of the materials inside the cell, while *exocytosis* is the reverse process, involving fusion of membrane-enclosed vesicles with the plasma membrane and deposition of their contents outside the cell. Endocytosis and exocytosis are also active transport mechanisms in that energy must be expended in the processes.

Transmembrane Transport

Diffusion. Diffusion is the net movement of molecules down a concentration gradient, that is, from high to low concentration. Thus, if a membrane separates two environments between which there is a difference in the concentrations of some particular molecule, there will be a tendency for movement from one side of the membrane to the other. In the case of electrically charged substances, or ions, another factor must be considered. Membranes maintain not only differences in concentrations of substances, but also differences in electrical potential. Hence both concentration and electrical potential differences, which together constitute an *electrochemical gradient*, will determine the driving force for simple diffusion. However, diffusion across a membrane requires that the substance in question be able to dissolve in the membrane, or that there exist pores through which the molecules can diffuse. The important substances that the cell must transport, mainly ions and organic substrates for metabolism such as sugars and amino acids, are not readily soluble in lipids. Indeed, it is this property that contributes to the capacity of membranes to maintain differences in concentration and to present barriers to free movement. It is unlikely, therefore, that simple diffusion, driven solely by differences in electrochemical potential, is important in getting materials into and out of the cells.

One possible exception is represented by the movement of water, which crosses the membrane much more rapidly that its solubility in lipid would predict, suggesting the presence of pores in the membrane. Although pores have not been seen directly by electron microscopy, it has been calculated that they must be very small and can occupy only a very small portion of the surface area of the membrane. Furthermore, it is possible that such pores are not permanent structural entities but represent transient and localized rearrangements in the molecular architecture of a dynamic membrane.

Osmosis and water balance. Let us now examine the specialized case of movement, into and out of cells, of water itself, the solvent in which other materials are dissolved, and how cells are able to maintain a proper water balance. A useful (and accurate) way to consider water movement is in terms of *water potential* differences, which determine the direction of movement of water from one region to another. The following example serves to illustrate how the two separate components of water potential, *osmotic potential* and *pressure potential*, interact. Consider two compartments separated by a membrane that is freely permeable to water but not to any solutes dissolved in that water, that is, a *differentially permeable* membrane (Figure 5.10). One compartment contains pure water, the other contains water in which some solute is dissolved. The presence of dissolved solute effectively lowers the concentration of water in the second compartment relative to the first, thereby reducing its osmotic potential. Thus there will be a net movement of water (Figure 5.10b) from the region of higher osmotic potential to the region of lower osmotic potential. (Note that osmotic potential is inversely related to the concentration of dissolved solutes.) Eventually, however, the resulting increase in pressure in the compartment with the solutes

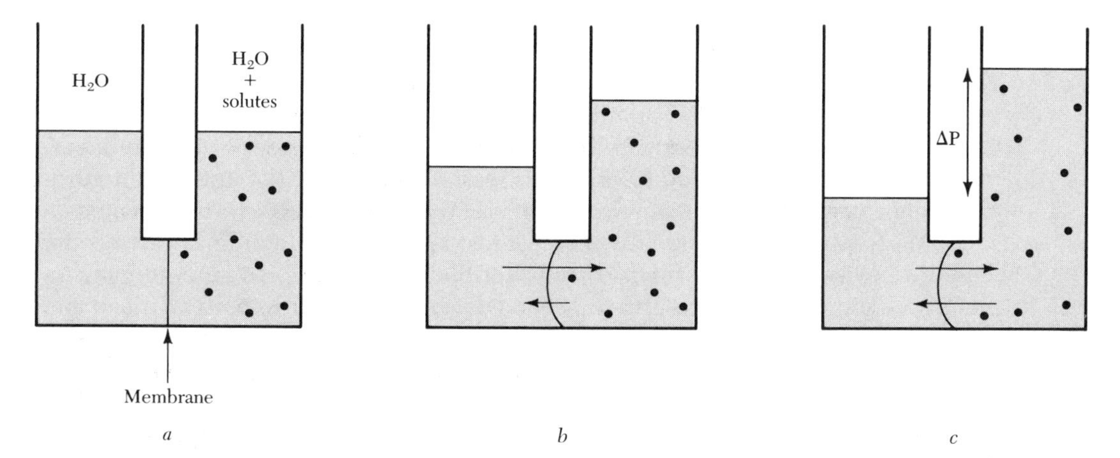

Figure 5.10 Illustration of how osmotic potential and pressure potential of water determine direction of water movement. (*a*) Two compartments, separated by a selectively permeable membrane, are filled with equal volumes of pure water and a solution of some substance in water respectively. The presence of dissolved solutes in the compartment on the right lowers the osmotic potential; hence water moves from left to right (*b*). As it does so it increases the pressure potential in that compartment to the point where the difference in pressure potential equals the difference in osmotic potential, and no more net water movement takes place (*c*). $\triangle$ P = pressure difference.

will force as many molecules back through the membrane (in response to the pressure potential difference) as are crossing into that compartment in response to the osmotic potential differences, with the result that no further net movement of water takes place (Figure 5.10c); that is, the osmotic potential difference is balanced by an equal and opposite pressure potential difference. The water potential difference is therefore zero.

We can now see how these phenomena apply to living cells, which are separated from their environment by a differentially permeable membrane. In the case of plant cells, the internal concentration of solutes is higher than in the medium outside; hence water will move into the cell, to the region of lower osmotic potential, and the increased pressure will cause the cell to tend to expand. However, the presence of a rigid wall outside the plant cell restricts such expansion, thereby maintaining a high pressure potential inside the cell, and there will be no more net movement of water across the membrane. Animal cells, on the other hand, have no cell walls, and can restrict the amount of water entering and leaving only by means of controlling osmotic potential differences between inside and outside.

Carrier-mediated transport. Many substances which are unable to diffuse freely across membranes can enter or leave the cell by interacting in specific ways with components of the membrane itself. Whether the driving force for such transport be an electrochemical gradient, resulting in what is known as "facilitated diffusion," or energy provided by cellular metabolism (active transport), the membrane components, almost certainly proteins, can act as carriers, either by ferrying the substrate through the hydrophobic interior of the membrane or by forming channels through which the substrate can pass (Figure 5.11). In addition to the movement through the membrane, such systems require that there be recognition and binding between substrate and carrier molecules on one side of the membrane, release of substrate at the other side, and reestablishment of the initial configuration of the carriers in the membrane in order that the process can be repeated.

Many current ideas on how membrane proteins can function as carriers are based on experiments in which certain antibiotics produced by microorganisms have been shown to increase greatly the movement of ions across both artificial and biological membranes. Such *ionophores*, which become inserted in the membrane, are of two types, depending on their mechanism of action in accelerating ion transport. The first type, exemplified by valinomycin, act as true carriers, binding cations at some position in the molecule, diffusing or rotating through the membrane, and releasing the ion on the other side (Figure 5.11a). The second type, such as gramicidin A, conduct ions through actual pores, which they form in the membrane (Figure 5.11b). Two molecules of gramicidin are required to form the channel, presumably by assuming a conformational state in the membrane that provides a water-filled transport route through which the ions can be passed.

Although the ionophoric antibiotics are not normal constituents of membranes, it is likely that some of the integral proteins of membranes can function

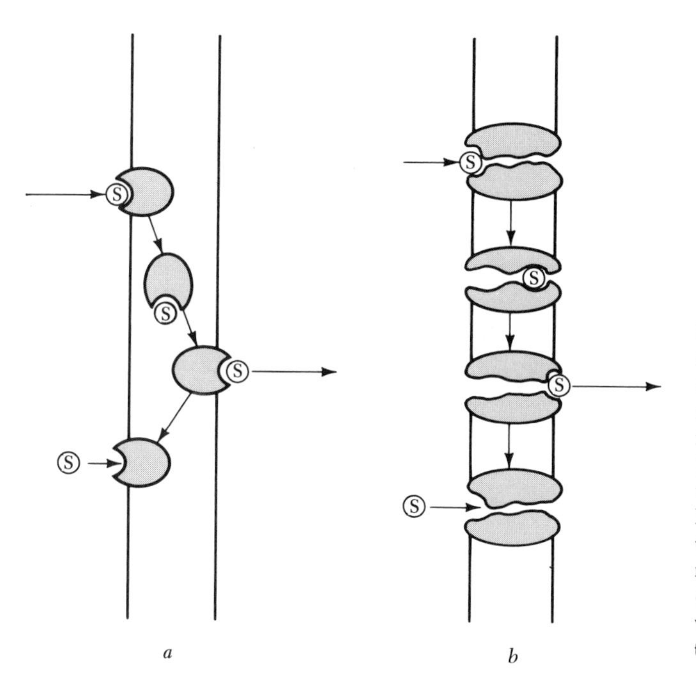

Figure 5.11 Two models of carrier-mediated transport across a membrane. (*a*) A protein binds to the solute, S, and rotates through the membrane, releasing the solute to the inside of the cell. (*b*) The protein binds to the solute, and undergoes conformational changes allowing the solute to be conducted through the pore. The driving force may be the electrochemical gradient (facilitated diffusion) or energy provided by cellular metabolism (active transport).

similarly, conferring upon the membrane the capacity to transport materials in specific and selective ways.

Active transport. While the driving force for facilitated diffusion is the difference in electrochemical potential across the membrane, the materials being transported by this mechanism moving from a higher to a lower free energy state, many substances, notably ions, are capable of being moved into and out of cells in the energetically unfavorable direction, that is, against an electrochemical gradient. Energy must therefore be expended in order to do the work represented by such uphill movement, or *active transport*. Much of the energy for active transport is supplied by the hydrolysis of ATP, and ATP-hydrolyzing enzymes (ATPases) are associated with the integral membrane protein systems involved in active transport.

One extensively studied active ion transport system of animal cells is the sodium/potassium/ATPase pump, in which sodium (Na^+) and potassium (K^+) transport against gradients is coupled to the hydrolysis of ATP (Figure 5.12). For most cells the concentration of Na^+ inside the cell is much lower than that outside. Potassium (K^+), on the other hand, is found in high concentrations inside the cell compared to the surroundings. Although membranes are usually fairly resistant to passive diffusion of these ions, some leakage would take place, tending to equalize concentrations inside and out. The differences in ionic concentrations are maintained, however, by an active transport system at the membrane which pumps Na^+ out of, and K^+ into, the cell.

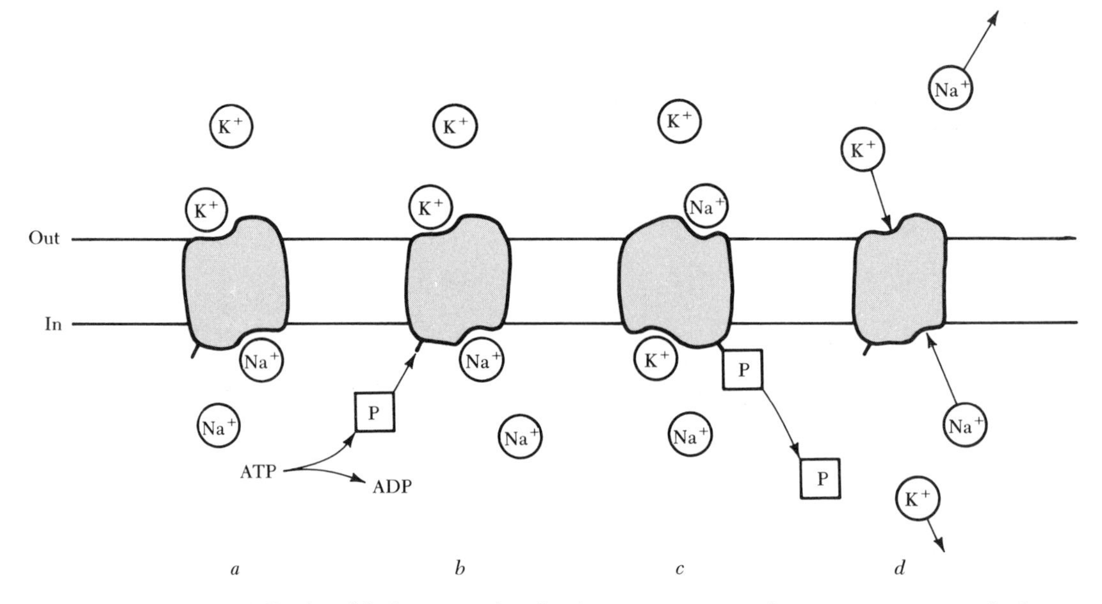

Figure 5.12 Simplified version of Na^+/K^+/ATP-ase pump. The transport protein, which is also an ATPase, binds Na^+ at a site inside the cell and K^+ at a site outside (*a*). The protein hydrolyses ATP (*b*), becomes phosphorylated, and undergoes conformational changes (*c*) which transport the ions and release them, Na^+ to the outside, K^+ to the inside. The protein is then dephosphorylated and returns to its original configuration (*d*). For each molecule of ATP hydrolyzed, three Na^+ are moved out of the cell and two K^+ come in.

The $Na^+ K^+$ ATPase protein has a binding site for Na^+ at the inner face of the membrane, which it spans, and a site for K^+ at the outer surface. Once the ions are bound, changes in the conformational state of the protein result in the Na^+ and K^+ being transported across the membrane and released, Na^+ to the outside and K^+ to the inside. The protein then returns to its original conformation. Some of the conformational changes that accompany the binding, transport, and release of the ions appear to involve phosphorylation and dephosphorylation of the transport protein, and the coupling of ATP hydrolysis by the protein itself provides the energy necessary for these conformational changes. The nature of the pump is such that three sodium ions leave the cell for each molecule of ATP hydrolyzed, and two potassium ions enter. Clearly, the ion–substrate binding interactions are complex, as must be the molecular nature and configuration of this and other membrane transport proteins.

The capacity of active transport pumps to maintain high electrochemical gradients across a membrane itself results in a source of energy that can be utilized for the work of the cell, including that of moving other materials across the membrane. Gradients in sodium concentration maintained by pumps such as that described above appear to be particularly important in this respect; the tendency of Na^+ to move back into the cells, that is, in the energetically favorable direction,

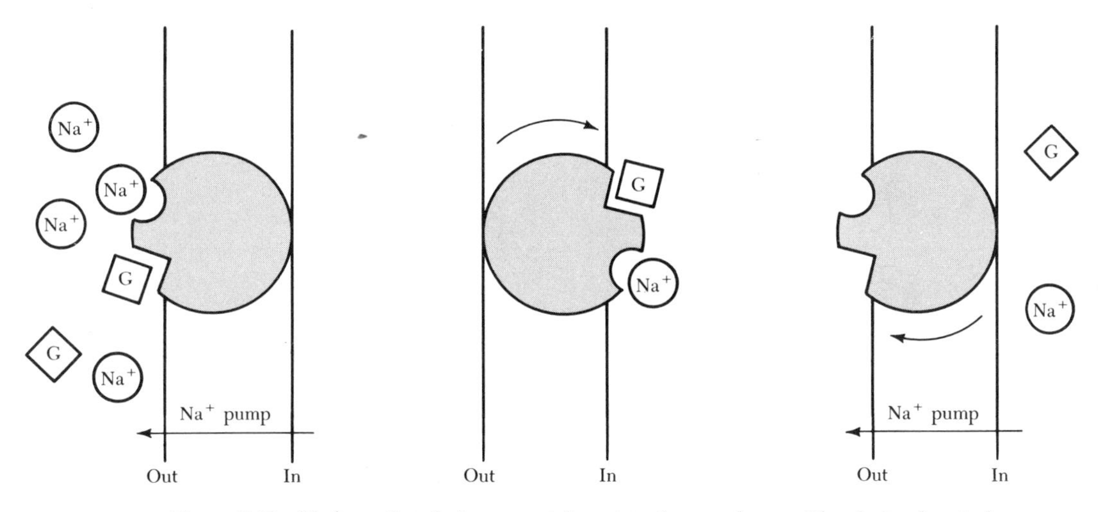

Figure 5.13 Na^+-mediated glucose uptake across the membrane. The electrochemical potential maintained by the pumping of Na^+ out of the cell provides the force for the "uphill" movement of glucose into the cell.

can be coupled to the uptake of various sugars and amino acids. For example, the absorption of glucose from the blood into the cells lining the lumen of the intestine occurs against a concentration gradient and is dependent on Na^+ (Figure 5.13). Again a carrier protein system is believed to be involved in the transport of both Na^+ and glucose; as the Na^+ is bound and transported across the membrane into the cell in the "downhill" direction, conformational changes are induced which facilitate the simultaneous binding and transport of glucose into the cell in the "uphill" direction. In the case of plant cells and microorganisms, proton gradients may be utilized in a similar fashion for transport work at the membrane.

Endocytosis and Exocytosis

Many types of cells may also take in material from the external medium by one of several *endocytotic* processes, in which the material is engulfed by the plasma membrane and deposited in the cytoplasm in pockets that are "pinched off" as vesicles toward the interior of the cell. The endocytotic vesicles may then be digested by the cell, generally following fusion with lysosomes (see later), small membrane-bound vesicles which contain degradative enzymes. In some cells, however, the fate of the endocytotic vesicles is not to be digested within the cell, but to be transported to, and fused with, another part of the plasma membrane. This process of *exocytosis*, which is essentially the reverse of endocytosis, allows the cell to deliver materials in bulk to its external environment (Figure 5.14).

The term *pinocytosis*, taken from the Greek words for "drink" and "cell," refers to the uptake of extracellular fluid containing various large molecules or ions incapable of passing through the membrane directly. In *fluid-phase* pinocytosis,

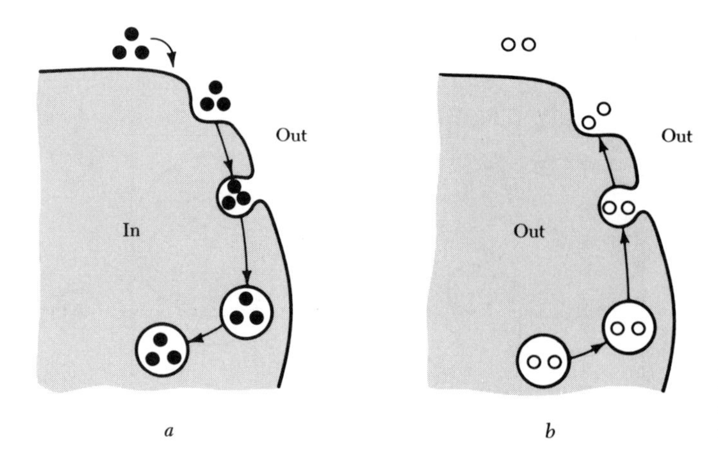

Figure 5.14 Bulk transport across membranes: (*a*) endocytosis, (*b*) exocytosis.

which appears to be nonspecific, materials enter the cell at rates which are proportional to their concentrations in the extracellular fluid, and there is no direct interaction between these materials and the membrane surface (Figure 5.15). *Receptor-mediated pinocytosis*, however, is much more selective and efficient, and depends not only on some form of recognition between the incoming substance and receptor molecules at the membrane surface, but also on a mechanism whereby the incoming material can be concentrated.

Receptor-mediated pinocytosis. Molecules which bind to receptors in the membrane are referred to as *ligands;* among the ligands which are incorporated into certain cell types by endocytosis are several proteins, including low-density lipoprotein (LDL), which carries cholesterol into cells, epidermal growth factor (EGF), a hormone protein, maternal yolk proteins, and even viruses. The pattern

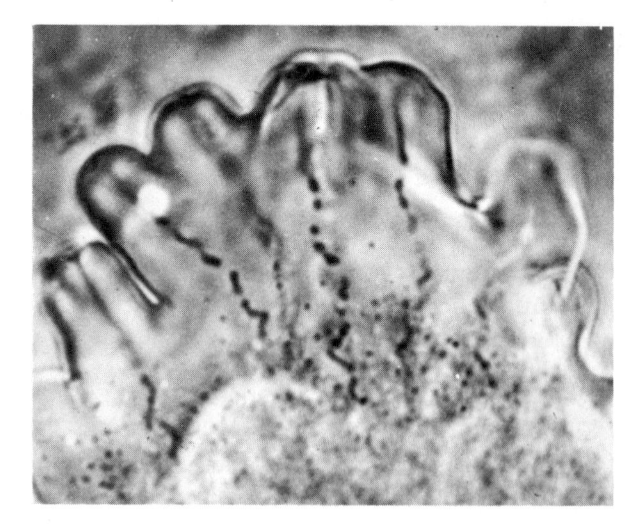

Figure 5.15 Photograph of the edge of a living amoeba, showing the pinocytotic channels (the dark lines converging toward the center of the cell). Liquids flow into the cell through these channels, to be pinched off as membrane-enclosed droplets; these eventually dissolve in the interior of the cell.

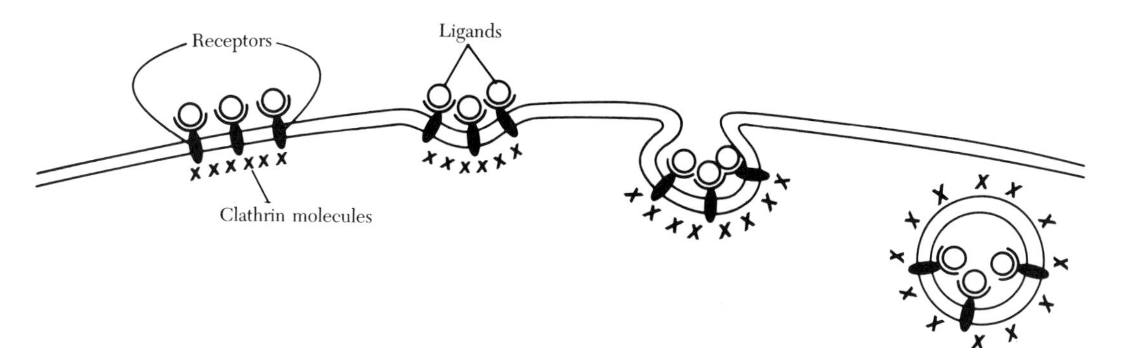

Figure 5.16 Receptor-mediated endocytosis. Ligands bind to receptors and become internalized by vesicle formation. Coated pits and coated vesicles have "cages" made of clathrin molecules at their cytoplasmic surfaces.

of binding of a protein ligand to its receptors in the membrane and subsequent incorporation has been followed by conjugating the protein in question to a substance such as ferritin or radioactively labeled iodine (^{135}I) which can then be detected by electron microscopy. Experiments using such techniques have shown that the ligands in the external fluids bind to receptors which are (or become) localized in particular domains in the membrane (Figure 5.16). The regions of the membrane in which the receptor molecules are aggregated have, attached to their cytoplasmic face, a network of molecules of a particular peripheral membrane protein called *clathrin*. These regions, which will subsequently invaginate to form the pinocytotic vesicles, are referred to as *coated pits* (Figure 5.16). While the mechanism by which the receptors are aggregated in the membrane is not understood, more than one type of receptor can be present in the same coated pit, and it is possible that the clathrin may somehow play a role in restricting their distribution in the membrane. Once the vesicles are formed, and the ligand internalized, the clathrin is lost from the vesicles, presumably to participate in the formation of another coated pit in the membrane. The resulting smooth vesicles, or *receptosomes*, then carry the enclosed material to wherever is its first destination within the cell (see Chapter 7).

Phagocytosis has traditionally been distinguished from pinocytosis on the basis of the nature of the materials incorporated into the cell, although the mechanisms involved may be similar to those described for pinocytosis. In phagocytosis (from the Greek, *phagein*, to eat) large particulate matter is tightly enclosed by membrane bound arms of cytoplasm, and most of the extracellular liquid appears to be excluded from the resulting invagination of the membrane. The solid material is then drawn into the cytoplasm, where it is eventually broken down by degradative enzymes.

As well as representing a feeding mechanism in cells such as *Amoeba* (Figure 5.17), phagocytosis is an important part of the body's defense mechanism. Specialized white blood cells are attracted to invading bacteria, which they

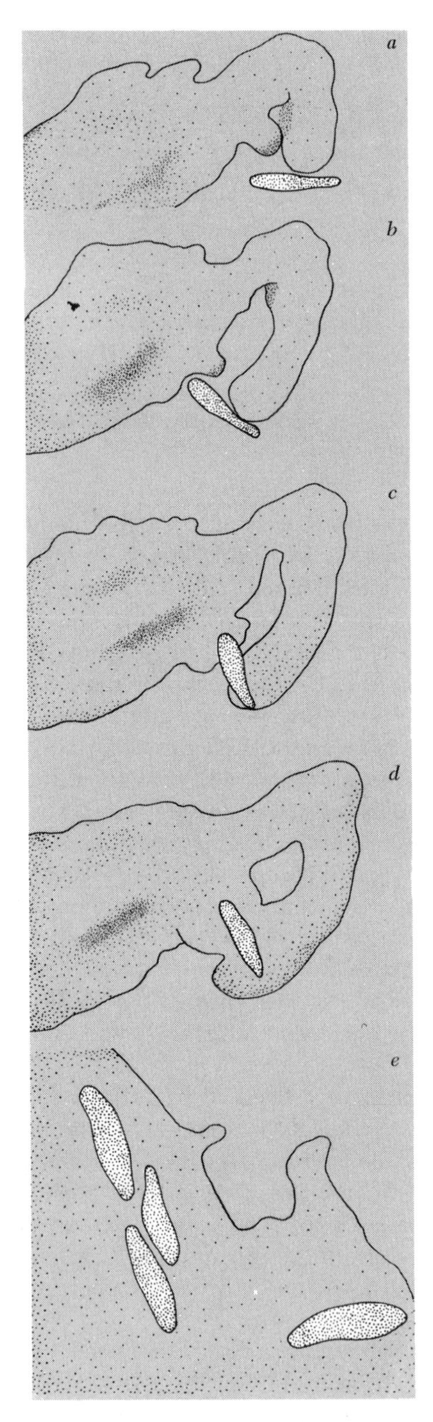

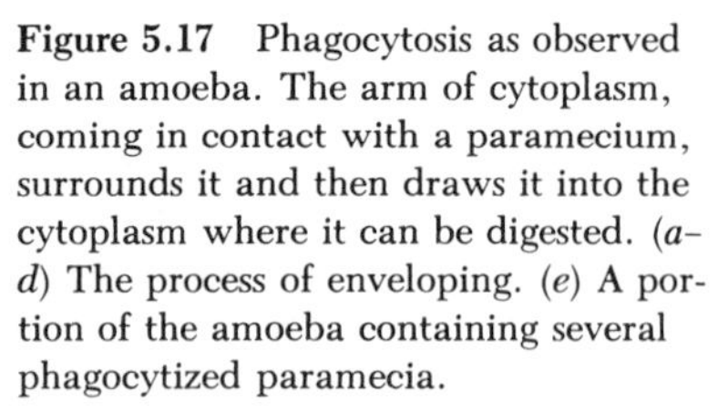

Figure 5.17 Phagocytosis as observed in an amoeba. The arm of cytoplasm, coming in contact with a paramecium, surrounds it and then draws it into the cytoplasm where it can be digested. (*a–d*) The process of enveloping. (*e*) A portion of the amoeba containing several phagocytized paramecia.

recognize by an interaction between specific proteins on the bacterial surface and receptor molecules in the white blood cell membrane. Following recognition, the white blood cell undergoes changes of shape, engulfs the foreign bacterium by phagocytosis, and destroys the invader by releasing degradative enzymes into the vesicle.

Such aspects of membrane behavior as endocytosis and exocytosis once again illustrate the dynamic nature of the plasma membrane, involving as they do membrane reorganization, movement, fusion, replacement, and turnover. The two latter aspects of membranes will be discussed in a later chapter, together with the relationships between the plasma membrane and the internal membrane systems of the cell.

MODIFICATIONS OF THE PLASMA MEMBRANE

From our discussion of the plasma membrane it should be clear that it is not a uniform, static structure, but rather that proteins are organized and distributed within the membrane in accordance with the specific functions of the membrane. Striking examples of this specialization are the modifications displayed at particular regions of very close association between parts of adjacent cells, which permit these cells to interact in specific ways. These intercellular junctions include *tight junctions* (Figure 5.18), which prevent movement through the intercellular spaces between one side of a cell layer and the other; *desmosomes* (Figure 5.19), which are involved in cellular adhesion; and *gap junctions* (Figure 5.20), which permit relatively free movement of ions and small molecules between adjacent cells.

Tight junctions are regions of very tight fusion of membranes of adjacent cells, and result from some sort of interaction of proteins of one membrane with similar proteins of the appressed membrane of the adjacent cell. Freeze-fracturing of tight junctions shows that a pattern of protein particles in one membrane matches up exactly with the pattern of particles in the adjacent membrane. Although the actual areas of contact are relatively small, the anastomosing network of connections forms an effective seal. Tight junctions are found between cells of the intestinal epithelium, a cell layer through which materials pass in a unidirectional fashion from the lumen of the intestine to the bloodstream at the other side. It has been shown that some molecules can move freely between adjacent cells of the epithelium only as far as the tight junctions, through which no further movement is possible.

The tight junctions may also play a role in maintaining the polarity of cells such as those of the intestinal epithelium. Since material enters the intestinal cell through the membrane facing the lumen and leaves it through the membrane facing the bloodstream, there are likely to be differences between the membranes on opposite sides of the cells. It can be shown that while lateral diffusion of proteins

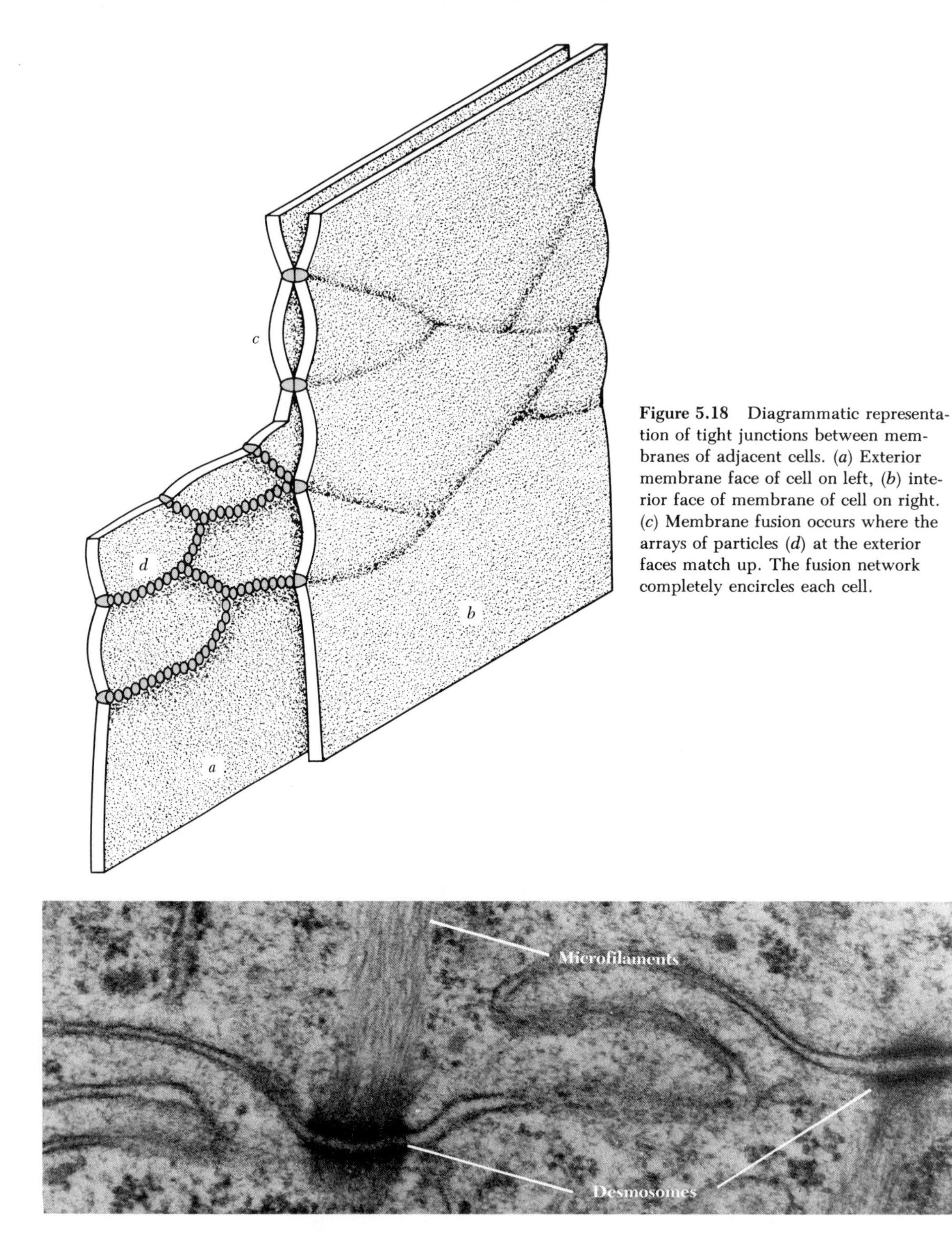

Figure 5.18 Diagrammatic representation of tight junctions between membranes of adjacent cells. (*a*) Exterior membrane face of cell on left, (*b*) interior face of membrane of cell on right. (*c*) Membrane fusion occurs where the arrays of particles (*d*) at the exterior faces match up. The fusion network completely encircles each cell.

Figure 5.19 The convoluted membranes of two adjacent columnar epithelial cells, showing two prominent desmosomes. Microfilaments extend from the desmosomes into the cytoplasm.

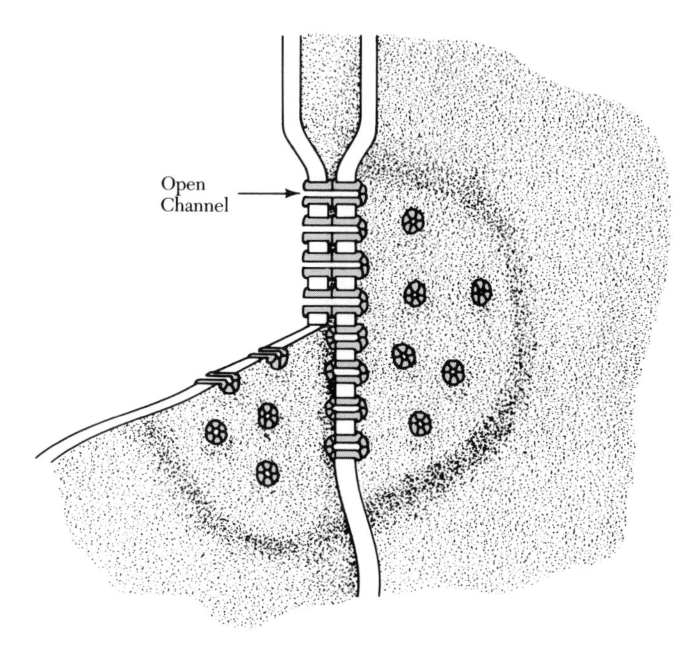

Figure 5.20 Diagrammatic representation of a gap junction. Hexagonal arrays of proteins surrounding channels in the membrane match up with similar arrays in an adjacent membrane, forming open channels, about 1.5–2.0 nm, in diameter, through which communication between cells can take place.

within the fluid matrix of the membrane can occur on either side of the tight junction, it cannot occur across the regions forming the junction (Figure 5.21).

Desmosomes are closely appressed regions of membranes of adjacent cells and appear to be involved in adhesion of cells to one another. They often show unique arrays of particles within the membrane, and are closely connected to bundles of filaments in the cytoplasm (Figure 5.19). It has been suggested that the desmosomes and underlying filaments may permit coordination of movement of groups of cells.

Like tight junctions and desmosomes, gap junctions represent much closer regions of association between adjacent membranes, although they do not form a barrier to movement through the intercellular space. The gap junctions are means of communication between cells, transmitting both chemical and electrical infor-

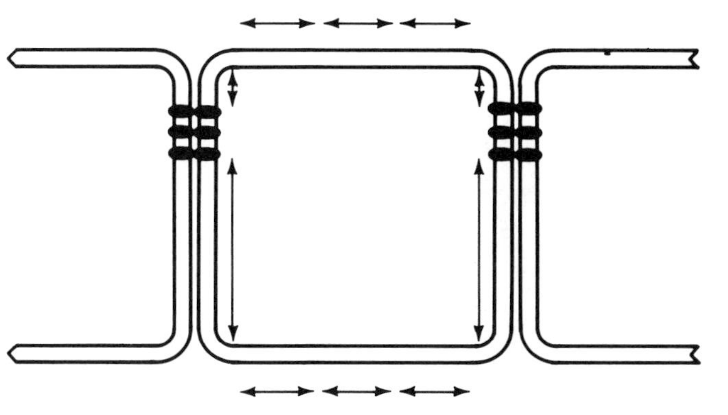

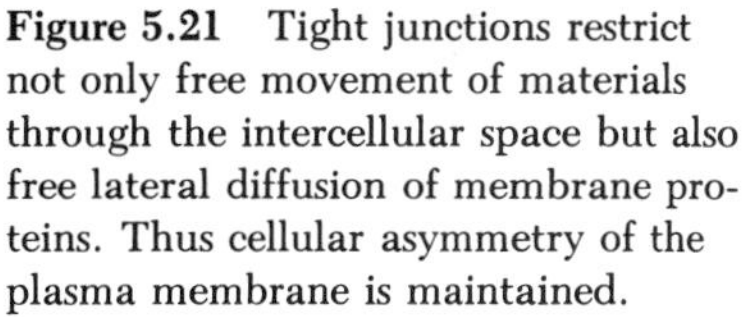

Figure 5.21 Tight junctions restrict not only free movement of materials through the intercellular space but also free lateral diffusion of membrane proteins. Thus cellular asymmetry of the plasma membrane is maintained.

mation. The particle distribution in the membranes is highly regular, and matching patterns are found on adjacent membranes. The proteins which make up the particles are thought to form pores through which intercellular communication can take place.

Plasmodesmata

A different type of intercellular association, found in plant tissues, is represented by plasmodesmata (Figure 5.22); these narrow channels run through the thick walls surrounding plant cells and are bounded by plasma membrane. The plasma membranes of adjacent cells, therefore, are continuous, and the plasmodesmatal channels represent cytoplasmic connections between adjacent cells. Most of the cytoplasmic volume of each plasmodesma is occupied by a tightly constricted tubular portion of endoplasmic reticulum (ER), part of the internal membrane system of cells (see Chapter 7). However, this ER tubule appears to be so compressed that the cisternal space, or lumen, is occluded (Figure 5.23); any in-

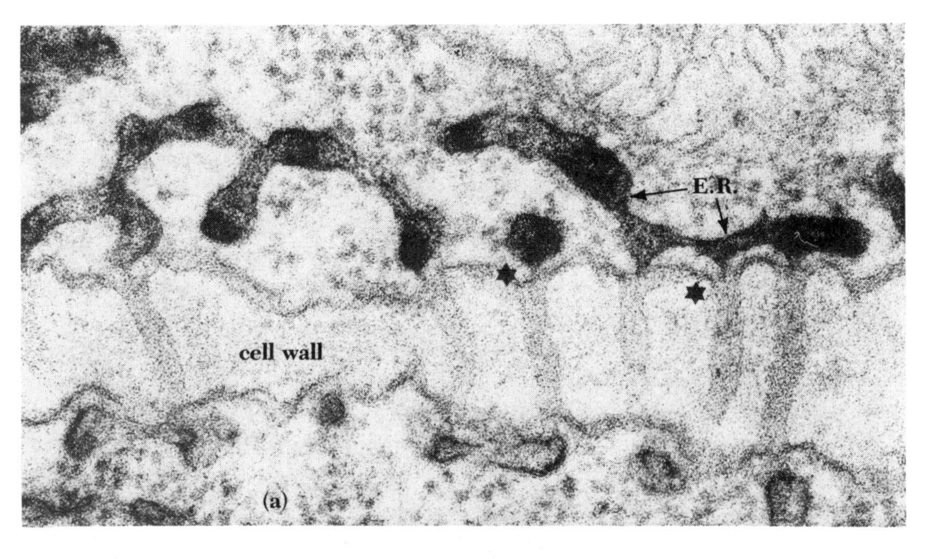

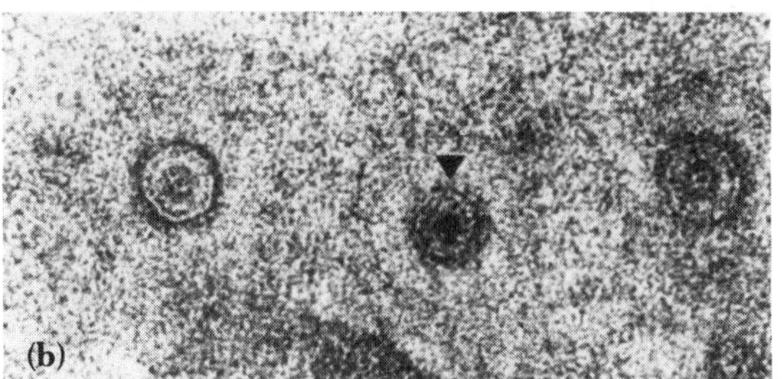

Figure 5.22 Plasmodesmata in plant cell walls. (*a*) Longitudinal section, showing close association of darkly-stained ER with the plasmodesmata between adjacent cells. (*b*) Oblique section through cell wall (see also Figure 5.23). (Courtesy of Dr. P. K. Hepler.)

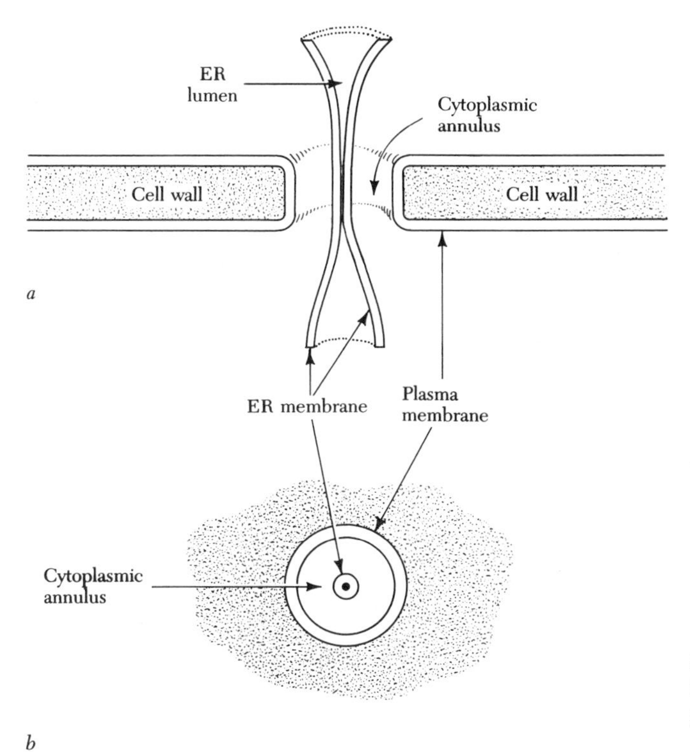

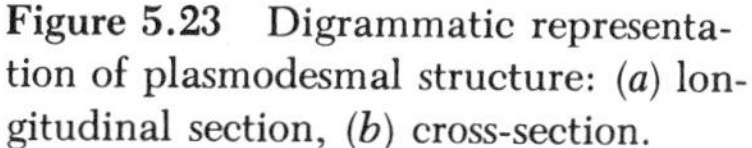
Figure 5.23 Digrammatic representation of plasmodesmal structure: (*a*) longitudinal section, (*b*) cross-section.

tercellular transport through plasmodesmata, therefore, is likely to occur in the cytoplasmic annulus surrounding the central ER "desmotubule."

Membrane Extensions

Highly convoluted regions of the membrane are also characteristic of many cell types. For example, columnar epithelium cells, which line the small intestine and are active in the absorption of digested food, form many projections, or *microvilli* (Figures 5.24 and 1.21), at their upper surface. These greatly increase the surface area of the cell, and hence provide a tremendous absorption area. Each such cell would have several thousand microvilli, while a square millimeter of the intestine would have as many as 200,000,000.

Another modification of the cell membrane can be seen in the light-receptor cells (rods and cones) of the vertebrate eye (Figure 5.25a). The outer segment of these cells is made up of a series of flattened discs, from 500 to 1000 in some cells, stacked one on top of the other like coins. The discs are derived from in-foldings of the cell membrane, but they tend to break free from the membrane and appear as free-floating structures, at least in sections prepared for electron microscopy (Figure 5.25b). The significance of these discs is that they represent the light-receptor surfaces of the eye. Embedded in the membranes of the rod cells are par-

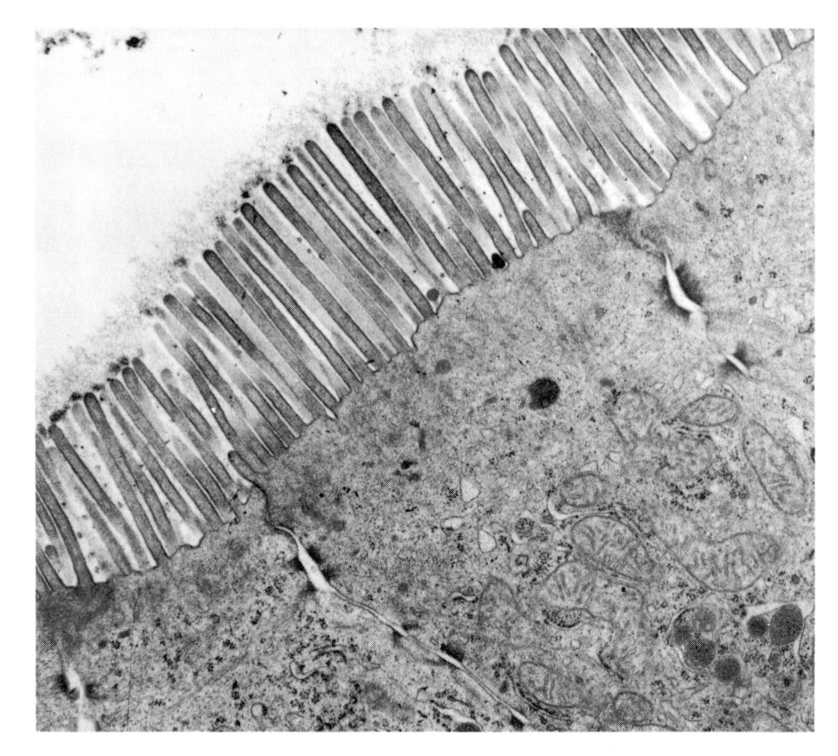

Figure 5.24 Microvilli projecting from the surface of a columnar epithelial cell. The glycocalyx is apparent at the tips of the microvilli.

ticles of rhodopsin, the only protein component of the disc membranes. These rhodopsin molecules change their molecular configuration and their orientation within the membrane in response to differences in light intensity, and these changes are somehow communicated to the nerve fibers associated with the rod cell and transmitted to the brain. These extensive in-foldings of the plasma membrane are responsible for an enormous increase in the light-receptive surface area of the retina cells.

The relation between a nerve cell and its associated Schwann, or satellite, cells presents another example of membrane flexibility. Here the entire plasma membrane of one cell forms an elaborate structural system around a portion of another cell. Figure 5.26 depicts a nerve fiber, or axon—the extended process of a nerve—with the associated Schwann cell wrapped around it. The development of the spiral proceeds as in Figure 5.27. The cytoplasm of the Schwann cells is largely squeezed to the outside, leaving the axon surrounded by a multilayered membranous system and isolated from its surroundings. The myelin sheath, as this layered structure is called, is thought to assist in the transmission of nerve impulses by insulating the nerve cell. It is unlikely that much enzyme activity is associated with these membranes, and in contrast to other membranes, few globular protein subunits can be detected at the surface.

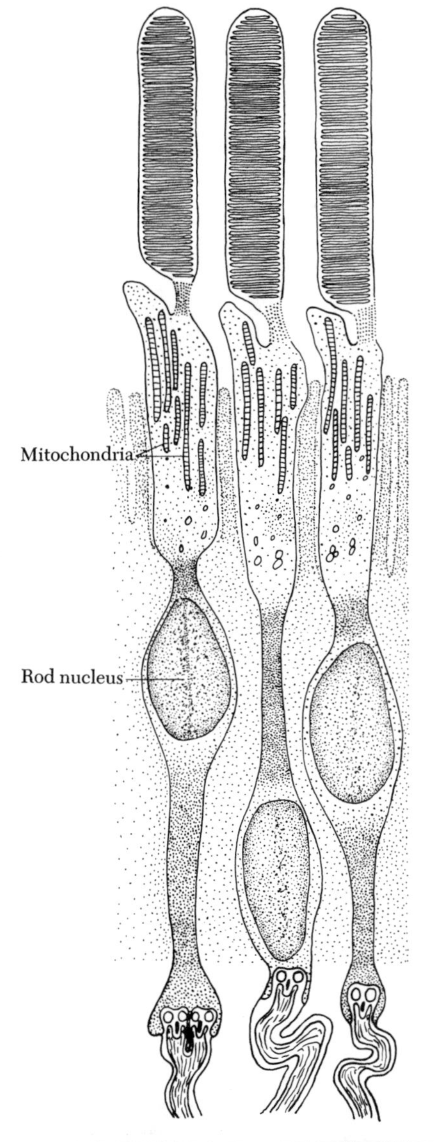

Figure 5.25 (*a*) Schematic representation of a rod (light-receptor) cell in the retina of the guinea pig. The discs at the top of the cell are folded and refolded to provide many layers of membranes, each of which contains light-sensitive pigments on its surface. The mitochondria are concentrated just below the light-sensitive area; the rod nucleus is also identified. At the base, each cell has an intimate connection with a nerve fiber. (*b*) A small portion of a retinal cell, with the folded membranes where the light-sensitive pigment is located. That the membranes originate from the cell membrane is not shown here.

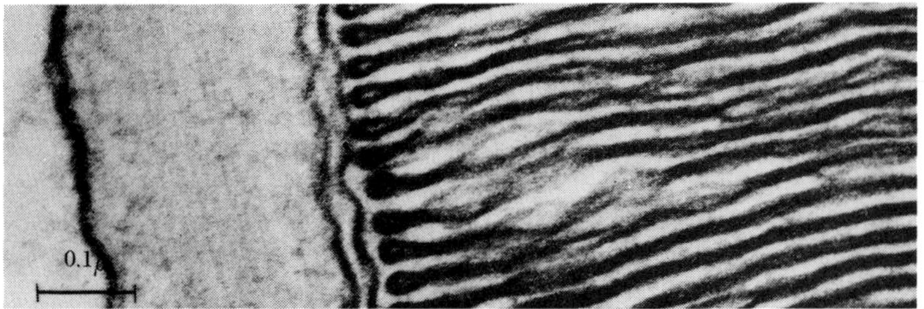

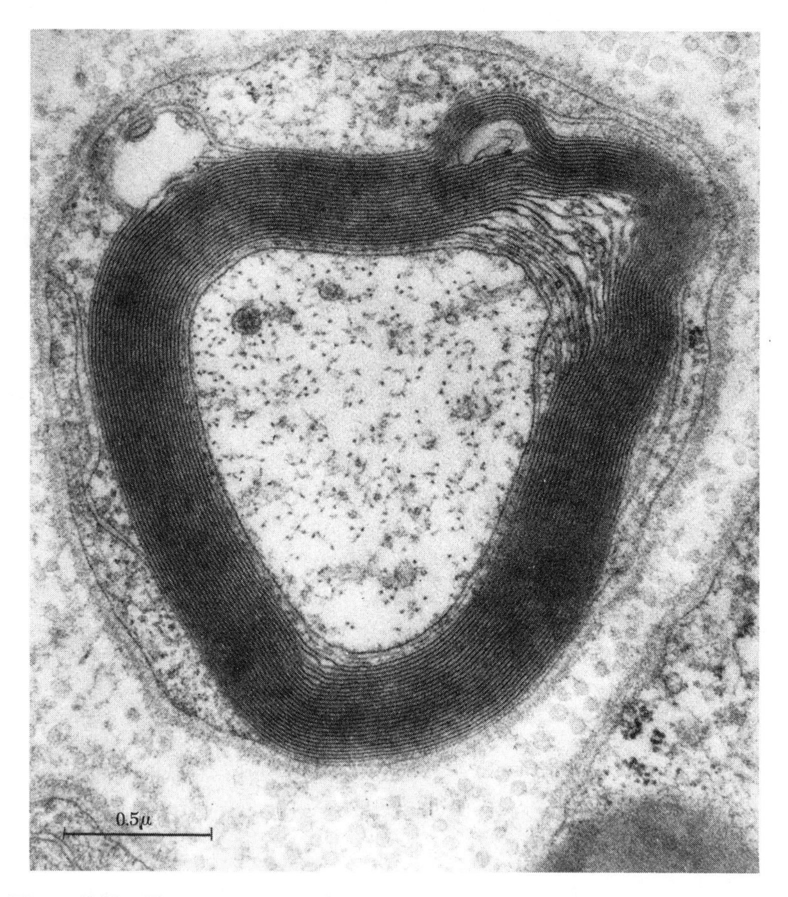

Figure 5.26 Electron micrograph cross section through a nerve fiber, with the membranes of the Schwann cell wrapped around the central axon. The cytoplasm of the Schwann cell can be seen outside the membranes and between the disrupted layers of membranes at the upper right.

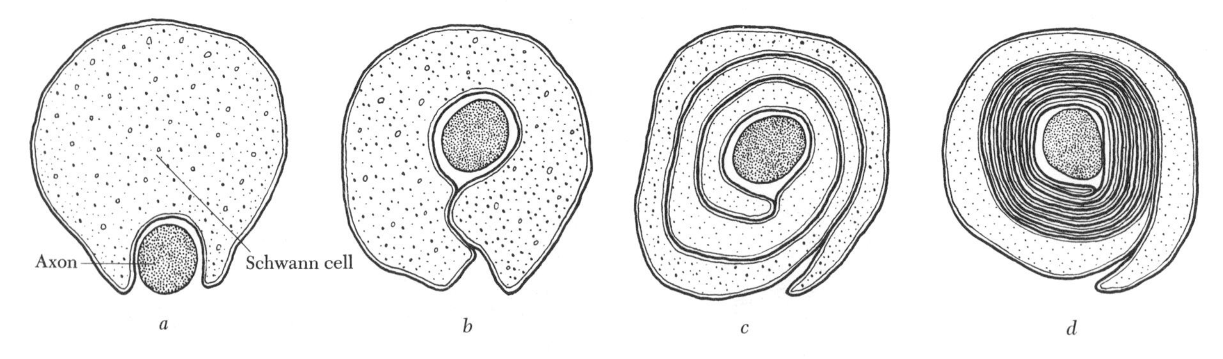

Figure 5.27 Schematic representation of the progressive envelopment of an axon by the membranes of the Schwann cell, as described by Dr. Betty B. Geren. Such an axon is said to be myelinated.

EXTRAMEMBRANE COMPONENTS OF CELLS

Although the plasma membrane is generally considered to be the outer living limit of the cell, most cells also have extracellular components associated with the membrane and these may contribute significantly to the surface properties and the functioning of the cells they enclose. We can see such outer boundaries most readily in plant cells, many of which possess heavy walls, but animal cells and many unicellular organisms also exhibit a number of somewhat comparable external substances, some of which are visible only at the level of electron microscopy, others being readily discernible with the light microscope. Many functions, varied and often multiple in nature, are performed by extracellular substances: *protection* as provided by tough, chitinous covering of insects; *support* as from the cellulose walls of plants, and from the collagen in cartilage and bone; *rigidity* and *hardness* as from the mineralized regions of bone, the dentine and enamel of teeth, the siliceous shells of diatoms, and chitin; *elasticity* as from the elastin fibers of artery walls; *locomotion* as in the pellicles of *Euglena* and *Tetrahymena;* and *adhesiveness* as from the middle lamellae of plant cells and the hyaluronic acid and chondroitin sulfate of animal cells. The slime secreted by amoebae and other aquatic organisms can provide adhesion as well as act as a lubricating material for gliding along surfaces, while the outer character of bacterial cells determines, among other things, their immunological and virulent properties. The adhesiveness of cell surfaces is particularly critical, because without it cells would fall apart after division, and multicellularity would be impossible. The permeability of cells may also be affected by the presence of extracellular substances, such as the waterproof waxes that are present in some plant cell walls, and the basal lamina of animal cells, which may act as a molecular filter.

The subject of extracellular substances, therefore, is large and varied, but we shall restrict our discussion to those commonly found in higher plants and animals.

Plant Cell Walls

Plant cells are characteristically enclosed by *cell walls*, which serve the mechanical functions of providing support for, and conferring rigidity on, the plant body. Related to this is the importance of the wall in preventing movement of water into the cell to the point where the protoplast would simply burst open. The tendency of water to enter the cell causes it to swell until pressure is exerted on the cell wall. It is this hydrostatic pressure that maintains the *turgidity* of plant parts. Furthermore, cell enlargement as a result of water moving into the cell is possible only as long as the cell wall remains sufficiently extensible.

The primary wall, the first layer to be laid down, consists of a framework of microfibrils (Figure 5.28) made up of the polysaccharide *cellulose* (Figure 5.29). The crystalline cellulose microfibrils are embedded in a matrix of several other polysaccharides and glycoproteins. Between two adjacent primary walls is the *middle lamella*, which consists mainly of another polysaccharide, pectin, and

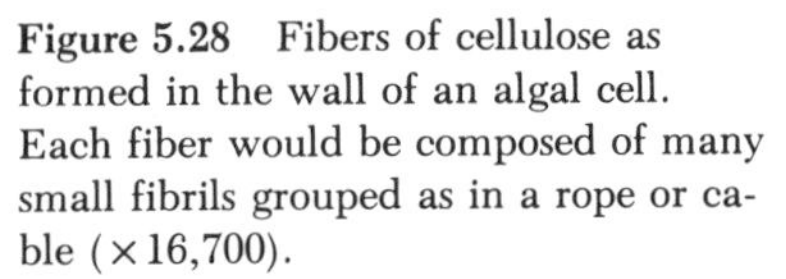

Figure 5.28 Fibers of cellulose as formed in the wall of an algal cell. Each fiber would be composed of many small fibrils grouped as in a rope or cable (×16,700).

which acts as an intercellular cement binding cells together. While a cell is enlarging, the primary wall is thin, elastic, and capable of great extension; although some thickening of wall can occur during elongation, in general this happens after the cell has reached its maximum size. After this time a *secondary wall* may be laid down between the primary wall and the plasma membrane. The secondary wall

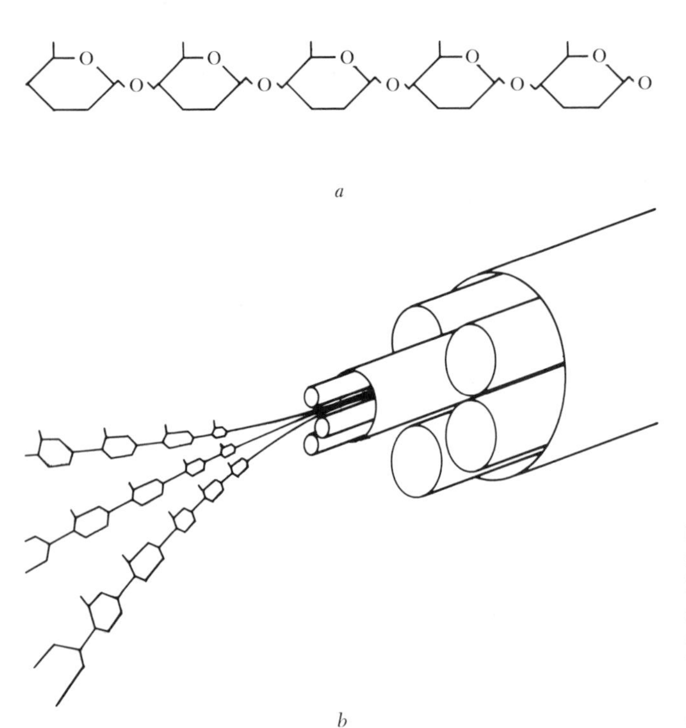

Figure 5.29 Cellulose and cellulose microfibril structure. (*a*) Cellulose is a long, unbranched molecule made up of repeating glucose sugar. (*b*) The long cellulose molecules are arranged in H-bonded crystalline bundles which together make up the microfibrils of the cell wall.

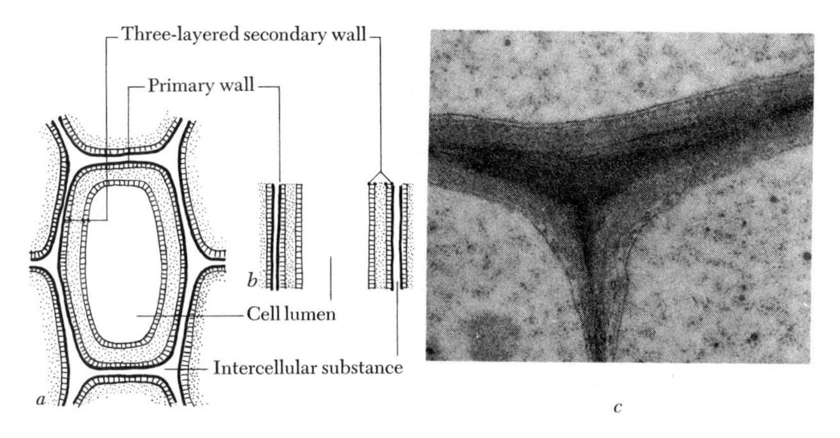

Figure 5.30 Typical wall structure of matured and lignified plant cells. (*a*) Cross section, showing arrangement of the various layers and the complex structure of the secondary wall. (*b*) Longitudinal section through a similar cell. (*c*) Electron micrograph of the cell walls of three adjacent cells: the darker middle area is the middle lamella; the lighter portions, the primary wall. (Reprinted with permission from K. Esau, *Plant Anatomy*. New York: John Wiley & Sons, Inc., 1953.)

may be thick or thin and of varying degrees of hardness or color. It is the part of the cell that gives various woods and plant fibers (cotton, flax, hemp) their particular character, and from which is derived the cellulose used in the manufacture of rayon, nitrocellulose, cellophane, and certain plastics. The organization of the mature cell wall is illustrated in Figure 5.30.

During secondary wall formation, other substances are deposited together with the cellulose. One of these is *lignin*, a complex, nonfibrous molecule, unrelated to the sugars, which forms in the spaces between the cellulose microfibrils. This arrangement is also the principle of the reinforced concrete that is used in many buildings; the cellulose provides rods of high tensile strength and the lignin is a hard substance that is resistant to pressures and acts as a glue or cement. When lignin is absent, as in balsa wood, the material is soft and brittle. Other substances, such as cutins and waxes, both derivatives of fatty acids, may replace lignin; in such instances the strength of the wall is less, but the cell surface is water-repellent, and helps prevent excessive water loss.

Let us examine the growth of the cotton fiber to illustrate the principles of cell wall formation. The mature fiber, or lint as it is called, may be 0.5 to 1.5 inches long. Located in the outermost layer or of cells, or epidermis, of the seed coat (Figure 5.31), each lint cell is attached to neighboring cells by a middle lamella and possesses a thin primary wall. Soon after fertilization of the egg has taken place, the lint cells begin to elongate, a process that takes 13 to 20 days and terminates when the cell is 1000 to 3000 times as long as it is wide. The primary wall then ceases to elongate, and the secondary wall forms as sugars in the cytoplasm are converted into cellulose fibrils and deposited on the inside of the primary wall. Deposition of

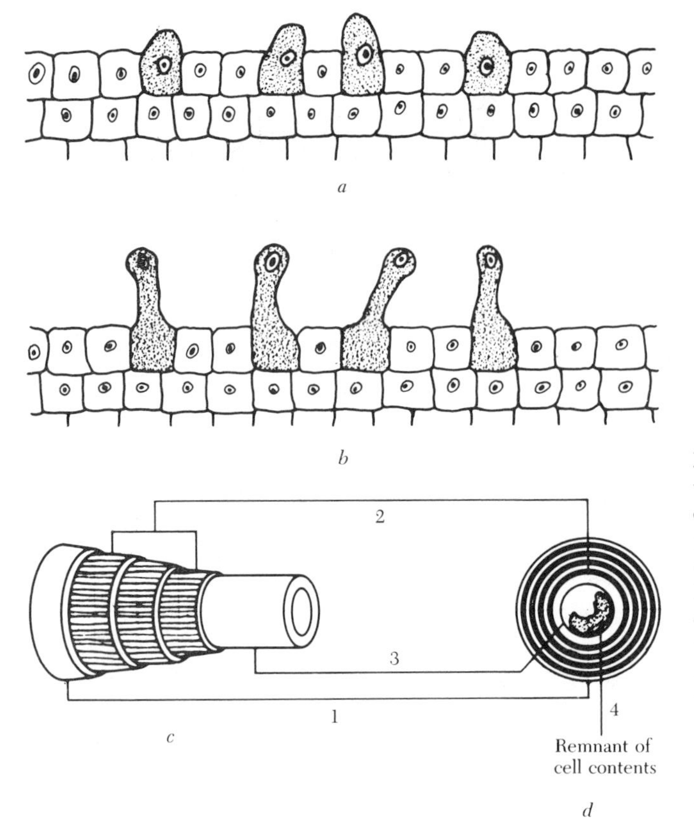

Figure 5.31 Growth and structure of the cotton fiber. (*a*) Outer layer of cells of young cotton seed, showing the beginning enlargement of the fibers on the day of flowering. (*b*) Same, 24 hrs. later. (*c*) Diagram of the various layers of cellulose laid down in a mature cotton fiber: (1) outer primary cell wall, (2) concentric inner layers, revealing the different orientation of the cellulose in the secondary thickenings, (3) last inner layer of the secondary wall. (*d*) Same, in cross section, with (4) representing the remains of cell contents.

cellulose continues until the fruit is mature. The cell then dies, collapses, and flattens to give the fiber used in the manufacture of cotton threads and cloths.

In cross section, the cotton fiber has an area of about 300 μm^2 and is made up of approximately 1 billion cellulose chains. These are grouped into fibrils of several orders of size, each one running the length of the entire fiber in a parallel, or sometimes helical (spiral), fashion. The spaces between the fibrils give the fiber its flexibility and allow for the complete penetration of dyes, whereas the parallel orientation of the fibrils accounts for its great tensile strength (nearly that of steel). It has been estimated that each cotton fiber contains about 10 trillion cellulose molecules, which are built up from approximately 60 quadrillion glucose molecules. Also, a single fiber is but one of many thousands growing on the surface of each cotton seed. From these rough calculations, we can gain some appreciation of the activity of plant cells in transforming carbon dioxide and water through photosynthesis (see Chapter 6) into organic molecules; joining many of these molecules together in specific ways, the cell then builds them into an elaborate structure, the cell wall. The cell carries out the process of polymerizing repeating molecules to form its structural elements—proteins, fats, nucleic acids, and

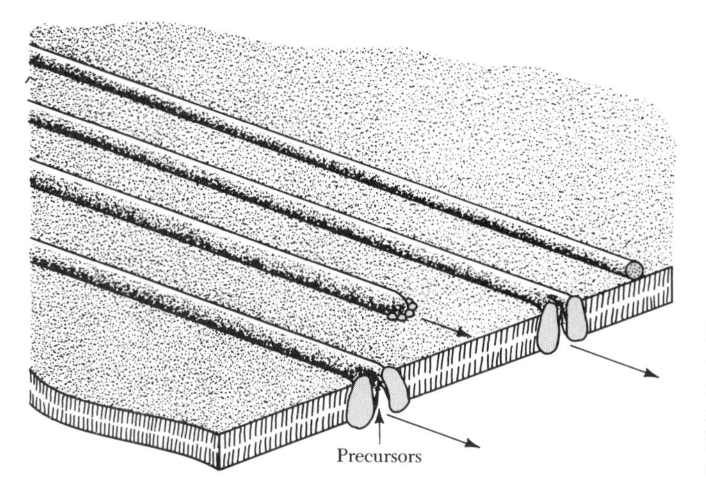

Figure 5.32 Schematic model of the cellulose synthesis/assembly complexes responsible for microfibril deposition. As complexes move through the membrane (arrows) the cellulose microfibrils are spun out behind them.

polysaccharides—in much the same way as we form plastics and synthetic fibers, although the chemical reactions are different.

The assembly of the sugars into cellulose and the aggregation of cellulose molecules into the crystalline microfibril takes place at the plasma membrane of the cell. Evidence from electron microscopy suggests that transmembrane protein complexes are associated with the ends of growing microfibrils, and the sugars from within the cell are polymerized into cellulose at these "terminal complexes." Extension of the basic 5-nm-diameter cellulose microfibril is presumably achieved by lateral movement of the complex in the fluid phase of the membrane, with the microfibril being "spun out" on the outer surface of the membrane, behind the moving complex (Figure 5.32). The direction in which the complexes move in the membrane and which determines the orientation of the microfibrils must in turn depend on some interaction between the membrane complex and the underlying cytoplasm. Cytoplasmic microtubules, long rodlike components of the cell, have been strongly implicated in how the polarity of cellulose microfibril deposition is determined.

It appears that the arrangement of cellulose fibers laid down in the early stages of wall deposition may be related to the kind of expansion a plant cell will undergo. When the fibers are randomly arranged, the cell expands uniformly in all directions. If the microfibrils are arranged in parallel fashion, they impose a restriction such that expansion of the cell occurs along an axis perpendicular to that of the microfibrils (Figure 5.33). Secondary walls, however, are deposited only after cell expansion is complete; thus the microfibril patterns in secondary walls, such as those of cotton hairs, are not related to the polarity of elongation.

Since most plant cells conduct water and dissolved substances as well as provide support, even after they have died (as in wood), the heavy wall must be interrupted at intervals to allow for passage from one cell to another. Interruption occurs in a variety of ways and as Figure 5.34 indicates, the secondary wall may exist

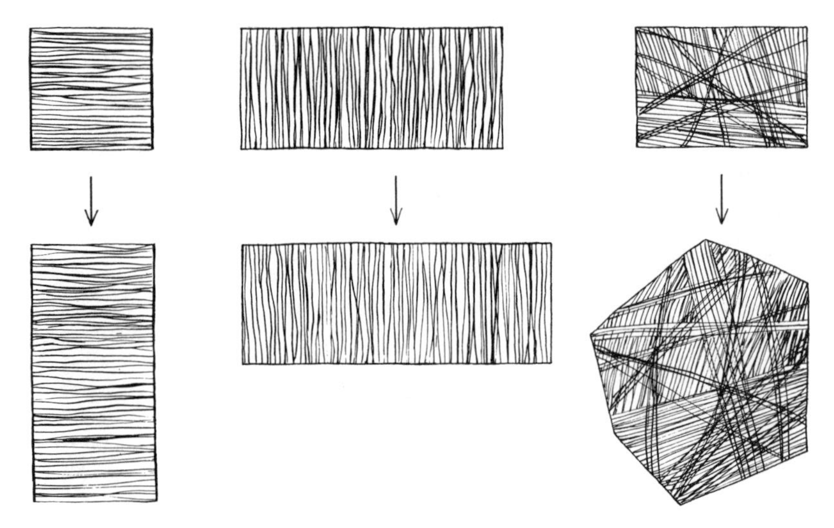

Figure 5.33 Schematic representation showing how the orientation of cellulose fibers in a young cell (top row) determines the axis of elongation of older cells. Elongation is essentially at right angles to the direction of the fibers. When the fibers exhibit no orientation (top right), the cell enlarges in all directions.

as rings, spiral bands, or sequences of thick and thin areas. Such gaps give flexibility as well as support and ease of conduction. In other cells, particular areas may be perforated by pits, or pit fields, or the end wall of a cell may be perforated or even missing to provide a connected column simulating a channel made up of short pieces of pipe (Figure 5.35).

Extremely elaborately sculptured walls form around pollen grains of many plants (see Figure 1.22, Chapter 1). These structures consist mainly of sporopollenin, which is lipoidal in nature, and are extremely tough, affording

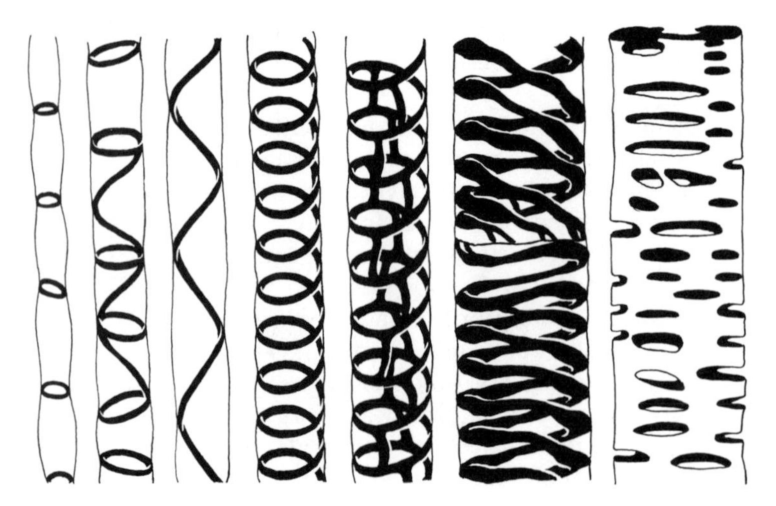

Figure 5.34 Cells in the wood of higher plants, exhibiting various patterns of secondary-wall formation. The interrupted areas are thin enough to permit the passage of water and dissolved materials.

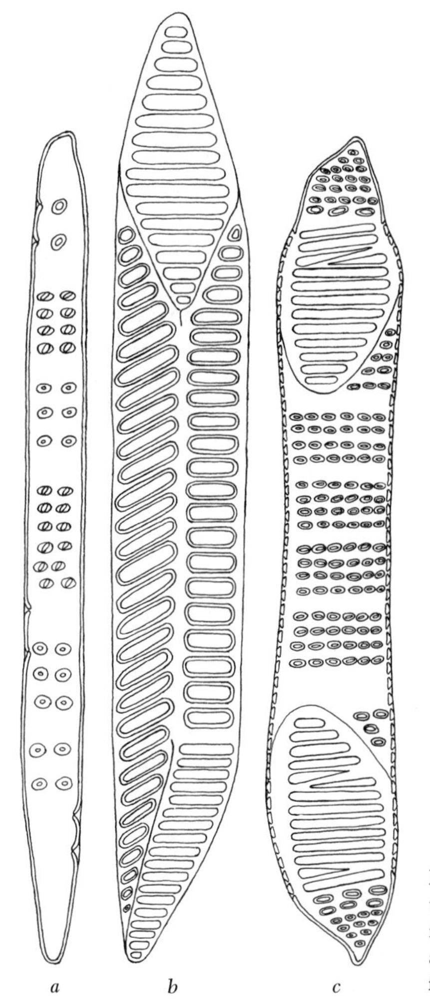

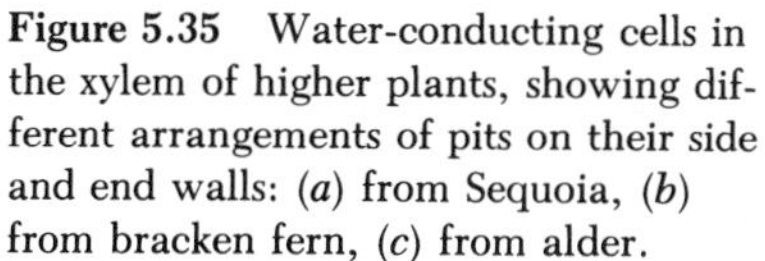

Figure 5.35 Water-conducting cells in the xylem of higher plants, showing different arrangements of pits on their side and end walls: (*a*) from Sequoia, (*b*) from bracken fern, (*c*) from alder.

much protection to the pollen grain. The walls can also contain substances that may play a role in determining whether the pollen grain is compatible or not with the stigma on which it lands. These substances are responsible for the pollen allergies that cause so much personal discomfort in humans, especially in spring and fall when pollen is shed in great quantities from the troublesome species.

Extracellular Substances of Animal Cells

The *extracellular matrix* is a complex array of macromolecules found predominantly beneath epithelial linings and surrounding cells of connective tissue. Originally believed to serve as no more than a supportive structure, it is becoming clear that cells interact with their matrix, and that the matrix determines various aspects of cell and tissue behavior. Cells aggregate into organs of definite shape and size, and organ systems are tied together to form the intact and functioning animal. Adhesiveness, lubrication, rigidity, and elasticity are but some of the required features of a developing and functioning organism that are governed by the quality and quantity of the extracellular matrix.

Several types of molecules constitute the extracellular matrix, and these interact with one another to determine the properties of the matrix. Three main classes of fibrous proteins can be present, and these are associated with the "ground substance" of the matrix, a mixture of complex polysaccharides and *proteoglycans*, protein molecules carrying polysaccharide side chains.

The most abundant of the fibrous proteins are the *collagens*, which may make up more than one-third of the total protein of a mammal. Collagen is located predominantly in those areas where a degree of firmness is needed—cartilage, bone, skin, tendons and so forth. The collagen molecules, which are very long and narrow, can aggregate spontaneously into fibers which show a very regular periodicity in their structure (Figures 5.36 and 1.19b). In cartilage the collagen fibers are interspersed with proteoglycans containing side chains of *chondroitin sulfate*. These proteoglycans are reversibly compressible molecules, and provide a firm gel which confers the necessary resilience and flexibility.

Another protein of the matrix is *elastin*, which, as its name suggests, has the capacity to stretch and snap back to its original state, much as a rubber band would. Consequently, it is prevalent in extracellular matrices of tissues where elasticity is required, as in ligaments and in the walls of major blood vessels. Unlike collagen fibers, the aggregates of cross-linked elastin molecules that are found in

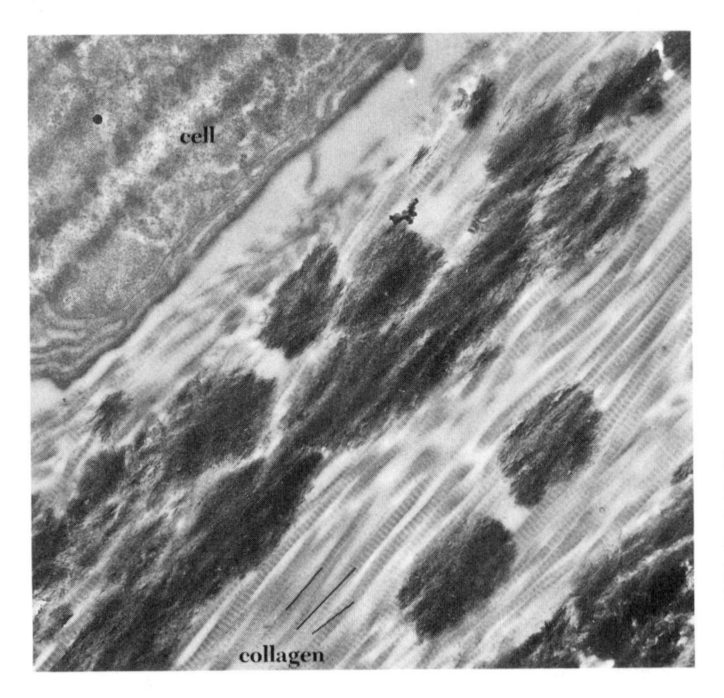

Figure 5.36 Electron micrograph of part of an osteoblast, or bone-forming cell, and the surrounding extracellular material. Note the characteristic axial periodicity of the collagen fibers every 64 nm.

the matrix show no periodicity, but rather form thick fibers with an amorphous appearance.

Several types of glycoprotein, of which *fibronectin* is the best known, are also present in extracellular matrices. Fibronectin can exist both as aggregates and fibers, and, like other components of the matrix, may be able to expand and contract. Furthermore, fibronectin can bind not only to other components of the matrix, but also to the plasma membrane itself. This interaction with the plasma membrane is important in the adhesion of cultured cells to the substrate on which they are growing, and in determining their morphology. It is possible that fibronectin binding to the membrane influences in some way the organization of the cytoplasm beneath, resulting in changes in cellular behavior.

The ground substance of the matrix, in which the various fibrous proteins are embedded, is composed mainly of polysaccharides. Sulfur-containing polysaccharides such as chondroitin sulfate are covalently bound to protein molecules to form the proteoglycans; *hyaluronic acid*, a jellylike, viscous, polysaccharide of high molecular weight forms long filaments with which the other components of the matrix are more loosely associated. Functionally, hyaluronic acid serves a number of purposes. As a "glue" it binds cells together at the same time that it permits flexibility; in the fluids of joints it acts as a lubricant and, possibly, as a shock absorber; in the fluids of the eye it acts to retain water and keep the shape of the eye fixed.

Basal Lamina

A specialized extracellular type of organization is found around surfaces of cells, other than connective tissue cells, which come into contact with the extracellular matrix. This *basal lamina* (or *basement* membrane) is prominent under epithelial cells and surrounding muscle cells. Attached to the cell membrane is a predominantly proteoglycan layer, outside of which is a regularly oriented sheet of bundles of collagen fibrils, separated from the connective tissue matrix by another proteoglycan layer. The basal lamina appears to function as an extracellular scaffold, helping to determine the shape and organization of the particular tissues or organs in which it is found.

Vitelline Coat

The true plasma membrane of most animal eggs is surrounded by another specialized, protective, extracellular layer, the vitelline coat. The thick, transparent surface coat of the mammalian egg is known as the *zona pellucida*, and is separated from the plasma membrane by a fluid-filled region, the *perivitelline space*. Before actual fusion of the sperm with the egg membrane proper, the sperm must attach to the exterior of, and then penetrate, this surface coat. The zona pellucida, which like the vitelline coat of other animal eggs contains polysaccharides, proteins, and glycoproteins, is thought to have on its surface specific receptors to which sperm cells bind. Hyaluronic acid is present in the coat, and passage of the sperm through the coat is facilitated by release of hyaluronidase and other enzymes which can dissolve away the protective barrier.

ADDITIONAL READING

ALBERSHEIM, P. 1975. The walls of growing plant cells. *Sci. Am. 232*(4): 80–95.

ANDERSON, R. G. W., BROWN, M. S., BEISIEGEL, U., and GOLDSTEIN, J. L. 1982. Surface distribution and recycling of the low density lipoprotein receptor as visualized with antireceptor antibodies. *J. Cell Biol. 93:* 523–531.

BECK, J. S. 1980. *Biomembranes.* Hemisphere Publishing Corp., Washington, D.C.

BRANTON, D., and PARK, R. B. (eds.). 1962. *Papers on Biological Membrane Structure.* Little, Brown and Company, Boston.

BRETSCHER, M. S. 1973. Membrane structure: some general principles. *Science 181:* 622–629.

CAPALDI, R. A. 1974. A dynamic model of cell membranes. *Sci. Am. 230*(3): 26–33.

DAUTRY-VARSAT, A., and LODISH, H. F. 1984. How receptors bring proteins and particles into cells. *Sci. Am. 250:* 52–58.

FINEAN, J. B., COLEMAN, R., and MICHELL, R. H. 1978. *Membranes and Their Cellular Functions,* 2nd ed. John Wiley & Sons, Inc., New York.

FOX, C. F. 1972. The structure of cell membranes. *Sci. Am. 226*(2): 30–38.

GIDDINGS, T. H., BROWER, D. L., and STAEHELIN, L. A. 1980. Visualization of particle complexes in the plasma membrane of *Microsterias denticulata* associated with the formation of cellulose fibrils in primary and secondary cell walls. *J. Cell Biol. 84:* 327–339.

GOLDSTEIN, J. L., ANDERSON, R. G. W., and BROWN, M. S. 1979. Coated pits, coated vesicles, and receptor-mediated endocytosis. *Nature 279:* 679–685.

HALL, J. L., and BAKER, D. A. 1977. *Cell Membranes and Ion Transport.* Longman, Inc., New York.

HAY, E. D. (ed.). 1981. *Cell Biology of the Extracellular Matrix.* Plenum Press, New York.

KAPLAN, J. 1981. Polypeptide-binding membrane receptors: analysis and classification. *Science 212:* 14–20.

MUELLER, S. C., and BROWN, R. M. 1980. Evidence for an intramembrane component associated with a cellulose microfibril-synthesizing complex in higher plants. *J. Cell Biol. 84:* 315–326.

PASTAN, I. H., and WILLINGHAM, M. C. Journey to the center of the cell: role of receptosomes. *Science 214:* 504–509.

QUINN, P. J. 1976. *The Molecular Biology of Cell Membranes.* University Park Press, Baltimore.

ROBERTSON, J. D. 1981. Membrane structure. *J. Cell Biol. 91:* 189s–204s.

SINGER, S. J., and NICOLSON, G. L. 1972. The fluid mosaic model of the structure of cell membranes. *Science 175:* 720–731.

UNWIN, N., and HENDERSON, R. 1984. The structure of proteins in biological membranes. *Sci. Am. 250:* 78–94.

6

Energy Transduction—Mitochondria and Chloroplasts

In earlier chapters we have pointed out that life is characterized by its capacity to assimilate, convert, and utilize energy, and have discussed some of the principles involved in how energy is manipulated in chemical reactions. We must now examine in more detail the processes by which energy is transduced, or changed from one form to another, within the cell. In so doing we shall also consider the specialized membrane-enclosed organelles, *mitochondria* and *chloroplasts*, which provide the structural basis for energy transduction.

RESPIRATION

Respiration is the process by which the energy stored in fuel molecules is extracted during the controlled breakdown of these molecules and conserved in the formation of ATP. As we discussed in Chapter 3, ATP represents the energy currency of the cell, and is the form in which metabolic energy is supplied to do the work of the cell. Although several types of fuel molecules can serve as substrates for respiration, carbohydrates are the most commonly used. We shall therefore consider the steps involved in the breakdown of glucose in our discussion of the main features of respiration.

The complete breakdown of glucose to CO_2, and hence maximum recovery of the energy stored in the molecule, requires oxygen, and is therefore known as *aerobic respiration*. The overall process of aerobic respiration represents the oxidation of glucose by oxygen, and can be summarized by the following equation:

$$C_6H_{12}O_6 + 6O_2 \longrightarrow 6CO_2 + 6H_2O + \text{energy}$$

$$\text{Glucose} + \text{oxygen} \longrightarrow \text{carbon dioxide} + \text{water} + \text{energy}$$

Fermentation, which represents only part of the process and which is much less efficient in energy recovery, occurs in cells which cannot use oxygen, or under conditions where oxygen is limiting.

We shall outline, without going into too much detail of the pathways, the

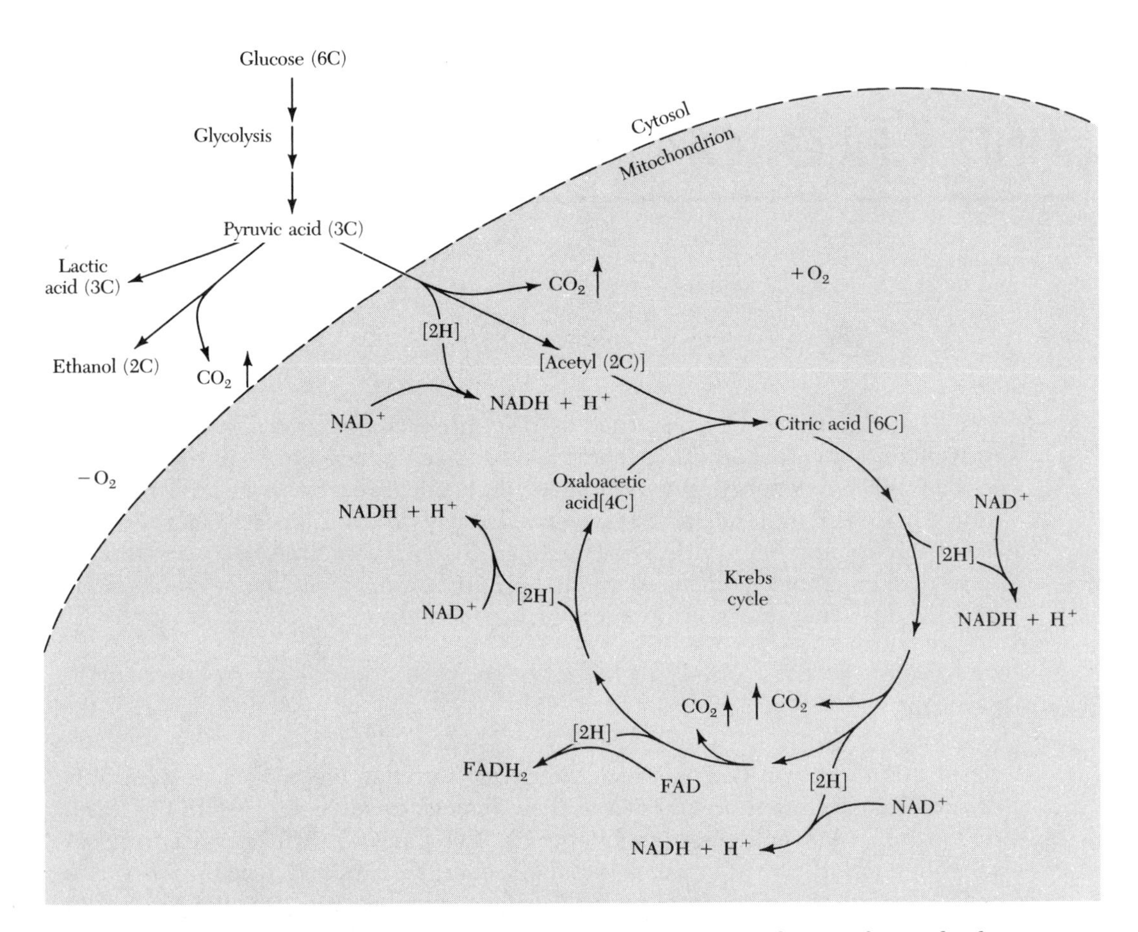

Figure 6.1 Pathway of oxidation of glucose in respiration. Glucose is first oxidized to pyruvic acid, which in the absence of oxygen is converted to either ethanol or lactic acid. In the presence of oxygen the pyruvic acid enters the mitochondria, is decarboxylated and fed into a series of reactions, the Kreb's cycle, being added initially to a 4-carbon acceptor molecule, oxaloacetic acid, to form the 6-carbon molecule, citric acid. Further oxidations regenerate oxaloacetic acid, and in these reactions electrons from the Kreb's cycle substrates are used to reduce NAD and the related compound FAD. Some of the energy in the original glucose molecule is thus conserved in the formation of reduced NAD and reduced FAD, and can be used to form ATP (see text and Figure 6.2).

three different complex sets of reactions by which glucose is oxidized, and then discuss in more detail the mechanisms believed to be involved in coupling the downhill flow of electrons from glucose to oxygen to the formation of ATP.

In the first series of reactions, the *glycolysis* pathway, the six-carbon glucose molecules are broken down only partially, yielding two molecules of pyruvic acid, a three-carbon compound (Figure 6.1). In the absence of oxygen or in cells that do not use oxygen, the pyruvic acid can be converted to either lactic acid or ethanol (Figure 6.1). This anaerobic process is involved in the production of wine and beer; fungi such as yeasts utilize the sugars of the grapes or malted barley as substrates for fermentation and convert them to ethanol. Cells of plants growing under anaerobic conditions may also produce this alcohol; in animal cells under oxygen stress, however, such as in active muscle, lactic acid is formed instead. Anaerobic respiration is very inefficient in terms of energy yield, since much of the energy is still locked up in the ethanol or lactic acid.

In the presence of oxygen the fate of the pyruvate formed in glycolysis is different. It enters the mitochondria (see below), where it is converted first to a two-carbon acetyl group; the acetyl group then combines with a molecule of oxaloacetic acid to form citric acid. The carbon is then further oxidized to CO_2 in a cyclic series of reactions (Figure 6.1) known as the Krebs cycle (= citric acid cycle, tricarboxylic acid cycle). The oxidation reactions of the Krebs cycle result in the transfer of electrons (hydrogen) to the electron carriers NAD (see Chapter 3) and FAD, thereby reducing them. Thus much of the energy originally locked up in the glucose molecule is conserved as the reducing power of NADH and $FADH_2$. To make this energy available for the formation of ATP, the electron carriers must pass their electrons to oxygen; however, the free energy change in the oxidation of NADH and $FADH_2$ by oxygen is so great that the electrons must be "dropped down" in a series of small steps in order that manageable amounts of energy are made available. The reduced electron carriers, therefore, donate their electrons to an electron transport system in which they are passed to successively lower energy levels, ultimately being accepted by oxygen. It is in the series of oxidation-reduction reactions that makes up the electron transport system that the free energy is made available and used to form ATP from ADP and inorganic phosphate (Figure 6.2). We shall discuss electron transport and oxidative phosphorylation in more detail later, but first we must consider the structural organization of the mitochondrion itself.

MITOCHONDRIA

While the reactions involved in the breakdown of the substrates of respiration to pyruvic acid, that is, glycolysis, occur outside the mitochondria, it is in these organelles that the major energy transfers of the Krebs cycle and the electron transport chain take place. Thus the mitochondria can be considered as the "power plants" of the cell, responsible for providing energy from the metabolic fuel.

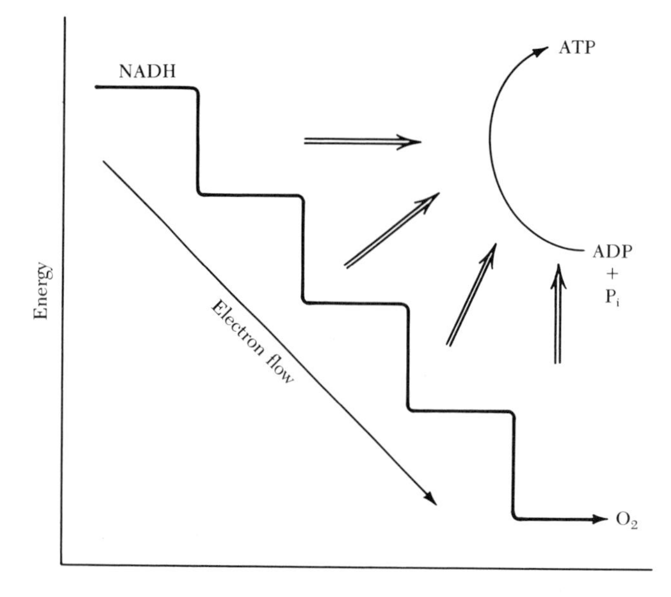

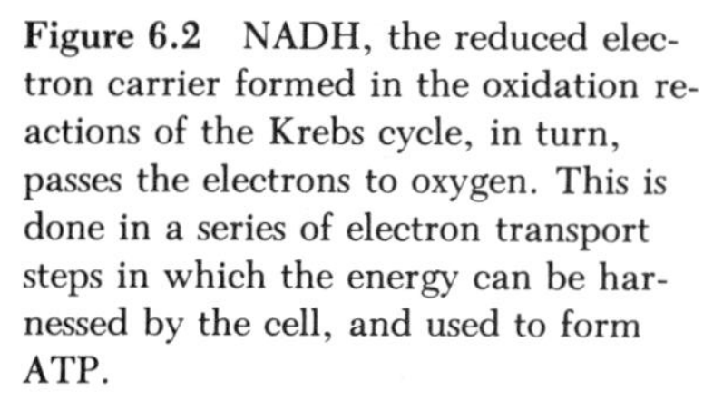
Figure 6.2 NADH, the reduced electron carrier formed in the oxidation reactions of the Krebs cycle, in turn, passes the electrons to oxygen. This is done in a series of electron transport steps in which the energy can be harnessed by the cell, and used to form ATP.

Mitochondria are found in all eukaryotic cells, although they may be lost in later stages of development of cells such as red blood cells or phloem sieve tube elements. The shape, size, and number of mitochondria vary, and may be characteristic of the cell type in which they are found. They range in size from about 0.2 to 7.0 μm and, in form, from spheres to rods to branching rods. The unicellular green alga, *Microsterias*, has a single mitochondrion per cell, while other cells have been reported to have many thousands. However, since most views of mitochondria are from electron microscopy, it may be difficult to tell if several different mitochondrial profiles seen in a thin section represent separate mitochondria, or different parts of the same mitochondrion.

Mitochondria, as seen in the electron microscope (Figure 6.3), consist of a smooth *outer membrane*, separated by a space from an *inner membrane*. The inner membrane is folded to form *cristae*, which extend into the *matrix* of the mitochondrion and separate it from the *intracristal* space (Figure 6.4). The cristae may be few in number, leaving most of the interior volume of the mitochondrion occupied by the matrix, or they may be many cristae packed closely together. Cristae may also assume tubular form, as in *Paramecium* and in some plant cells, or reticulate form, as in flight muscle cells of some insects. These various ways of increasing the available surface area of the inner membrane system appear to be related to the degree of metabolic activity of the mitochondrion. Dramatic conformational changes of the inner membrane are also known to occur as mitochondria change from an inactive to an active state.

It is possible to separate the inner and outer membranes from the internal matrix, and to analyze these components for activities of various enzymes. None of the major respiratory enzymes are located in the outer membrane, although other enzymes may be present there. Most of the enzymes that catalyze the reactions of

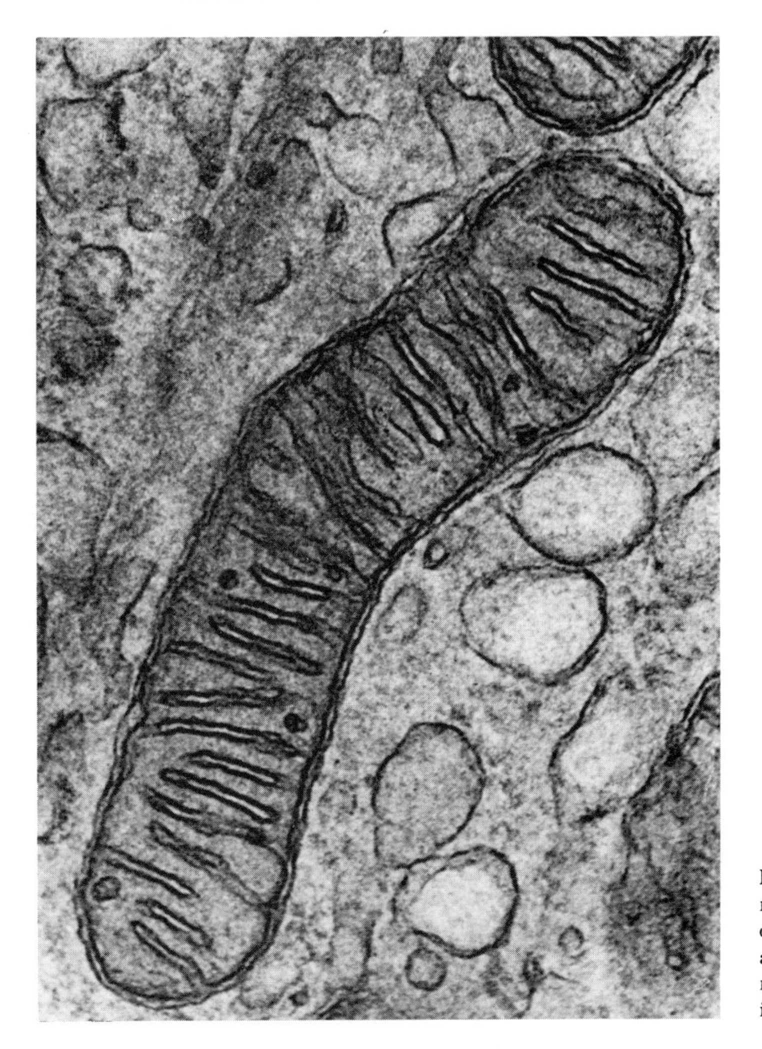

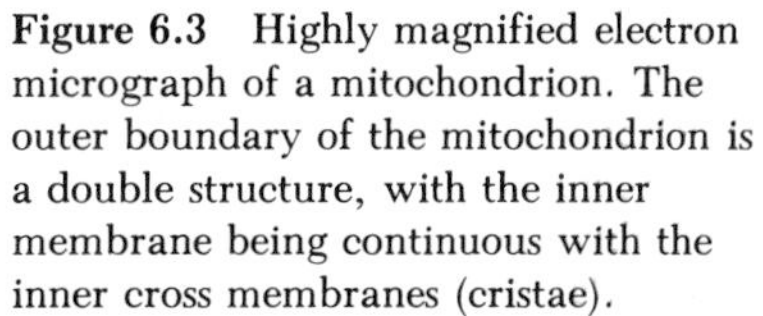

Figure 6.3 Highly magnified electron micrograph of a mitochondrion. The outer boundary of the mitochondrion is a double structure, with the inner membrane being continuous with the inner cross membranes (cristae).

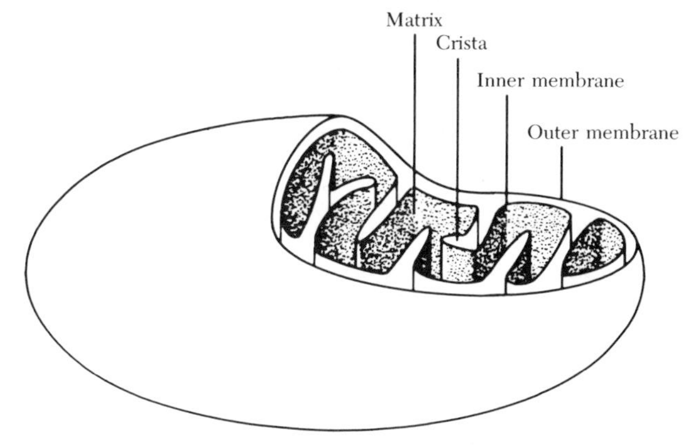

Figure 6.4 Schematic drawing of a typical mitochondrion.

the Krebs cycle are in the matrix compartment, and it is here that the energy-rich electron carrier NADH is formed. The inner membrane of the mitochondrion contains the assemblies of molecules, including FAD, that constitute the electron transport chain, as well as the enzymes necessary for the electron transport-dependent formation of ATP.

Electron Transport and Oxidative Phosphorylation

As we have already mentioned, oxidative phosphorylation is the process by which the energy made available by the oxidation of the electron carriers NADH and $FADH_2$ is conserved in the formation of ATP. We shall now consider the mechanism by which this energy transfer is accomplished, that is, the mechanism by which phosphorylation is coupled to electron transfer. To do so, we must examine in more detail the role of the inner membrane in aerobic respiration, since it is in the chemical composition and structural integrity of this membrane that we find the basis of the capacity of the cell to harness the energy of electron transport, and then to conserve it in the formation of ATP.

The electron transport system of mitochondria is made up mainly of electron-carrying proteins that are firmly embedded in the inner membrane. These proteins, which include flavoproteins, metalloproteins, and cytochromes, all have attached to them atoms that can be alternately oxidized or reduced. For example, the cytochromes, which constitute the major class of electron transport proteins, contain an iron atom which can exist in either the oxidized ferric (Fe^{3+}) or the reduced ferrous (Fe^{2+}) state, and can therefore accept or donate electrons. The oxidation and reduction of other carriers, such as the flavoproteins and the nonprotein ubiquinone, involve both protons and electrons.

The electron carriers, with their oxidizable and reducible atoms, are arranged in complexes that maintain a specific spatial configuration and that can conduct electrons, with or without accompanying protons, in an orderly fashion from one carrier to the next. The major components of the system, as well as the sequence in which they alternately accept and donate electrons, are shown in Figure 6.5a. The orientation of the electron transport components within the inner membrane is asymmetrical, and some of the electron transfer steps actually span the membrane; thus the electrons move from side to side of the membrane as they pass to successively lower energy levels (Figure 6.5b).

Small fragments of mitochondrial inner membranes are capable of carrying out electron transport, indicating that the membrane must consist of many repeating subunits of the orderly array of enzymes necessary for these reactions. Such fragments can also make ATP from ADP and inorganic phosphate, but only if the membrane fragments form closed vesicles; thus separation of two compartments by the inner membrane is necessary for the coupling of electron transport and ATP synthesis.

Another feature of the inner membrane is the presence of stalked particles attached to its matrix face (Figure 6.6). When these are present on membrane

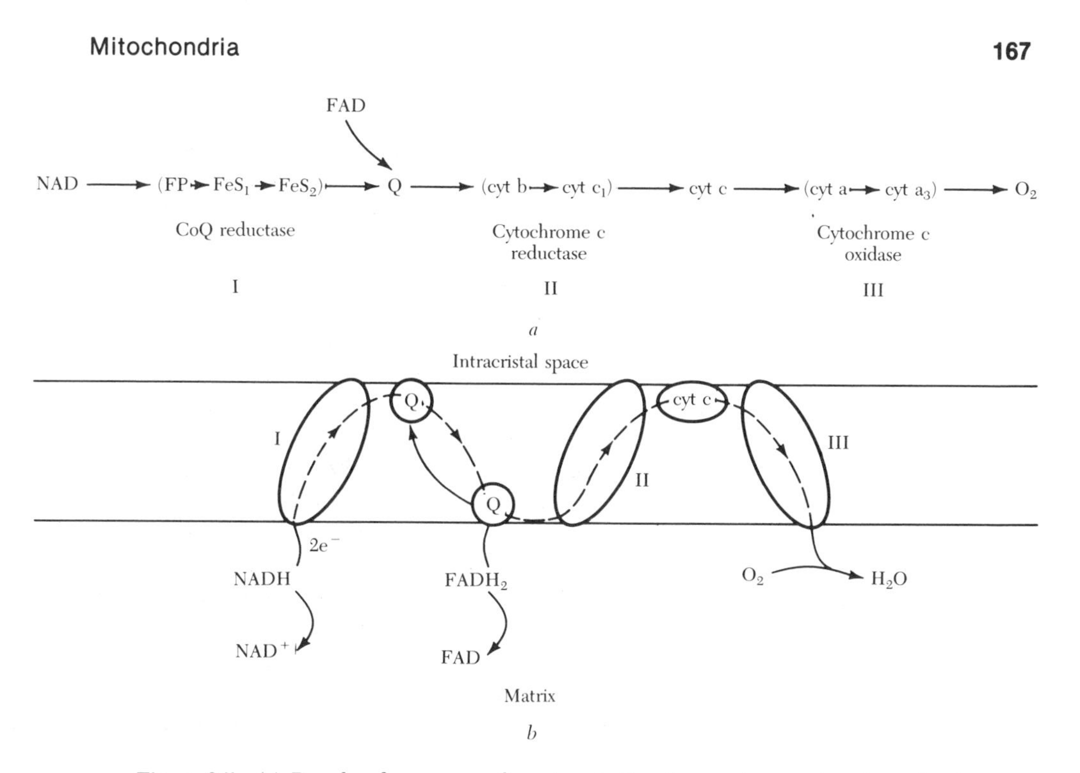

Figure 6.5 (*a*) Postulated sequence of carriers in the electron transport system of mitochondria. Enzyme complex I, coenzyme Q reductase, consists of a flavoprotein (FP) and two iron-sulfur (FeS) proteins, and conducts electrons from NADH to ubiquinone (Q). Complex II, cytochrome c reductase, includes cytochromes b and c, and in turn reduces cytochrome c. Complex III is cytochrome-oxidase, and includes cytochromes a and a_3; this terminal oxidase transfers the electrons to their final destination, oxygen. (*b*) Simplified scheme of the arrangement of electron carriers in the inner membrane, showing the vectorial nature of the electron transport pathways. Ubiquinone actually moves back and forth across the membrane.

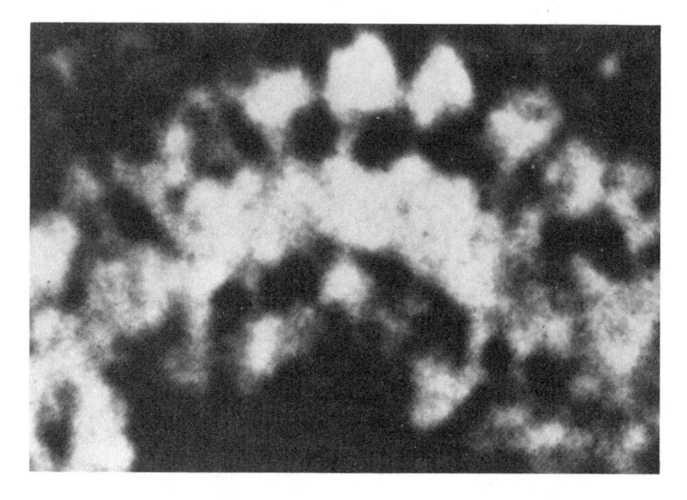

Figure 6.6 Portion of inner mitochondrial membrane showing stalked particles that contain ATP-ase activity.

preparations, both electron transfer and ATP formation occur; upon removal of these particles, electron transport continues but no ATP is formed. However, the isolated particles show ATPase activity, and when the inner membrane is reconstituted by addition of the particles, the capacity to make ATP is recovered. Thus the particles, of about 10nm diameter, represent the ATP-forming enzyme complex associated with electron transport, and are known as *coupling factors*. These observations show that although ATP formation is tightly coupled to electron transport under normal conditions, the two processes are separable. This is also shown by the effect of *uncouplers*, molecules which allow the normal flow of electrons from NADH to oxygen to occur but which prevent ATP formation. (Under such conditions the free energy of oxidation is dissipated as heat.)

What, then, is the nature of the connection between electron transport and oxidative phosphorylation? The most widely accepted (but, of course, not universally so) explanation of how oxidation and phosphorylation are coupled is the *chemiosmotic* hypothesis proposed by Mitchell. In this model (Figure 6.7) electron flow through the electron transport system of the inner membrane is used to pump protons across the membrane, from the matrix side to the intermembrane (or intracristal) space. Thus gradients of both proton concentration and electrical potential are built up across the inner membrane. These gradients represent a source of energy, which can be tapped as protons flow back across the membrane through

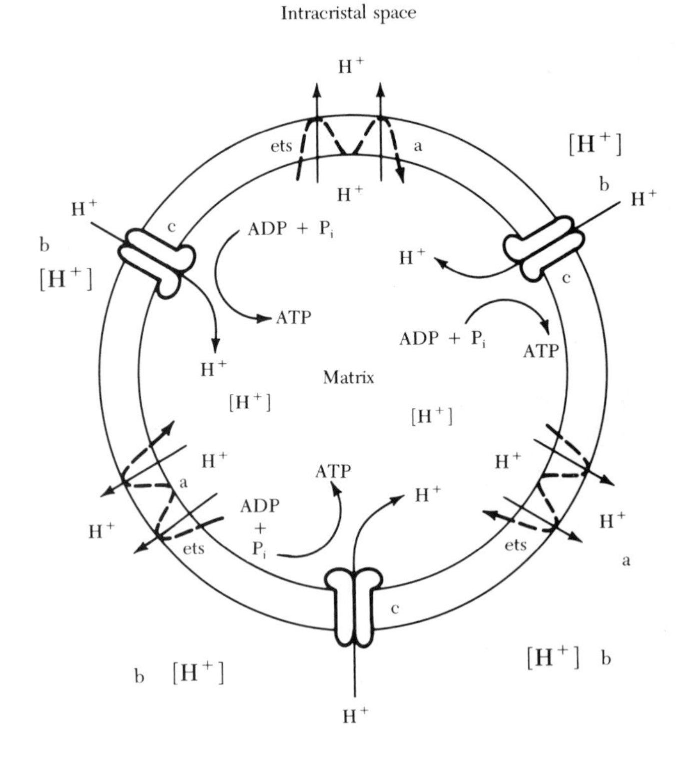

Figure 6.7 The Chemiosmotic Hypothesis. (*a*) As electrons pass along the electron transport system (ets), energy is used to transport protons (H^+) across the membrane. (*b*) As a result a proton gradient is established, which represents potential energy. (*c*) Protons flow back through proton-conducting channels in the membrane that also have ATP-synthetase activity; as they do so the energy is used to form ATP.

specific channels, giving up their energy as they do so to an ATP-generating system in the membrane, presumably the ATPase represented by the stalked particles.

This model, of course, accounts for the above-mentioned requirement for closed, intact inner membrane vesicles if oxidative phosphorylation is to occur; no proton gradient could be established without a membrane separating two compartments. Other evidence in favor of the chemiosmotic hypothesis is that during electron transport the compartment outside the membrane becomes both more acid (higher H^+ concentration) and more electrically positive than the matrix. Furthermore, the uncouplers of oxidative phosphorylation already referred to act by transporting protons back across the membrane, thereby dissipating the proton gradient and hence the force used to form ATP.

Let us consider how the energies of oxidation in the electron transport system reactions might transport protons from the matrix to the outside of the inner membrane. As we mentioned earlier, the electron transfer complexes are asymmetrically oriented within the membrane, with the result that electrons traverse back and forth as they move from NADH and $FADH_2$ to oxygen. This polarity of electron flow across the membrane is believed to establish a polarity of proton flow. For example, in some of the oxidation-reduction reactions of electron transport, the transfer of electrons is accompanied by proton transfer, that is, hydrogen atoms are used in the reduction of the acceptors. In others, however, electrons only are transferred. Thus, if electron transfer from the inner to outer side of the membrane is accompanied by proton transfer, while that back from outer to inner is not, protons will be picked up from the matrix, and released to the medium outside (Figure 6.8). Additional protons may also be pumped across the membrane in a less direct way; it is possible, for example, that the energy of oxidation available from electron transport may induce conformational changes in the proteins, enabling them to transport protons across the membrane and to contribute to the proton gradient.

Although it is clear that electron transport within the membrane establishes an electrochemical gradient across the membrane, it is not yet known exactly how the transient store of potential energy in the form of the proton gradient is converted to the chemical energy of ATP. The stalked particles representing the ATP-forming complex are attached to other protein complexes, which are embedded within the membrane and through which protons can move across the membrane. It is believed that as the gradient forces protons to flow back to the matrix through these channels, or gates, somehow the energy is used to drive the synthesis of ATP at the catalytic site of the ATP-forming complex.

Finally, we should point out that ATP formation driven by proton gradients, themselves established across membranes as a consequence of electron transport, is not unique to mitochondria. As we shall see, ATP formation in the process of photosynthesis is accomplished by a similar mechanism within chloroplasts, while electron transfer and phosphorylation in bacteria are coupled by a proton gradient across the cell membrane. Indeed, transmembrane electrochemical proton gradients represent stores of energy which can be used for other kinds of work, including transport across the inner membrane of the mitochondrion. This mem-

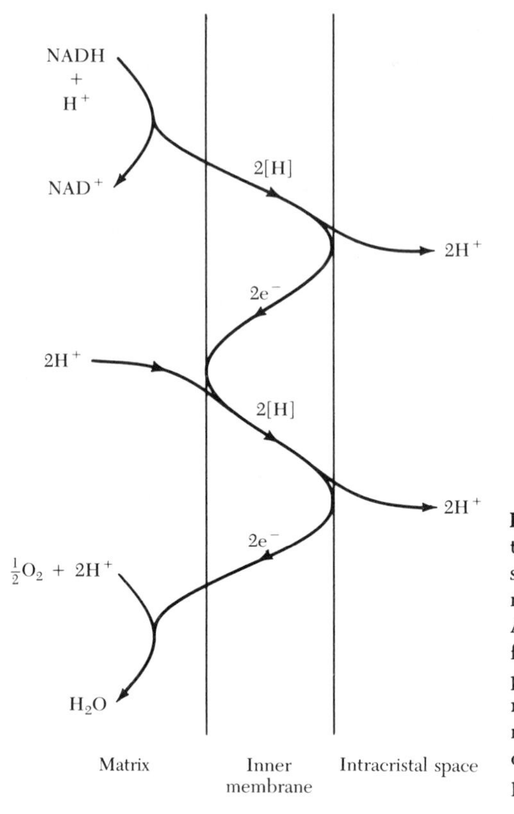

Figure 6.8 Possible mechanism of proton pumping across a membrane, resulting from the asymmetric arrangement of the electron transport system. As electrons move from the matrix surface of the membrane they are accompanied by protons; in the oxidation-reduction reactions in which electrons move back towards the matrix side, only electrons are transferred, and the protons are left in the intracristal space.

brane is extremely impermeable to many of the metabolites found within the matrix. For example, the ATP formed within the mitochondrion must be transported to the surrounding cytoplasm. The mitochondrial inner membrane proton gradient is believed to provide not only the power to drive ATP synthesis, but also that which allows this ATP to be transported to the surrounding cytoplasm. Other substances, including ions, must be transported across the extremely impermeable inner membrane; again, the proton gradient can provide the free energy for such transport, which presumably takes place via carrier systems in the membrane itself.

PHOTOSYNTHESIS

All living organisms are able to oxidize organic food molecules in the process of respiration, and most do so aerobically, thereby extracting the maximum amount of free energy available. However, the supply of these foods (or fuels) depends almost entirely on the green plants (and some bacteria) that carry out *photosynthesis*. The importance of photosynthesis to life in general and to humankind in particular cannot be overstated. Not only does photosynthesis capture and transform energy, and thereby provide the fuel on which all forms of life depend, it has also been responsible for providing, in the form of fossil fuels, most of the energy reserves currently available to humankind and our increasingly industrialized society. Approximately 840 trillion kilowatt-hours a year are funneled through photosynthesis, several times more than our current world energy consumption. Furthermore, much of the atmospheric oxygen on which higher forms of life depend is produced as a by-product of photosynthesis, liberated by the oxidation of water.

Just as respiration can be considered as the oxidation of carbohydrates to CO_2, so photosynthesis represents the reduction of CO_2 to carbohydrates. The overall process of photosynthesis can be represented by the following equation:

$$nCO_2 + nH_2O + \text{energy} \longrightarrow (CH_2O)n + nO_2$$

Since the free energy level of the carbohydrates formed in photosynthesis is much higher than that of the CO_2, energy must be supplied. The source of energy to drive photosynthesis is, of course, sunlight; photosynthesis, then, involves the absorption and retention of light energy, its conversion into chemical energy, and the storage of this chemical energy in the final products of photosynthesis.

We can consider photosynthesis as the sum of two sets of reactions, one set of which can occur only in the light, while the other set can be carried out in the light and in the dark and depend on light only indirectly. In the *light reactions*, some of the energy of solar radiation is captured and conserved as chemical energy in the form of two high-energy compounds, ATP and NADPH (a reduced electron carrier similar to the NADH formed in respiration). The phosphate-group potential of the ATP and the reducing power of NADPH are then used in the *dark reactions* for the synthesis of carbohydrates from CO_2.

Light Reactions

The formation of ATP and NADPH in the light reactions can be represented by the following equation:

$$ADP + P_i \longrightarrow ATP + H_2O$$

$$NADP + H_2O \longrightarrow NADPH + H^+ + 1/2O_2$$

These reactions, the formation of ATP from ADP and inorganic phosphate, and the transfer of electrons from water to NADP are highly endergonic. We must now consider the way in which light energy is used to drive these uphill reactions.

Chlorophyll

Chlorophyll, the pigment responsible for the green color of plants, is characterized by its capacity to absorb light quanta of particular wavelengths. When light strikes the chlorophyll molecule, one of the electrons in it is raised to a higher energy level, and is said to be in an "excited state." If a solution of chlorophyll extracted from a plant is illuminated it will fluoresce; that is, it will give off light. The light represents the reemission of the absorbed energy as the electron returns to its "ground state." In the cell, however, the chlorophyll molecules are arranged in chloroplast membranes in such a way that as the excited electrons return to their ground state, the energy of excitation is transferred from one chlorophyll molecule to another until it reaches a *reaction center*. The reaction centers represent chlorophyll molecules complexed with specific proteins in such a way that the energy of excitation is sufficient to transfer an electron from the chlorophyll to another kind of molecule. Thus the reaction center chlorophyll is oxidized, and an electron acceptor molecule is reduced, thereby increasing the free energy of the latter. In this way the light energy "pumps" electrons from a relatively low energy level in chlorophyll to a relatively high energy level in the acceptor molecule (Figure 6.9). Although the actual charge separation takes place at the reaction center chlorophyll, the large excess of non-reaction center chlorophyll molecules acts as an "antenna" system, harvesting as much light as possible and funneling the energy to the reaction center.

Two separate photosystems of this kind, known as Photosystems I and II, are used in the light reactions of photosynthesis. In Photosystem I (PS I) the energy of excitation is funneled to the reaction center chlorophyll, where it is sufficient for the excited electrons to be transferred to the electron acceptor ferredoxin. Thus the reaction center chlorophyll of PS I is oxidized and the ferredoxin is reduced. The ferredoxin is, in turn, able to pass the electrons on to NADP, thereby reducing it (Figure 6.10). In this way some of the light energy absorbed by chlorophyll is converted to chemical energy in the form of the reducing power of NADPH.

However, if the process is to continue, the electrons lost from the reaction center chlorophyll of PS I must be replaced. These electrons come from the second

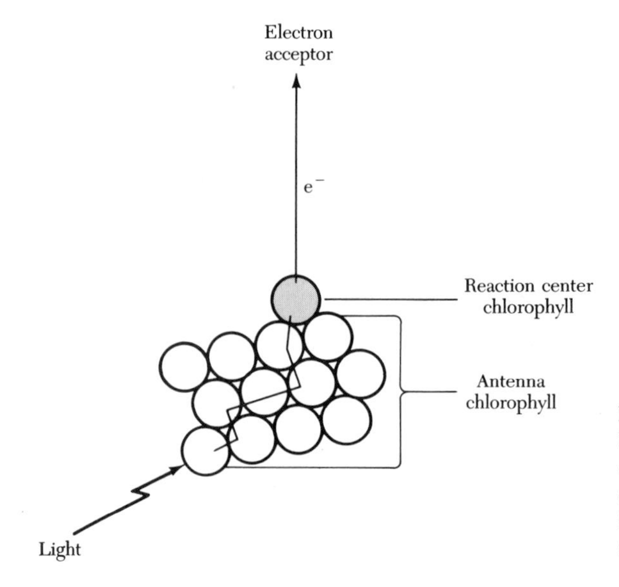

Figure 6.9 Absorption of light energy by antenna chlorophyll. The energy of excitation is transferred to the reaction center chlorophyll, which now has sufficient potential to reduce an electron acceptor.

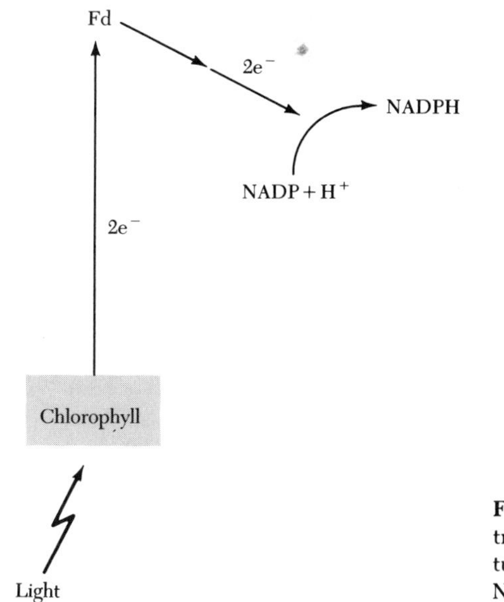

Figure 6.10 Photosystem I. The electron acceptor is ferredoxin, which in turn can reduce NADP, forming NADPH.

Photosystem, PS II, in which light energy is used to pump electrons from the reaction center chlorophyll of PS II to a still-unidentified electron acceptor, called Q. The oxidized chlorophyll of PS II is a strong oxidant and is able to replace its electrons from water, which is oxidized to molecular oxygen (Figure 6.11). The two photosystems are linked in such a way that electrons can be passed from Q, the reduced electron acceptor of PS II, to the oxidized chlorophyll of PS I, via an elec-

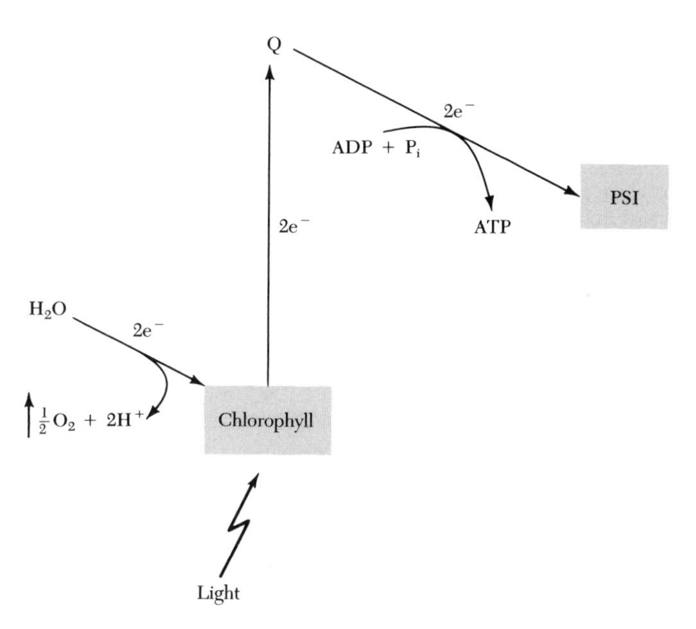

Figure 6.11 Photosystem II. The electron acceptor Q becomes an electron donor when reduced, and passes electrons along an electron transport chain to replace those lost by photosystem I. ATP is formed in the process.

The oxidized chlorophyll is a sufficiently strong oxidant to be able to replace its electrons from water, liberating protons and molecular oxygen.

tron transport system. This system, like that of mitochondria, involves a series of oxidation-reduction reactions in which the electrons are passed to successively lower energy levels before replacing those lost by PS I (Figure 6.11). Some of the energy given up in this series of electron transfer is available for the formation of ATP from ADP and inorganic phosphate in the process of photophosphorylation (see later).

Thus the light reactions of photosynthesis result in the removal of electrons from water and their transfer to NADP (Figure 6.12). The electrons are pumped from a low-energy status in water to a high-energy status in NADPH. Two light-driven steps, mediated by chlorophylls, are involved; these "uphill" steps are connected by a downhill flow of electrons in which energy is made available to form ATP. The flow of electrons from water to NADP is a one-way process and is referred to as *noncyclic* flow. However, if sufficient reducing power in the form of NADPH is already present, the electrons of PS I can be returned from the acceptor to the chlorophyll via electron transport molecules. Such *cyclic* electron flow can also make energy available for the formation of ATP (Figure 6.12).

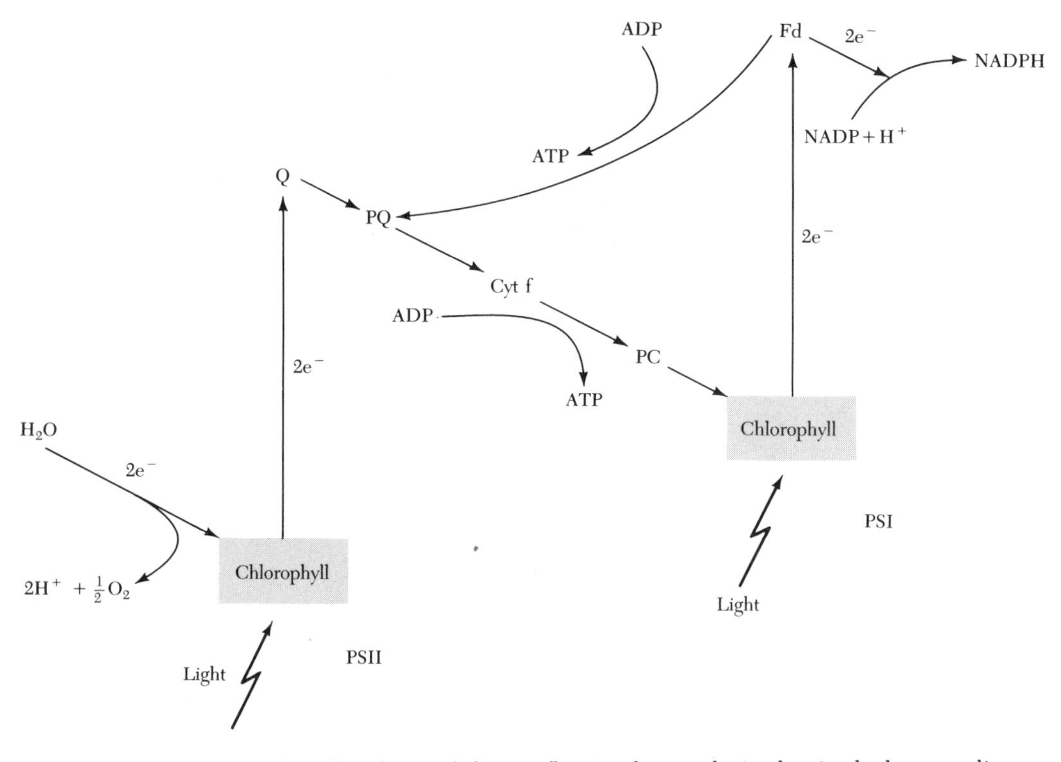

Figure 6.12 Overall pathway of electron flow in photosynthesis, showing both non-cyclic and cyclic electron flow and phosphorylation. PQ = plastoquinone, cyt f = cytochrome f, PC = plastocyanin. While overall the scheme is correct, some details of some of the electron transfer steps and sequences are still unknown.

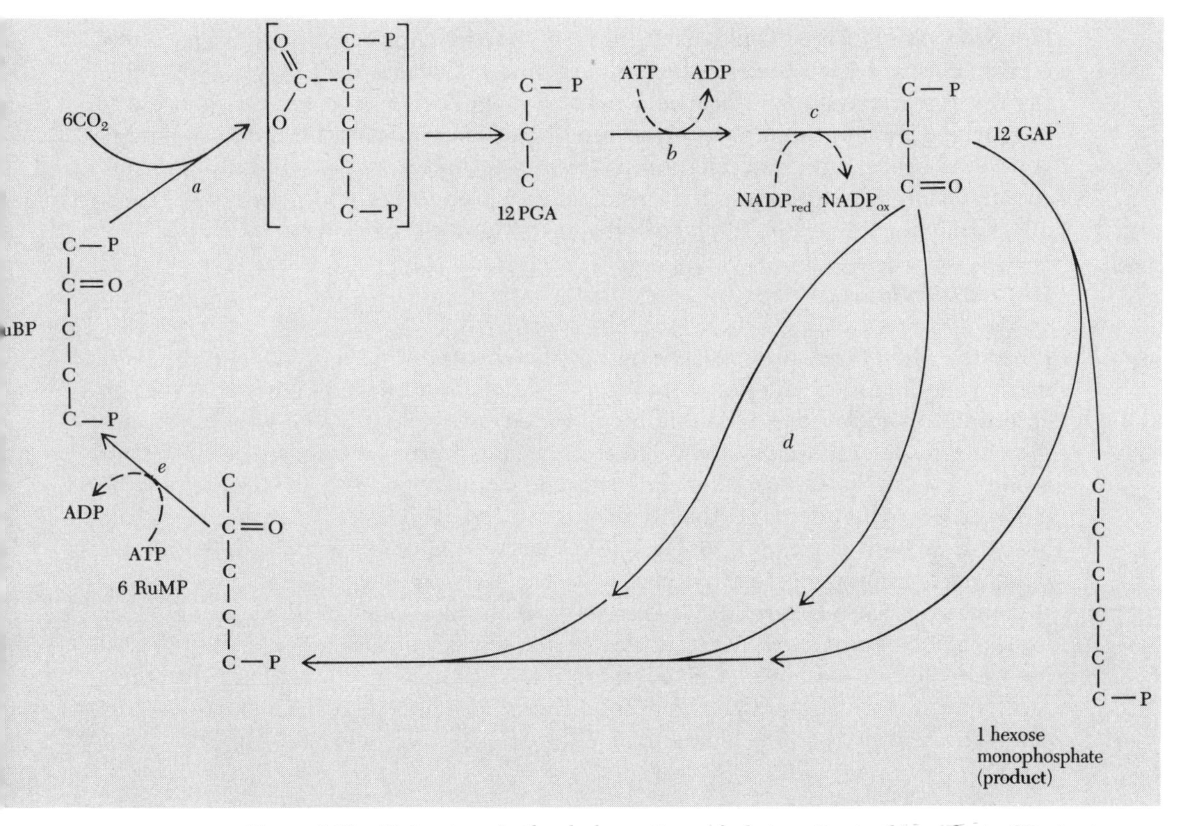

Figure 6.13 Major steps in the dark reactions of photosynthesis. Atmospheric CO_2 is attached to a 5-carbon sugar, ribulosediphosphate (RuBP), in a reaction catalyzed by the enzyme RuBP carboxylase (*a*). The unstable 6-carbon intermediate breaks down to form phosphoglyceric acid (PGA), which is in turn converted to glyceraldehyde phosphate (GAP). In a complex cyclic series of reactions (*d*), the 5-carbon acceptor sugars are regenerated from GAP, and a final product, glucose, is formed. The ATP and reduced NADP produced by the light reactions provide the energy and the electrons necessary for the dark reactions (*b*, *c*, and *e*). (The -H and -OH groups present on the carbon atoms have been omitted for simplification.)

Dark Reactions

The two high-energy compounds, ATP and NADPH, formed in the light reactions provide the energy for the subsequent, non-light-requiring reactions, the reduction of atmospheric CO_2 to sugars. While we need not describe in detail the very complex cyclic series of chemical reactions involved (Figure 6.13), several main features should be recognized. The first step in the fixation and reduction of CO_2 is accomplished by the addition of CO_2 to acceptor molecules of the five-carbon sugar ribulose bisphosphate (RuBP), catalyzed by the enzyme RuBP carboxylase. The resulting three-carbon molecules of phosphoglyceric acid are then reduced, using the ATP and NADPH provided by the light reactions, to phosphoglyceric aldehyde. Most of the PG aldehyde (or GAP) participates in a series of sugar interconversions in which the five-carbon acceptor molecules are regenerated, while the rest is converted to a six-carbon (hexose) sugar, either glucose monophosphate or fructose monophosphate. It is in these dark reactions that the chemical energy generated by the light reactions is stabilized in the hexose sugars. Finally, the hexose sugars are converted into the principal end products of photosynthesis, generally starch or sucrose, in which both the energy and the carbon can be stored.

Chloroplasts

Both the light and dark reactions of photosynthesis take place in discrete, membrane-bound organelles within the cell, the chloroplasts (Figure 6.14). Each chloroplast is enclosed by a double-membraned envelope, inside of which is a complex internal membrane system. These inner membranes are organized into flattened sacs, or *thylakoids*, embedded in the aqueous matrix, or *stroma*, of the chloroplast. In higher plants the thylakoids are arranged in stacks, rather like a pile of coins, as shown in Figure 6.14. These stacked thylakoids constitute the *grana*, which are interconnected by channels enclosed by membranes, the stroma thylakoids (see also Figure 6.22). Thus within the chloroplast itself we see a further example of compartmentalization, the thylakoid space, or *lumen*, being separated by its enclosing membranes from the stroma. Starch grains and lipid bodies, representing storage products of photosynthesis, also may be present (Figure 6.15).

The pattern of development of chloroplasts illustrates the importance of

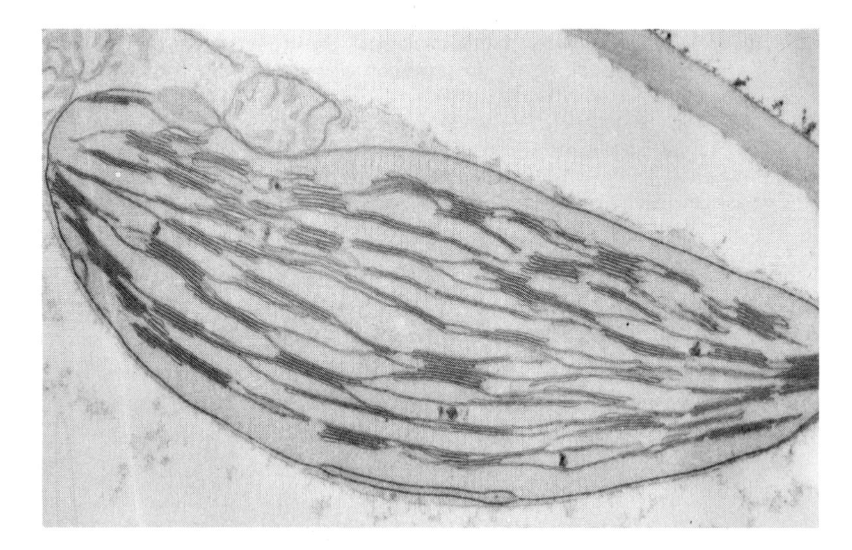

Figure 6.14 Electron micrograph of a chloroplast from the duckweed, *Lemna minor*. An outer membrane surrounds the structure, and the grana do not consist of many layers. Compare with Figure 6.15.

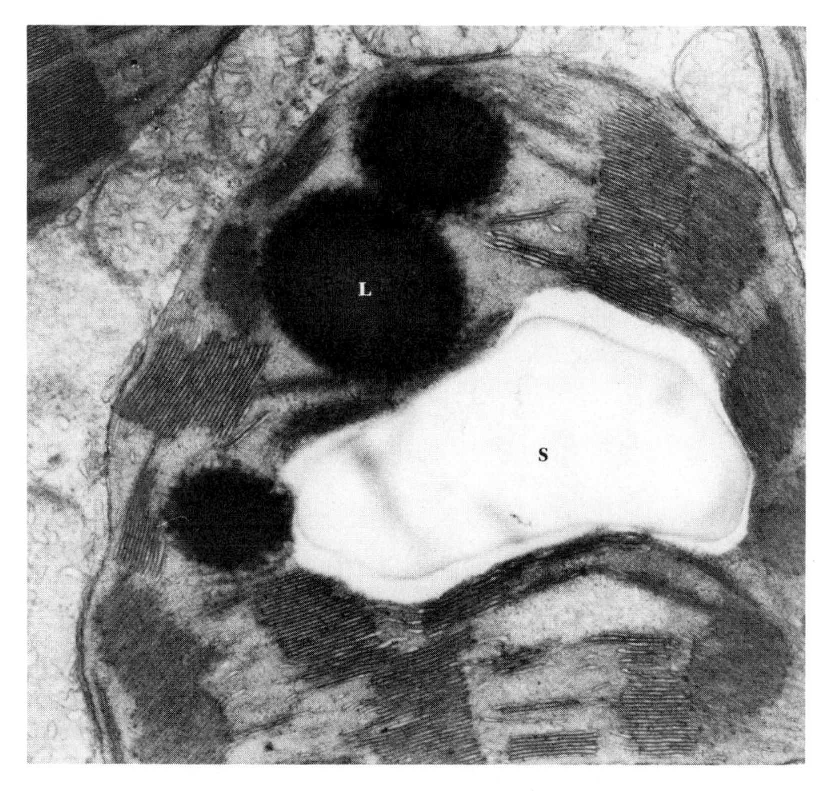

Figure 6.15 Portion of a chloroplast of a leaf of the rubber plant, *Ficus elastica*, showing grana, starch grain (S), and lipid bodies (L). (Courtesy of Dr. J. G. Duckett.)

thylakoid organization in photosynthesis, and the importance of both light and chlorophyll to that organization. Chloroplasts (as well as other types of plastids) develop from *proplastids*, small, double-membrane-bound structures seen in the immature cells of very young leaves. In the presence of light, proplastids develop into normal chloroplasts; invaginations of the inner membrane form thylakoids, which subsequently become organized into functional grana. At the same time, synthesis of chlorophyll and the rest of the photosynthetic machinery takes place. In the dark, however, no chlorophyll is synthesized and very few thylakoids form; instead, an aggregate of membraneous tubules, the prolamellar body, appears (Figure 6.16). When such dark-grown cells are exposed to light, chlorophyll is formed, and the prolamellar body is converted into functional thylakoids, which themselves form grana.

Chloroplasts may assume many forms and vary widely in number per cell in different plants. In some algae, such as the filamentous *Spirogyra*, only a single spiral chloroplast is present in each cell; when the cell divides, so also does the chloroplast. In contrast, a cell in the photosynthesizing part of a grass leaf may

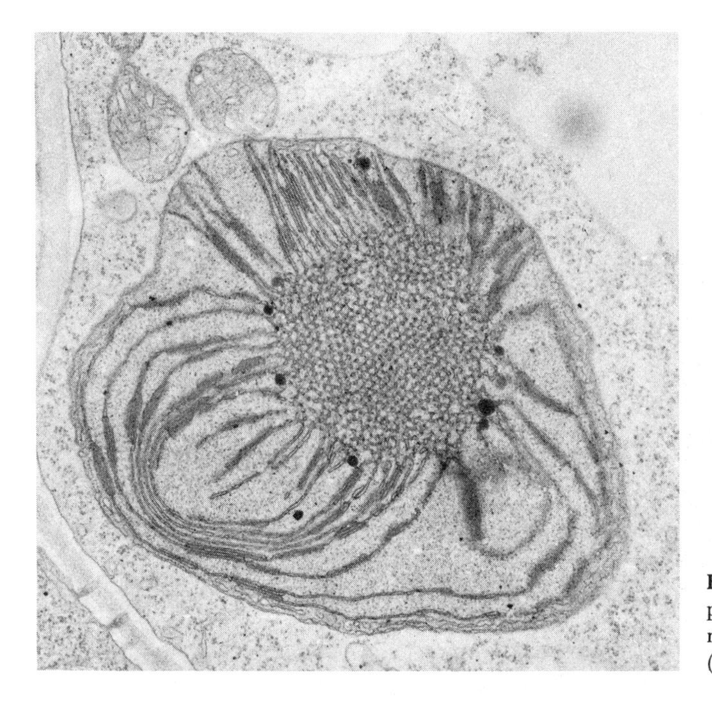

Figure 6.16 Electron micrograph of prolamellar body. The photosynthetic membranes are beginning to form. (Courtesy of Dr. C. L. F. Woodcock.)

have 30 to 50 chloroplasts; their division, which occurs in the immature, or proplastid, state, is not correlated with cell division in any exact way. The stacked grana are missing in some chloroplasts, as in some red and brown algae, to be replaced by long membranes running the length of the chloroplast, but these presumably function in the same manner as the grana. The blue-green algae, or *Cyanobacteria*, on the other hand, lack definite chloroplasts; instead they possess loosely arranged membranes in the cytoplasm on which the photosynthetic pigments are layered (Figure 6.17). In some bacterial cells vacuolelike chromatophores (Figure 6.18), which are bounded by membranes, are the photosynthetic units. Other bacteria may have beautifully organized layers of membranes comparable to grana, but as in that depicted in Figure 6.19, the membranes are chemosynthetic rather than photosynthetic, the energy for synthesis being derived from the breakdown of sulfur compounds.

Not all plastids contain chlorophyll and function photosynthetically. Some, as in the potato tuber, are for starch storage (Figure 6.20); others may contain oil or protein. These lack the lamellar construction of the chloroplast. However, they, too, are derived from proplastids, and what each becomes is related to the kind of cell in which it is found at maturity. For example, starch-containing plastids found in root cap cells (Figure 6.21) are thought to be involved in the perception of gravity, to which most roots respond. Several such structural and functional modifications of plastids are possible, but it is only in the chlorophyll-containing, thylakoid type of organization of the chloroplast that occur the reactions of photosynthesis,

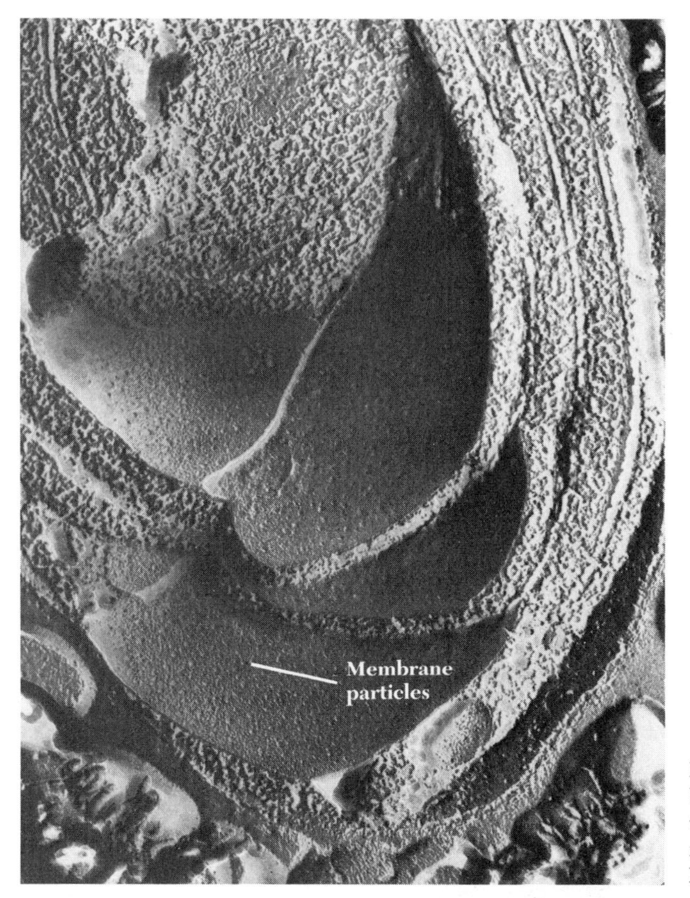

Figure 6.17 Electron micrograph of a blue-green algal cell showing photosynthetic membranes. Particles in the membranes can be seen. (Courtesy of Dr. S. C. Holt.)

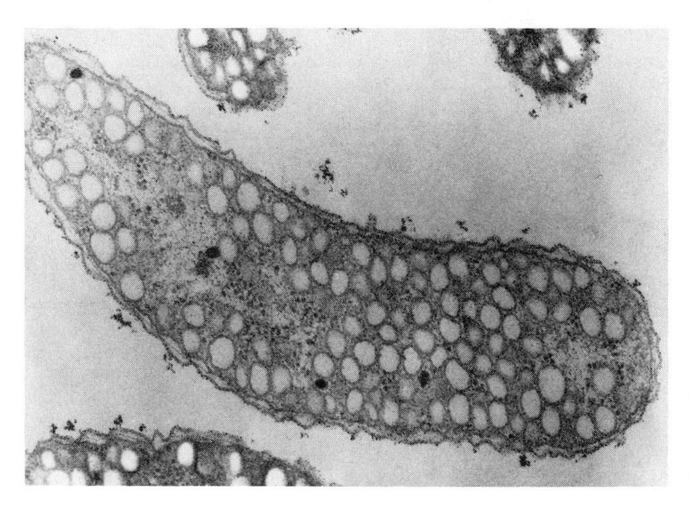

Figure 6.18 Electron micrograph of photosynthetic bacterium showing membrane-enclosed chromatophores. (Courtesy of Dr. S. C. Holt.)

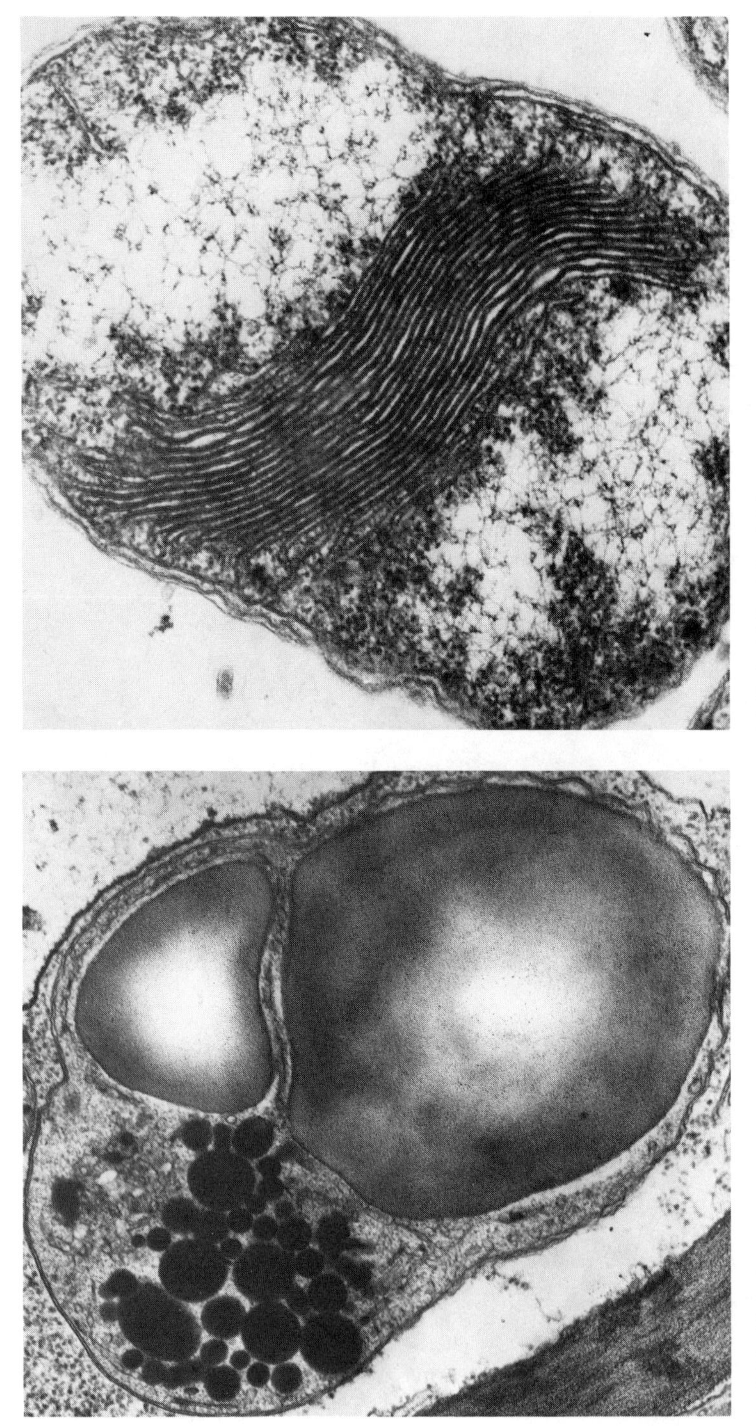

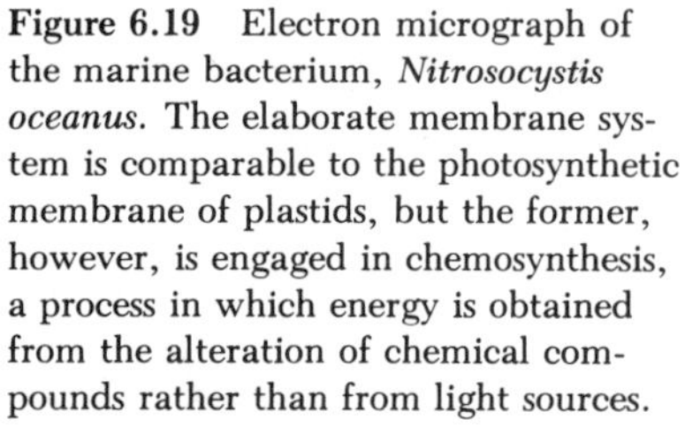

Figure 6.19 Electron micrograph of the marine bacterium, *Nitrosocystis oceanus*. The elaborate membrane system is comparable to the photosynthetic membrane of plastids, but the former, however, is engaged in chemosynthesis, a process in which energy is obtained from the alteration of chemical compounds rather than from light sources.

Figure 6.20 A starch-storage plastid from the tuber of the sweet potato. The large light masses are stored starch; no grana are present.

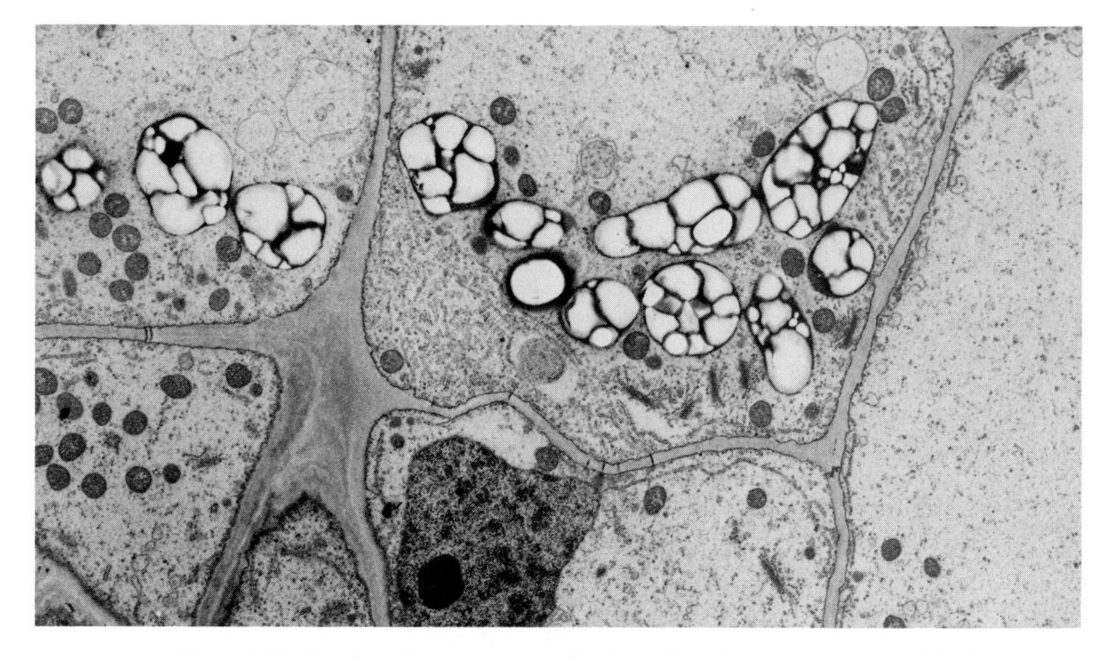

Figure 6.21 Starch grains in root cap cells of pea. Note the orientation toward the bottom of the cells.

the process which allows life to reverse temporarily the trend toward thermodynamic equilibrium.

Thylakoid membranes. It is relatively easy to fractionate chloroplasts by differential centrifugation; in this way it can be shown that while the enzymes that catalyze the dark reactions are localized in the stroma, the membrane fraction contains the chlorophyll and is capable of carrying out the light reactions of photosynthesis. The molecular composition and organization of the thylakoid membrane, therefore, must be such that Photosystems I and II, the electron transport chain, and the enzymes necessary for the formation of ATP and reduced NADPH are capable of functioning in the membrane in an integrated manner.

Various aggregates of proteins can be extracted from thylakoid membranes and shown to carry out different parts of the overall sequence of events that constitute the light reactions. One of these complexes also contains much of the thylakoid chlorophyll and represents the light-harvesting system in which the energy of excitation is funneled to the reaction center chlorophyll. Two others, which contain the rest of the chlorophyll, display Photosystem I and Photosystem II activities and include the respective reaction centers. Yet another, consisting of several iron-containing electron carriers, including cytochromes, appears to represent the electron transport system through which electrons are transferred from plastoquinone to plastocyanin, that is, from PS II to PS I. Finally, a coupling factor complex, which can synthesize ATP from ADP, has been isolated from the thylakoid membranes.

Electron microscopic examination of freeze-fractured and freeze-etched thylakoid membranes clearly demonstrates their structural complexity. As we have already mentioned, this technique provides not only surface views of membranes, but also representations of internal organization, since the membranes can be "cleaved" along the plane of the relatively weak bonds that hold the two lipid layers of the membrane together.

Two major size classes of particles are embedded within the internal membrane of the chloroplast. The larger class, although spanning most of the width of the membrane, is more closely associated with the side of the membrane facing the lumen, while the smaller particles are oriented toward the stromal matrix (Figure 6.22). This structural asymmetry of the photosynthetic membrane is related to a corresponding functional asymmetry; it is known that the reaction centers of the photosystems are oriented toward the lumen, while the electron acceptor complexes face the outer side. Thus electron flow in the light reactions, from water to NADPH, is vectorial in that electrons are transported from the inner lumen to the surrounding matrix, while in the electron transport system connecting the two photosystems, the electrons move in an outside-to-inside direction (Figure 6.23).

Whatever the exact nature of the two types of proteinaceous particles, they appear to be associated with the two separate photosystems. The appressed mem-

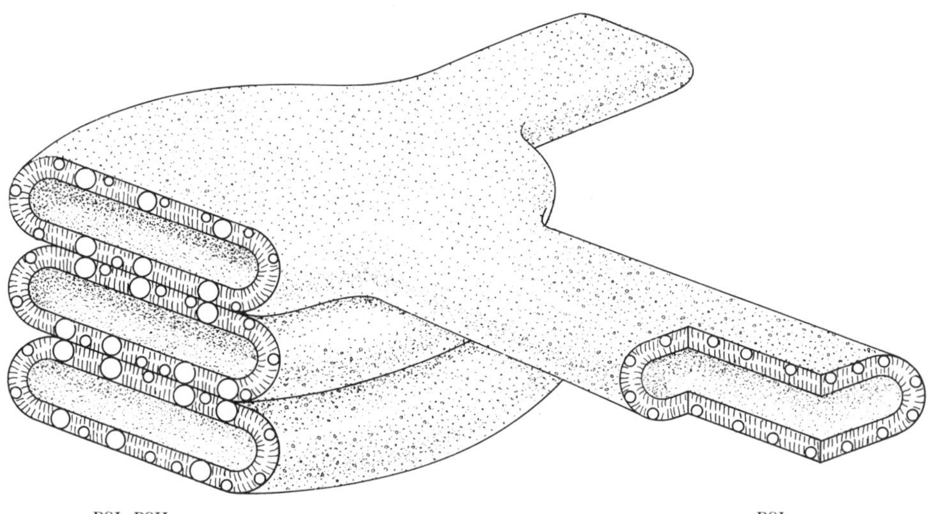

Figure 6.22 Model of the structural organization of grana thylakoids (stacked) and stroma/thylakoids (unstacked). The different particles in the membrane represent different complexes of the light-harvesting and electron transfer components of the light reactions of photosynthesis.

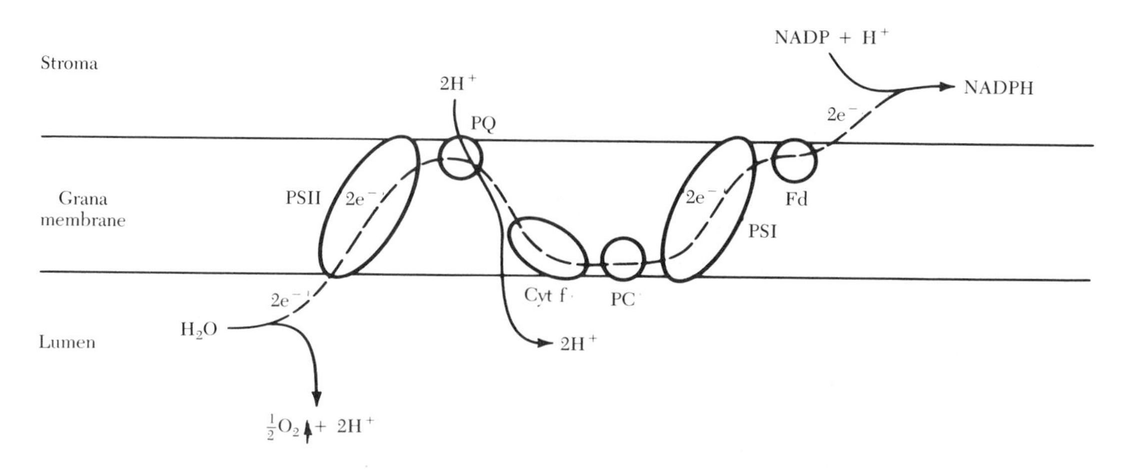

Figure 6.23 Possible arrangement of photosystems I and II and the electron transport chain. The directional flow of electrons can generate a proton gradient across the membrane by increasing the H^+ concentration in the lumen over that in the stroma matrix.

branes of the grana, which contain both types of particle, can carry out both Photosystem I and Photosystem II, while the stroma thylakoids, which lack the large particles, show only Photosystem I and not Photosystem II activity. Indeed, the stacking phenomenon itself may depend on the presence, in the grana thylakoid membranes, of the light-harvesting chlorophyll-protein complexes of Photosystem II, which have been shown to mediate adhesion between membranes.

Another class of particles is present on the outside surface of the unstacked membranes. These are similar in size and appearance to the stalked particles on the matrix surface of the mitochondrial inner membrane and, like these particles, represent the ATP-forming complex of the organelle. The absence of the ATP-synthesizing enzymes from the appressed faces of the grana thylakoid membranes shows that the site of ATP formation in the chloroplast can be distinct from that of the electron transport system, a possibility allowed for in the chemiosmotic hypothesis of ATP formation.

Photophosphorylation. Photophosphorylation is the process by which energy is made available for ATP synthesis as electrons pass to successively lower energy levels in the oxidation-reduction reactions of the electron transport system in thylakoid membranes. In noncyclic photophosphorylation the energy comes from electron flow between Photosystem II and Photosystem I, while in cyclic photophosphorylation, energy is provided as electrons from Photosystem I, instead of being used to form NADPH, are returned from the electron acceptor to the reaction center chlorophyll of Photosystem I (Figure 6.12).

As we saw to be the case for oxidative phosphorylation in mitochondria, the coupling of electron transport to ATP formation in photosynthesis is not direct, but is mediated by an electron transport-generated proton gradient across a membrane, in this case the thylakoid membrane of the chloroplast. Two ways by which the proton gradient is established are illustrated in Figure 6.23, and can be seen to depend on the asymmetrical arrangement of the photosynthetic apparatus within the membrane. The enzyme system which oxidizes water to provide electrons for Photosystem II is located at the inner (lumen) side of the thylakoid membrane. For each molecule of water oxidized, therefore, two protons are released into the lumen. Similarly, the reduction of NADP by the electron acceptor of Photosystem I occurs at the outer, or matrix, surface of the membrane. Since this reduction requires two electrons and one proton, a proton is taken up from the stromal matrix. The second way in which the proton concentration inside the lumen is increased relative to that in the matrix is in one of the steps in the electron transport system. The reduction of plastoquinone by Photosystem II requires not only two electrons from Photosystem II, but also two protons, which must be taken up from the stroma. The subsequent reduction of the cytochromes by plastoquinone, however, does not require these protons, which therefore can be released into the lumen. The cyclic flow of electrons associated with Photosystem I (Figure 6.12) can also contribute in this way to the establishment of the proton gradient across the membrane.

The free energy represented by the proton gradient can be used in the formation of ATP. The mechanism by which this is accomplished is very similar to that described for mitochondria. As the protons are permitted to flow back into the stroma through proton gates in the membrane, the energy is made available for ATP synthesis, again occurring at the ATP-forming particles attached to the stroma surface of the thylakoid membranes (Figure 6.24).

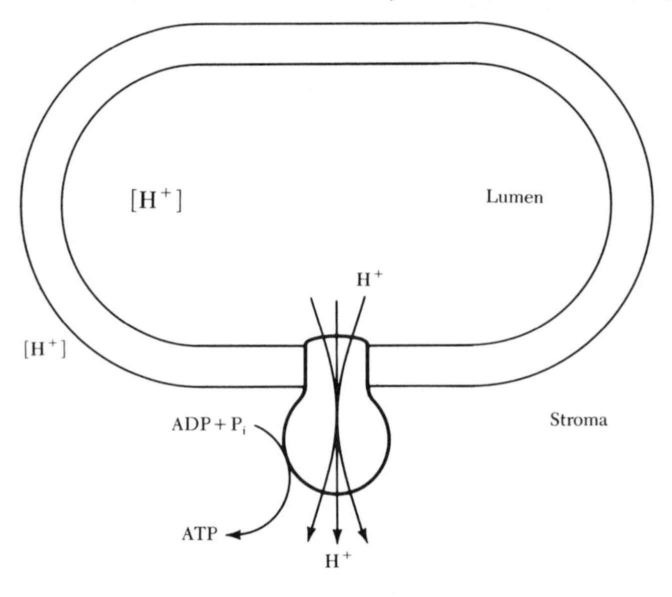

Figure 6.24 The energy of the proton chemiosmotic gradient generated by electron transport can be harnessed as protons flow back across the membrane via the ATP-synthesizing complexes in the membrane.

AUTONOMY OF MITOCHONDRIA AND CHLOROPLASTS

These energy-converting organelles are integral components of eukaryotic cells, but certain features indicate that they display a considerable amount of structural and functional independence. For example, both their internal membrane systems, which are organized in specific and orderly ways to carry out specific and vital reactions of making energy available to the cell, and their aqueous matrices are separated from the rest of the cytoplasm by other membranes. Furthermore, neither mitochondria nor chloroplasts arise *de novo*, but from growth and division of other mitochondria and plastids. That mitochondria and chloroplasts might also contain the informational machinery to determine at least certain aspects of their own development and activity was first suggested on the basis of inheritance patterns of organelle traits. Such patterns were consistent with the view that the traits, unlike those showing Mendelian inheritance, were under the control of nonnuclear genes.

We now know that indeed mitochondria and chloroplasts contain a genetic system similar, but not identical, to that of the rest of the cell. Both organelles contain DNA molecules that are usually closed circular double helices. Protein-synthesizing apparatus, including ribosomal RNA and ribosomes, transfer RNA, and activating enzymes, is also present. The DNA replicates in a semiconservative manner, and messenger RNA transcribed from the DNA is translated on the organelle ribosomes. The proteins coded for by the organelle DNA include some of the critical enzymes of photosynthesis and respiration.

Chloroplasts and mitochondria, however, are not totally autonomous in this respect; indeed, most photosynthetic and respiratory enzymes are coded for by nuclear DNA and synthesized on cytoplasmic ribosomes. The close integration between the organelle and nonorganelle genetic systems is illustrated by the ATP-synthesizing enzyme complex of mitochondria. This complex, which, as we pointed out earlier, couples the proton gradient established by electron transport to the formation of ATP, consists of several different subunits; some of these subunits are coded for by mitochondrial DNA and synthesized on mitochondrial ribosomes, while others are coded for by nuclear genes, synthesized on cytoplasmic ribosomes, and transported into the mitochondrion. Some of the other organelle products known to be under organelle control are listed in Table 6.1.

The process by which mitochondrial and organelle proteins made on cytoplasmic ribosomes enter the organelles appears to be similar in principle to that by which proteins enter the endoplasmic reticulum (see Chapter 7). For example, the smaller of the subunits of RuBP carboxylase, the enzyme involved in fixing CO_2 in the dark reactions of photosynthesis, is synthesized in the cytoplasm, the other in the chloroplast. The small subunit is formed, on free ribosomes in the cytoplasm, as a polypeptide with a signal sequence of amino acids at one end. Once the completed polypeptide is released from the ribosomes, the signal sequence is somehow

TABLE 6.1 ORGANELLE COMPONENTS KNOWN TO BE CODED FOR BY ORGANELLE DNA.

Mitochondria	Chloroplasts
Ribosomal RNAs:	Ribosomal RNAs:
12S, 16S—mammals, yeast	5S, 16S, 23S
5S, 18S, 26S—plants	
Transfer RNAs	Transfer RNAs
Cytochrome oxidase subunits	Cytochrome oxidase subunits
Cytochrome *b*	ATP synthase subunits
ATP synthase subunits	RuBP carboxylase large subunit
Several (8–15) unidentified proteins	Several to many unidentified proteins

recognized by the chloroplast and allows the polypeptide to be transported across the chloroplast membrane. Once inside, the signal sequence is removed and the functional RuBP carboxylase forms by association of the two subunits.

Both the degree of autonomy exhibited by chloroplasts and mitochondria, and the structural and biochemical features they share with blue-green algae and bacteria, have led to the suggestion of a symbiotic origin of these organelles; that is, that they are actually invaders that have become not only established in but also essential to the host cell. During the course of evolution, according to this hypothesis, their role in the symbiotic association has become refined to the point where their function is now crucial. Whether this is correct or if the organelles have some different evolutionary history (see Chapter 12), it is clear that a fine and coordinated balance of development and activity exists between the energy transducers and the rest of the cell.

ADDITIONAL READING

BASSHAM, J. A. 1962. The path of carbon in photosynthesis. *Sci. Am. 206*(6): 88–100.

GOVINDJEE, and GOVINDJEE, R. 1974. The absorption of light in photosynthesis. *Sci. Am. 231*(6): 68–82.

GRIVELL, L. A. 1983. Mitochondrial DNA. *Sci. Am. 248*(3): 78–89.

HINKLE, P. C., and MCCARTY, R. E. 1978. How cells make ATP. *Sci. Am. 238*(3): 104–123.

LEHNINGER, A. L. 1971. *Bioenergetics*, 2nd ed. W. A. Benjamin, Inc., Menlo Park, Calif.

MALONEY, P. C. 1982. Energy coupling to ATP synthesis by the proton-translocating ATPase. *J. Membrane Biol. 67*: 1–12.

MCELROY, W. D. 1971. *Cell Physiology and Biochemistry*, 3rd ed. Prentice-Hall, Inc., Englewood Cliffs, N.J.

STRYER, L. 1981. *Biochemistry*, 2nd ed. W. H. Freeman and Company, Publishers, San Francisco.

7

The Endomembrane System

As we saw earlier, the information necessary for the manipulation of energy in directed metabolism flows through the enzymes of the cell. Each cell has many thousands of enzymes, catalyzing the various reactions involved in the activities in which the available energy is utilized. These reactions, which are carried out in a controlled and precise manner, do not occur in a homogeneous milieu, but in one that is highly structured and compartmentalized. In discussions of the nucleus, the plasma membrane, and the chloroplasts and mitochondria, the components whose activities confer on the cell its characteristic autonomy as the unit of life, we saw how the structural basis for these activities represents a high degree of molecular organization and of compartmentation. Such structural complexity, which certainly requires information and energy for its maintenance, is itself necessary for the manipulation of information and energy.

Eukaryotic cells also contain several interrelated membrane-bound compartments, known collectively as the *endomembrane system*, that play both an architectural and a biochemical role in many cellular activities. Once again, the significance of the ubiquitous type of macromolecular organization represented by membranes lies in its capacity to restrict diffusion, and therefore to compartmentalize the cell, and to provide surfaces in which specific orientation of multienzyme systems can be maintained and on which specific types of reactions can be localized.

While the membranes of the nuclear envelope can be considered as part of the endomembrane system, the major components we shall discuss here are the *endoplasmic reticulum* and the *Golgi apparatus*, along with the other membrane-delimited compartments derived from them. The latter include small membranous vesicles which serve various roles in the cell: exocytotic vesicles destined for the plasma membrane, where they can contribute material to that membrane as well as depositing their contents outside the cell; *lysosomes*, which serve to compartmentalize specific types of enzymes; and various storage compartments for proteins which subsequently may be secreted or mobilized for internal use. As we shall see, the endomembrane system functions as a distribution system, with proteins synthesized in association with the endoplasmic reticulum being sorted and packaged, usually but not always while in transit through the Golgi apparatus, and delivered to their ultimate destinations in the cell.

ENDOPLASMIC RETICULUM

The endoplasmic reticulum (ER) consists of a system of cytoplasmic membranes, often continuous with the outer membrane of the nuclear envelope (Figure 7.1), that delimit a series of often interconnected regions of the cell. This vacuolar system, which can vary considerably in amount from cell to cell, can also assume various configurations within the cell, forming flattened sacs (cisternae), tubules, and vesicles (Figure 7.2). The ER is also classified according to whether ribosomes are attached to the outer surfaces of the membranes (rough ER) or not (smooth ER). Thus the ER can be thought of as a relatively labile membrane system that can display intracellular differentiation, becoming modified in different ways to facilitate the various activities associated with it.

Rough ER

Rough ER (Figure 7.3) is particularly abundant in cells that are actively synthesizing proteins, which might be expected in view of what is known of the role of the ribosomes in protein synthesis (see Chapter 4); however, ribosomes are also present

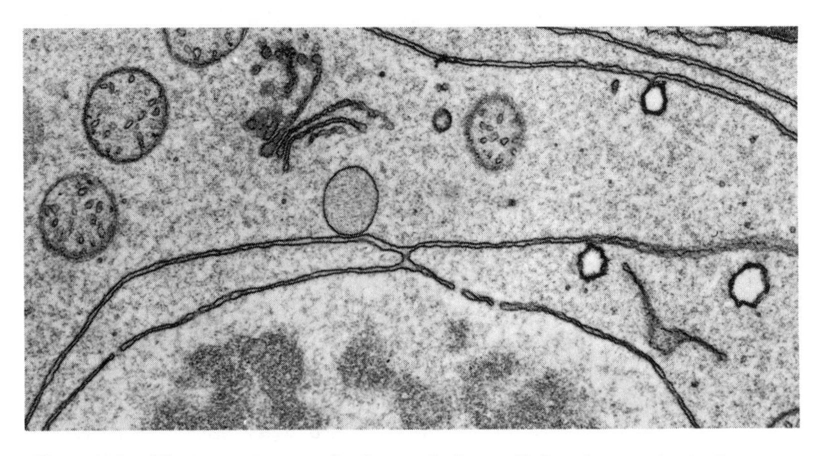

Figure 7.1 Electron micrograph of part of plant cell showing continuity between the outer membrane of the nuclear envelope and the ER.

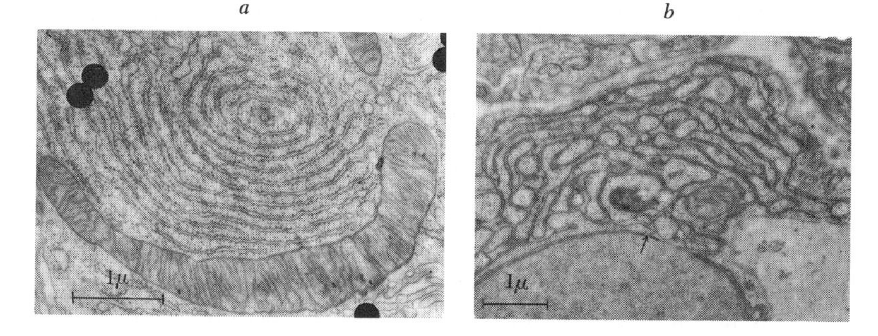

Figure 7.2 The endoplasmic reticulum (ER) in parotid (salivary-gland) acinous cells of the mouse. (*a*) The ER above the mitochondrion is of the rough, or granular, variety, containing ribosomes, and is much more highly organized than in the area immediately below; many of the membranes end blindly in the cytoplasm. (*b*) The mass of rough ER membranes is a continuously branching and interconnected system, which is also connected with the outer portion of the nuclear membrane (arrow).

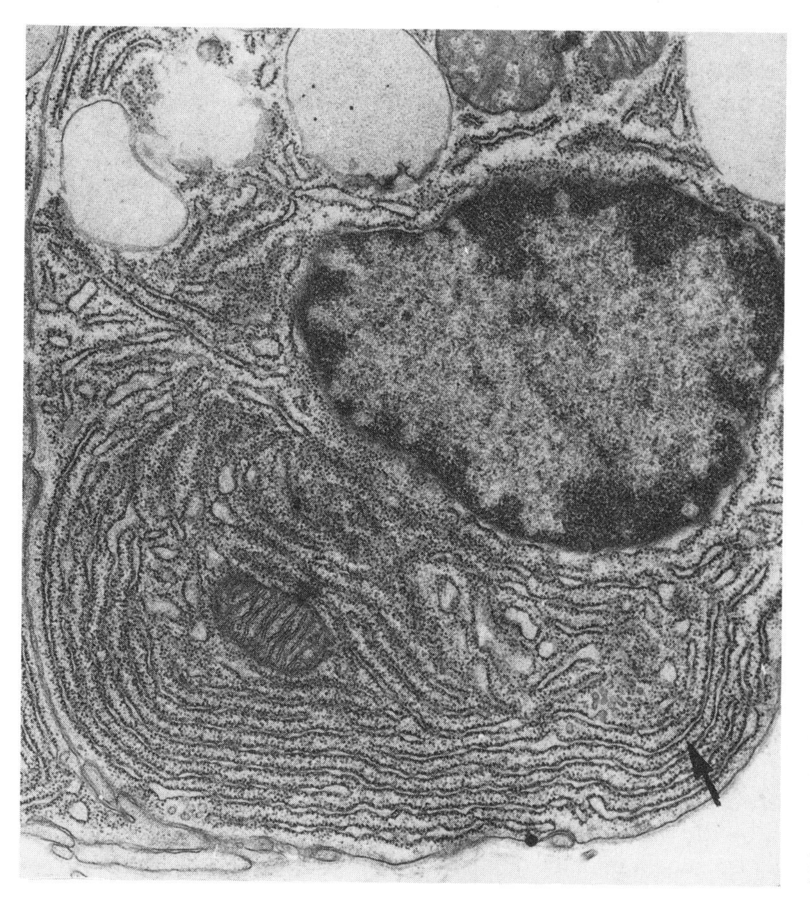

Figure 7.3 Electron micrograph of portions of "chief" cells of the bat, showing the extensive rough ER. The ER forms flattened saclike arrangements, with the ribosomes (arrow) attached to the outer surfaces of the membranes enclosing the sacs, or cisternae. These cells are active in synthesis and secretion of the digestive enzyme, pepsin. (Courtesy of Dr. K. Porter.)

in the cytoplasm free of any associations with membranes. The bound ribosomes, which are often arranged to form spiral polysomes, are attached to the ER membrane by their large subunits.

At least part of the significance of the attachment of ribosomes to the ER is related to the types of proteins synthesized on these ribosomes and their subsequent fate and distribution within the cell. Among such proteins are those that are destined to be secreted from the cell, such as extracellular matrix proteins, digestive enzymes, and hormones; others, for example hormone receptors and antigens, become integral proteins of the plasma membrane. Yet others, to be used elsewhere in the cell, end up in membrane-bound vesicles of the endomembrane system, such as lysosomes (see later). All of these types of protein are bound for specific destinations within the cell; also, they must be modified in various ways before reaching these destinations. Synthesis of these proteins on rough ER represents the first step in a system of communication that, as we shall see, permits segregation of the products formed, their modification, and their delivery to the sites where they will ultimately function.

Much of what we know of these subsequent steps comes from work on both secreted and integral membrane proteins of animal cells; although the following description is based on this work, it is likely that similar, if not identical, mechanisms operate in determining the fate of other types of ER-synthesized proteins.

The way of ensuring that the proteins to be delivered to the endomembrane system are synthesized on ER-bound ribosomes and then incorporated into the ER depends on a built-in feature of the proteins themselves. The messenger RNAs coding for the proteins first associate with free ribosomes, and attachment of the polyribosome complex to the membranes takes place only after translation has begun. The first-formed 15 to 29 amino acid sequence of the growing polypeptide chain acts as a *signal sequence. Signal recognition particles,* consisting of ribonucleoproteins, are believed to bind to these sequences and to receptors in the ER membrane; the hydrophobic signal sequence is then inserted into the membrane, and the ribosomes bind to the membrane surface (Figure 7.4). As synthesis of the protein proceeds, the signal sequence passes into the lumen of the ER, and the elongating chain is somehow threaded into the membrane. It is not yet clear if specific proteins in the ER membrane are responsible for guiding the nascent polypeptide chains into the membrane, or if conformational changes resulting from interactions between hydrophobic and hydrophilic regions of the chains and the membrane drive the process spontaneously.

In the case of the secreted proteins, the completed polypeptide chain passes right through the membrane and is released into the lumen, while the integral proteins remain embedded, in their proper conformation, within the membrane (Figure 7.4). In both cases, however, the leading signal sequence peptide, no longer necessary for the distribution or functioning of the protein, is enzymatically removed in the lumen by the time synthesis of the protein is completed.

The ER itself is also biochemically competent to participate in other post-

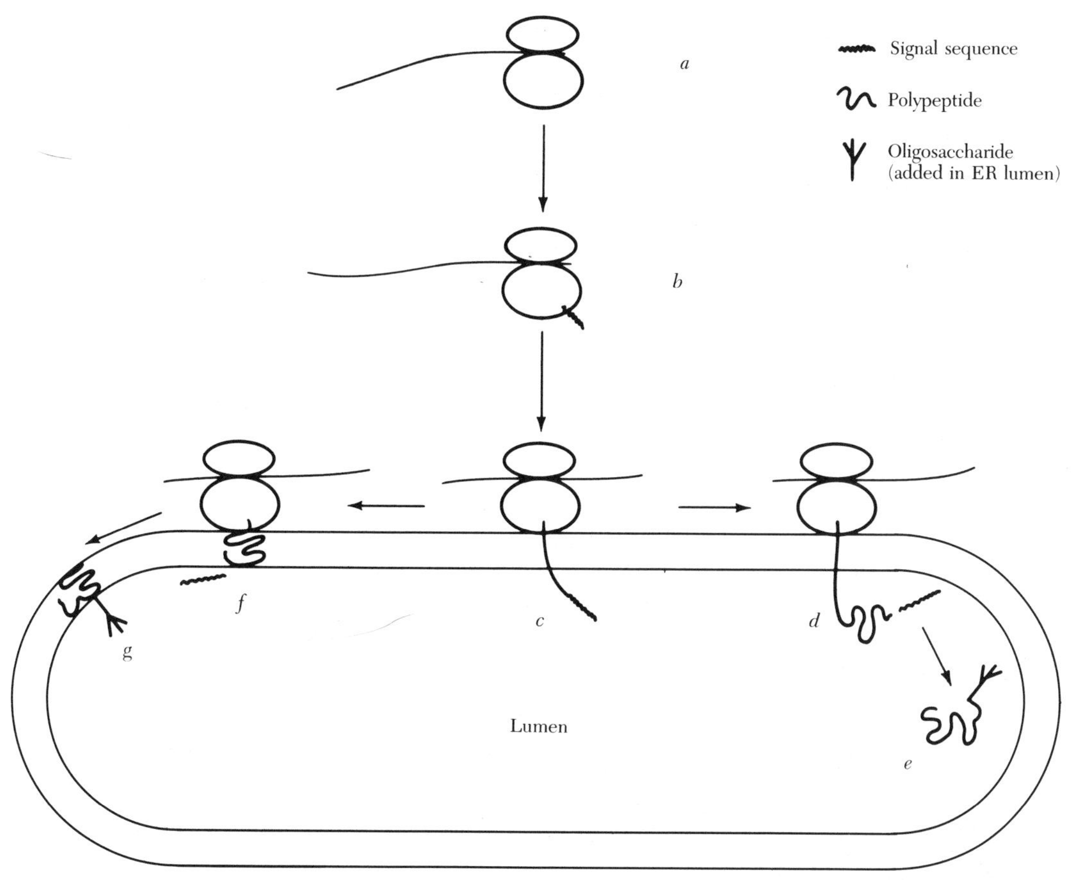

Figure 7.4 Signal hypothesis: (*a*) the mRNA attaches to the ribosome; (*b*) the sequence of the first few amino acids coded for is synthesized; (*c*) the signal sequence is recognized and attached to the membrane of the ER, along with the ribosome. (*d*) As synthesis of secretory protein proceeds, the signal peptide is removed, and the protein is inserted into the lumen of the ER (*e*). (*f*) and (*g*) Synthesis of integral membrane proteins by a similar pathway results in the proteins being inserted into the ER membrane.

translational modifications of proteins. Many secreted and membrane proteins are glycoproteins; that is, they have chains of sugar molecules attached to them. Part of this glycosylation process takes place in the ER while the proteins are being synthesized, and the enzymes involved are associated with the ER membranes. In at least one class of glycosylations, the sugars are first assembled into oligosaccharides by attachment to one of the lipids in the ER membrane. The 14-sugar oligosaccharides are subsequently transferred within the lumen from the lipid carrier to one of the asparagines in the growing polypeptide chain.

Once the proteins have been sequestered in the lumen or in the membrane of the ER, they must be transferred to the plasma membrane. By following the fate of the proteins synthesized on the ER, it has been possible to show that many of them first pass from the ER to the other major component of the endomembrane system, the Golgi apparatus. Small vesicles, called *transition vesicles*, are often seen lying between the ER and the Golgi; these are believed to be "budded off" from the ER and to represent the path by which both the membrane and the secretory proteins are delivered to the Golgi.

As we have already mentioned, not all proteins synthesized on ER-attached ribosomes are secretory or plasma membrane proteins. Furthermore, although transfer of proteins from the ER to the Golgi apparatus can be demonstrated for these and other proteins, this may not always be so. The subsequent destination of ER-synthesized proteins will be discussed shortly, as well as how they may be sorted and forwarded to these destinations. First, however, we shall consider some of the other functions associated with the endoplasmic reticulum.

Smooth ER

Smooth ER (Figure 7.5), whose membranes usually form tubular elements free of attached ribosomes, has also been implicated in a range of functions, including synthesis, storage, degradation, and transport of materials.

The spatial continuity and morphological similarities between membranes of rough and smooth ER systems suggest that these may be parts of the same basic membrane system. There is also evidence that smooth ER may be derived from rough ER, the constituent proteins of the former being synthesized on ribosomes that become detached once the protein has been asserted into the membrane.

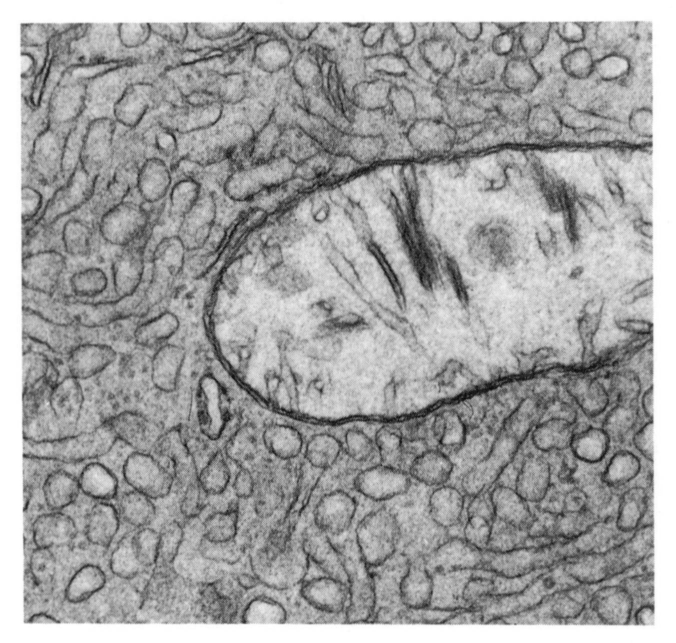

Figure 7.5 Electron micrograph of a portion of a cell from the testis of an opossum, showing the smooth, or agranular, form of ER.

Among the cell types in which smooth ER is prevalent are those that synthesize and secrete steroid hormones, such as the cells of the testis and of the adrenal gland, and those in which stored glycogen is broken down to glucose, such as liver cells. The respective enzymes involved in these activities can be localized in the smooth ER membranes of the cells. Smooth ER in liver cells also contains an electron transfer system that appears to be involved in detoxification; administration of certain drugs can result in dramatic increases in both the amount of smooth ER and of the detoxifying enzymes. Even within one cell the homogeneous appearance of the smooth ER in electron micrographs may mask a considerable amount of functional heterogeneity.

Another function associated with smooth ER is the sequestering and release of calcium ions, thereby regulating Ca^{2+} levels in different parts of the cell. This activity is most clearly demonstrated in a specialized form of smooth ER, the *sarcoplasmic reticulum* of skeletal muscle cells. Following stimulation by a nerve impulse, the sarcoplasmic reticulum, the lumen of which contains very high Ca^{2+} levels, becomes permeable and releases the Ca^{2+} to the cell, where it plays a critical role in the regulation of muscle contraction (see Chapter 8). Calcium-rich smooth ER has also been detected in nonmuscle cells, and implicated in other Ca^{2+}-regulated processes, such as chromosome movement during cell division (Figure 7.6), vesicle fusion during cell plate formation in plant cells (Figure 7.7), and light adaptation in photoreceptor cells.

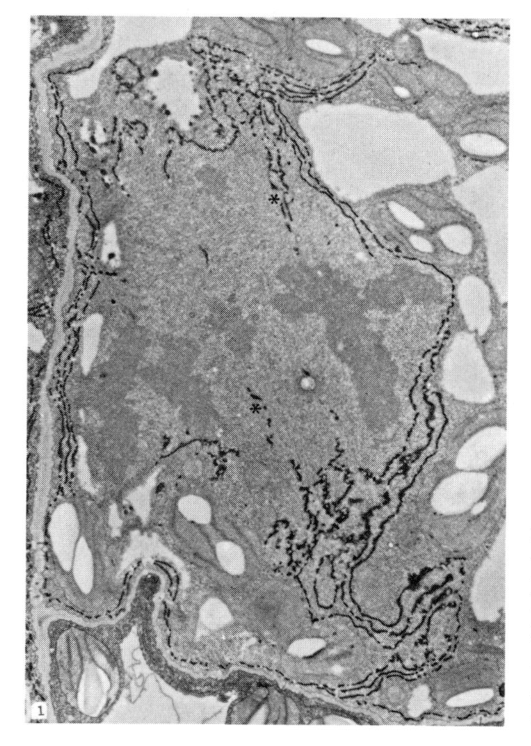

Figure 7.6 Smooth ER associated with the mitotic apparatus in a dividing cell. Darkly-stained ER is seen at the spindle poles and extending into the spindle toward the chromosomes (*). The ER is thought to contain Ca^{++} involved in spindle function (see Chapter 9). (Courtesy of Dr. P. K. Hepler.)

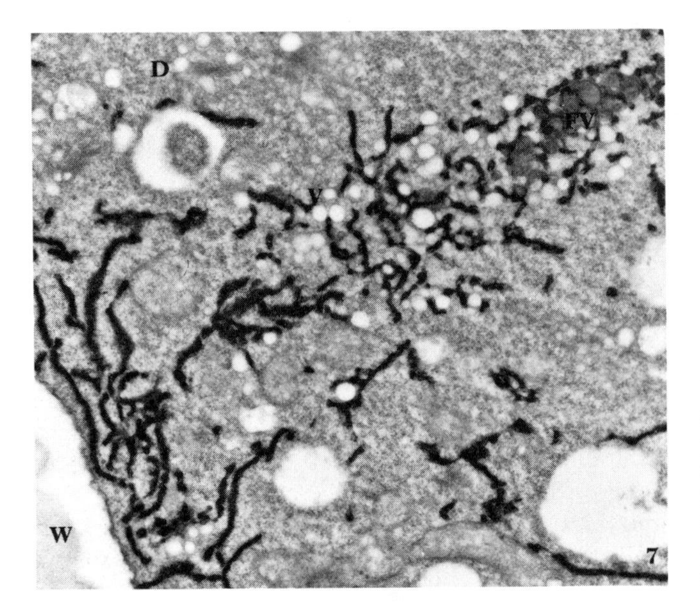

Figure 7.7 Darkly-stained tubular ER, possibly containing Ca^{++}, is closely associated with the formation of the cell plate from fusion vesicles (FV) themselves derived from dictyosome (D) vesicles (V) (see also Figure 7.12). (Courtesy of Dr. P. K. Hepler.)

THE GOLGI APPARATUS

Like the endoplasmic reticulum, to which it can be intimately related, the Golgi apparatus is a system of membranous compartments found in almost all eukaryotic cells. These membranes play a central role in the processing, packaging, and distribution of material to and from other regions of the cell and to and from the outside of the cell.

The Golgi system of a cell frequently consists of several apparently distinct Golgi *complexes*, or *dictyosomes*, each of which is made up of a parallel array of membrane-enclosed, flattened cisternae (Figure 7.8) ranging in number from 3 to 20 depending on the cell type. The cisternae, which are often distended at their periphery, are sometimes curved, giving the Golgi complex a distinct polarity with both a convex, or *cis*, face, often oriented toward the nucleus and a concave, or *trans* face. The terms "forming" face and "maturing" face are used synonymously with "cis" and "trans," and reflect the idea that both the contents and the membranes of the Golgi flow through the organelle. A network of tubular and vesicular membranous components radiates out from the cisternae (Figure 7.9). While it is difficult to tell from the static images seen in electron micrographs what the relationship is between these and the cisternae, there is evidence that they represent the means by which material is transported into and out of the Golgi. Some of the vesicles fuse with Golgi membranes, thereby delivering both their contents and at least some of their membrane components to the Golgi; others are budded off the Golgi, carrying with them materials for elsewhere in the cell. We shall now look at the nature of some of this intracellular traffic, and how it is handled as it flows through the Golgi.

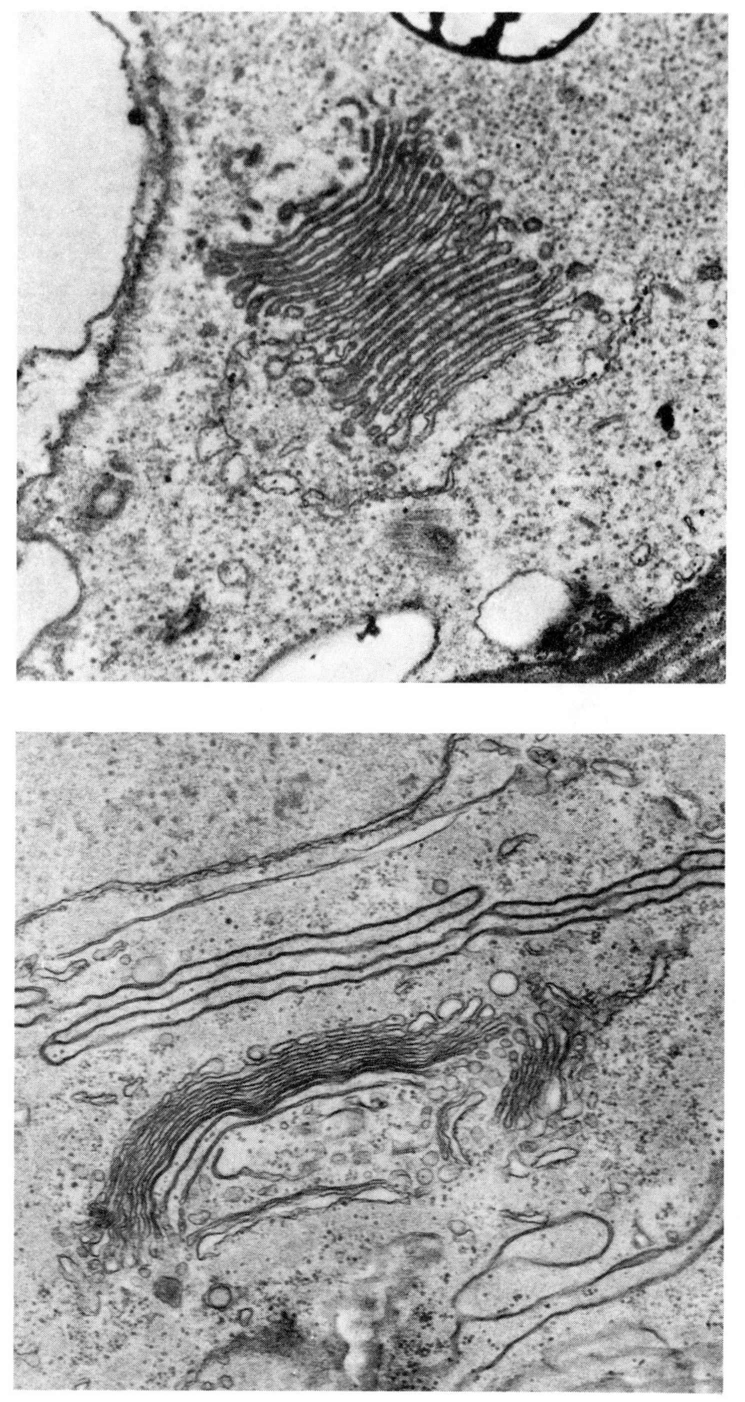

Figure 7.8 Electron micrograph of a Golgi complex in *Euglena*. Note the large number of cisternae. (Courtesy of Dr. S. C. Holt.)

Figure 7.9 Electron micrograph of a Golgi complex in an animal cell. The parallel flattened sacs are usually curved slightly, and they tend to bud off vesicles at the ends of the sacs.

Modification and Distribution of Proteins

As we have already mentioned, proteins synthesized on rough ER are destined for secretion, for the plasma membrane, or for other compartments within the cell. By radioactively labeling these proteins, their subsequent fate can be followed by electron microscopy or by cell fractionation techniques. In this way many of these proteins have been shown to become associated with the Golgi apparatus soon after being synthesized, presumably as a consequence of fusion of transition vesicles derived from the ER with some component of the Golgi. The exact pathway of flow of both membrane and lumen proteins through the Golgi is not yet known, but in at least some cases appears to take place in a cis-to-trans direction. However, it is possible that different subcompartments may exist and that not all proteins follow the same route. Such heterogeneity within the Golgi would not be surprising, since different proteins must be segregated, processed in specific ways, and distributed to specific sites.

Processing. The oligosaccharide components of ER-synthesized glycoproteins are extensively modified during transit through the Golgi. This modification involves initial removal of sugar from, and subsequent addition of other sugars to, the oligosaccharide core, and the necessary enzymes for both trimming and glycosylation are localized in the Golgi cisternae. One such enzyme, galactosyl transferase, which catalyzes a terminal step in glycosylation, is present only in the trans cisternae, indicating that the different reactions involved in modification of oligosaccharides may be spatially as well as temporally separated.

Lysosomal enzymes, which are also glycoproteins, undergo similar processing in the Golgi. One additional modification of lysosomal enzymes is the phosphorylation of one of the mannose sugars of the oligosaccharide, and this phosphorylated sugar may play a role in how these enzymes are recognized, segregated, and directed to lysosomes.

Another type of post-translational processing of proteins that occurs during the time spent in the Golgi is proteolytic cleavage of secretory peptide hormones. For example, both insulin and parathyroid hormone are synthesized on rough ER as inactive forms; these peptides are subsequently found in the Golgi, where they are converted to the active form by removal of a sequence of amino acids (Figure 7.10).

Distribution. Once the proteins have been modified by the varied and complex biochemical activities of the Golgi, they can be delivered to their intracellular and extracellular destinations by one of several different pathways (Figure 7.11). Some secretory proteins leave the Golgi in small vesicles which fuse to form *secretion granules* in which the proteins are concentrated and stored. At the appropriate time, the secretion granules are induced to fuse with the plasma membrane, and the contents are released to the outside. Other secretion products, as well as integral plasma membrane proteins, are carried in Golgi-derived vesicles directly to

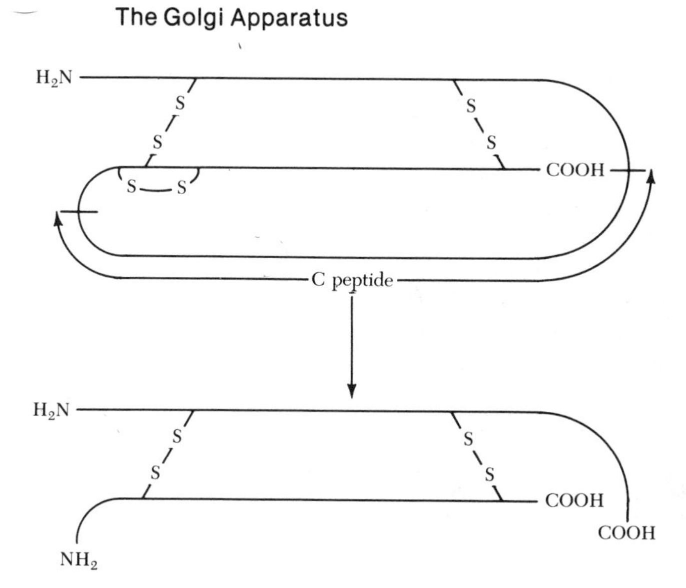

Figure 7.10 Post-translational processing of insulin in Golgi apparatus. The proinsulin precursor is a polypeptide consisting of 84 amino acids. A central peptide, C, is removed in the Golgi to form the active insulin, consisting of the remaining A and B peptides, held together by disulfide bonds and H-bonds (not shown).

the cell surface. Fusion of these *exocytotic* vesicles with the plasma membrane results in the deposition of the fluid contents to the exterior of the cell, or the incorporation of the membrane-embedded proteins into the plasma membrane. Lysosomal enzymes are also thought to be transported to lysosomes in small vesicles, although the exact route followed by their enzymes is not yet clear.

Polysaccharide secretion in plant cells. Several proteins that are synthesized on rough ER membranes in plant cells are either secreted or accumulated in other membrane-bound compartments in the cell. However, it has not yet been

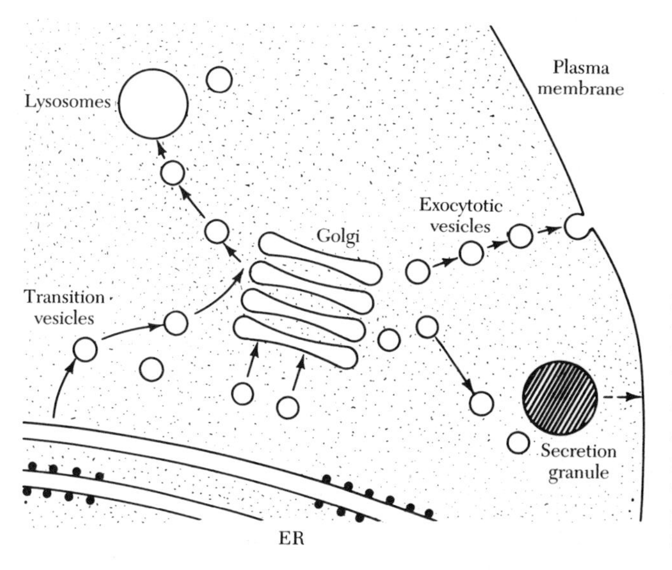

Figure 7.11 Diagrammatic representation of fates of Golgi-derived vesicles. Transition vesicles from the ER deliver both membrane and contents to the Golgi apparatus. The proteins of the membrane and of the contents are sorted and processed before being delivered to lysosomes, directly to the plasma membrane, or to secretion granules where proteins are concentrated.

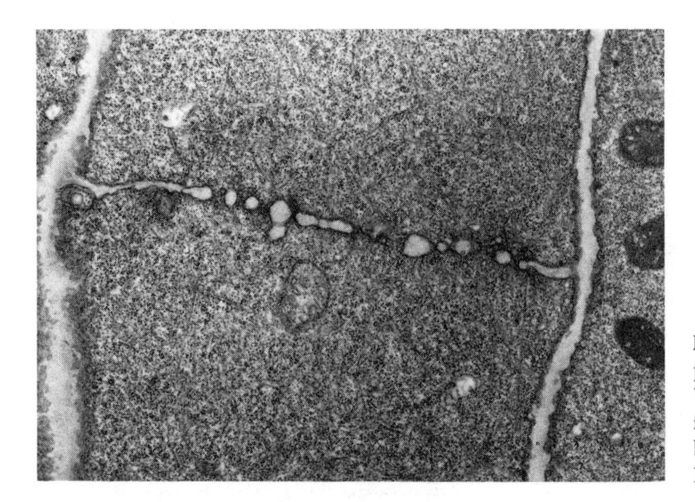

Figure 7.12 Formation of phragmoplast following mitosis in a root tip cell. Vesicles derived from Golgi complexes fuse with each other to form new membranes and wall. (Courtesy of W. McDaniel.) See also Figure 7.7.

clearly established that they all follow the route through the Golgi described for animal cells. Rather, the major role of the Golgi apparatus in plant cells is in the secretion of polysaccharides, either as extracellular mucilaginous slime or as the matrix materials of newly forming or rapidly extending cell walls. In some cases actual synthesis of these materials, which are delivered to their destinations in secretory vesicles, may occur within the Golgi complex.

The final stage in cell division in higher plants is the partitioning of the original cell into two by the formation of a new cross wall, bounded by membranes, between the two nuclei resulting from mitosis. Following mitosis, vesicles derived from the Golgi complexes move to the region where the new wall will form and fuse with each other (Figure 7.12). As a result of this fusion, two sheets of membrane form, with the former contents of the Golgi-derived vesicles between them. The fused membranes then become the adjacent plasma membranes of the newly formed cell, and the material between them forms the matrix of the new cell walls.

Golgi vesicles also contribute matrix material, pectins and hemicellulose, to preexisting walls of cells that show tip growth, such as pollen tubes and root hair cells. In such rapidly growing regions of the cell surface, it is likely that the membranes of the secretory vesicles are incorporated into the newly forming plasma membrane.

THE LYSOSOMAL SYSTEM

Lysosomes are small, often-spherical structures surrounded by a single membrane. They are characterized by the presence within them of *acid hydrolases*, degradative hydrolytic enzymes which act most effectively under conditions of relatively low pH. Over 50 different hydrolases have been shown to be associated with the lysosome compartment, including proteases, nucleases, and glycanases (enzymes that hydrolyze polysaccharides). These hydrolases are glycoproteins, and are synthesized on rough ER then delivered, via the Golgi, to the lysosomes.

The contents of the lysosomes provide a clue to their function, which is the breakdown and degradation both of material brought into the cell from the outside, and of other cellular components already present. *Primary lysosomes* are those which contain the degradative enzymes but which have not participated in any digestive process. They give rise to *secondary* lysosomes, into which the material to be digested has already been incorporated.

In the case of materials brought into the cell by endocytosis (see Chapter 4) the primary lysosomes fuse with the incoming endocytotic vesicles to form secondary lysosomes, in which digestion of the incorporated material takes place. The low pH necessary for the functioning of the acid hydrolases is maintained inside the lysosome by an ATP-driven pump in the membrane which translocates protons to the inside of the organelle.

In autophagic processes, that is, self-digestion, other cellular components, such as mitochondria, may be incorporated into autophagic vacuoles (Figure 7.13), where they are degraded by the lysosomal enzymes. The significance of such autodigestion is not clear, although presumably some cellular processes require controlled breakdown of materials. The fusion of lysosomes with certain secretory granules, for example, is believed to represent a selective process by which excess secretory products are eliminated. Regular turnover of various intracellular components, involving their destruction by lysosomes, may also be required for the successful functioning of the cell. Even cell death can be part of normal developmental processes, and indeed dying cells often appear to contain many lysosomes; release of the lysosomal hydrolytic enzymes would certainly accelerate the dissolution of a cell. Lysosomes can be thought of as disposal units of the cell, being involved in digestion of material brought from outside and in removing foreign bodies or elements of cellular architecture no longer needed. The necessity of keep-

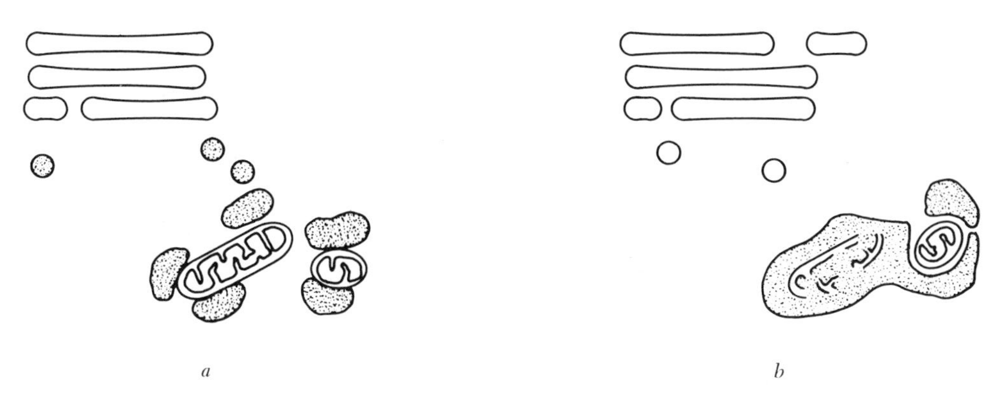

Figure 7.13 Autodigestion of cellular material (in this case mitochondria) by lysosomally-derived vacuoles. The lysosomes surround the mitochondria, coalesce, then digest the enclosed materials by means of hydrolytic enzymes.

ing the powerful lysosomal enzymes packaged in a membrane, isolated from the rest of the cell, which they could digest, and in a medium in which their pH requirements can be met, is obvious. How the cell controls the time and place of activity of these enzymes, however, is not known.

PLANT CELL VACUOLES

No organelles comparable to the lysosomes found in animal cells have been unambiguously demonstrated in plant cells. However, there is evidence that some of the intracellular digestive processes characteristic of lysosomes are associated with the plant cell vacuoles.

Vacuoles are characteristic of mature plant cells and consist of an aqueous solution surrounded by a single membrane, the *tonoplast*. As much as 90 percent of the internal volume of the cell is likely to be occupied by the vacuole, with the cytoplasmic components and the nucleus being appressed to the plasma membrane and cell wall (Figures 7.14 and 7.15).

The vacuole functions in the maintenance of *turgor* of plant cells and in providing an aqueous environment for the accumulation of water-soluble compounds.

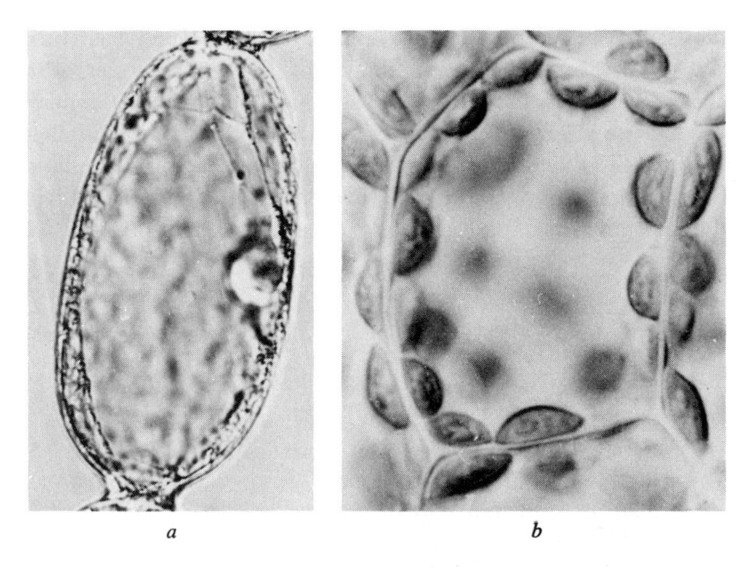

Figure 7.14 Two examples of plant cells, illustrating the manner by which the formation of a vacuole pushes the cytoplasm to the outside, thus increasing the exchange of materials between the cytoplasm and the exterior of the cell. (*a*) Cell from the stamen hair of the spiderwort, *Tradescantia*, showing nucleus appressed to cell wall, and strands of cytoplasm traversing the vacuole. (*b*) Cell from "leaf" of a moss, showing the cytoplasm and chloroplasts forced to the outside.

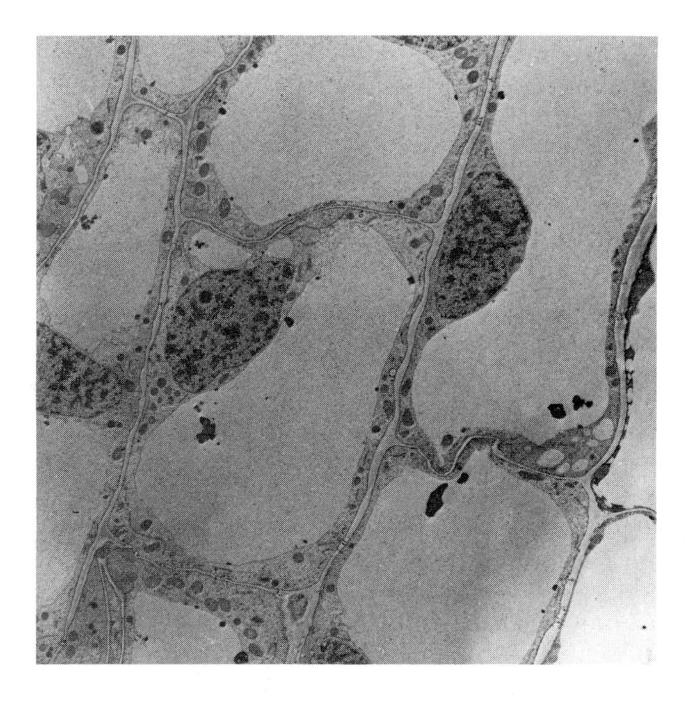

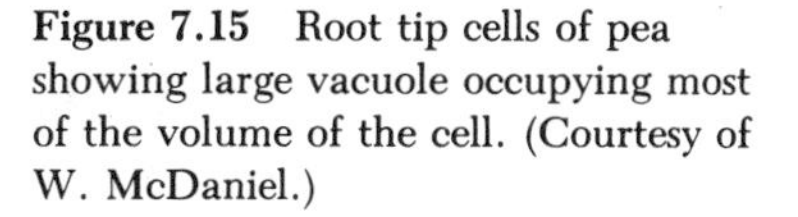

Figure 7.15 Root tip cells of pea showing large vacuole occupying most of the volume of the cell. (Courtesy of W. McDaniel.)

The tonoplast membrane, like the plasma membrane, is differentially permeable and hence can maintain concentrations of materials very different from those found in the cytoplasm. High concentrations of salts, sugars, and organic acids are found in most vacuoles, and water-soluble pigments, such as the red pigment of beets and many flower color pigments, also may be present. The tendency of water to move into the hypertonic vacuole exerts pressure on the surrounding cytoplasm and hence on the cell wall. This *turgor pressure* not only allows plant cells to remain relatively rigid, but is also responsible for enlargement of the cell before the outer wall becomes too restricting.

Lysosomal Functions of Vacuoles

The vacuole forms from the enlargement and fusion of smaller vacuoles present in meristematic cells; these provacuoles, which are believed to be derived from the ER and possibly the Golgi, contain acid hydrolases similar to those of animal cell lysosomes. Acid hydrolases are also associated with the tonoplast of the large vacuoles of differentiated cells. Occasionally, evidence of autophagic activity can be seen inside the vacuole; various membranous components, including mitochondria and plastids, can be detected in electron micrographs, and these appear to be digested in the vacuole itself.

A specialized case of lysosomal activity may be represented by the fate of the storage proteins formed in the seeds of legumes. These proteins are synthesized on the ER and, via a subsequent pathway that includes the Golgi, are eventually deposited within the vacuole. The vacuole then splits up to form membrane-enclosed protein bodies, which accumulate more of the storage protein. The breakdown, or digestion, of the proteins to provide food reserves to the developing embryo starts to take place in the vacuole and the protein bodies, and results from acid hydrolase activity. Thus lysosomelike activity appears to be associated with these organelles and is necessary for the early stages in seed germination.

MICROBODIES

Another class of membrane-enclosed organelles found in both plant and animal cells is that comprised of the *microbodies*. This term is used to describe structures that, like lysosomes, are bounded by a single membrane and characterized by the presence within them of specific enzymes. The function of microbodies in general is to compartmentalize specific enzymes and, therefore, specific biochemical reactions in various regions of the cell. These microbodies frequently contain proteinaceous crystallinelike bodies, which may represent an orderly array of some of the enzymes present.

Two main types of microbody have been described, depending on which reactions are catalyzed by the enzymes they contain; some enzymes, however, are common to both types. *Peroxisomes* were first described in liver and kidney cells of the rat, where they carry out various oxidations in which oxygen is taken up and hydrogen peroxide formed; the potentially dangerous hydrogen peroxide is subsequently destroyed by catalase, the enzyme generally used as the identifying feature of microbodies. Plant peroxisomes contain similar enzymes, and one major role of the peroxisomes in leaf cells is in the oxidation of certain intermediates formed in the dark reactions of photosynthesis; close associations between leaf peroxisomes and chloroplasts are commonly encountered. The second class of microbody, known as *glyoxysomes*, is also present in plants, particularly in cells of some germinating seedlings. In addition to the oxidases and catalases, the glyoxysomes contain enzymes involved in converting storage fats of the seed to carbohydrates; following germination of the seedling and the utilization of the storage fats, the glyoxysomes are lost from the cells.

Although microbodies are often seen in electron micrographs to be closely associated with the ER, it is not clear what relationship, if any, exists between the two types of membrane. One suggestion is that they are derived, by a budding process, from the ER membranes. However, most of the peroxisomal and glyoxysomal enzymes studied appear to be synthesized on free ribosomes rather than on those of the rough ER, and to be post-translationally inserted through the membranes of preexisting vesicles. Furthermore, the protein composition of the microbody membranes themselves may be very different from that of ER membranes, and it is possible that the microbodies represent a separate compartment within the cell.

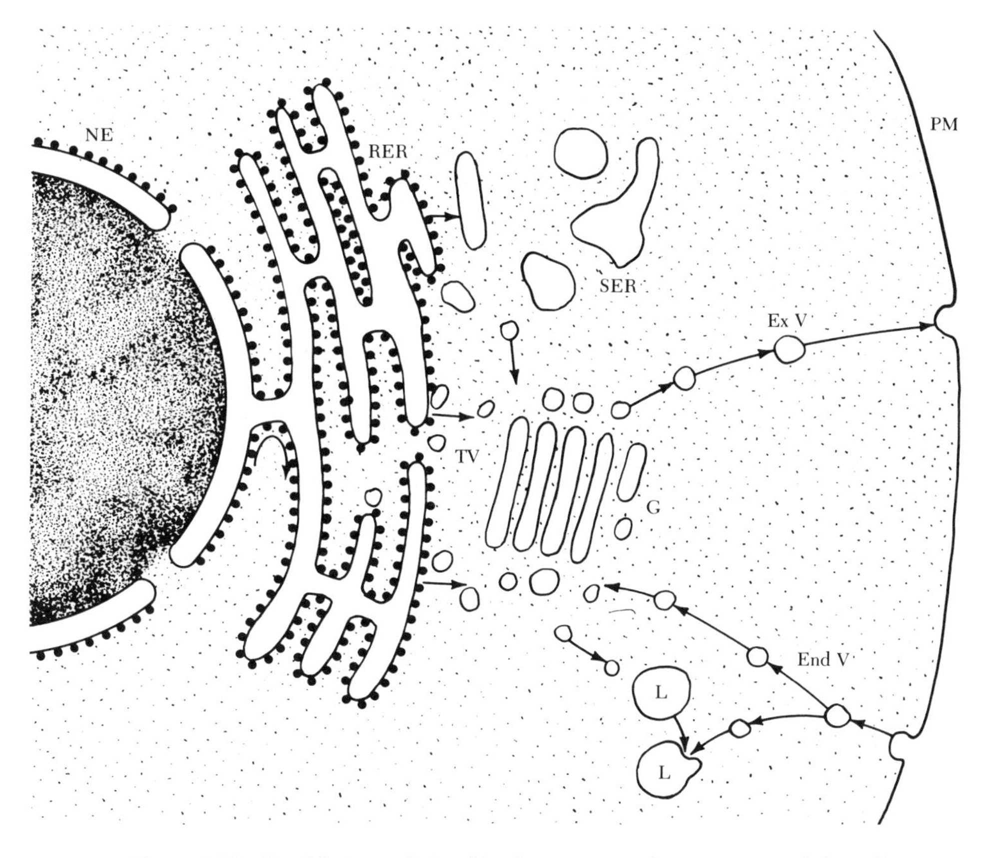

Figure 7.16 Possible interrelationships between membrane systems of the cell. *NE:* nuclear envelope; *RER:* rough endoplasmic reticulum; *SER:* smooth endoplasmic reticulum; *Ex V:* exocytotic vesicles; *End V:* endocytotic vesicles; *L:* lysosomes; *TV:* transition vesicles; *G:* golgi.

MEMBRANE FLOW AND SORTING

From our discussion of ER, Golgi, lysosomes, microbodies, and vacuoles, it is apparent that all of these represent membrane-enclosed compartments of the cell, specialized in different ways related to their various functions. Although direct physical continuity between these different compartments is only rarely observed in the electron microscope, it is certain that there must be temporal continuity between the different intracellular membranes, and between them and the plasma membrane. Thus we can think of all of these membrane systems as being in a state of dynamic equilibrium in the cell, with a basic membrane organization being modified at different times and in different regions of the cell to fulfill various roles. In this way the distribution of materials and labor in the cell would be accomplished by a flow of membranes through the different cellular compartments (Figure 7.16). For example, the integral proteins of the plasma membrane are in-

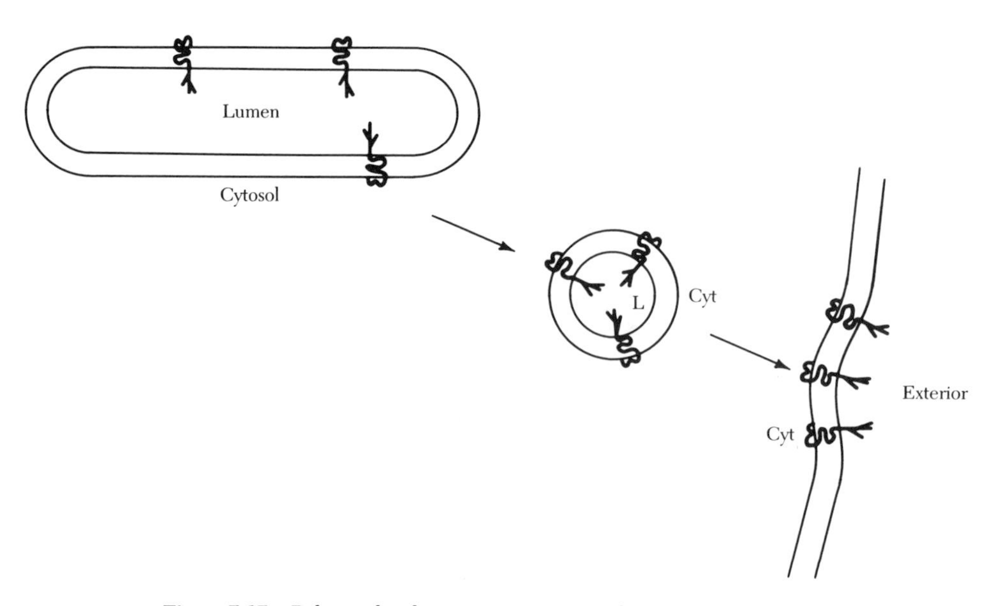

Figure 7.17 Relationship between orientation of integral proteins in ER, vesicles, and plasma membrane, as the proteins pass through these compartments. L, lumen; C, cytosol.

serted into the membranes of the ER in a specific orientation; this orientation is maintained in a membrane as it passes through the Golgi and up to the time it becomes part of the plasma membrane (Figure 7.17). Also, endocytotic vesicles remove membrane from the plasma membrane and return at least some of it to the internal membrane complement of the cell, including the Golgi apparatus. It should be clear, however, that since different membranes are of different and specialized compositions, some mechanism must exist to ensure that the appropriate compositions are maintained. Furthermore, since the nonmembrane proteins within the ER lumen also have specific destinations, the cell must have some means of sorting out these proteins and delivering them to different sites.

Two examples may illustrate how the specificities of membrane composition and membrane-compartment contents could be determined. In our earlier discussion of the plasma membrane (Chapter 5), we pointed out that receptor-mediated endocytosis involved the clustering of ligand-binding receptor molecules, which were then internalized by the formation of a vesicle. The role of the clathrin coat that underlies the clustered receptors and that at least initially surrounds the endocytotic vesicle is not known; it is conceivable that the clathrin can recognize the cytoplasmic domains of the receptors and induce them to cluster, as well as contributing to the formation of the vesicle itself. Once the vesicle and its contents have been internalized, the clathrin coat is lost. The receptor molecules, minus the ligands, are then returned to the plasma membrane. One possible route is by the

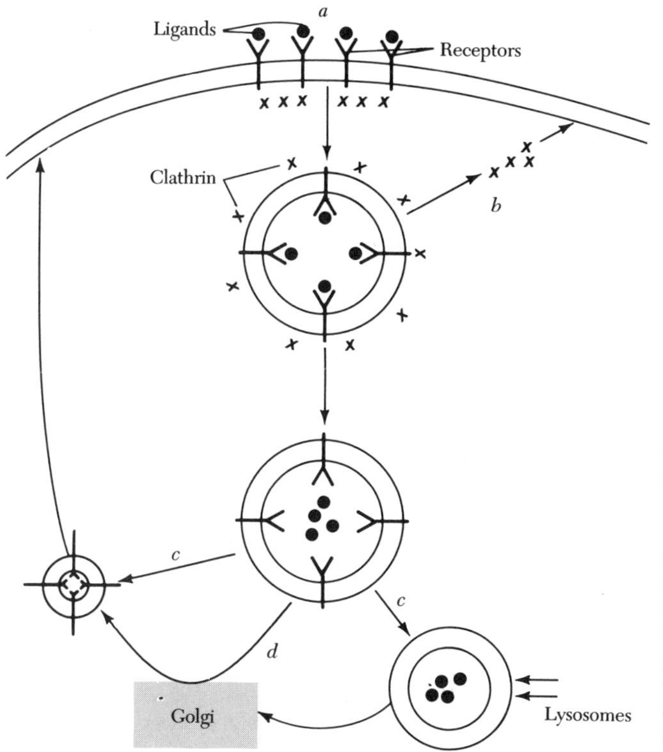

Figure 7.18 Some possible pathways of recycling of plasma membrane receptors. (*a*) The receptors, clustered at clathrin-coated pits in the membrane, bind their ligands. (*b*) Following endocytosis, the clathrin returns from the coated vesicle to the plasma membrane. (*c*) The receptors release their ligands and are budded off and returned in a small vesicle to the membrane, while the endocytosed materials are delivered to lysosomes, Golgi, or other destinations. (*d*) Alternatively the receptors may be delivered with the endocytotic vesicle to the Golgi, before being returned, via the normal secretion pathway, to the membrane.

"budding off" from the vesicle, either before or after fusion with primary lysosomes, of small segments of membrane containing the receptors and the refusion of these segments with the plasma membrane. A second possibility is that the receptors may be delivered to the Golgi and then returned via exocytic vesicles to be incorporated in the membrane again (Figure 7.18).

Similar receptors may be involved in directing intracellular traffic; we have already pointed out that lysosomal enzymes are modified by the phosphorylation of a mannose residue. This apparently allows the hydrolases to be recognized by a mannose-6-phosphate receptor that is present both in the ER and the Golgi-associated vesicles. Thus the lysosomal enzymes can be sequestered by this receptor and concentrated in the vesicles that deliver them to the lysosomes (Figure 7.19). The mannose-6-phosphate receptors must also be recycled, since they are not found in the lysosomes themselves. Clathrin-coated vesicles have also been implicated in the process by which lysosomal enzymes are compartmentalized and channeled to their proper destinations, although again how they may be involved is not understood.

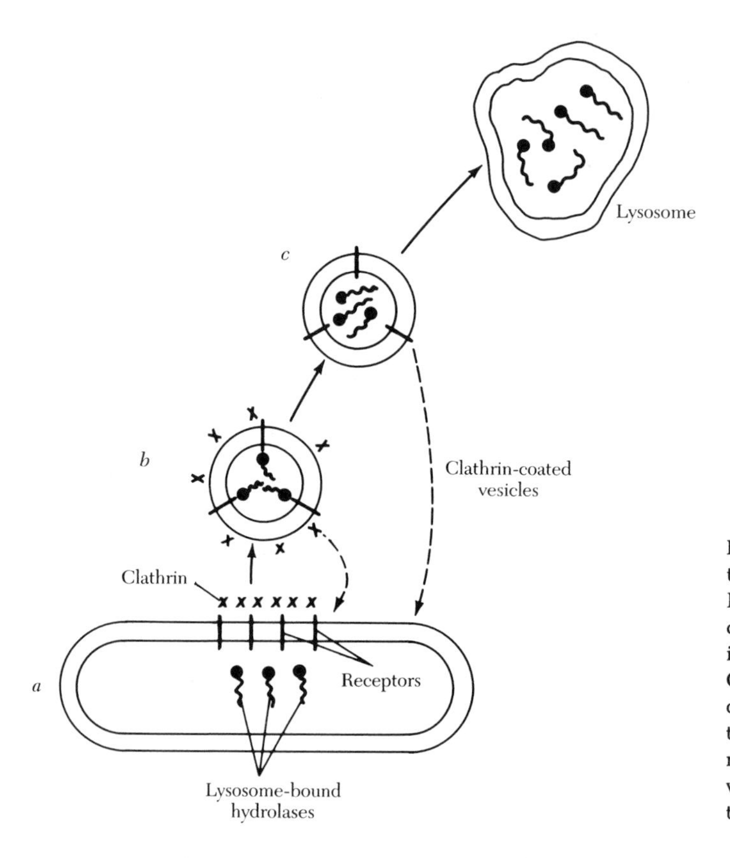

Figure 7.19 Segregation and delivery to lysosomes of lysosomal enzymes. (*a*) ER and/or Golgi compartment with clustered receptors for 6-mannose-P residue of lysosome-bound hydrolases. (*b*) Clathrin-coated vesicle containing hydrolases bound to 6-mannose-P receptors in membrane. (*c*) The clathrin and receptors are lost (recycled?) from the vesicle before it deposits its contents in the lysosome.

ADDITIONAL READING

Bainton, D. F. 1981. The discovery of lysosomes. *J. Cell Biol. 91:* 66s–76s.

Chrispeels, M. J., Higgins, T. J. V., Craig, S., and Spencer, D. 1982. Role of the endoplasmic reticulum in the synthesis of reserve proteins and the kinetics of their transport to protein bodies in developing pea cotyledons. *J. Cell Biol. 92:* 5–14.

deDuve, C. Microbodies in the living cell. 1983. *Sci. Am. 248*(5): 74–84.

Farquar, M. G., and Palade, G. E. 1981. The Golgi apparatus (1954–1981)—from artifact to center stage. *J. Cell Biol. 91:* 77s–103s.

Glaumann, H., Ericsson, J. L. E., and Marzella, L. 1981. Mechanisms of intralysosomal degradation with special reference to autophagocytosis and heterophagocytosis of cell organelles. *Int. Rev. Cytol. 73:* 149–182.

Goldman, B. M., and Blobel, G. 1978, Biogenesis of peroxisomes: intracellular site of synthesis of catalase and uricase. *Proc. Natl. Acad. Sci. USA 75:* 5066–5070.

Hepler, P. K. 1982. Endoplasmic reticulum in the formation of the cell plate and plasmodesmata. *Protoplasma 111:* 121–133.

Lodish, H. F., and Rothman, J. E. 1979. The assembly of cell membranes. *Sci. Am. 240*(1): 48–63.

NORTHCOTE, D. H. 1979. The involvement of the Golgi apparatus in the biosynthesis and secretion of glycoproteins and polysaccharides. *Biomembranes 10:* 51–76.

PASTAN, I. H., and WILLINGHAM, M. C. 1981. Journey to the center of the cell: role of the receptosomes. *Science 214:* 504–509.

ROTHMAN, J. E. 1981. The Golgi apparatus: two organelles in tandem. *Science 213:* 1212–1219.

SABATINI, D. D., KREIBICH, G., MORIMOTO, T., and ADESNIK, M. 1982. Mechanisms for the incorporation of proteins in membranes and organelles. *J. Cell Biol. 92:* 1–22.

STEINMAN, R. M., MELLMAN, I. S., MULLER, W. A., and COHN, Z. A. 1983. Endocytosis and the recycling of the plasma membrane. *J. Cell Biol. 96:* 1–27.

TOLBERT, N. E., and ESSNER, E. 1981. Microbodies: peroxisomes and glyoxysomes. *J. Cell Biol. 91:* 271s–283s.

WICKNER, W. 1980. Membrane assembly. *Science 210:* 861–868.

WILLINGHAM, M. C., PASTAN, I. H., and SAHAGIAN, G. 1983. Ultrastructural immunocytochemical localization of the phosphomannosyl receptor in Chinese hamster ovary (CHO) cells. *J. Histochem. Cytochem. 31:* 1–11.

8

The Cytoskeleton—Motility and Cell Shape

The properties and activities of cells are many and varied, and as we have pointed out, depend on the manipulation of energy. One feature of all living cells is that movement occurs: single cells swim, muscle cells contract, chromosomes move, cytoplasm streams, some cells change shape, and so on. Such cellular and intracellular motility represents mechanical work, and therefore requires that force be generated and energy expended. The structural basis for motility is a complex network of rods and filaments present in the cytoplasm of eukaryotic cells. This architectural framework, known as the *cytoskeleton*, can also be responsible for the determination of cell shape and possibly for the nonrandom spatial distribution of other components of the cytoplasm.

The three major components of the cytoskeletal framework are *microtubules*, *microfilaments*, and *intermediate filaments*. These components, each of which is comprised of a fairly small number of specific proteins, can be distinguished on the basis of their chemical composition as well as their morphologies and dimensions. Indeed, the specificity of the proteins has facilitated the detection of cytoskeletal elements and their distribution in the cell. If fluorescent antibodies against the specific proteins are allowed to bind to cells, the distribution of the antibodies, as detected by fluorescence microscopy, indicates the distribution of the particular type of cytoskeleton protein (Figure 8.1). Interactions within and between the cytoskeletal constituents, and between them and other components of the cell, such as the plasma membrane, are important in generating movement and in determining cell shape; as we shall see, the molecular mechanisms involved in both motility and shape may be very similar.

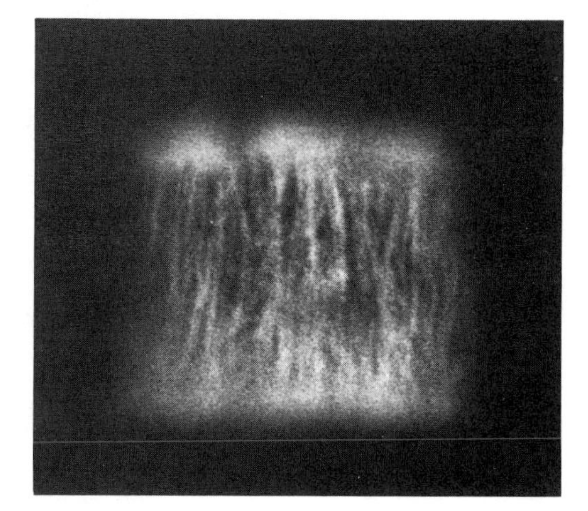

Figure 8.1 A roottip cell showing, by means of immunofluorescence microscopy, the distribution of tubulin and microtubules at interphase. Tubulin, the protein of the microtubules, is identifiable since it has been tagged with a rabbit antibody (antitubulin) to which a fluorescent dye has been attached. Although not obvious in single isolated cells, the microtubules are known to be aligned perpendicular to the root axis. (Courtesy of Dr. Sue Wick.)

MICROTUBULES

Microtubules are long, narrow, hollow cylindrical rods about 25nm in diameter, consisting of a wall approximately 5.0nm in thickness (Figure 8.2). The walls are composed of *protofilaments;* usually 13 protofilaments are present, as seen in cross-sectional views. Occasional linkages, or cross-bridges, between adjacent microtubules are also observed. Microtubules may be arranged in the cell to form well-organized structures, such as in cilia and flagella, the hairlike projections involved in cell propulsion, or in the spindle which is assembled during cell division and on which chromosome separation occurs. Other microtubules are less regularly distributed in the cytoplasm, sometimes extending in a radial fashion from around the nucleus toward the periphery of the cell, sometimes lying just beneath, and parallel to the plane of, the plasma membrane.

The main component of microtubules is *tubulin*, a polypeptide with a molecular weight of 55,000. Tubulin can exist free in the cytoplasm in dimer form, each dimer consisting of two slightly different monomers, α-tubulin and β-tubulin, or in the highly polymerized form characteristic of the microtubules. Assembly of microtubules takes place by polymerization of tubulin dimers to their ends, while disassembly can occur by depolymerization and the consequent release of free tubulin from the microtubule ends (Figure 8.3). Thus microtubules can grow or shrink in length, and such changes may be important to their functions in the cell. Many cellular activities are impaired by drugs that interfere with the polymerization and depolymerization of tubulin, and indeed the presumption of involvement of microtubules is often based on the effects of such drugs on these activities.

Several nontubulin proteins, known as microtubule-associated proteins (MAPs), have been identified as components of microtubules. Some of these appear to play a role in the regulation of polymerization and depolymerization of tubulin,

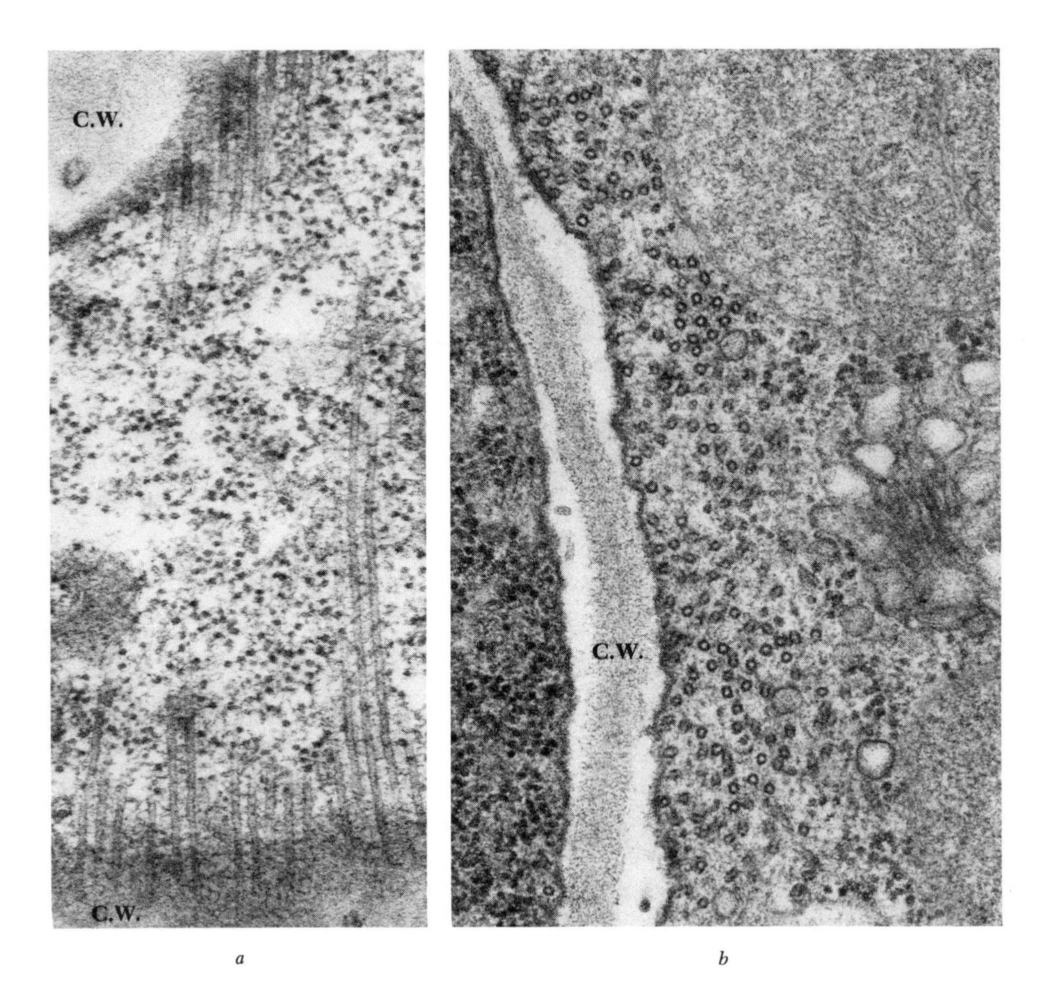

Figure 8.2 Electron micrographs of microtubules in plant cells. (*a*) Oblique section, cut parallel to the cell surface; the microtubules, seen in longitudinal section, lie at the periphery of the cytoplasm just beneath the plasma membrane. (C.W. = cell wall) (Courtesy of W. McDaniel). (*b*) Section cut perpendicular to the cell wall, showing many microtubules in cross-section. This bundle of microtubules constitutes the pre-prophase band (see also Figure 8.5), which determines the position of the cross wall formed when the cell divides. (Courtesy of P. K. Hepler.)

while others are thought to serve to cross-link microtubules to each other or to other cellular components. There is also evidence that different MAPs may be associated with microtubules from different cell types, and it is possible that some of the MAPs may be involved in determining specificity of both organization and function of microtubules.

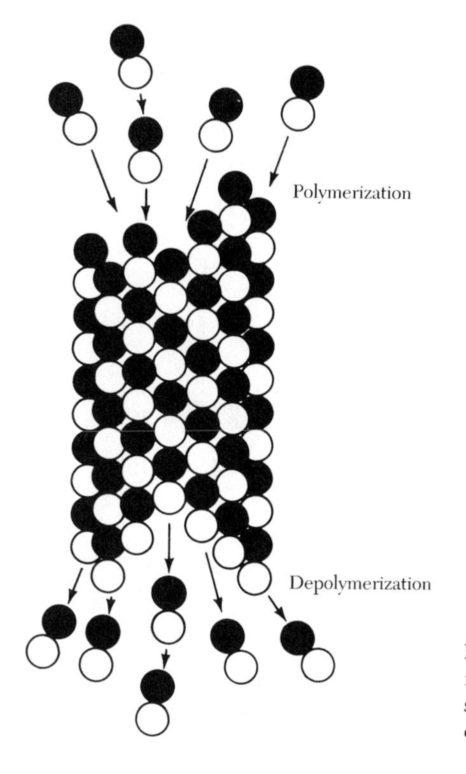

Figure 8.3 Assembly or disassembly of microtubules occurs by the addition or subtraction of sub-units, each of which exists as a dimer of tubulin.

Microtubules and Cell Shape

As we have already mentioned, the various elements of the cytoskeleton can be responsible for determining the shapes of cells. A striking example of the role of microtubules in this respect is the formation of *axopodia* in certain heliozoan protists. These single-celled organisms form long, needlelike projections, through each of which runs a spirally arranged bundle of microtubules (Figure 8.4). If the cells are treated with agents that destroy microtubules, such as *colchicine*, a drug that binds to tubulin and inhibits polymerization, the axopodia collapse and are retracted into the cell. Reestablishment of axopodia occurs only when the microtubules reform, and elongation of the axopod is accompanied by elongation of the microtubules toward the tip. Thus not only do the microtubules confer the rigidity necessary to maintain the shape of these cell extensions, but also microtubule elongation appears to provide the force necessary for their formation.

Nerve cells also display a morphology which is highly asymmetric and which appears to depend on specific arrays of microtubules. Long processes, or *axons*, which conduct nerve impulses and transmit them to adjacent cells, extend from initially rounded cells. In human nerves these axons can reach many centimeters in length, and contain bundles of microtubules running along their axes. As was the

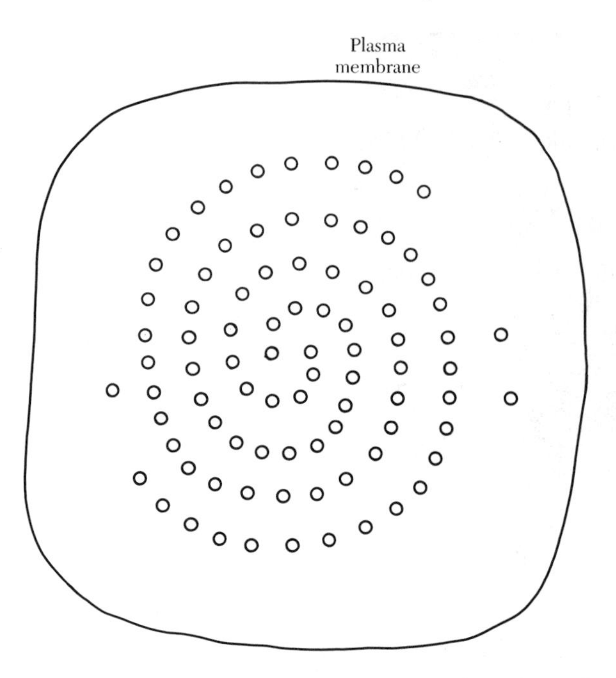

Figure 8.4 Spiral arrangement of microtubules as would be seen in a cross-sectional view of an axopod of a heliozoan cell.

case for axopodia, treatment of cultured nerve cells with microtubule-depolymerizing drugs inhibits extension of these axons and causes them to retract. However, both extension and retraction seem to require more than microtubule polymerization and depolymerization, since inhibitors of microfilament function can also interfere with changes in axon shape; it is likely, therefore, that interactions between microtubules and other cytoskeletal fibers are involved in modulating axon shape. It should also be apparent that such changes in shape as we have described require movement; as we shall see later, microtubules are also required for a different kind of motility, that involving transport of materials along the axon.

Cell Wall Formation

Microtubules also play an important, though not yet fully understood, role in influencing the shape of plant cells. Plant cells, because of their external walls (Chapter 5), are less dependent on an internal cytoskeleton for providing the rigidity necessary to maintain their shape. However, the orientation in which the cellulose microfibrils of the growing wall are deposited influences the polarity of cell expansion (Chapter 5) and hence cell shape. In growing cells the orientation of microtubules in the peripheral cytoplasm, just beneath the plasma membrane, is frequently correlated with the orientation of the newly laid down cellulose fibrils in the wall on the other side of the membrane. If cells that are about to elongate are treated with colchicine, the new microfibrils are deposited randomly in the wall instead of in the normal orientation, and the cell expands isodiametrically instead

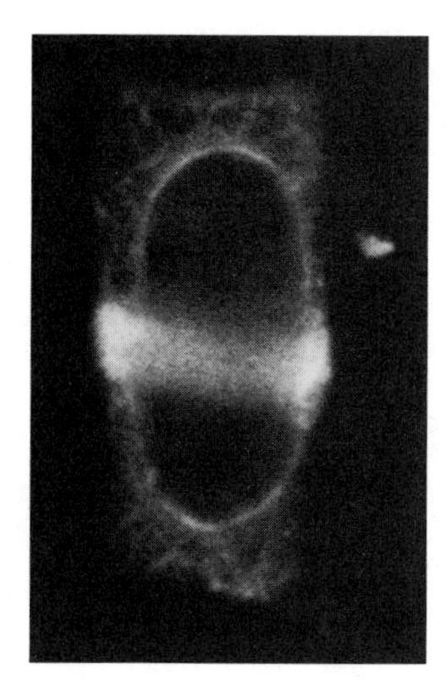

Figure 8.5 A meristematic roottip cell, treated with antitubulin to reveal the arrangement of the microtubules to form a pre-prophase band. The band is also perpendicular to the root axis. The fluoresence of the nuclear envelope is characteristic of pre-prophase, and a network of microtubules can be seen to fill the cytoplasm between the nucleus and the plasmalemma. (Courtesy of Dr. Sue Wick.)

of in a polar fashion. The nature of the connection between cytoplasmic microtubules and cellulose microfibrils of the wall is not known, but may involve some interaction between microtubules and the membrane by which the cellulose synthesizing assembly complexes would be guided in the appropriate direction.

The pre-prophase band. Microtubule orientation can also influence plant cell shape by determining the plane in which new walls are formed during division of the cell (Figure 8.5). Prior to the onset of cell division a band of microtubules becomes organized in a specific position around the circumference of the cell. This band disappears before mitosis begins, presumably to contribute to the formation of the microtubular spindle on which the chromosomes will be separated (Chapter 9). However, the plane demarcated by the circumferential pre-prophase band is the same as that in which the cell will be partitioned, that is, that in which the new cross-wall is laid down following mitosis (Figure 8.5). Clearly, some aspect of microtubule orientation at pre-prophase must influence the cytoplasm and/or membrane in a way that predetermines the sites of activities leading to the subsequent establishment of the new cell walls.

Microtubules and Motility

Cilia and flagella. A direct involvement of microtubules in generating motion is most clearly demonstrated, and the mechanism most fully understood, in the cilia and flagella of motile cells. The beating of cilia and flagella provides the force for the movement of material past the cell surface and for propulsion of the cell, and is in turn a consequence of interactions between the constituent microtubules.

Structure. The microtubules of cilia and flagella are arranged in a specific and characteristic manner. Two central microtubules are surrounded by an outer

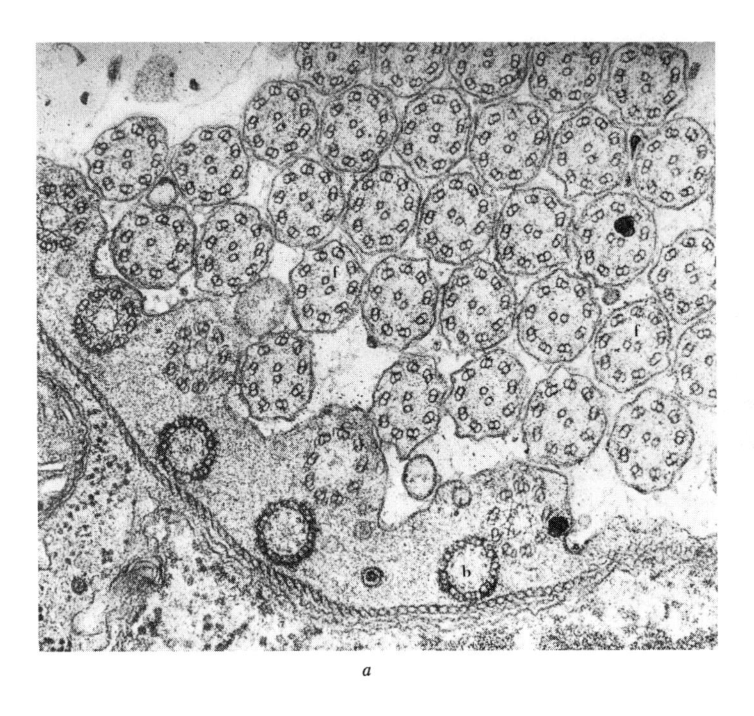

a

Figure 8.6 (*a*) Electron micrograph of spermatid of *Equisetum*, showing cross sections of many flagella (f), with the typical 9 + 2 arrangements of microtubules. The basal bodies (b) are also shown. (*b*) Cross-sectional view of microtubule arrangement of flagellum at higher magnification.

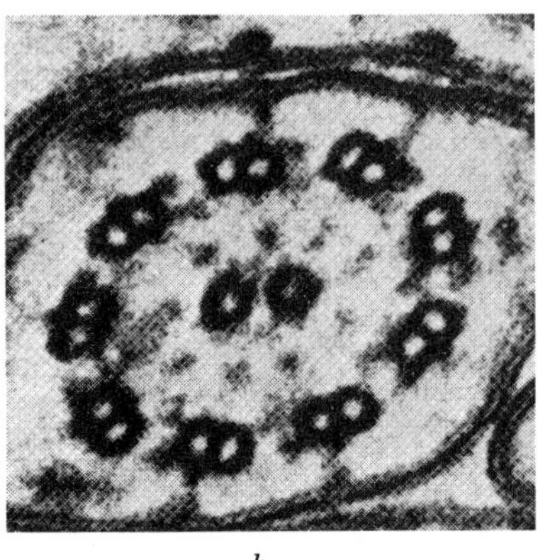

b

circle of nine pairs, or "doublets," as shown in Figure 8.6. These microtubules and associated structures constitute the *axoneme*, which is embedded in a matrix enclosed by an extension of the plasma membrane. Anchoring the cilia and flagella in the cytoplasm are structures called *basal bodies* (Figure 8.7a), the microtubules of which are arranged in nine triplets (Figure 8.7b). The basal bodies are also involved in the formation of cilia and flagella, and are thought to provide a template on which the 9 + 2 arrangement of axoneme microtubules is assembled.

The details of the structural organization of the axoneme are shown in Figure 8.8. Each outer "doublet" consists of two types of microtubule. The A microtubule consists of the typical 13 protofilaments while the B microtubule is incomplete, and consists of only 11 protofilaments. Attached to each A tubule is a double row of arms extending toward the B tubule of the adjacent doublet, plus a row of radial spokes extending toward the central tubules. The doublets are connected to each other by a row of links between A and B tubules of adjacent doublets, while the central pair are also connected by cross-bridges. As we shall now see, these interconnections play an important role in ciliary and flagellar movement.

Mechanism of beating. The beating of cilia and flagella results from the wavelike propagation of bends along the length of the organelles. This bending is based on the capacity of adjacent doublets of the axoneme to slide past one another. If sliding is restricted at some region along the axoneme, the displacement of doublets relative to one another can be accommodated only by bending.

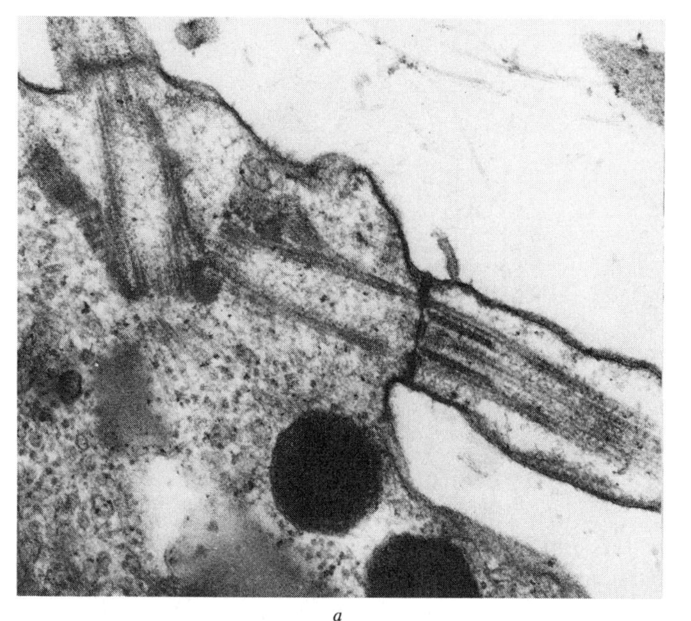

a

Figure 8.7 (*a*) Longitudinal section through basal bodies and flagella of motile zoospore of *Phytophthora*, the potato blight fungus. (*b*) Cross-sectional view of microtubule arrangement of basal body.

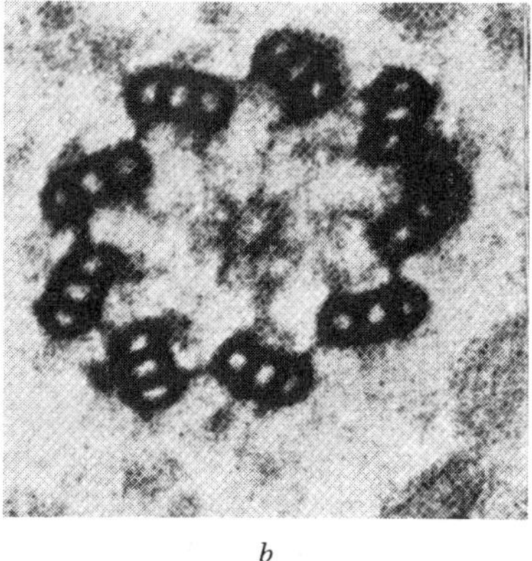

b

The sliding motions are generated by the activity of the two rows of arms attached to the A microtubule. These arms are made of the ATP-hydrolyzing protein, *dynein*. ATP-driven conformational changes in the dynein arms allow them

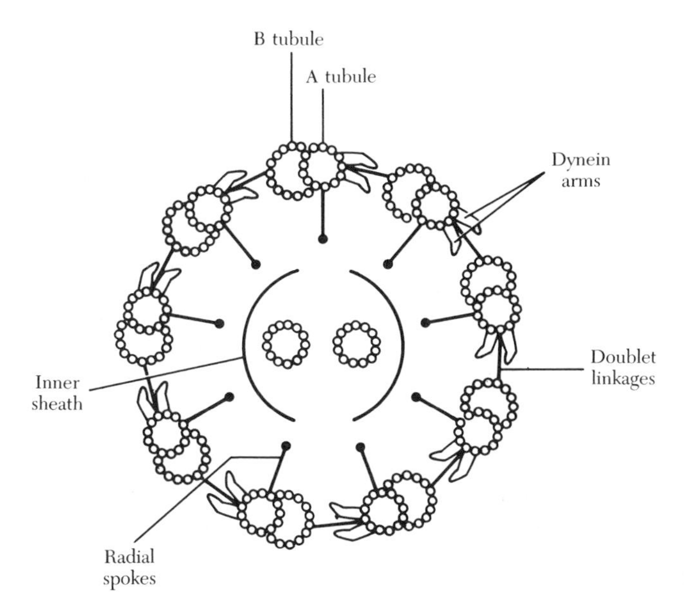

Figure 8.8 Schematic representation of flagellar organization (cf. Fig. 8.6b). See text for description.

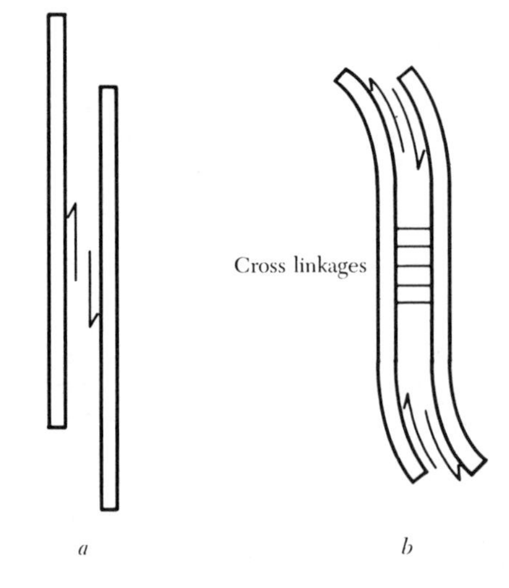

Figure 8.9 Diagrammatic representation of the "sliding filament" mechanism of ciliary and flagellar bending. (*a*) The sliding force is not resisted, and the filaments (doublets) are free to slide past one another. (*b*) The sliding force is resisted by cross-links between the filaments at one region; hence the displacement of doublets in non-linked regions must be accommodated by, and result in, binding.

to interact with successive binding sites on the B microtubule of the adjacent doublet, and produce a shearing force between the doublets. The resulting sliding, however, is restricted by the other cross-bridges, including the radial spokes linking the outer and central doublets, causing the axoneme to bend (Figure 8.9). The bending can then be propagated as a wave as both the shearing forces and breakage and re-formation of cross-bridges move along the axoneme in a coordinated fashion.

Nonflagellar microtubules. Microtubules also appear to be involved in the generation of internal motive forces in nonflagellated cells, although the mechanism(s) by which they do so are not fully understood.

The segregation and distribution of chromosomes on the spindle during division of the cell is one of the most dramatic displays of movement associated with microtubules. However, we shall delay discussion of chromosome movement until later, when we can consider the overall process of mitosis and cell division.

Two other examples of microtubule-based motility systems are the extended axons of neurons, along which mitochondria, membrane-enclosed vesicles, and certain chemicals are rapidly transported, and pigmented cells known as chromatophores, in which color changes are brought about by the aggregation or dispersal of pigment-containing granules in the cell. In both cases, movement of materials occurs in close association with, and is prevented by disruption of, the microtubule arrays present in the cells. Such microtubules, therefore, must at least provide a framework upon which movement can occur; that they can themselves generate movement is suggested by the presence of associated dynein-like proteins. While it is not clear to what extent the latter may contribute to the driving forces, dynein activity similar to that of cilia and flagella may be a part of the force-transducing mechanisms of all microtubule-based motility systems.

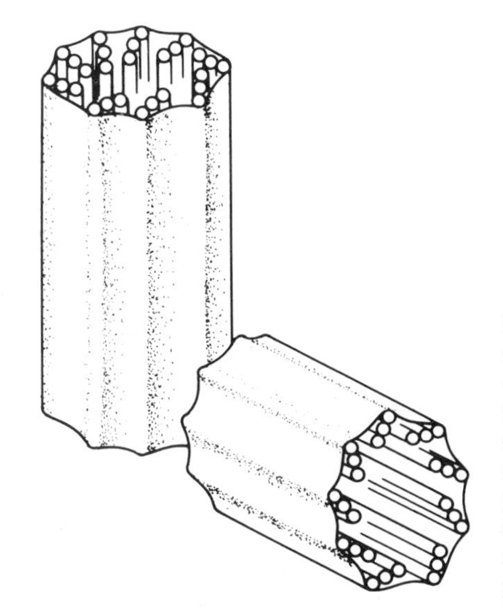

Figure 8.10 Diagrammatic representation of a centriole "pair". The sister centrioles of the pair lie perpendicular to one another; one already-duplicated centriole pair, consisting of a parent and a daughter centriole, lies at each pole of the metaphase spindle. The two members of each pair will separate and replicate during the subsequent interphase period.

Microtubule Organizing Centers, Centrioles, and Basal Bodies

As we have already described, microtubule assembly results from polymerization of tubulin dimers. In order for microtubules to be assembled in specific patterns and in appropriate orientations in the cell, organizing centers from which the microtubules can grow must be present at specific sites in the cell. These microtubule organizing centers (MTOCs) may be associated with identifiable structures in the cell, such as the basal bodies of cilia and flagella (Figure 8.7), the *centrioles* associated with the mitotic spindle apparatus of animal cells (Figure 8.10), and the *kinetochore* of chromosomes (Figure 8.11). In other cases, however, notably in plant cells, microtubule assembly proceeds from less easily detectable, diffuse regions of the cytoplasm.

Basal bodies are frequently derived from centrioles, remarkably similar organelles found in cells of organisms that have a motile stage in their life cycles.

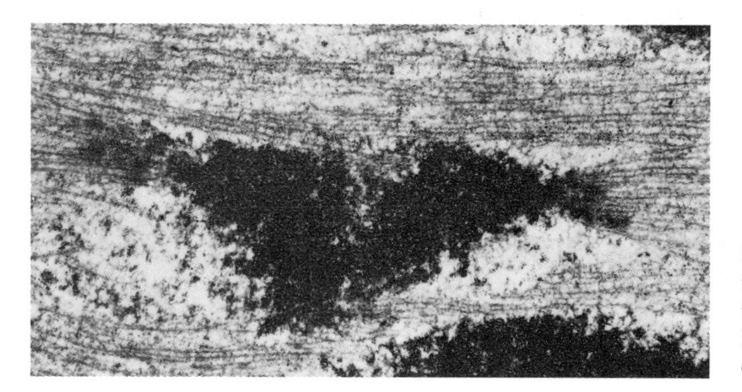

Figure 8.11 Microtubules of the mitotic spindle; the microtubules attached to the chromosomes emanate from the kinetochores (arrows), which act as microtubule organizing centers.

The centriole microtubules are also present in nine sets of three. Centrioles are found in close association with the nucleus and can replicate by the formation of a new centriole perpendicular to the old. Centriole duplication is frequently coordinated with cell duplication, and prior to cell division the centrioles move to the regions to be occupied by the poles of the spindle. Centrioles were formerly thought to be necessary for spindle formation; this is not the case, however, since in cells of higher plants, which do not have centrioles, and in animal cells from which the centriole has been removed, spindle formation and cell division occur normally. The behavior of centrioles during cell division may simply reflect a mechanism to ensure the distribution of centrioles to all of the daughter cells. Lower plants also have centrioles, although usually in cells that are motile, such as sperm cells of ferns and zoospores of algae. The absence of centrioles from other cells of the organism suggests that they may arise de novo during formation of the motile cells.

Studies of microtubule polymerization show that indeed isolated basal bodies and centrioles can act as nucleating centers. Microtubules start to form in close association, but not in direct contact, with these organelles. Extension of the microtubules occurs by addition of tubulin to the distal end.

Microtubules can also be seen radiating out from regions close to centrioles in intact animal cells, and from very similar-appearing regions in plant cells. These presumed MTOCs also are redistributed in cells undergoing dynamic changes in microtubule organization, such as those that occur before, during, and after cell division.

Although the molecular composition of MTOCs is not known, it is clear that their distribution and activity in the cell are important in maintaining the spatial organization of the microtubule networks. Thus regulation of MTOC distribution must represent a critical aspect of how the cell organizes its cytoplasm and maintains its shape.

MICROFILAMENTS

Microfilaments are typically smaller than microtubules, ranging for the most part from 5 to 7 nm in diameter. They are also made up of protein subunits, consisting mainly of actin molecules. Actin, which is often one of the most abundant proteins of animal cells, can exist as a free monomeric form, called G-actin, or it can be polymerized in a polar fashion into filaments, in which form it is referred to as F-actin (Figure 8.12).

Individual actin filaments may be associated in well-defined bundles, or they may form less regularly oriented networks. The arrangement of actin filaments in the cytoplasm appears to depend on interactions with various actin-binding proteins; such interactions can therefore influence both the physical properties of the cytoplasm and the activities in which actin filaments participate.

Figure 8.12 Actin filaments are made up of two helically-twisted rows of polymerized actin (F-actin). The unpolymerized actin is G-actin.

Muscle Contraction

The role of actin filaments which is most fully understood is in the movement represented by the rapid, coordinated contraction of striated muscle, in which the microfilaments are organized in highly specific and regular ways in relation to each other and to other filamentous proteins. We shall therefore describe the structure and activities of striated muscle, in order to demonstrate how actin filaments can participate in movement.

Skeletal muscle tissue is made up of individual muscle fibers that are elongate and tapered. The fibers, which are multinucleate as a result of fusion of individual muscle cells, may be from 1 to 40 mm in length and from 10 to 40 μm in width. Each fiber is enclosed by a membrane, the *sarcolemma*, from which tubular invaginations, *T tubules*, extend into the cell (Figure 8.13). Internally, the cytoplasm consists primarily of filaments arranged into longitudinal bundles called *myofibrils* and surrounded by a specialized type of ER, the *sarcoplasmic reticulum* (Figure 8.13). The sarcoplasmic reticulum contains high levels of Ca^{2+}, which, as we shall see, plays an important role in muscle movement.

The highly organized filaments are arranged such that a repeating pattern of bands is observed along the length of each myofibril (Figure 8.14). Each repeating unit between successive narrow *Z bands* is known as a *sarcomere;* as we shall see, it is the shortening of the sarcomeres that causes the muscle fibers to contract. Actin microfilaments are attached to the Z bands, and extend into the sarcomere from each end. In the center of the sarcomere is the *A band*, which contains an array of thicker filaments made of another protein, *myosin*. The end regions of the myosin filaments overlap with the ends of the actin filaments, giving the outer regions of the A band a darker appearance than the central region, where only myosin filaments are present. The overlapping actin and myosin filaments are interdigitated in such a way that each myosin filament is surrounded by, and can be cross-linked to, six actin filaments (Figure 8.15).

During contraction of the muscle, the actin and myosin filaments slide past one another, thereby drawing the Z bands closer together and causing the sarcomere to shorten (Figure 8.16). This shortening requires that the distal ends of the actin filaments be firmly anchored in the Z discs; an actin-binding protein, α-actinin, is present in the Z-disc regions, and may be responsible for the attachment of the actin to the disc.

Now let us look at the nature of the interactions between the actin and myosin filaments that are responsible for the sliding, and hence the contraction.

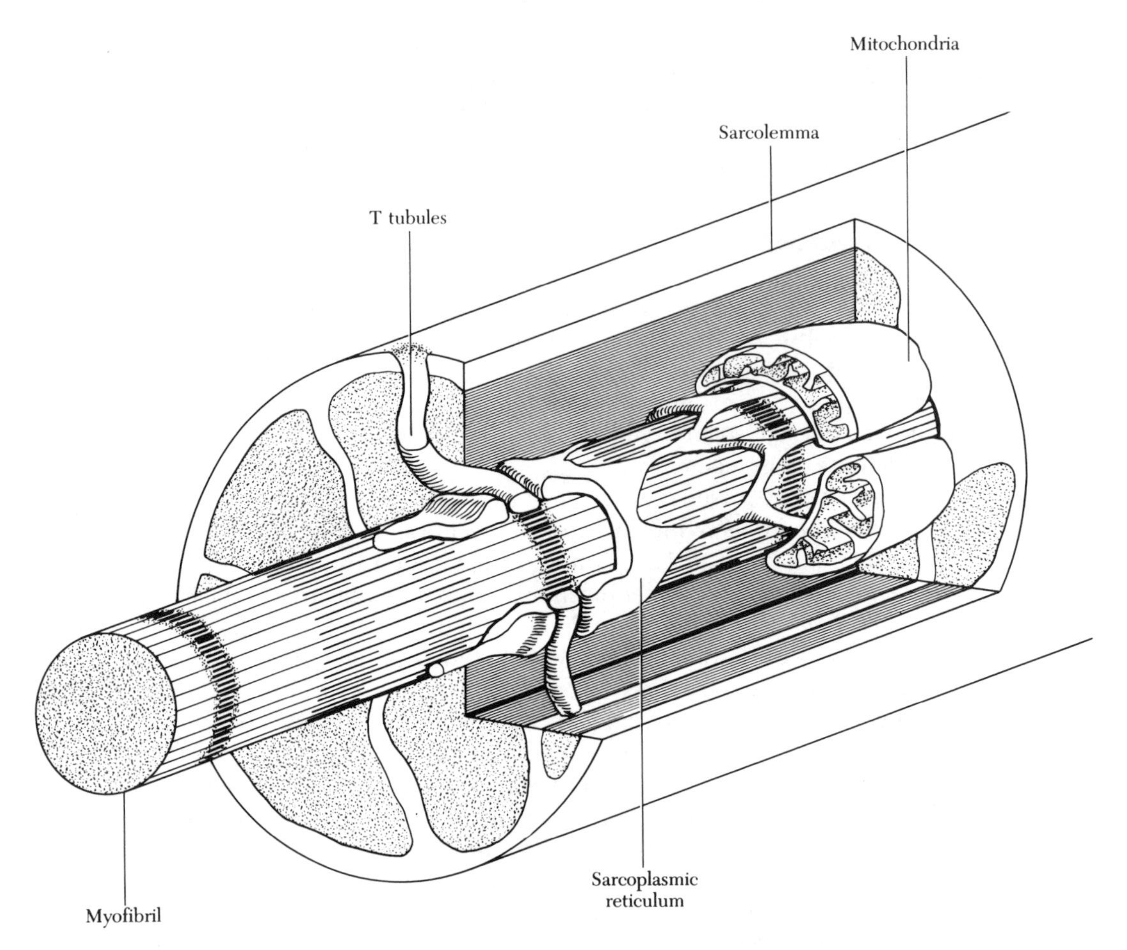

Figure 8.13 Schematic representation of a muscle fiber, showing the relationship of the myofibrils with the other components of the cell. Several myofibrils are enclosed by the cell membrane, the sarcolemma, from which invaginations called T tubules extend toward the myofibrils. Each myofibril is surrounded by a sarcoplasmic reticulum system; and associated with mitochondria.

Actin–myosin interactions. The arrangement of the globular actin molecules in the thin filament is such that two rows of molecules are twisted around one another in a helical fashion (Figure 8.12). Two other proteins, *troponin* and *tropomyosin*, are associated with the actin in the filament (Figure 8.17) and are involved in regulating actin–myosin interactions.

The myosin filaments, which are about 10 nm in diameter, are aggregates of individual molecules of the protein myosin. This protein consists of a filamentous portion and a club-shaped end with two globular heads (Figure 8.18a); the filamentous portions are wound into the filament, while the heads turn outward to

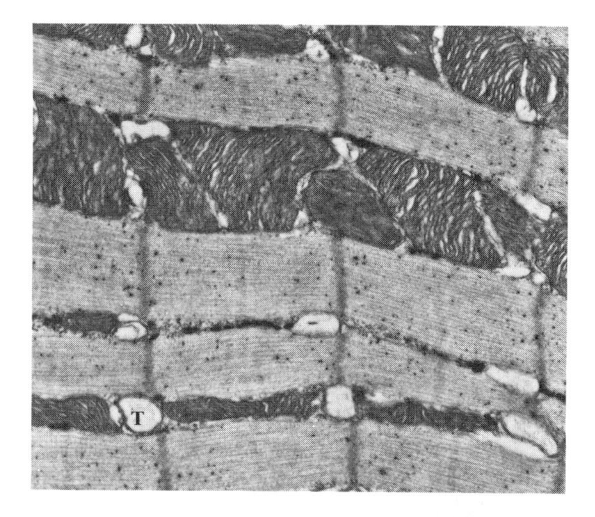

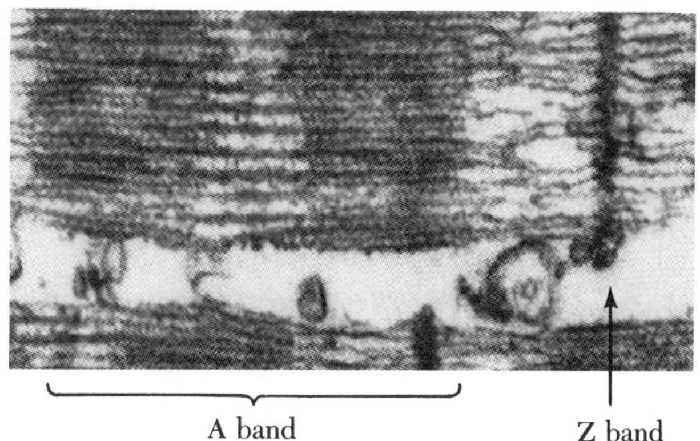

Figure 8.14 Electron micrographs of portions of striated muscle. Top: a sarcomere is a region between two Z bands (the dark vertical lines), and lines of sarcomeres are separated from each other by closely packed masses of mitochondria T (T tubule). Bottom: a higher magnification of a region between two Z lines, with the larger, darker lines of myosin and the thinner, lighter lines of actin.

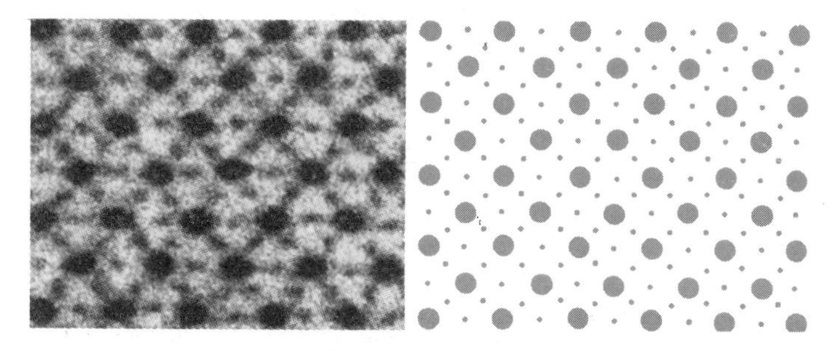

Figure 8.15 An electron micrograph and a diagrammatic representation of a cross section through the sarcomere of a striated muscle.

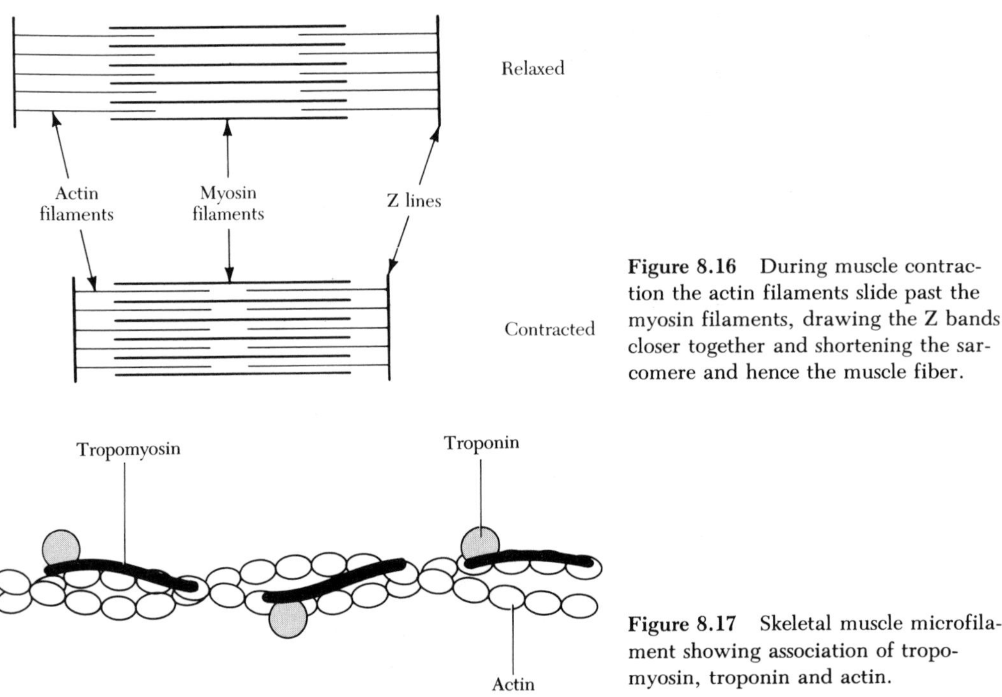

Figure 8.16 During muscle contraction the actin filaments slide past the myosin filaments, drawing the Z bands closer together and shortening the sarcomere and hence the muscle fiber.

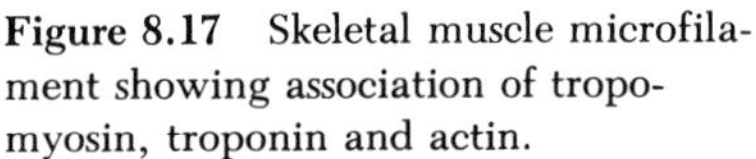

Figure 8.17 Skeletal muscle microfilament showing association of tropomyosin, troponin and actin.

form bridges that can connect with the actin filaments. As can be seen in Figure 8.18b, the two ends of the myosin filaments are of opposite polarity with respect to the orientation of the myosin heads.

Two properties of the myosin heads are important in the proposed actomyosin sliding mechanism. One is that they can bind to actin; the other is that they can hydrolyze ATP to ADP plus phosphate in the presence of actin. The significance of these features is shown in Figure 8.19, which is a diagrammatic representation of how sliding is thought to occur. When ATP is hydrolyzed by a myosin head, the products of hydrolysis, ADP and phosphate, remain associated with the myosin. The myosin head can then bind to a site on the actin filament, forming a bridge. The ADP and phosphate are released from the head, which

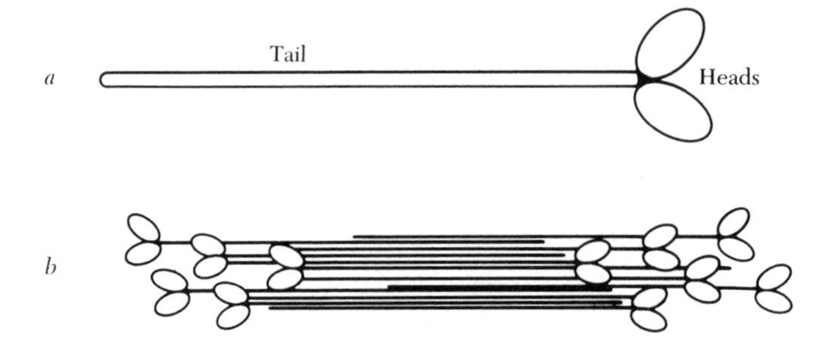

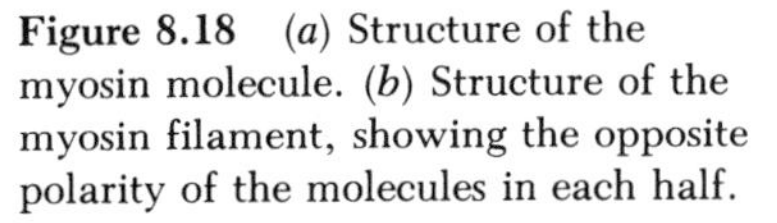

Figure 8.18 (*a*) Structure of the myosin molecule. (*b*) Structure of the myosin filament, showing the opposite polarity of the molecules in each half.

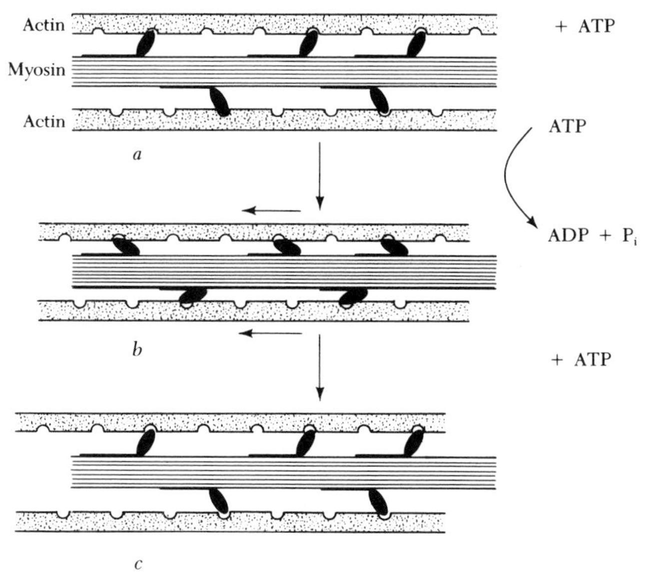

Figure 8.19 Sliding filament mechanism of muscle contraction. (*a*) The hydrolysis of ATP by the myosin heads allows them to bind to the actin filaments, and then undergo a conformational change that tugs the actin past the myosin. (*b*) New bridges are formed (*c*) when the myosin heads bind and hydrolyze more ATP.

undergoes a conformational change that generates a tugging force between the filaments. The combined forces of many such bridges along the length of the filaments is sufficient to move them past one another. The spent actin–myosin bridge is broken when another ATP molecule binds to the myosin head and is hydrolyzed; a new bridge can then be formed at a different position on the actin filament and the process repeated. Thus the actin filaments are moved along the myosin filaments by a ratchet-type mechanism, in which the myosin heads "walk" along the actin filaments. In this way the energy of ATP is used for the mechanical work represented by muscle contraction.

In order for myofilaments to be actually shortened as a result of the actin–myosin interaction, the distal ends of the actin filaments must be firmly anchored in the Z disc. As we pointed out earlier, α-actinin may be involved in the attachment of the actin to the Z disc.

Regulation of muscle contraction. The two proteins associated with actin in the microfilaments, troponin and tropomyosin, play a role in how muscle movement is regulated. In the relaxed state, these proteins prevent binding of the myosin heads to the actin molecules (Figure 8.20). However, when a nerve impulse which signals contraction reaches the muscle fiber membrane, the permeability of the sarcoplasmic reticulum is changed, and the calcium it contains is released into the cytoplasm. The Ca^{2+} binds to the troponin, causing a conformational change that displaces the tropomyosin, thus allowing the interaction between the myosin heads and the actin sites that are now exposed. As long as the muscle is being stimulated the Ca^{2+} level will be sufficiently high for the cycle of cross-bridging and movement to occur. However, if the signal from the nerve is no longer being transmit-

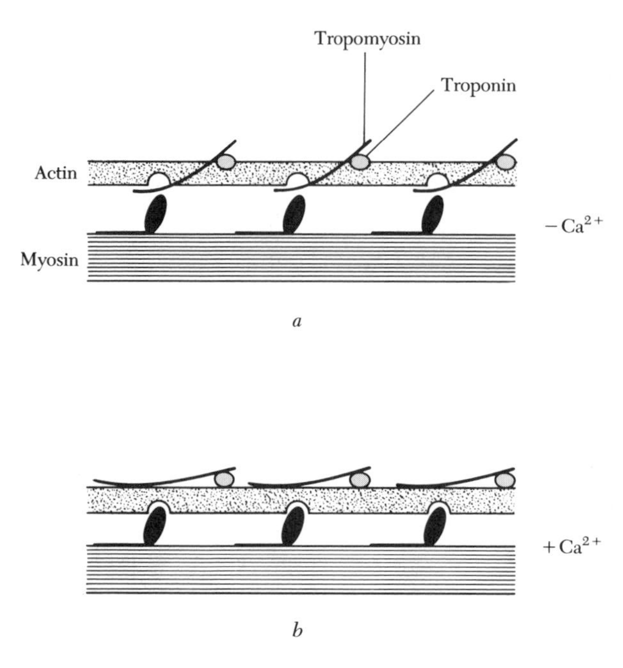

Figure 8.20 Stimulation of muscle contraction by calcium. In the unstimulated muscle, tropomyosin prevents binding of the myosin heads to the actin (*a*). Upon nervous stimulation Ca^{++} is released by the sarcoplasmic reticulum, binds to the troponin and causes tropomyosin to be displaced, making the actin sites accessible to the myosin heads.

ted, the Ca^{2+} is rapidly pumped back into the SR, the troponin–tropomyosin interaction with actin prevents cross-bridge formation, and the muscle returns to its relaxed state.

Actin in Nonmuscle Cells

Actin-containing microfilaments are also found in a wide range of nonmuscle cells, in some of which actin may be the most abundant protein present. The arrangement of the filaments is less highly organized and more labile than that of striated muscle fibers; they may be associated in bundles, or may form intermeshing networks, and these arrangements can assemble and disassemble depending on the particular activities of the cell. Although they are generally associated with cytoplasmic motility, in certain cells the role of the microfilaments may be more structural than dynamic.

Examples of motile processes in which actin filaments participate are cytoplasmic streaming in plant cells, amoeboid locomotion, and the formation of the ever-narrowing constriction which pinches a dividing animal cell into two. In the giant cells of the green algae *Chara* and *Nitella*, an inner layer of cytoplasm, the endoplasm, moves continuously around the cell past an outer static layer of ectoplasm. Parallel bundles of microfilaments are embedded in the ectoplasm at the interface between the static and streaming layers, and are oriented in (and presumably determine) the direction of streaming (Figure 8.21). Bundles of actin filaments are also present in amoebae, both in the advancing *pseudopods*, lobe-shaped projections of the cell pushed out in the forward direction, and in the

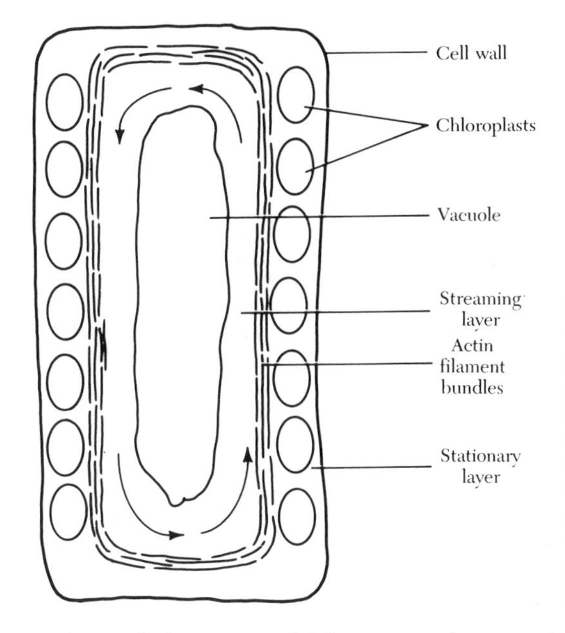

Figure 8.21 Relation of actin filaments to cytoplasmic streaming in the green alga *Chara*. A stationary layer of cytoplasm just beneath the cell wall includes chloroplasts, and, possibly attached to the chloroplasts, bundles of actin filaments. The cytoplasmic layer between the actin filaments and the central vacuole is in constant motion within the cell, and is believed to contain myosin molecules that slide past the stationary actin filaments.

retracting tail, between which regions the cytoplasm is moved. In dividing animal cells, an encircling ring of microfilaments lies just beneath the plasma membrane in the region of the constriction; as this ring contracts, the constriction is narrowed until the cells are separated. In all three of the cases above, the movement involved is prevented by treatment with the drug cytochalasin B, which is known to inhibit the formation of actin filaments.

The molecular bases of such types of actin-mediated movement are less well understood than is that of muscle contraction; however, as in striated muscle the activity of microfilaments appears to be regulated by interactions between actin and various other proteins. For example, myosin, tropomyosin, and α-actinin, three of the proteins that participate in sarcomere contraction, are associated with actin filaments in many types of cells. Although myosin filaments are rarely detected as readily as in sarcomeres, myosin undoubtedly can contribute to movement by functioning in a musclelike ATP-dependent actomyosin contraction mechanism.

Actin may also contribute to both movement and stability of the cytoplasm in other ways, involving interactions with proteins other than myosin. Several proteins are known which can bind to actin, and by affecting the way in which the filaments are organized can produce changes in the viscosity and stiffness of the cytoplasm. Some actin-binding proteins can affect the length of actin filaments, either by regulating the degree of polymerization of actin or by fragmenting the filaments, while others can cross-link actin filaments to form gel-like bundles or networks. For example, in the brush-border microvilli of intestinal epithelium cells a regularly oriented core bundle of microfilaments extends from the tip of the microvillus into a terminal web of filaments in the underlying cell (Figure 8.22).

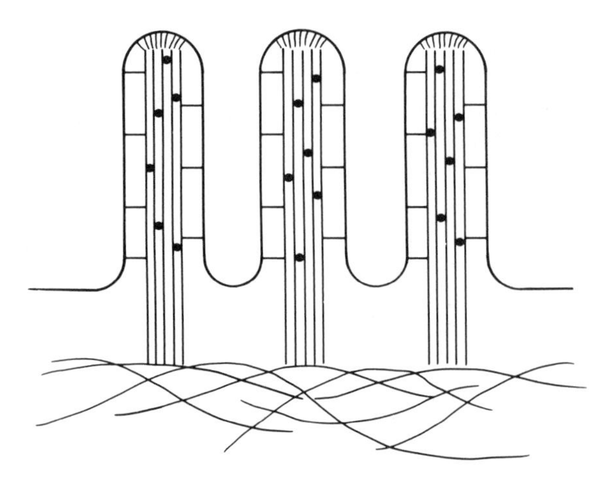

Figure 8.22 Microfilament organization in brush border microvilli. The parallel microfilaments are cross linked by bridges to each other and to the membrane of the microvillus. The bundles end in a more diffuse terminal web of cytoskeletal components.

The core microfilaments have several proteins other than actin associated with them; two of these bind to actin and cause the microfilaments to be packed into bundles, while another forms periodically spaced side arm connections between the filaments and the membrane of the microvillus. Such interactions are necessary to maintain the stability of the microvillus, and presumably the stability and activity of similar types of microfilament organization.

A different type of cross-linking is exemplified by *spectrin*, the flexible membrane-associated protein of red blood cells (Chapter 5). Spectrin cross-links small actin filaments to form the cytoskeletal framework underlying the membrane, and is itself attached to the membrane by the protein ankyrin. This filamentous framework is able, therefore, to maintain the shape of the cell and to interact with the membrane to regulate the distribution of integral membrane proteins. Similar submembrane actin-containing networks are likely to be involved in other cytoplasm/membrane interactions, and to participate, together with spectrin-like proteins, in cell shape determination, movements of the membrane, and movement of integral proteins within the membrane.

Cross-linked actin networks are also constituents of the *microtrabecular lattice*, an interconnecting three-dimensional network of 3- to 6-nm filaments revealed by special EM techniques. This lattice, found in both plant and animal cells, also includes intermediate filaments. It is likely not only to contribute to cell shape but also to control the distribution of organelles and other components of the cell. For example, in fibroblasts polyribosomes are clustered close to the nucleus; they are not, however, membrane-bound polyribosomes, but rather are attached to the fibrous, anastomosing cytoskeletal lattice. Messenger RNA, including actin mRNA, can also be associated with and translated on these ribosomes, which may be sites of synthesis of some of the cytoskeletal proteins themselves.

The versatility of actin clearly allows it to participate in different ways to the structural organization of the cytoplasm and its activities. These activities depend on, and are determined by, the various interactions that occur between actin and other proteins in the cell.

INTERMEDIATE FILAMENTS

The third component of the cytoskeletal framework is made up of filaments ranging in diameter from 7 to 11 nm, intermediate in size between microtubules and microfilaments. The functions of these intermediate filaments (IFs) are not understood; although they probably play a mainly structural role, they may also participate in some types of movements.

Unlike the microfilaments and microtubules, the constituent proteins of intermediate filaments are cell-type specific. Five major classes of IF can be recognized, each composed of characteristic proteins. These are the *cytokeratin* filaments of epithelial cells, *vimentin* filaments of various types of mesenchyme cells, *desmin* filaments of muscle cells, the neurofilaments of nerve cells, and finally the filaments found in glial cells. The particular type of intermediate filaments appears, therefore, to be related in some way to the differentiated state of these cells.

Although the proteins from each type of filament are not identical, they are related, and have certain sequences of amino acids in common. The proteins all form rodlike particles, which aggregate in a helical ropelike arrangement to form the filaments. Cross-linkages and binding between IFs and other components of the cell, including microfilaments, microtubules, plasma membrane, and even mitochondria, can be observed, although the significance of these and their relation to possible IF functions is unknown.

Although we have discussed the three components of the cytoskeleton separately, it should be recognized that they can be integrated both structurally and functionally. The cytoskeleton is a complex assembly of various proteins, an assembly that contributes to the maintenance both of the shape of the cell and of spatial organization within the cell. However, the cytoskeleton is clearly a dynamic structure, and interactions between cytoskeletal and associated proteins are also responsible for the changes in this organization that can generate both intracellular motility and cellular locomotion.

ADDITIONAL READING

BECKERLE, M. C., and PORTER, K. R. 1983. Analysis of the role of microtubules and actin in erythrophore intracellular motility. *J. Cell Biol. 96:* 354–362.

DUSTIN, P. 1980. Microtubules. *Sci. Am. 243*(2): 66–76.

LAZARIDES, E. 1980. Intermediate filaments. *Nature 283:* 249–256.

LAZARIDES, E., and REVEL, J. P. 1979. The molecular basis of cell movement. *Sci. Am. 240*(5): 100–113.

LLOYD, C. W. (ed.). 1983. *The Cytoskeleton in Plant Growth and Development.* Academic Press, Inc., New York.

MURRAY, J. M., and WEBER, A. 1974. The cooperative action of muscle proteins. *Sci. Am. 230*(2): 58–71.

PORTER, K. R., and TUCKER, J. B. 1981. The ground substance of the living cell. *Sci. Am. 244*(3): 57–67.

SATIR, P. 1974. How cilia move. *Sci. Am. 231*(4): 44–52.

STEBBINS, H., and HYAMS, J. S. 1979. *Cell Motility.* Longman, Inc., New York.

WEEDS, A. 1982. Actin-binding proteins—regulators of cell architecture and motility. *Nature 296:* 811–816.

9

The Cell in Reproduction

One of the important tenets of the cell theory is that cells do not arise de novo but are formed by reproduction of preexisting cells. The billions of cells that comprise an adult human being are all derived by successive cycles of cell reproduction, starting with the single cell, the zygote, formed by fertilization of an egg by a sperm. The successive duplications that take place produce new cells, which become differentiated and aggregated to form the many cell types and the recognizable structures of a multicellular embryo. This embryo continues to grow by cell reproduction and cell differentiation, these processes occurring in a coordinated manner throughout early life, and leading eventually to the development of a mature adult.

Both the rates of cell reproduction occurring in an organism during its development, and the localization and restriction of reproduction to particular tissues and cells are normally under strict control and determine the characteristic form of members of a species. Cell reproduction also continues throughout the lifetime of an organism at rates demanded by its inherent needs and in response to internal and external environmental conditions. The indeterminate growth of many higher plants is a result of the continuous production of new cells in localized regions, such as the *apical meristems*, which contribute to the extension of roots and shoots, and the *lateral cambium*, which contributes to their increasing girth. Most animals, on the other hand, exhibit determinate growth, and after attainment of mature proportions the active regions of cell duplication in bone marrow, skin, intestinal epithelium, and in some glands are necessary for the replacement of cells that die or are lost rather than for addition of cells to the total number in the organism.

Most of the cells of plants and animals do not reproduce, however, although some may retain the capacity to do so under certain conditions. For example, the production of scar tissue in wound-healing responses in both plants and animals involves reproduction of cells that otherwise would not divide, while cancer is itself a consequence of the uncontrolled reproduction of cells that have become free of the restraints normally imposed upon them.

We know that cell reproduction does not only result in the formation of two cells from one, but also that it produces two cells whose information contents are identical to each other and to that of the original cell. We must look at the processes involved in cell duplication, therefore, as means of ensuring exact duplication of cellular information and its equal distribution to the two daughter cells. Since we also know that the nucleus is the control center of the cell, containing the informational DNA, we must concentrate primarily on the behavior of the nucleus during cell reproduction.

The overall process of cell reproduction is basically similar in all eukaryotic cells, although modifications do occur that reflect some degree of evolutionary divergence. The stages most obvious under the microscope involve the condensation of the nuclear material into visible chromosomes, followed by the equal distribution of the already duplicated chromosomal material into two daughter nuclei. The cytoplasm also participates in this process and is itself partitioned into the daughter cells. Prior to these visible stages of nuclear and cell division, a complex series of biochemical events occurs, resulting in the exact duplication of the chromosomal material to be segregated. The entire process of cell duplication, therefore, can be thought of as a cyclical affair, since the products of one duplication can themselves go on to divide. This *cell cycle* can be divided into three main stages: *mitosis*, which refers to those cytologically detectable events of chromosome behavior that result in equal distribution of the nuclear material; *cytokinesis*, during which the two newly formed nuclei are enclosed in their own mass of cytoplasm by the formation of cell membranes; and *interphase*, the interval between two successive mitoses during which the cell is preparing to divide.

We shall describe mitosis and cell division as they typically occur in cells of higher plants and in animal cells, pointing out the differences that do exist between the two essentially similar processes.

ROOT TIP CELLS IN MITOSIS AND DIVISION

As pointed out, cell reproduction in plants is confined to localized meristems. The apical meristems of young root tips are particularly suitable for studies of mitosis, since most of the cells of the meristem undergo duplication, and as many as 10 to 15 percent of the cells may be in mitosis at any one time.

The most commonly used technique for examining the behavior of the nucleus as it divides is to fix the cells, stain them with a DNA-specific stain (Feulgen reagent is commonly used), and "squash" them on a slide to obtain a monolayer of

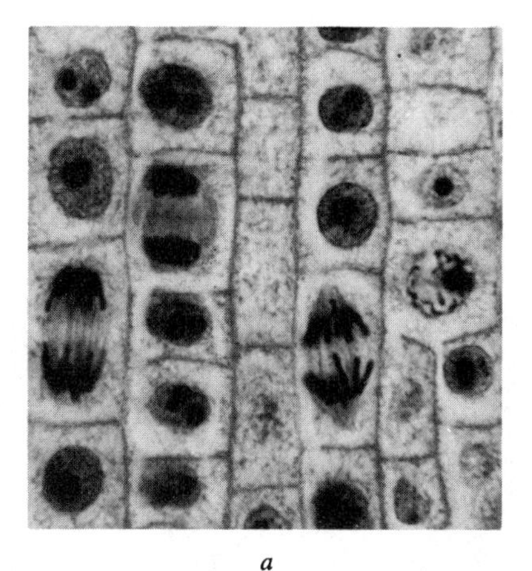

a

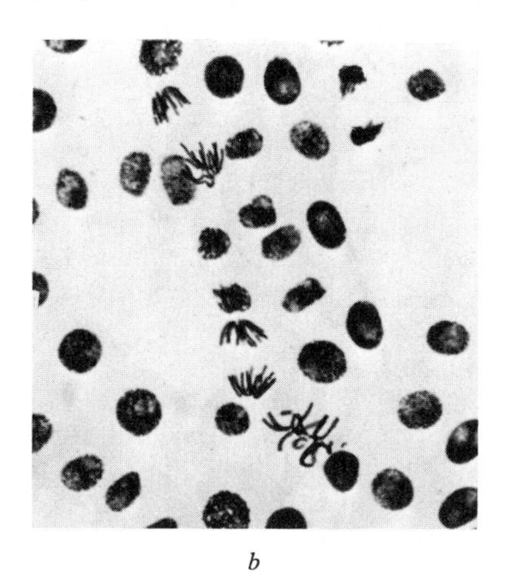

b

Figure 9.1 Panoramic view of sectioned and smeared root tip cells. (*a*) Sectioned view of dividing cells in the onion root, stained with iron hemotoxylin to show chromosomes, spindle, walls, and cytoplasm; the various stages of division range from interphase to telophase. (*b*) Smeared cells from the root of the broad bean, *Vicia faba;* smearing disrupts the arrangement of the cells, while the Feulgen stain used in this instance is specific for chromosomes and stains no other part of the cells.

flattened cells. Root tips also can be embedded in paraffin wax after being fixed, and then sectioned on a microtome before staining. Root tip cells prepared in both ways are illustrated in Figure 9.1.

Most of the cells in a root tip are in interphase, because that stage usually lasts much longer than mitosis and cytokinesis combined. The interphase nucleus is readily visible and is enclosed by the nuclear envelope. The chromatin appears fairly homogeneous, although some densely stained regions of heterochromatin are visible (Figure 9.2). One or more nucleoli can be seen within the nucleus, although they do not stain with DNA-specific dyes. During interphase, the nucleus and cytoplasm increase in volume, and, as we shall see later, active synthetic processes take place.

The first stage of mitosis is *prophase,* and its onset is marked by the chromatin becoming resolved into visibly distinct, long, thin threads, the chromosomes (Figures 9.3 and 9.4). Each chromosome consists of two longitudinal subunits, the

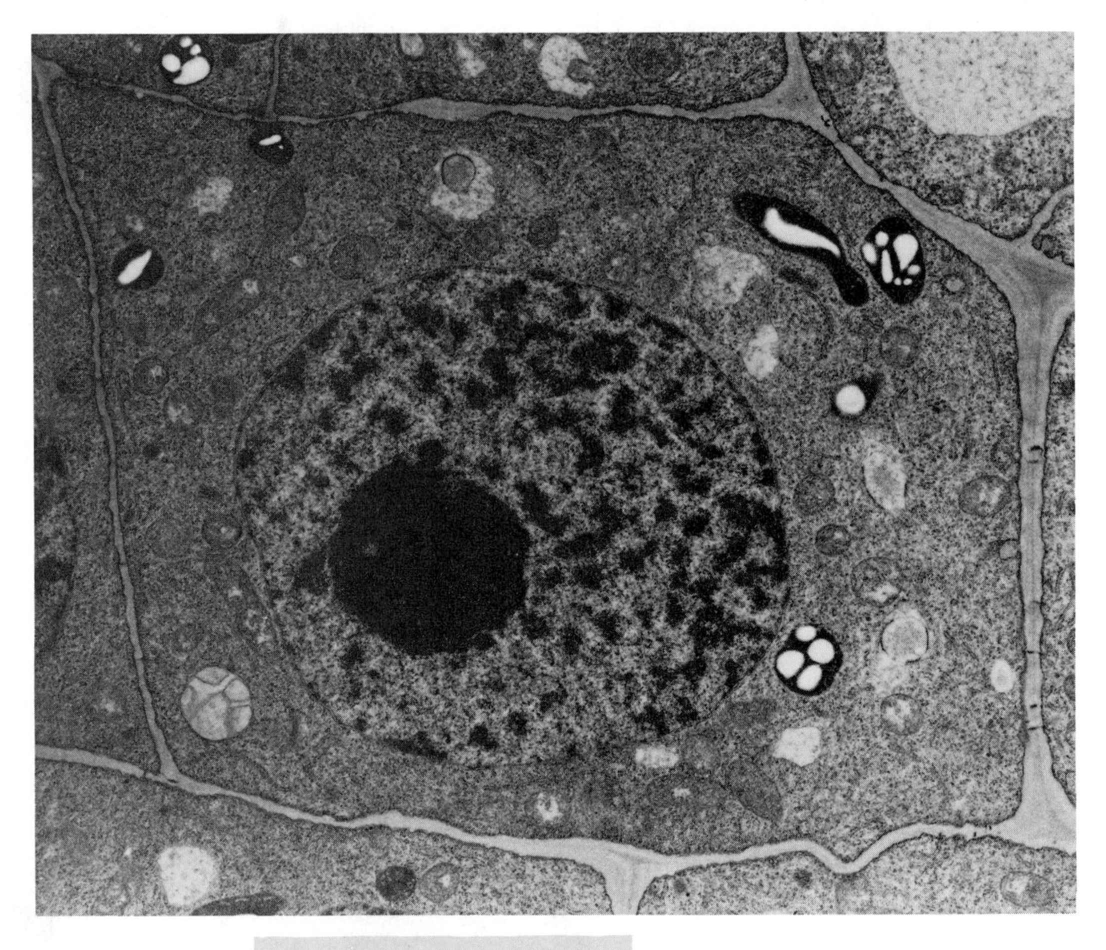

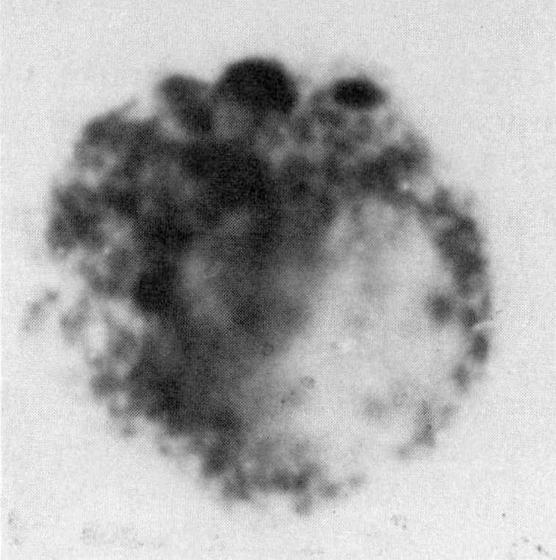

Figure 9.2 (*a*) Electron micrograph of root meristem cell in interphase. The darkly stained nucleolus is prominent, but little definite structure can be seen. (*b*) Light micrograph of interphase nucleus in a root meristem cell—the nucleolus is the central region that does not stain with DNA-specific stains and condensed regions of heterochromatin can be seen. (Courtesy of W. McDaniel.)

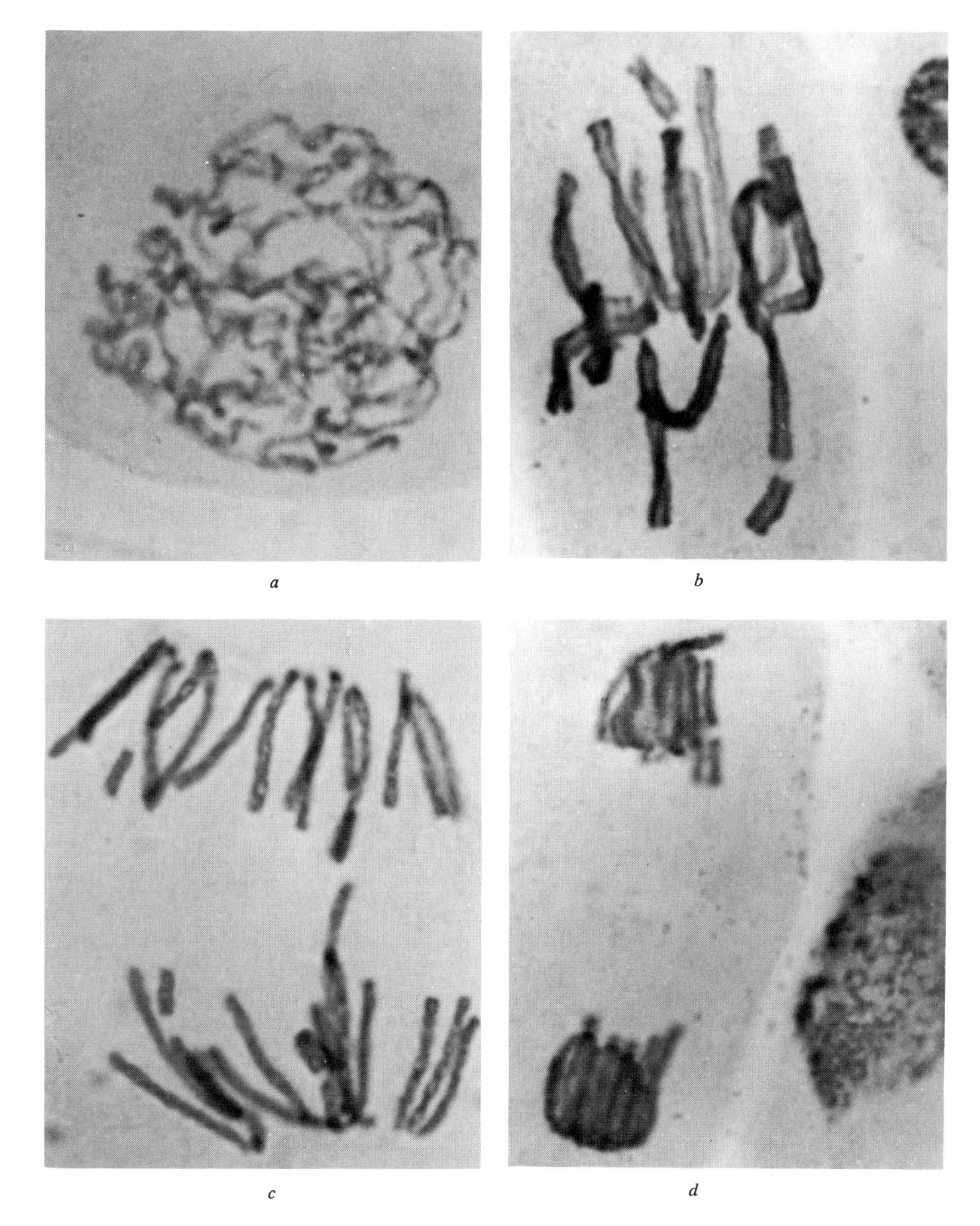

Figure 9.3 Stages of mitosis in root tip cells of the broad bean, *Vicia faba:* (*a*) prophase, (*b*) metaphase, (*c*) anaphase, (*d*) telophase.

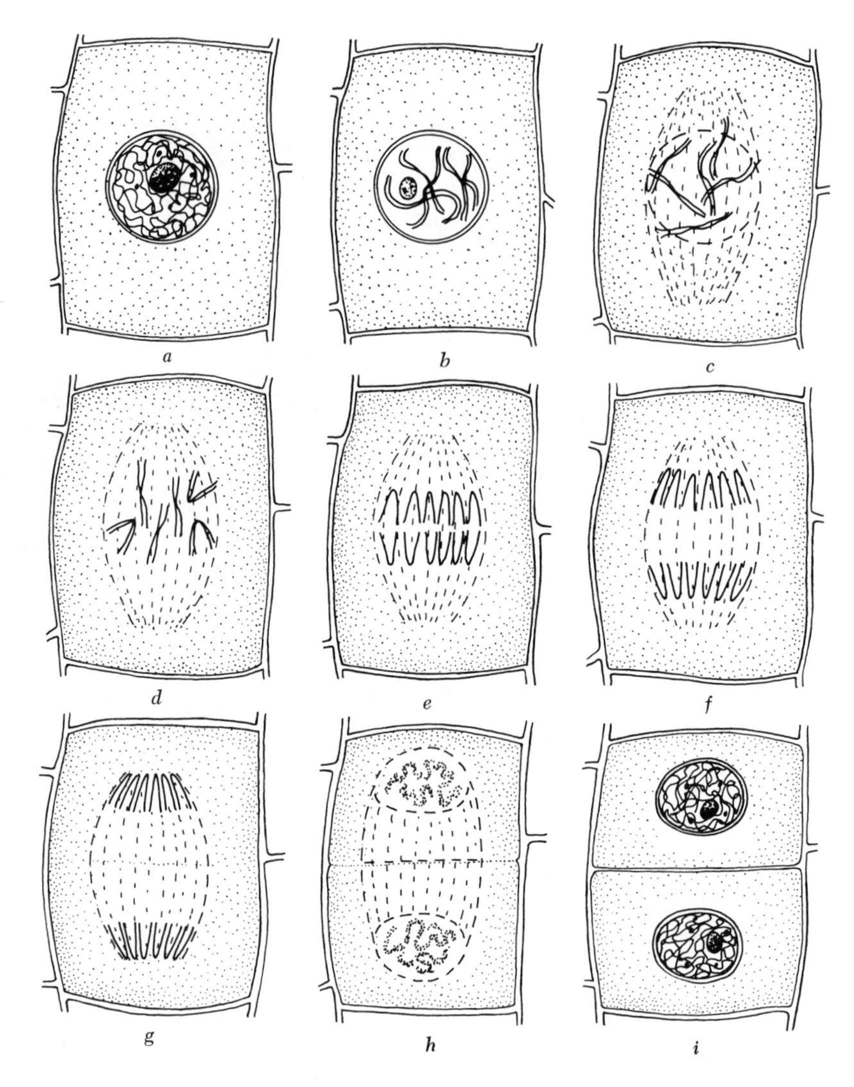

Figure 9.4 The progress of cell division, outlined in schematic form. As the cell prepares to divide, the chromosomes appear as distinct bodies in the nucleus, with a split along their length. The spindle appears at metaphase and separates the two chromatids of each chromosome at anaphase, after which the cell plate cuts the cell into two new cells. Karyokinesis, or mitosis, refers to the nuclear events of cell division; cytokinesis refers to the division of the cytoplasm by the cell plate.

chromatids, which are in very close association with each other all along their length. These sister chromatids are relationally coiled (twisted) around each other. As prophase continues, the chromosomes continue to condense, each chromatid becoming shorter and fatter (therefore, more distinct) by some controlled process

Figure 9.5 Late-prophase stage in a spermatogonial cell of the amphibian *Amphiuma*. Each chromosome is longitudinally split into two chromatids, and the centromeres are indicated by the constricted region in each chromosome. The fuzziness of the chromosomes is due to projecting loops of fine chromatin; these would be withdrawn into the body of the chromosome by full metaphase.

of internal coiling. A lightly stained, less contracted region of the chromosome, the *centromere*, can be detected, each chromosome having its centromere located at the particular position (Figure 9.5).

During prophase the nucleoli, which can now be seen to be associated with specific regions of specific chromosomes, the nucleolar organizer regions (NORs), gradually diminish in size, and eventually the nucleolar material becomes dispersed. The nuclear envelope also breaks down, to form membranous vesicles, and the chromosomes are "released" into the cytoplasm. However, also during prophase a *spindle* forms in the cytoplasm around the nucleus. The spindle is bipolar and consists of bundles of microtubules oriented longitudinally between the poles

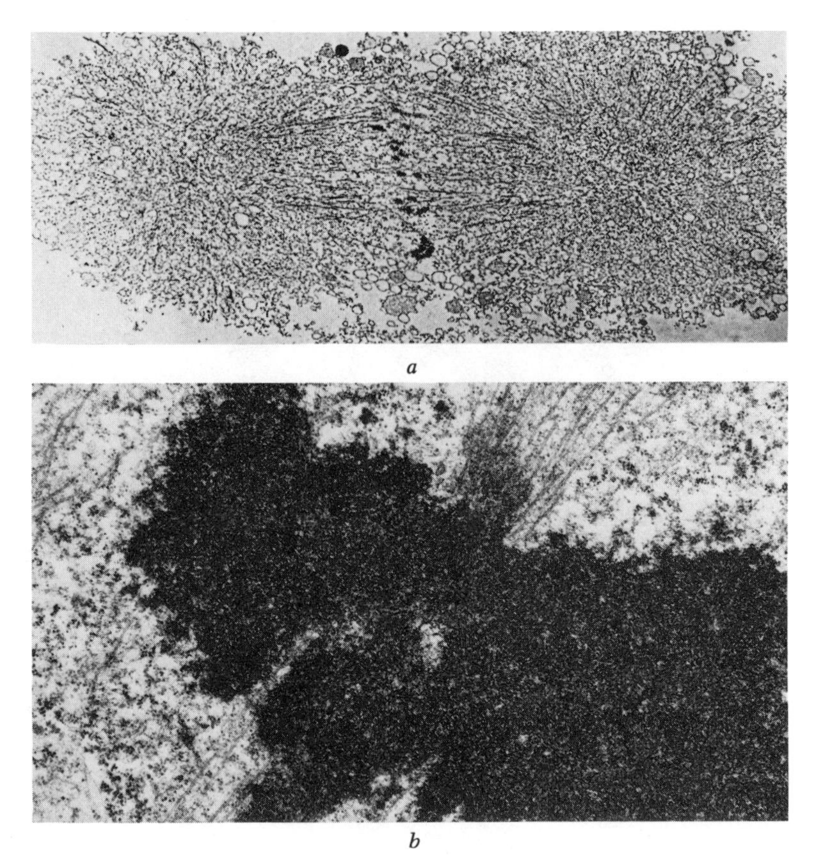

Figure 9.6 Electron micrographs of spindles and spindle structure in sea-urchin eggs. (*a*) Isolated spindle at low magnification (× 2,100), with the chromosomes appearing dark on the metaphase plate, a vague region across center of spindle; (*b*) Metaphase chromosome of *Haemanthus* showing the attachment of spindle microtubules to the kinetochores of sister chromatids.

(Figure 9.6). Once the nuclear envelope disintegrates, each chromosome becomes attached by its centromere region to certain spindle fibers. The term "kinetochore" is often used to refer to that part of the centromere to which the microtubules are attached, and which appears to be a disc or platelike structure.

Following attachment of the chromosomes to the spindle, they become aligned in such a way that all of the centromeres lie in a plane equidistant from the spindle poles. This stage is *metaphase* and the plane is sometimes referred to as the *metaphase plate.* Metaphase is also the stage at which the chromosomes have reached the point of maximum contraction (Figures 9.3 and 9.4).

The next stage, *anaphase*, begins when the centromeres, each of which up until now has behaved as a functionally single unit, divide and begin to separate (Figure 9.4). The sister centromeres move apart as the spindle fibers shorten, thus separating the sister chromatids from each other. Thus, during anaphase two

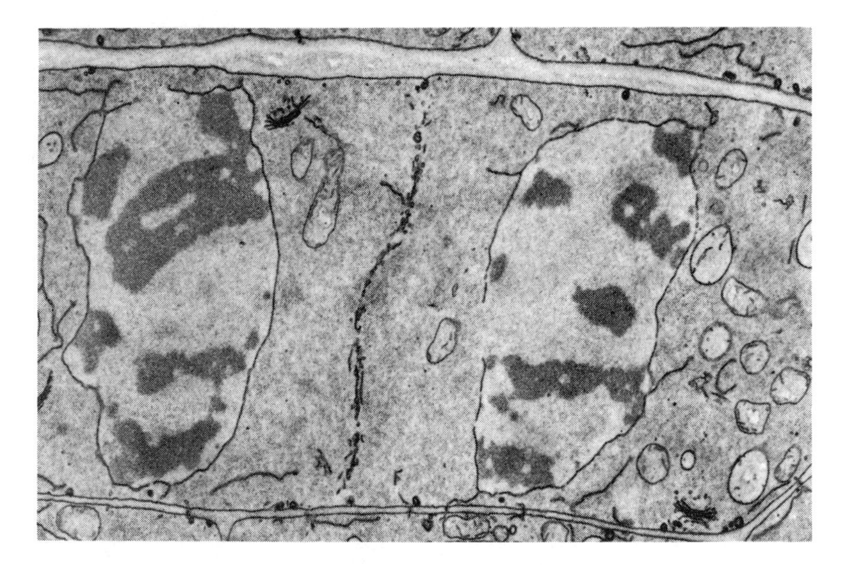

Figure 9.7 Electron micrograph of a maize cell in late telophase, with the cell plate forming across the center.

groups of chromatids, each chromatid representing half of an originally double chromosome, are moving toward opposite poles.

Telophase (Figures 9.3 and 9.4) begins when the centromeres reach the poles of the spindle. During telophase the chromatids, each of which can now be considered as chromosomes, become decondensed and fuzzy in outline. A new nuclear envelope reforms around each daughter group, nucleoli reappear, and the daughter nuclei assume a typical interphase appearance.

Finally, true cell division, or *cytokinesis*, occurs, involving the formation of a cell wall between the daughter nuclei. This begins as a *cell plate*, or *phragmoplast*, formed by aggregation of secretion vesicles from the Golgi complexes (Figure 9.7, see also Figure 7.12, Chapter 7). These vesicles which contain precursors of cell wall material, fuse with each other to form cell membranes and cell walls, thereby segmenting the cytoplasm into two parts. The spindle, which is also involved in cell plate formation, then disintegrates, and cell reproduction is completed. Thus, two new cells, each with a nucleus containing a full complement of information, have been formed from the original cell.

MITOSIS AND DIVISION IN ANIMAL CELLS

The end result of cell reproduction is the same in both plant and animal cells: the formation of daughter cells of like genetic constitution. This stems from the fact that the chromosomes in each behave similarily. Differences do exist, however, and the division of cells in the embryo of the whitefish reveals these in striking fashion (Figure 9.8). The first major difference is in the appearance of the spindle

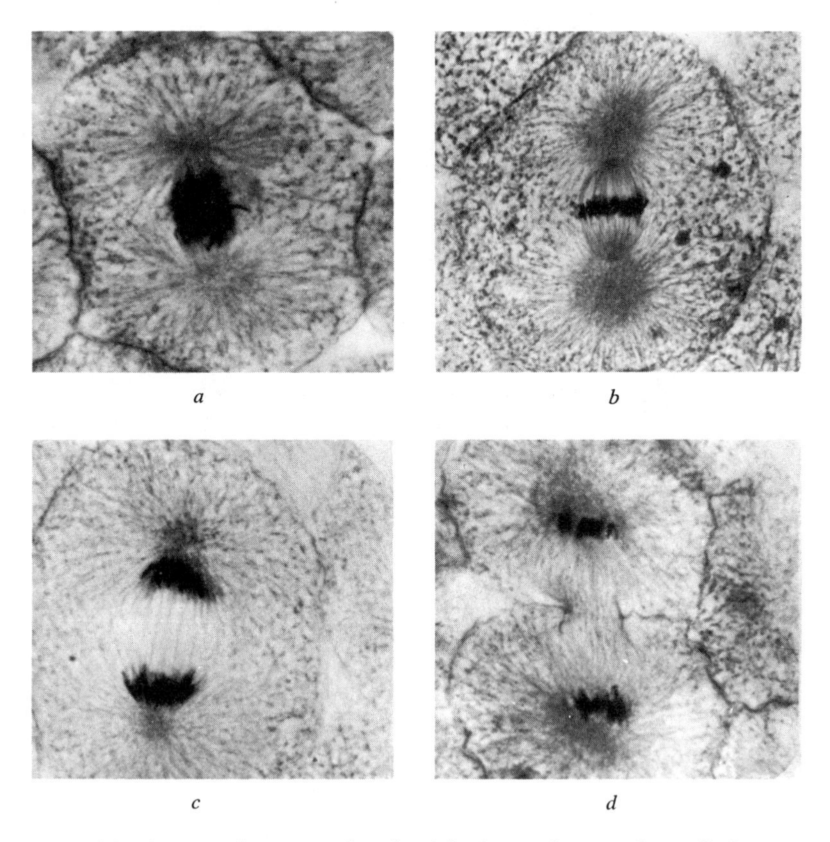

Figure 9.8 Stages in division in the whitefish. (*a*) prophase, with spindle beginning to form; (*b*) metaphase; (*c*) anaphase; (*d*) telophase, with the furrow cutting the cell into two new daughter cells.

apparatus. During late interphase a structure known as the *aster* can be detected; this consists of microtubules radiating out from a central area, or *centrosome*, within which are two centrioles. The centrioles, each of which replicates during interphase, separate until they lie opposite each other outside the nuclear envelope. As the centriole pairs separate from each other, spindle microtubules form between them, until a bipolar spindle with an aster at each pole is formed (Figure 9.8). As in the root tip cells, the chromosomes become attached to the spindle fibers and are aligned with their centromeres in the equatorial plane; anaphase movement then segregates sister chromatids to opposite poles.

The actual division of the animal cell into two daugher cells differs from that of a root tip cell. Instead of a cell plate forming, a process of *furrowing*, beginning at the outer edges of the cell and midway between the poles, cleaves the cell in two. High concentrations of actin are present around the cleavage furrow, suggesting that microfilaments as well as microtubules are important for the movements involved in cytokinesis in animal cells.

COLCHICINE

If we wish to examine the morphology of chromosomes, we can interrupt the normal sequence of events during mitosis by use of the drug *colchicine*. This plant alkaloid prevents the polymerization of tubulin subunits into microtubules, and hence inhibits spindle formation. The chromosomes then lie free in the cell and undergo further contraction, making them more distinct.

Figure 9.9 shows the chromosomes of the broad bean, *Vicia faba*, as seen in a colchicine-treated cell. Each one has a distinct morphology that is characteristic. The location of the centromere is constant and is identified by the constriction it forms, dividing each chromosome into two *arms* of characteristic length. The centromere, as pointed out, is the structure concerned with movement of the chromosome. Without it, a chromosome cannot orient on the spindle, and the chromatids cannot segregate from each other. In the longest chromosome found in the broad bean, another constriction is also present. This chromosome formed a nucleolus at that point, and the constriction, or gap, which is an uncoiled region of the chromosome, is the site occupied by the nucleolus before it disappeared. We can also see that the 12 chromosomes in the broad bean complement consist of one pair of long chromosomes with median centromeres and a secondary constriction plus five pairs of short chromosomes with subterminal centromeres. Colchicine is frequently used to facilitate an analysis of chromosome abnormalities, such as those induced by radiation and certain chemicals; many of these induced changes in the chromosome complement are readily recognizable and available for analysis.

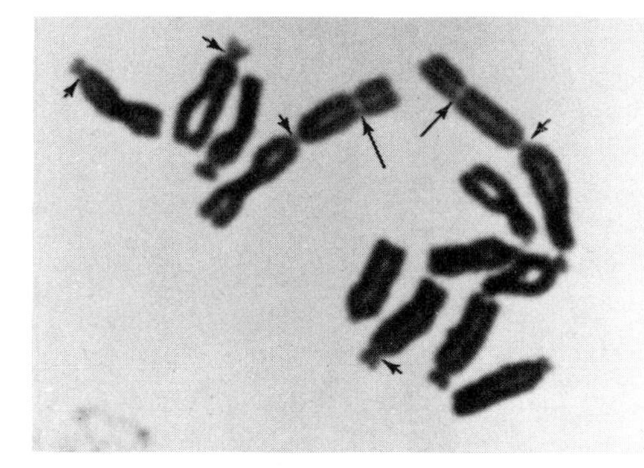

Figure 9.9 Colchicine metaphase in root tip cell of *Vicia faba*. The centromeres, which are not aligned in an equatorial plane, are evident (short arrows), as are the secondary constrictions on the two long chromosomes (long arrows).

THE MITOTIC SPINDLE

By using a technique of light microscopy in which orderly arrays of macromolecules can be made to appear brighter than other areas, the fibers can be shown to be regularly organized in the long axis of the spindle. Electron microscopy reveals the fibers to be bundles of microtubules which, during late prophase, become organized from a cytoplasmic pool of protein subunits. As we discussed in Chapter 8, the major protein involved is *tubulin;* however, other proteins are also present in the spindle, and are probably crucial to spindle form and function. The tubulin is not necessarily synthesized just prior to metaphase since it appears that the microtubules in the cell are in equilibrium with their subunits, with assembly into tubules and disassembly into subunits depending upon the circumstances at the moment.

As many as several thousand tubules may be present in a single spindle, and at least two classes of microtubules can be recognized. Polar microtubules extend from the polar regions toward the equatorial plane, where they interdigitate, or overlap, while the kinetochore microtubules extend from the centromere region of each chromosome toward the opposite spindle poles (Figure 9.10). There are also, in many but not all cells, two components to anaphase separation of chromatids; one of these involves separation of the two spindle halves, thereby increasing the distance between the poles, and the other involves shortening of the kinetochore-to-pole microtubules, thereby moving the attached chromatids toward the poles.

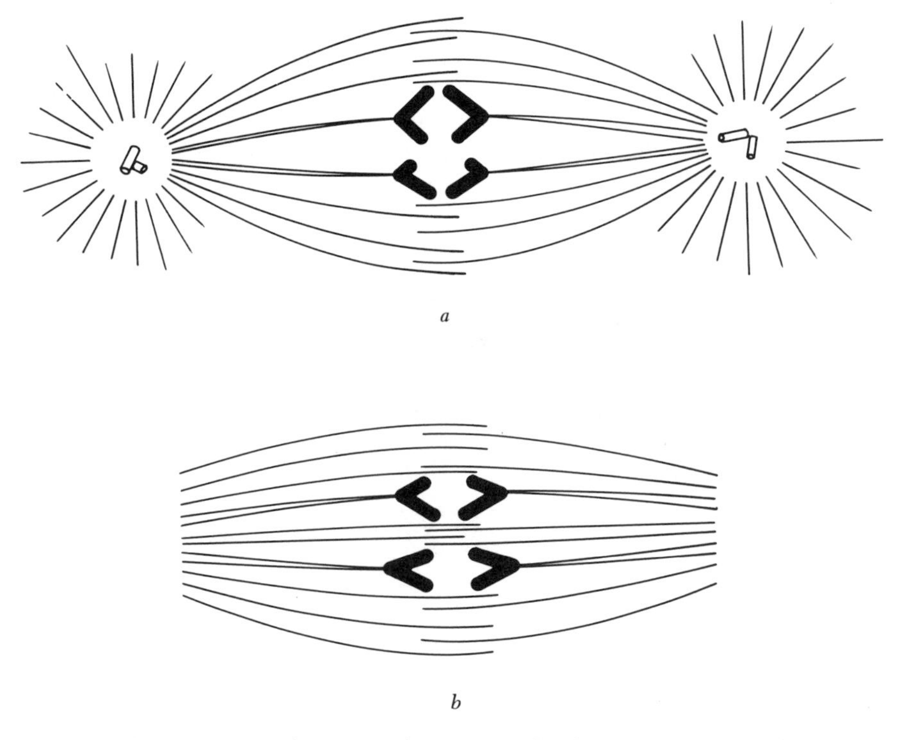

Figure 9.10 Diagram of structure of mitotic spindles showing polar microtubules overlapping at equatorial region and kinetochore microtubules to which the separating chromatids are attached. (*a*) Astral spindle with centrioles, typical of animal cells. (*b*) Anastral spindles usually found in higher plants.

The forces that move the chromatids, however, are still not understood. The simultaneous shortening of some tubules and a lengthening of others would seem to disprove the idea that chromosome movement is due to microtubule contraction. Another hypothesis is that depolymerization of the kinetochore tubules into monomeric subunits of tubulin occurs at the poles, somehow resulting in the shortened tubules with their attached chromatids being pulled toward the poles. This implies, of course, that as depolymerization takes place, the tubules remain constantly attached in some fashion to the polar area. This same hypothesis also considers the lengthening of the polar tubules to be a consequence of adding subunits to the existing microtubules, resulting in a further separation of the poles by an elongation of the tubules. Another hypothesis to account for chromatid separation suggests that somehow the two classes of tubules can slide past one another in specific directions, the force being generated by the breakdown and formation of cross-links between tubules. There is some evidence for such bridges and for ATPase activity associated with these. Such a mechanism bears some relation to the sliding filament model for muscle contraction (page 223). Of special interest with respect to the sliding filament mechanism for anaphase movement is that actin molecules have been found to be associated with the kinetochore tubules, but how such a relation can account for the behavior of spindle microtubules is not clear.

INTERPHASE

Although the fascinating dance of the chromosomes during mitosis has attracted the attention of cytologists for many years, it is now clear that, like any complicated dance, mitosis takes place only after far longer periods of preparation. It is during interphase, that period between the end of telophase and the onset of the subsequent prophase, that most of the preparation for mitosis and cytokinesis takes place. Enlargement of the cells occurs during this period, and the cellular constituents to be distributed to the daughter cells are synthesized in interphase. This can readily be demonstrated through the use of radioactive molecular tracers. If the precursor molecules of nucleic acids, proteins, lipids, and so on, are supplied to the cells, they are readily incorporated into these cellular macromolecules. If these same precursors were supplied during the period between the onset of prophase to the end of telophase, a very different pattern of incorporation, and at a very low level, would be found. However, since we are concerned with the cell in reproduction, we can point to the replication of DNA as the most significant event of interphase, a process which ensures that the daughter cells will possess the exact genetic information as the mother cell from which they arose. Looked at from the point of view of the chromosome, we can ask how the single, longitudinally undivided, chromosome of telophase becomes transformed into the two-chromatid chromosome of the following prophase.

DNA Synthesis

As indicated in Figure 4.20, DNA replicates in such a manner that each of the two original strands remains intact and acts as a template upon which a new complementary strand is formed. This method of replication is known as *semiconservative*, that is, when a double helix is being replicated each one of the polynucleotide strands is conserved (old DNA) while its complementary strand (new DNA) is synthesized from a pool of cellular constituents. Before this was made clear, however, there was debate as to whether the replication pattern was *conservative*, *semiconservative*, or *dispersive*. Conservative replication would mean that the original double helix remained intact, while the copy was totally new, while dispersive replication postulates that the original double helix is broken down, and each new double helix reassembled *de novo*. That DNA replication is indeed semiconservative was elegantly demonstrated by the experiment of Meselson and Stahl. They cultured cells of *E. coli* in a medium in which all of the nitrogen was ^{15}N, a heavy isotope of the more commonly found ^{14}N. This procedure made the DNA heavier than that from cells grown in a ^{14}N-containing medium, and these two kinds of DNA can readily be distinguished from each other by centrifugation in a cesium chloride (CsCl) density gradient (Figure 9.11). The cells containing only ^{15}N DNA were then transferred to a ^{14}N medium, and allowed to go through one round of DNA replication. When the DNA from these cells was extracted and centrifuged, its location in the density gradient was halfway between the ^{14}N and ^{15}N positions. Had the replication process been conservative, one-half of the DNA double helices should have been heavy (that is, ^{15}N DNA), the other half light, or ^{14}N DNA. After another round of replication in the ^{14}N-containing medium, the DNA was again extracted and centrifuged. Two bands were observed, one corresponding to that found after one round of replication, that is, it was half ^{14}N – half ^{15}N, while the other band corresponded to that for ^{14}N DNA. Had the replication process been dispersive, only one band would have been obtained, representing DNA that was a complete mixture of ^{14}N and ^{15}N. Figure 9.11 shows how the results are consistent with semiconservative replication.

The mechanism by which DNA is exactly replicated is extremely complicated, and requires the coordinated activity of many enzymes. The actual polymerization of nucleotides into a polynucleotide sequence specified by the template strand is accomplished by the enzyme DNA polymerase III. This enzyme adds nucleotide-5′-phosphate residues to the 3′-OH end of the growing chain. Thus the new polynucleotide sequences are always synthesized in the 5′ → 3′ direction (Figure 9.12). However, the opposite polarity of the template strands of a double helix presents a problem; while the new strand copied from the 3′ → 5′ template strand can be synthesized continuously, the strand copied from the 5′ → 3′ template of the same double helix cannot. Instead, this other new strand, which *overall* is formed in the 3′ → 5′ direction, is synthesized in short segments, each formed in the 5′ → 3′ direction (Figure 9.13). These *discontinuously* synthesized segments, called *Okazaki* fragments, are about 200 nucleotides long in

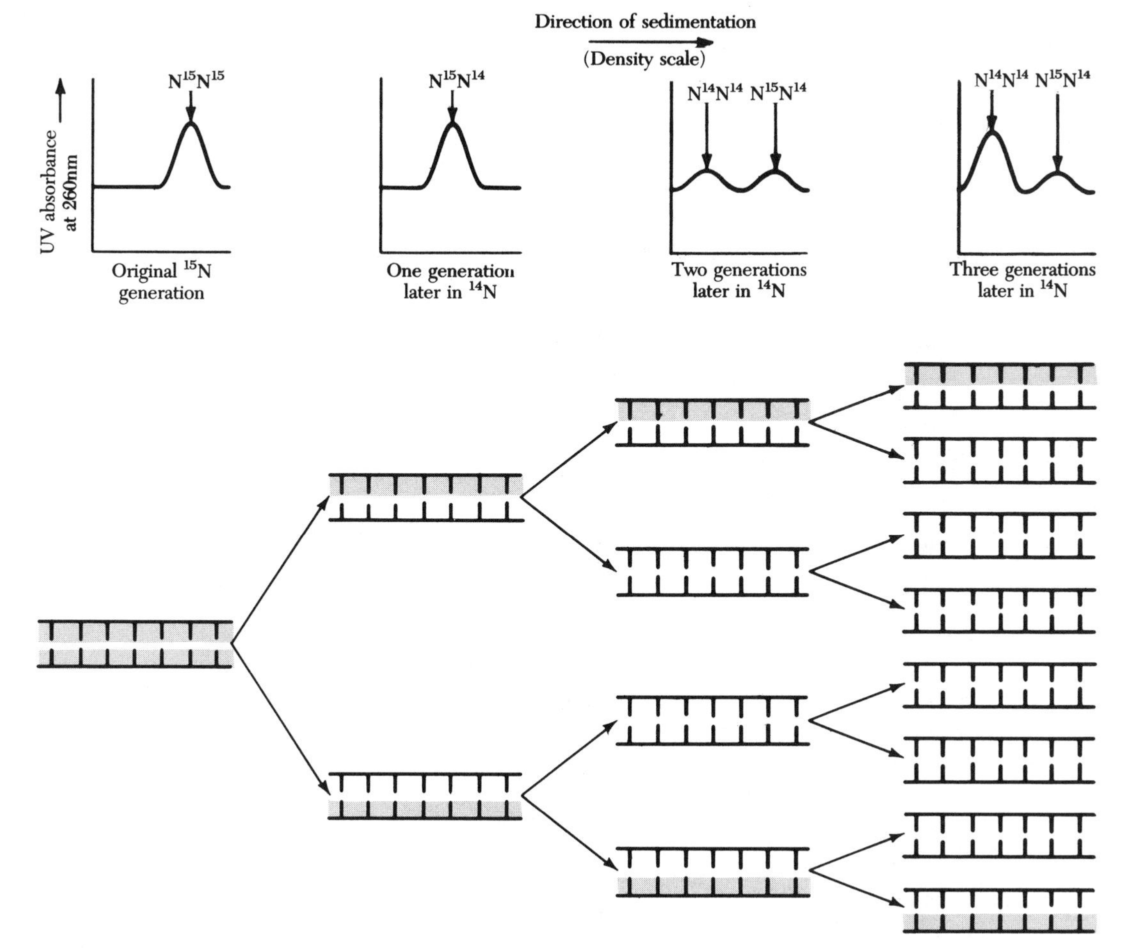

Figure 9.11 Semiconservative replication in *E. coli* as demonstrated by Meselson and Stahl, and based on the fact that DNA formed in the presence of heavy nitrogen (^{15}N as opposed to ^{14}N) can be distinguished from that containing ^{14}N by centrifugation in a CsCl density gradient. The original heavy nitrogen polynucleotide strands (shaded) are conserved, and although replication thereafter is only in the presence of $^{14}N/^{14}N$ helices can be distinguished from the ^{14}N, the $^{15}N/^{14}N$ helices. The sedimentation profiles are given above the appropriate double helices.

eukaryotes, and are primed by a short stretch of RNA; that is, initiation of synthesis involves the formation of a short sequence of ribonucleotides rather than deoxyribonucleotides. Once the chain has been initiated, it is completed by the successive addition of deoxyribonucleotides, and the short polyribonucleotide sequence is subsequently removed. As the ribonucleotides are excised from one *Okazaki* fragment, deoxyribonucleotides are added to the end of the adjacent fragment until the gaps are filled in with DNA (Figure 9.14). Another enzyme, DNA polymerase I, is

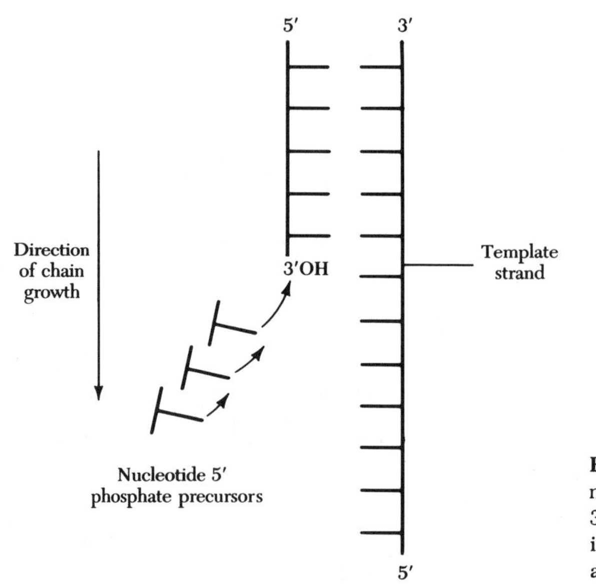

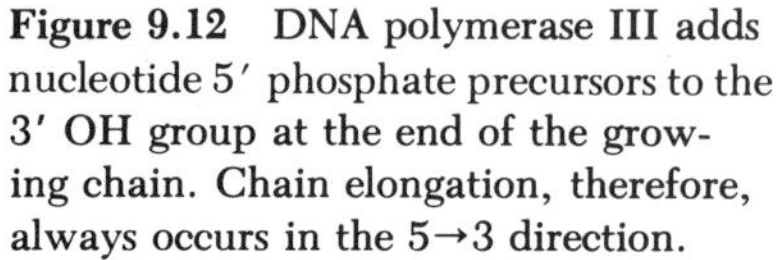
Figure 9.12 DNA polymerase III adds nucleotide 5′ phosphate precursors to the 3′ OH group at the end of the growing chain. Chain elongation, therefore, always occurs in the 5→3 direction.

responsible both for removal of the RNA and for insertion of the DNA. Finally, the ends of the completed DNA sequence are joined together by yet another enzyme, DNA ligase (Figure 9.14).

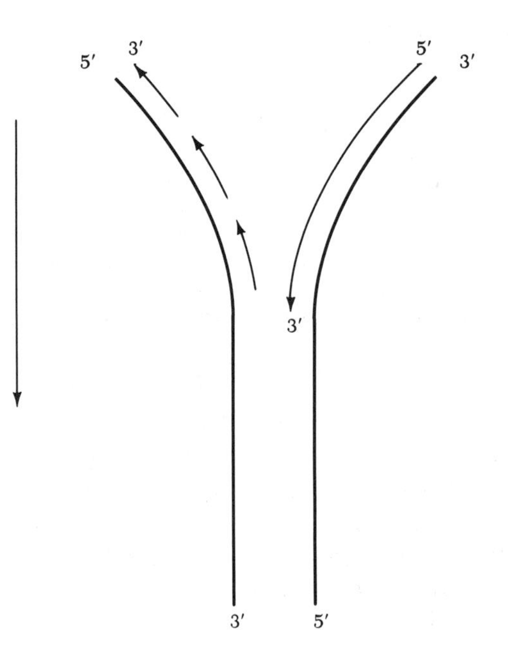

Figure 9.13 Discontinuous synthesis of one strand of DNA. The strand being copied from the template on the right is synthesized continuously in the 5→3 direction. The strand being copied from the template on the left is synthesized as successive short fragments, each formed in the 5→3 direction, and subsequently joined together.

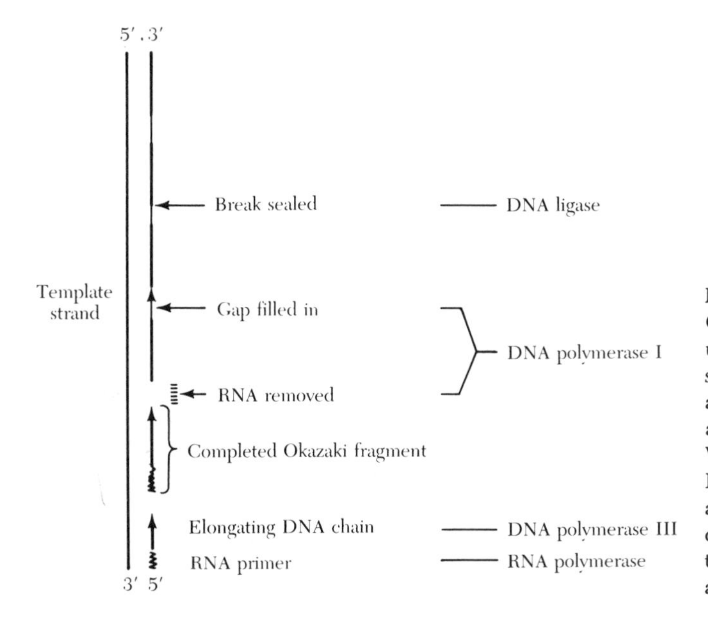

Figure 9.14 Formation and joining of Okazaki fragments in the discontinuously synthesized strand of DNA. A short stretch of RNA is formed and acts as a primer to which the DNA polymerase can add deoxyribonucleotides. When the fragment is completed the RNA primer is removed from the 5′ end and the gap is filled by addition of deoxyribonucleotides to the 3′ end of the adjacent fragment. The fragments are then joined together by DNA ligase.

Replicons

The process we have described above is what happens at the *replication fork*, which of course moves along the stretch of double helix being replicated. In eukaryotic cells the rate of fork movement is about 0.5 to 2.5 μm per minute. However, mammalian chromosomes, for example, may contain DNA molecules of about 20,000 to 30,000 μm in length. If replication were initiated at one end of the molecule and proceeded to the other end, it would take many days for the completion of replication instead of the known 6 to 8 hours. However, the problem is solved by each chromosome having a number of points of replication initiation, and therefore different stretches of the same molecule are replicated simultaneously. This was demonstrated by the use of tritiated thymidine ([^{3}H]TdR) a radioactively labeled precursor which is incorporated only into DNA. When replicating cells are exposed to [^{3}H]TdR for a short period of time (a pulse), and the DNA is then gently extracted in a relatively unfragmented state and stretched out on a slide, the radioactive portions of DNA can be seen to alternate with nonradioactive portions (Figure 9.15). Longer pulses result in longer stretches of radioactivity than do shorter pulses, indicating that the replicated segments of DNA are gradually as being extended. Thus the DNA of a chromosome can be thought of being composed of short stretches, each capable of initiating replication; these are the *replicons*, and a number of them are joined together to form a chromosome. Each replicon has a point at which replication begins, but an examination of figures such as those in Figure 9.15 shows that a replicon grows in both directions from the point of origin (Figure 9.16). Eventually, the replicons meet each other to complete the replication of the chromosomal DNA.

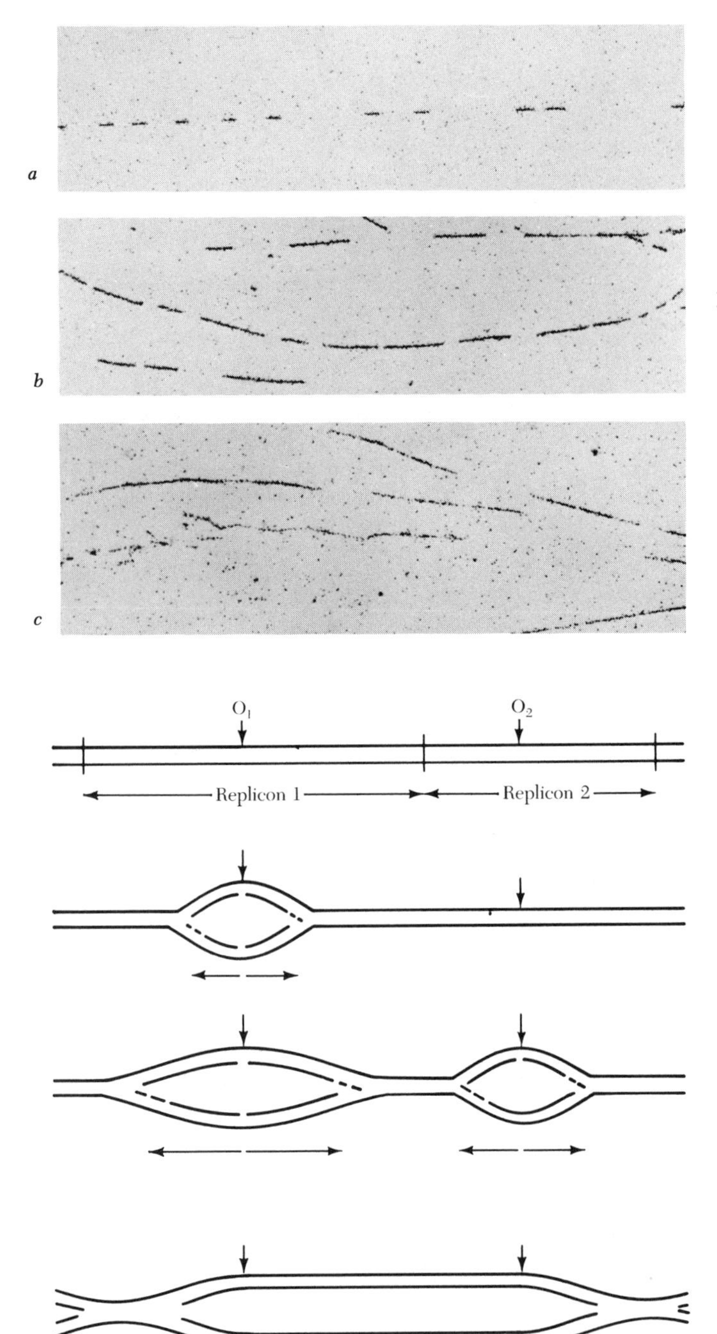

Figure 9.15 Autoradiographs of replicons in the DNA of sunflower, *Helianthus annuus*, labeled with ^{3}H-TdR. (*a*) Only short segments of the bidirectional replicon have been replicated during the pulse label. (*b*) Longer labeling times permit more of each replicon to be replicated. (*c*) DNA labeled first with high specific activity ^{3}H-TdR, followed by exposure to low specific activity ^{3}H-TdR. The fading "tails" of silver grains point in the direction of replication. (Courtesy of Jack Van't Hof.)

Figure 9.16 Adjacent replicons in a linear chromosome. Replication begins at the origins (O) and proceeds bidirectionally in each replicon until adjacent replicons merge.

Segregation of DNA in Chromosome Duplication

We must now consider how the newly replicated DNA double helices are distributed to the two chromatids of the replicated chromosome. By supplying tritium-labeled thymidine to cells that are synthesizing DNA, and then following the distribution of the radioactive DNA by radioautography, it can be demonstrated that each chromatid of a mitotic chromosome contains some "old" DNA, present before replication, and some "new" DNA, formed during the preceding interphase. At the first mitosis following synthesis, both chromatids of the chromosome are labeled, indicating that both chromatids contain newly synthesized DNA (Figure 9.17a). If the daughter cell progeny are allowed to complete another interphase in the *absence* of radioactive thymidine and enter a second mitosis, only one of the two chromatids will be labeled (Figure 9.17b). These results are best explained (Figure 9.18) as follows: a single unreplicated chromatid consists of the two strands of a DNA double helix. Replication in the presence of radioactive thymidine results in the formation of two chromatids, each consisting of one strand of the DNA double helix of the original chromatid and one newly synthesized strand (Figure 9.18). Thus both chromatids will appear radioactive at mitosis. During the next interphase, the single chromatids, each consisting now of a double helix with one radioactive and one nonradioactive strand, replicate again, this time in the absence of radioactive thymidine. Thus one of the two chromatids formed will consist of a double helix with a radioactive strand and a

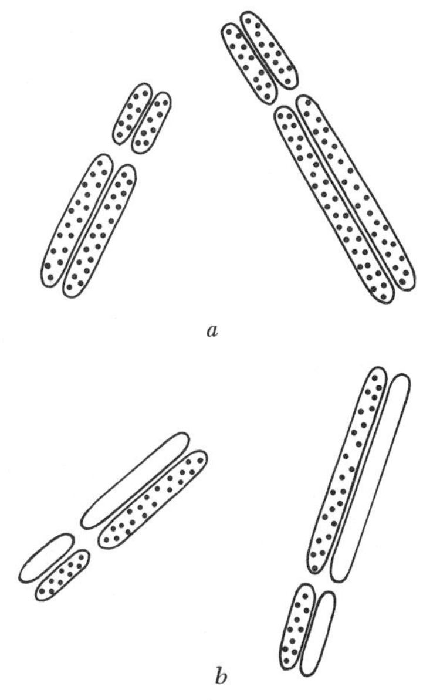

Figure 9.17 Appearance of chromosomes in autoradiographs of cells in (*a*) first metaphase and (*b*) second metaphase following DNA synthesis in the presence of ^{3}H-thymidine. In (*a*) both chromatids are labeled, in (*b*) only one of the two chromatids of each chromosome is labeled.

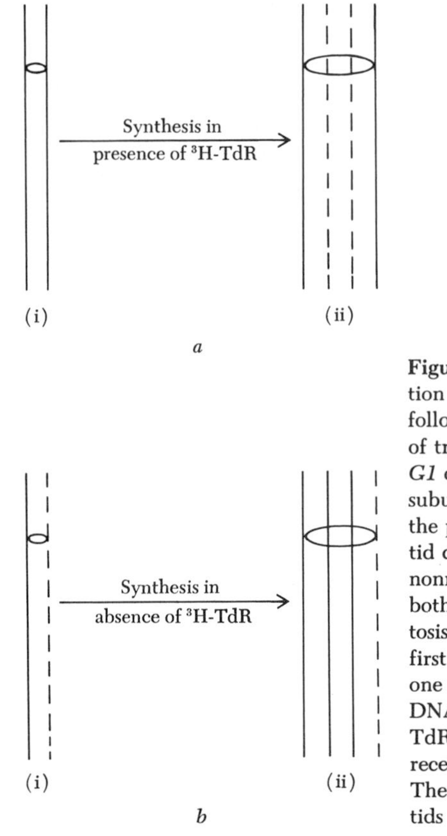

Figure 9.18 Explanation of distribution of radioactivity in chromosomes following DNA synthesis in the presence of tritiated thymidine (^{3}H-TdR). (*a*) (i) *G1* chromosome consisting of two intact subunits. Following DNA synthesis in the presence of ^{3}H-TdR, each chromatid consists of one radioactive and one nonradioactive subunit (ii). Therefore, both chromatids are labeled at first mitosis. (*b*) (i) *G1* chromosome following first mitosis, consisting of two subunits, one of which is radioactive. Following DNA synthesis in the absence of ^{3}H-TdR, only one of the two chromatids receives the radioactive subunit (ii). Therefore, only one of the two chromatids is labeled at the second mitosis.

newly formed nonradioactive strand; this chromatid will appear labeled. The other chromatid, however, will be unlabeled, since it consists of a double helix with two nonradioactive strands (Figure 9.18).

TIMING OF DNA SYNTHESIS AND THE MITOTIC CYCLE

DNA synthesis occurs in the S (for synthesis) period of interphase, a discrete and limited portion of time which is preceded by a G1 phase, and followed by a G2 phase (G stands for *gap*). This permits the mitotic cycle to be represented as in Figure 9.19, with the S period occupying a middle position covering about one-third of the interphase period in rapidly dividing cells. Some chromosomes begin replication early in S, others later, and among human chromosomes, at least, some chromosomes can be identified by their timing of DNA synthesis. Heterochromatin DNA, in particular, is known to be replicated late in S, and a pulse of [^{3}H]TdR given to human cells late in S will result in radioactivity showing in the centric

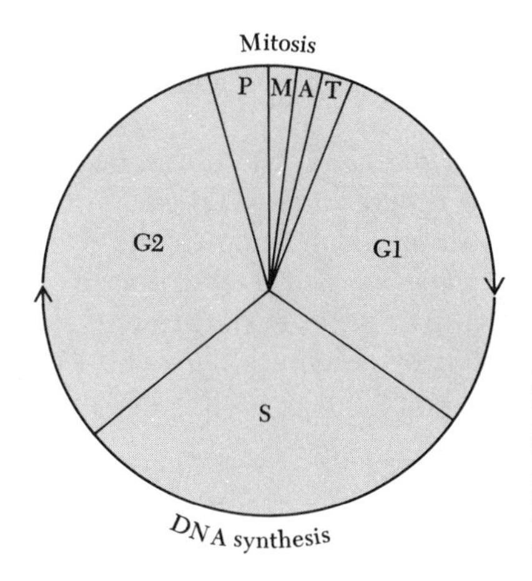

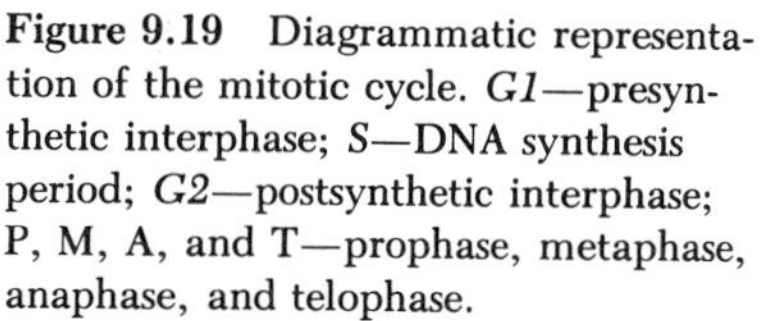
Figure 9.19 Diagrammatic representation of the mitotic cycle. *G1*—presynthetic interphase; *S*—DNA synthesis period; *G2*—postsynthetic interphase; P, M, A, and T—prophase, metaphase, anaphase, and telophase.

regions of most chromosomes as well as in the almost totally heterochromatic Y chromosome.

The discreteness of the S period permits us to measure the duration of the cell cycle and its constituent phases by an ingenious technique involving the use of [^{3}H]TdR and autoradiography. If root meristems, for example, are supplied with [^{3}H]TdR in a pulse of 30 minutes, only those cells in S at the time will incorporate the precursor in their DNA. Samples of root tips are then taken at different times following such a "pulse" label and prepared as autoradiographs. By counting the percentage of mitotic figures that are labeled at these different times, the progression through mitosis of the cells that were in S at the time of labeling can be followed (Figure 9.20). For a period equal to the duration of G2 no labeled mitotic

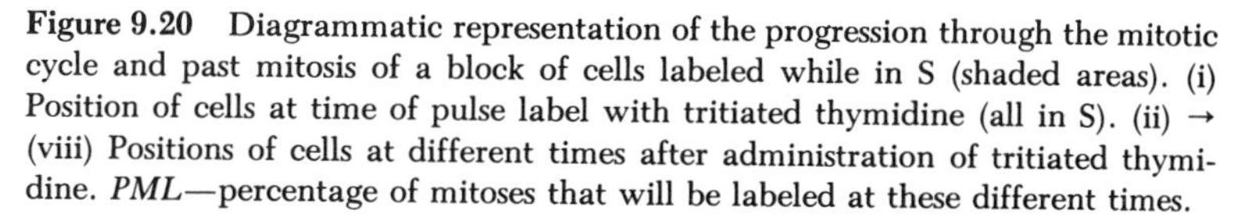
Figure 9.20 Diagrammatic representation of the progression through the mitotic cycle and past mitosis of a block of cells labeled while in S (shaded areas). (i) Position of cells at time of pulse label with tritiated thymidine (all in S). (ii) → (viii) Positions of cells at different times after administration of tritiated thymidine. *PML*—percentage of mitoses that will be labeled at these different times.

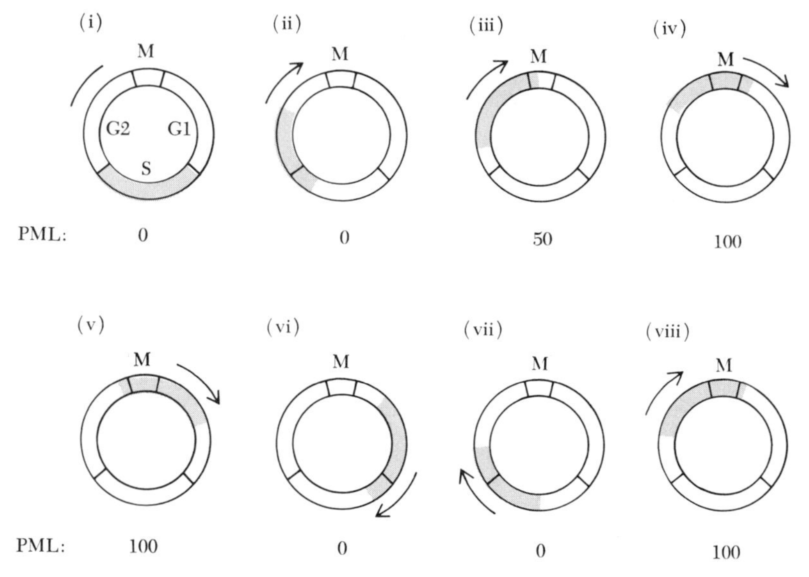

figures will be seen. The first labeled cells to reach mitosis will be those cells completing DNA synthesis (that is, cells that were at the end of S) during administration of the tritiated thymidine. These cells will reach mitosis after a time interval equal to the duration of G2, which can, therefore, be measured directly (Figures 9.20 and 9.21a). After this, the percentage of labeled mitotic figures will reach 100 percent, this level being maintained as long as the former S cells are reaching mitosis. Eventually, the cells that were at the beginning of S during the pulse label will be followed into mitosis by the unlabeled, former G1' cells, and the percentage of labeled mitotic figures will drop (Figures 9.20 and 9.21a). Thus, the time period between the appearance of labeled mitoses and their subsequent disappearance is

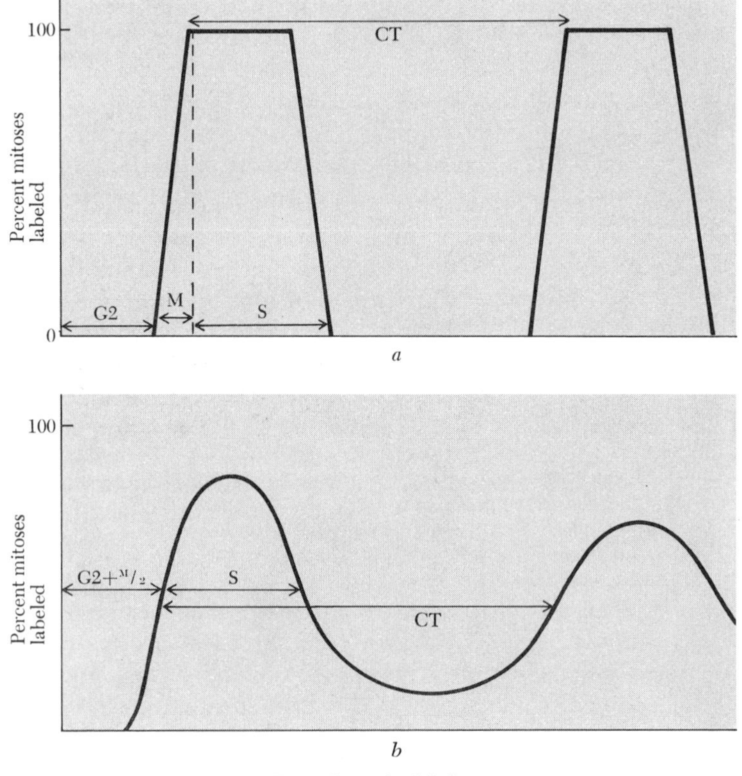

Figure 9.21 Percentage mitosis labeled curve to measure time parameters of the mitotic cycle. The percentages of mitoses which are labeled are determined at different times after administration of a pulse label of tritiated thymidine. The durations of some of the phases of the cycle can be measured from the plotted data. (See text and Figure 7.13). (*a*) Theoretical curve. (*b*) Actual curve obtained in practice and used for estimates. G2—Duration of *G2*. M—Duration of mitosis. S—Duration of *S* period. CT—Duration of mitotic cycle. G2 + M/2—Average duration of *G2* plus half of mitosis.

equivalent to the duration of S, which can, therefore, be determined. The next time labeled mitotic figures will be seen is when the daughters of the former S cells, which are themselves still labeled, complete interphase and enter mitosis (Figures 9.20 and 9.21a). Thus the duration of the complete cell cycle can be estimated by measuring the time interval between successive appearances of labeled cells in mitosis (Figure 9.21a). The duration of mitosis itself can be estimated in one of several ways, and hence the duration of G1 determined by subtraction. Figure 9.21b shows how average values of various phase durations are obtained in practice from such data.

The durations of the complete cell cycle and of its component phases, G1, S, G2, and M (mitosis) vary from cell type to cell type. For example, in lateral root tips of *Vicia faba*, the average cycle duration of most of the meristematic cells is about 14 hours at 22°C. About $2\frac{1}{2}$ hours are spent in G1, 6 hours in S, $3\frac{1}{2}$ hours in G2, and mitosis itself lasts 2 hours. Cells of the crypts of the intestinal epithelium of the mouse divide on average once every 19 hours, with G1 lasting about $9\frac{1}{2}$ hours, S taking $7\frac{1}{2}$ hours, and G2 plus mitosis together lasting 2 hours.

Even between cells of the same tissue, variation exists in the time parameters of the cycle. Most of this variation is in the G1 phase, suggesting that once a commitment to DNA replication is made, cells proceed through the rest of the cycle at some fixed rate. This is not always the case, however, since cells can be arrested in G2 during normal development.

Although DNA synthesis and mitosis are the most easily recognizable processes occurring during the cell cycle, as well as being the most significant, they themselves depend on other events that must occur. Progression through the cell cycle does require the synthesis of specific enzymes at certain times, and the production and utilization of these enzymes must be under strict temporal control. The cell cycle is a complicated affair; nucleus and cytoplasm interact in an orderly sequence of events in which all of the elements are in a state of readiness at the proper time. Each event depends on preceding events and is in turn necessary for the events to follow.

Presumably, the G1 and G2 phases represent stages in the cycle during which preparations for DNA synthesis and mitosis respectively are taking place, but we still are unfamiliar with the exact nature of these preparations. We do know, however, that during the cycle certain internal conditions that come to exist in the cell can signal the beginning of DNA synthesis and mitosis. This can be shown by fusing cells from different stages of the cycle. For example, if cells in early G1 are fused with cells in S, the G1 cells are stimulated to initiate DNA synthesis much earlier than normal. Furthermore, fusion of interphase cells with cells in mitosis results in the premature condensation of the chromosomes of the interphase cells; if the prematurely condensed chromosomes are from a G2 cell, they consist of two chromatids, but those from a G1 cell are single, not yet having undergone replication. Such experiments demonstrate clearly that specific internal environments during the cell cycle are responsible for determining progression of cells through the cycle.

Preparation for nuclear division			Nuclear division	
Interphase	Prophase	Metaphase	Anaphase	Telophase
RNA and protein synthesis			Separation of sister centromeres	
Replication of chromosomes	Organization of spindle proteins	Organization of spindle proteins	Movement of chromatids to poles	Uncoiling of chromosomes
	Shortening of chromosomes	Shortening of chromosomes	Spindle elongation	Disappearance of spindle
	Disappearance of nucleoli			Reappearance of nucleoli
	Disappearance of nuclear envelope			Reappearance of nuclear envelope
		Connection of centromeres to spindle		Division of cell
		Movement of centromeres to equatorial plane		
	Replication of	Replication of	centrioles	

Figure 9.22 Table of events taking place in preparation for, and during, nuclear division.

Some of the important events we know to take place during the cell cycle are presented in Figure 9.22. Other changes do occur, including alterations in cell membrane configuration and permeability, modification of chromosomal proteins, and changes in sensitivity to radiation. However, the significance of these and other changes, and their relationship to chromosome replication and segregation, is still not fully understood.

SIGNIFICANCE OF CELL REPRODUCTION

Cell reproduction is, of course, part of the process of growth. Although the dance of the chromosomes and the formation of the spindle and the daughter cells are the more obvious parts of the drama, it is also involved with the assimilation of materials from the outside, their transformation through breakdown and synthesis into new cellular parts, and the utilization of energy. Cell enlargement also takes place. We know of no cells, except the fertilized egg and a few of its derivative cells, that simply divide from one large cell into two others of half size and again into four cells of quarter size. This is not the usual way in which cell reproduction proceeds, for interspersed in the process are periods of growth; each division is a tumultuous affair, from which the cells must recover before proceeding again through the cycle.

The great significance of cell reproduction is the fact that it ensures a continuous succession of identically endowed cells. Chaos would result if only a random array of cells of varying qualities and capacities were to reproduce themselves; organized growth must proceed from cells of similar nature that subsequently can be molded according to the demands of the species. The species could not otherwise persist. We mentioned earlier that the chromosome is an intricate fabric composed of nucleic acids and proteins. Since the nucleus is the control center of the cell, and since the nucleus contains little else but chromosomes, the chromosomes must be the regulators of cellular metabolism and the structural characteristics of the cell. Therefore, if two cells are to behave similarly, they must have the same amount and type of nucleic acids and proteins. The longitudinal duplication of the chromosomes into identical chromatids and their segregation to the poles at anaphase must be exact to the minutest degree; the kind of cell reproduction described provides the mechanism needed. From the time a particular species was formed, this process has gone on with an exactitude that almost defies the imagination. Accidents and variations do occur—and indeed they must if evolution is to take place—but they are relatively few in number.

Cell reproduction in unicellular organisms is, of course, equivalent to exact reproduction of individuals and is responsible for increase in number of individuals of like type. Cell reproduction is, therefore, an act of survival, since cells that do not reproduce must eventually die. Reproduction and the growth associated with it bring fresh substance into the cell and effectively prevent aging, giving the cell potential immortality. In multicellular organisms, the production of new cells by cell duplication allows for division of labor, different cells and groups of cells becoming specialized to carry out different functions.

Viewed in this manner, cell division is, therefore, a first step toward cell differentiation. But this is the antithesis of survival. Since differentiated cells lose their capacity to divide, differentiation is also a first step toward eventual death. The significance of cell division, then, depends not only on the phenomenon itself but also on the kind of cell that is dividing and the consequences of division to an organism.

Although the exact process of cell reproduction ensures that all of the cells of a multicellular organism contain the same type and amount of information, different types of cells in an organism manipulate energy in different ways, and not all of the information they contain need be used. Exact duplication and segregation of information in cell reproduction, however, allows more flexibility to be retained by cells, since they always will have the potential to change their function should such be required. The kind of cell reproduction we have described, which does ensure that different cells do contain the information they need as well as some they may not, is also probably more efficient to the organism than an unequal but highly specific distribution of only those different types of information required by specific cells.

ADDITIONAL READING

INOUE, S. 1981. Cell division and the mitotic spindle. *J. Cell Biol. 91:* 132s–147s.

KORNBERG, A. 1980. *DNA Replication.* W. H. Freeman Company, Publishers, San Francisco.

LEWIN, B. 1980. *Gene Expression,* Vol. 2: *Eukaryotic Chromosomes,* 2nd ed. John Wiley & Sons, Inc., New York.

MAZIA, D. 1974. The cell cycle. *Sci. Am. 230*(1): 54–64.

MCINTOSH, J. R. 1979. Cell division. In *Microtubules,* K. Roberts, and J. S. Hyams, (eds.). Academic Press, Inc., New York.

MESELSON, M., and STAHL, F. W. 1958. The replication of DNA in *E. coli. Proc. Natl. Acad. Sci. USA 44:* 671–682.

MITCHISON, J. M. 1971. *The Biology of the Cell Cycle.* Cambridge University Press, Cambridge.

PICKETT-HEAPS, J. D., TIPPIT, D. H., and PORTER, K. R. 1982. Rethinking mitosis. *Cell 29:* 729–744.

SLOBODA, R. D. 1980. The role of microtubules in cell structure and cell division. *Am. Sci. 68:* 290–298.

TAYLOR, J. H. 1974. Units of DNA replication in chromosomes of eukaryotes. *Int. Rev. Cytol. 37:* 1–20.

10

Meiosis—The Transmission of Information

If, over the course of several generations, we examine the individual members of any species, we recognize a range of certain morphological and behavioral characteristics that are shared by all individuals, and that allow us to distinguish them from members of other species. This distinction is readily made with human beings, for our nearest relatives, the chimpanzees and gorillas, are sufficiently different from us to eliminate the possibilities of mistaken identity. But such distinctions become more difficult when we are dealing with species of bacteria, moths, or oaks. Even here, however, close scrutiny reveals that these species produce individuals true to form in much the same way as does the human species. There is, consequently, from generation to generation, a *continuity* of species, but we find on even closer scrutiny that the individuals do differ among themselves, a fact more readily observable among human beings but nonetheless equally detectable in other species.

The mechanisms that keep the species going, generation after generation, must be able to account for both the conservation of type (continuity) as well as the diversity among the members of the species. Since we recognize that the cell is the basic unit of organization, that eggs and sperm, among sexually reproducing species, are the slender cellular bridges connecting one generation to the next, and that DNA is the molecular basis of heritable information, we must turn to the cellular level for an understanding of these phenomena.

The continuation of any species, man or amoeba, oak tree or bacteria, depends on an unending succession of individuals. No individual is immortal, so the members of a population must reproduce if the species is to escape extinction. In unicellular organisms such as the amoeba, somatic cell division, as described in the preceding chapter, serves this function; it is a reproduction device that leads to the continued formation of new individuals. And, since somatic cell division and the mitotic cycle are mechanisms by which DNA replicates itself and chromosomes are distributed in a precise manner to daughter cells, all offspring arising through mitosis have the same number of chromosomes as the original parent—barring, of course, any accident of mitotic segregation.

The amoeba, however, like many unicellular and some multicellular organisms, is asexual; it does not produce sexual cells—*eggs* and *sperm*—and all descendant individuals have a unilateral or uniparental inheritance. But other unicellular and most multicellular organisms reproduce by sexual means; at some time during their life cycles they produce *gametes* (a general term applied to any type of sexual cell), which unite in pairs to form a single new cell called a *zygote*. From this cell a new individual develops, the product of biparental inheritance. The union of gametes is called *fertilization* or *syngamy*.

It is important to recognize that when two gametes unite through fertilization, the principal event is the fusion of gametic nuclei. Let us consider what this means in terms of chromosome number. The cells of the human being, for example, contain 46 chromosomes (Figure 10.1). If we assume for the moment that mitosis is the only type of nuclear division, the human egg and sperm would each contain 46 chromosomes since they would be mitotic descendants of the original zygote. The zygote formed by their union would then contain 92 chromosomes, and so, too, would the eggs and sperm produced by the individual developing from the new zygote. The grandparents of this individual would have had 23 chromosomes in each cell; the individuals of the next generation would possess 184 chromosomes. Starting with parents each possessing 46 chromosomes, each individual, by the end of the tenth generation, would have cells containing 23,552 chromosomes.

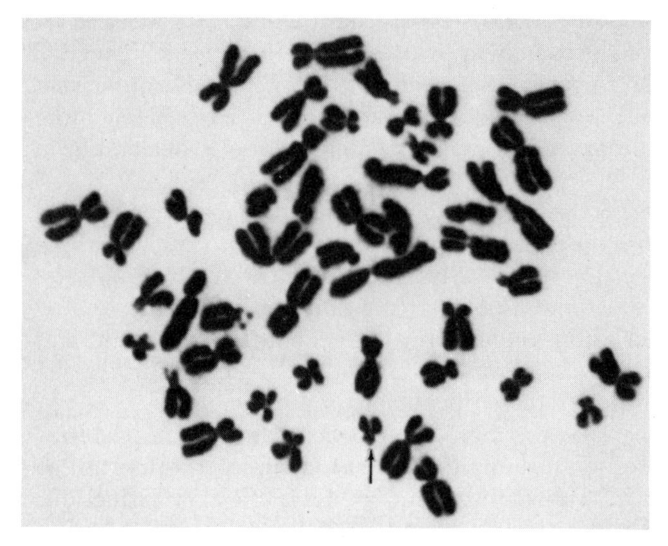

Figure 10.1 The somatic chromosomes of the human male, derived from cultured leucocytes. The Y chromosome is probably the small one identified by an arrow; the X is a medium-sized chromosome and difficult to identify with certainty.

Obviously, this would be a ridiculous state of affairs. We are merely emphasizing that, in a sexually breeding population, the increase in chromosome number resulting from fertilization cannot and does not go on indefinitely. At some time during the life cycle of an individual, a compensatory mechanism must reduce this number, for we know that the cells of individuals belonging to the same species have a striking constancy of chromosome number. Thus, normal human cells have 46 chromosomes; those of maize, 20; of the mouse, 40; of the rat, 42; and so on.

This numerical constancy for each species is repeated generation after generation. The gametes, therefore, of sexually reproducing species must have one-half the number of chromosomes found in the zygote and in the other cells of the body (since the latter arise from the zygote by mitosis). The reduction in number of chromosomes is accomplished by a special type of cell division called *meiosis*, which, in its barest essentials, consists of *two nuclear and cytoplasmic divisions but only one replication of chromosomes.* It is a type of division peculiar to sexually reproducing eukaryotes.

Before considering the details of meiosis and the features that distinguish this type of cell division from mitosis, we need to recognize certain terms that conveniently describe the chromosomal states of eukaryotes (Figure 10.2). The chromosomes in the nuclei of gametes are variously said to be of a *reduced, gametic, haploid*, or *n* number; those in the zygote and all cells derived from it by mitosis are termed *unreduced, zygotic, diploid*, or *2n* number. Thus, a human egg or sperm, prior to fertilization, possesses 23 chromosomes, in contrast to the 46 in the zygote. Furthermore, the 46 chromosomes are not all individually different; they exist as 23 pairs, as indicated in Figure 10.3, the members of each pair being similar in shape, size, and genetic content. This is because the gamete contributions to the zygote are equivalent not only in terms of chromosome number, but also in terms of chromosome types. Thus the zygote contains two of each type of chromosome found in gametes, one from each parent. The members of each pair are said to be *homologous* to each other and nonhomologous with respect to the

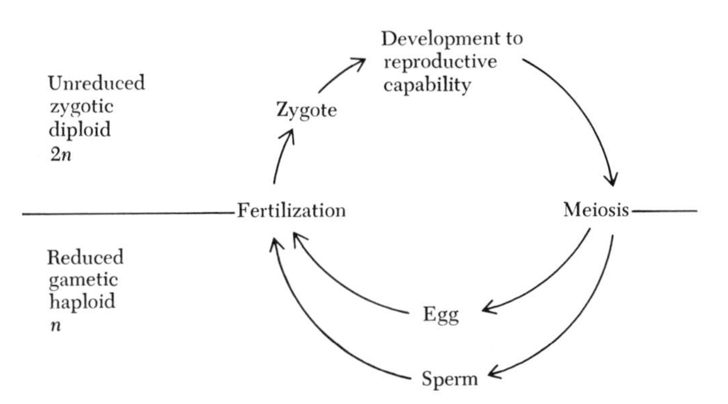

Figure 10.2 The life cycle of sexually breeding animals, with meiosis and fertilization being the events that govern chromosomal states. In higher plants, meiosis would produce haploid spores, which germinate to form the gametophytic generation, which, in its turn, would produce the eggs and sperm.

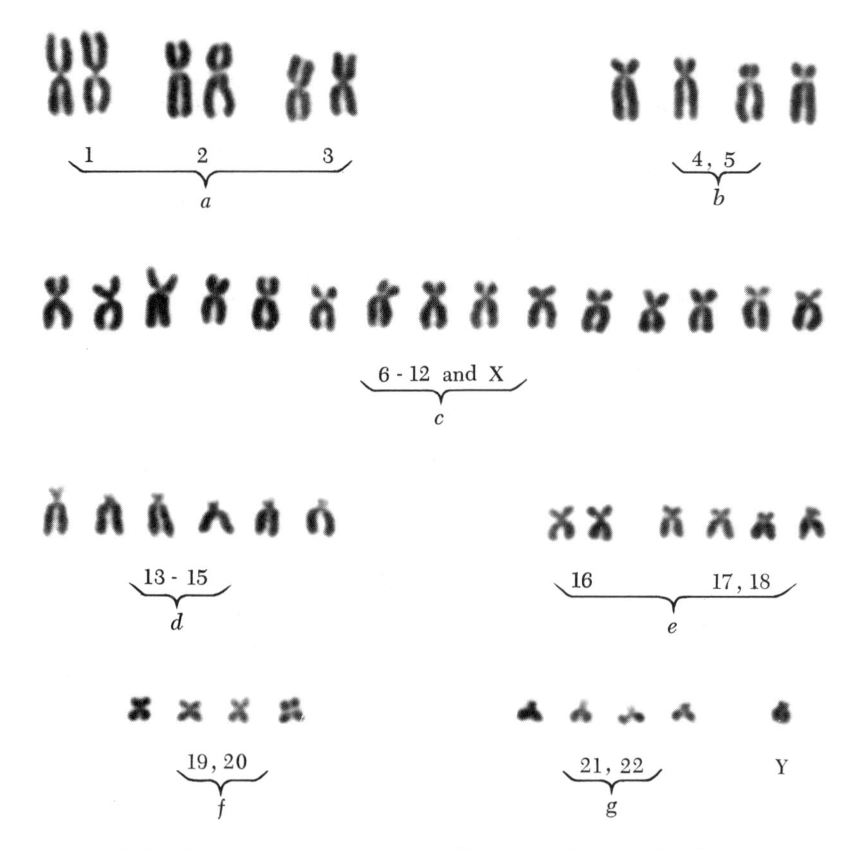

Figure 10.3 The chromosomes of a normal human male, with the chromosomes arranged in homologous pairs, numbered according to size, and also designated by letter groups. The male has an XY sex-determining system; the small Y chromosome is indicated at the lower right in group g, while the X, which is difficult to identify positively, is one of those included in group c.

other chromosomes. Each diploid cell derived from the zygote by mitosis also contains two homologues of each type, one a descendant of a chromosome contributed by one parent, the other a descendant of the equivalent chromosome contributed by the other parent. As we shall see, when meiosis takes place in a diploid cell to produce haploid cells, the homologous chromosomes are distributed in such a way that each haploid cell receives one of each type to contribute to the next generation.

Just as there is an alternation of haploid and diploid chromosome numbers during the life cycle of an individual (Figure 10.2), so is there a comparable alternation in the amounts of DNA per nucleus. These amounts are expressed as C values. If C represents the amount of DNA in a haploid sperm or egg, a diploid cell such as a zygote or a cell derived from the zygote by mitosis has a $2C$ value. As the diploid cell prepares to divide by mitosis, the DNA is replicated to a $4C$ value during the S period of interphase, and then reduced to $2C$ during the anaphase separation of chromosomes. Prior to meiosis in a diploid cell there is a comparable increase in DNA from the $2C$ to the $4C$ value, concomitant with chromosome replication. However, meiosis, unlike mitosis, consists of two consecutive divisions. In the first of these the homologous chromosomes, each consisting of two chromatids, separate from each other; this division produces two nuclei, each with the haploid number of chromosomes and with the $2C$ amount of DNA. The haploid nuclei then divide by mitosis, the chromatids of each previously replicated chromosome separating from one another; the overall result is the formation of four haploid nuclei, each with the C amount of DNA. The relationships between DNA content and chromosome number during the life cycle are shown in Figure 10.4.

One exception to the similarity of paired homologues in shape and size is the pair of chromosomes characterizing the two sexes. Figure 10.3 shows this pair of chromosomes in a human male, although the size and shape of the X makes identification difficult. The human female is XX, and consequently her sex chromosomes are similar and homologous; the male is XY, but the two chromosomes, while differing in size and genetic content, are sufficiently homologous to pair in the prophase of meiosis (see meiosis stages below). Not all sexually reproducing species possess recognizable sex chromosomes, and the XX-XY system is only one variation among many that are known (Figure 10.5). All function in some manner to bring about the expression and functioning of the sexual state.

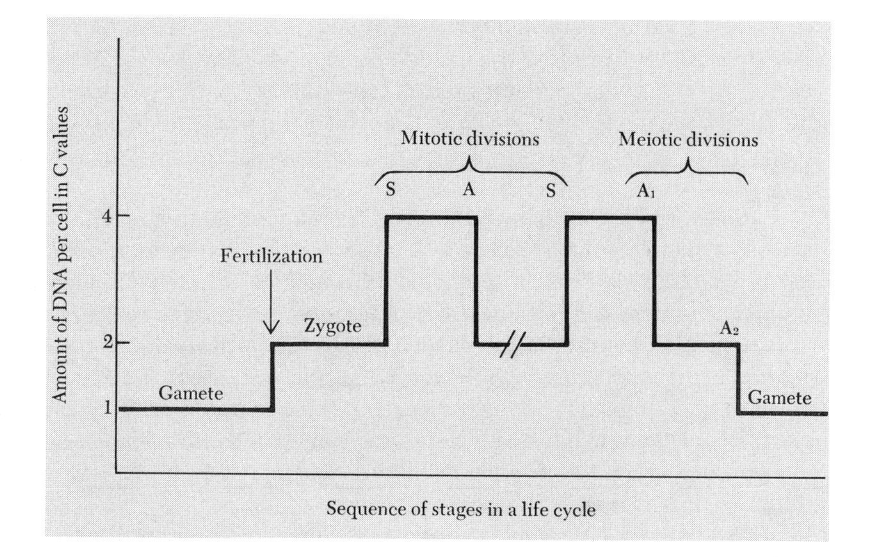

Figure 10.4 Sequences of stages in the life cycle of a sexually reproducing animal correlated with the changes in the amount of DNA per cell. S = period of DNA synthesis in interphase; A = mitotic anaphase; A_1 = first meiotic anaphase; A_2 = second meiotic anaphase.

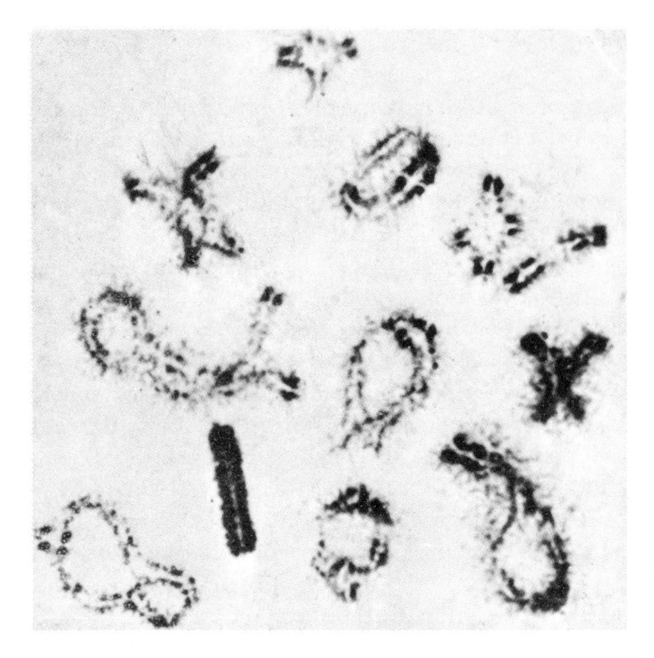

Figure 10.5 The chromosomes in meiosis in the spermatocyte of the grasshopper, *Schistocerca gregaria.* This insect has an XO sex-determining system, and the unpaired X chromosome is seen as the deeply stained rod.

STAGES OF MEIOSIS

Meiosis is a rather complicated type of cell division, yet, remarkably, like somatic cell division and mitosis, the crucial nuclear events and end results are essentially the same wherever encountered. Consequently, a single account of it applies equally well to a fungus, an insect, a flowering plant, or a human. Except for the type of cell resulting from meiosis, the process is basically similar in both sexes as well.

Some of the features that should be recognized in the detailed description of meiosis which follows are so crucial that they deserve to be pointed out in advance. First, we shall see that homologous chromosomes pair with each other. Second, while paired, the homologues participate in a reciprocal exchange of parts. Third, homologous chromosomes subsequently separate from one another. The significance of these three features is not only that haploid cells, with one chromosome of each type, are produced, but also that each haploid set consists of a mixture of maternally derived and paternally derived chromosomes and chromosome parts. Thus each individual passes to his or her offspring a mixture of the traits inherited from his or her parents.

We can separate meiosis into a sequence of steps similar to those in mitosis (Figures 10.6 and 10.7). Prophase, however, is a more leisurely process and hence longer in duration, and the modifications introduced affect the character of the resultant cells. Five separate prophase stages are recognizable, even though the prophase progression is continuous.

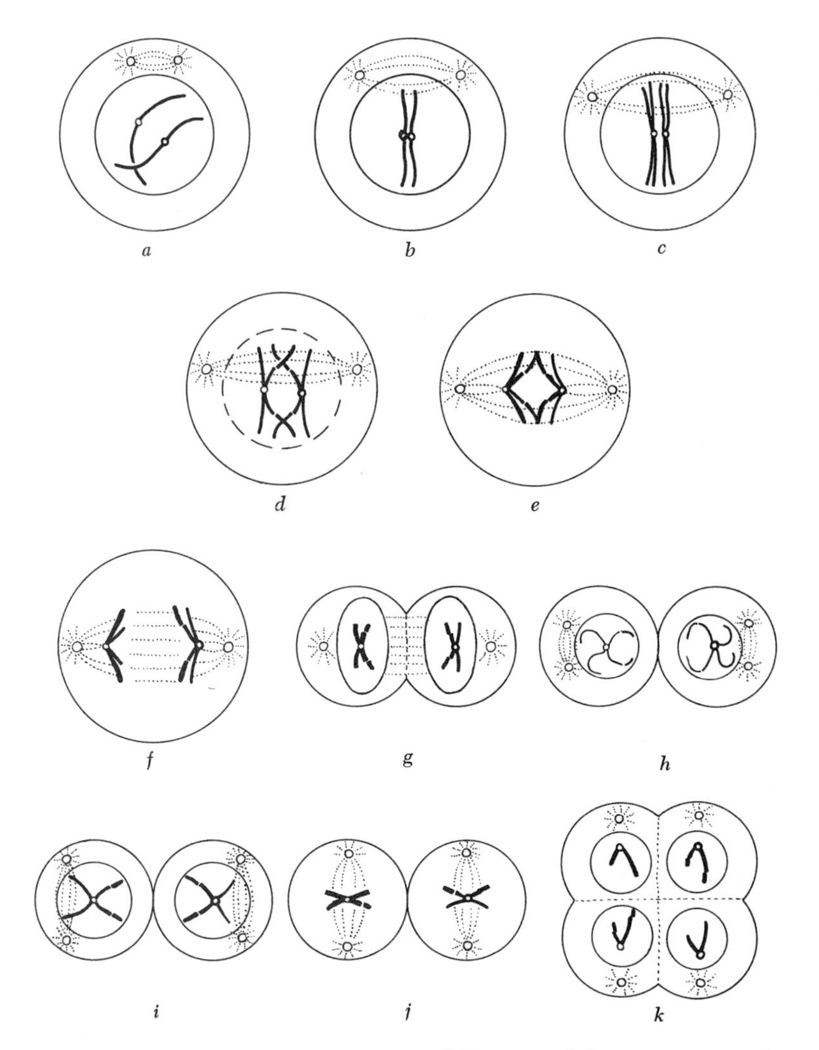

Figure 10.6 Diagrammatic representation of the stages of division in meiosis I and II: (*a*) leptotene, prior to synapsis; (*b*) beginnings of synapsis at zygotene; (*c*) pachytene; (*d*) diplotene; (*e*) metaphase I; (*f*) anaphase I; (*g*) telophase I; (*h*) interphase between the two meiotic divisions; (*i*) prophase II; (*j*) metaphase II; (*k*) telophase II. For simplification, only one pair of homologues has been included in this figure.

The *leptotene* stage initiates the first visible steps of meiosis (Figures 10.8 and 10.9). Meiotic cells and their nuclei are generally larger than those of the surrounding tissues. The chromosomes, present in the diploid number, are thinner and longer than in mitosis and are, therefore, difficult to distinguish individually. Leptotene chromosomes, however, differ from those in ordinary mitotic prophase in two ways: (1) they *appear* to be longitudinally single rather than double, although

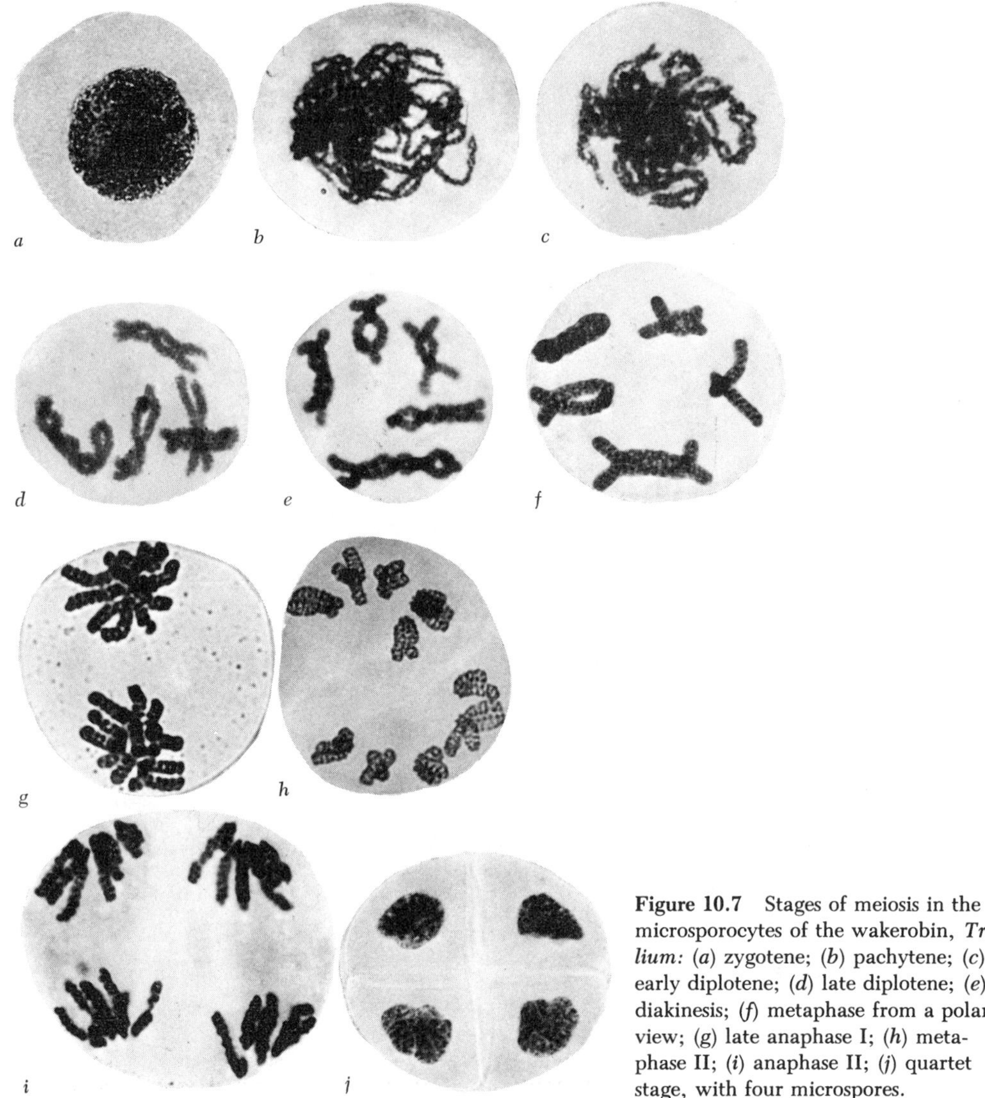

Figure 10.7 Stages of meiosis in the microsporocytes of the wakerobin, *Trillium:* (*a*) zygotene; (*b*) pachytene; (*c*) early diplotene; (*d*) late diplotene; (*e*) diakinesis; (*f*) metaphase from a polar view; (g) late anaphase I; (*h*) metaphase II; (*i*) anaphase II; (*j*) quartet stage, with four microspores.

DNA synthesis has already occurred, indicating that they are in fact double; and (2) their structure is more definite, with a series of dense granules, or *chromomeres*, occurring at irregular intervals along their length. These bodies are not to be confused with the beaded nucleosomes of chromatin seen in Figure 4.34, Chapter 4; the level of magnification is quite different, with chromomeres of meiosis being hundreds of times larger than the nucleosomes.

The chromomeres of any given organism are characteristic in number, size, and position; they can, as a result, be used as landmarks for the identification of particular chromosomes, especially at the succeeding zygotene and pachytene

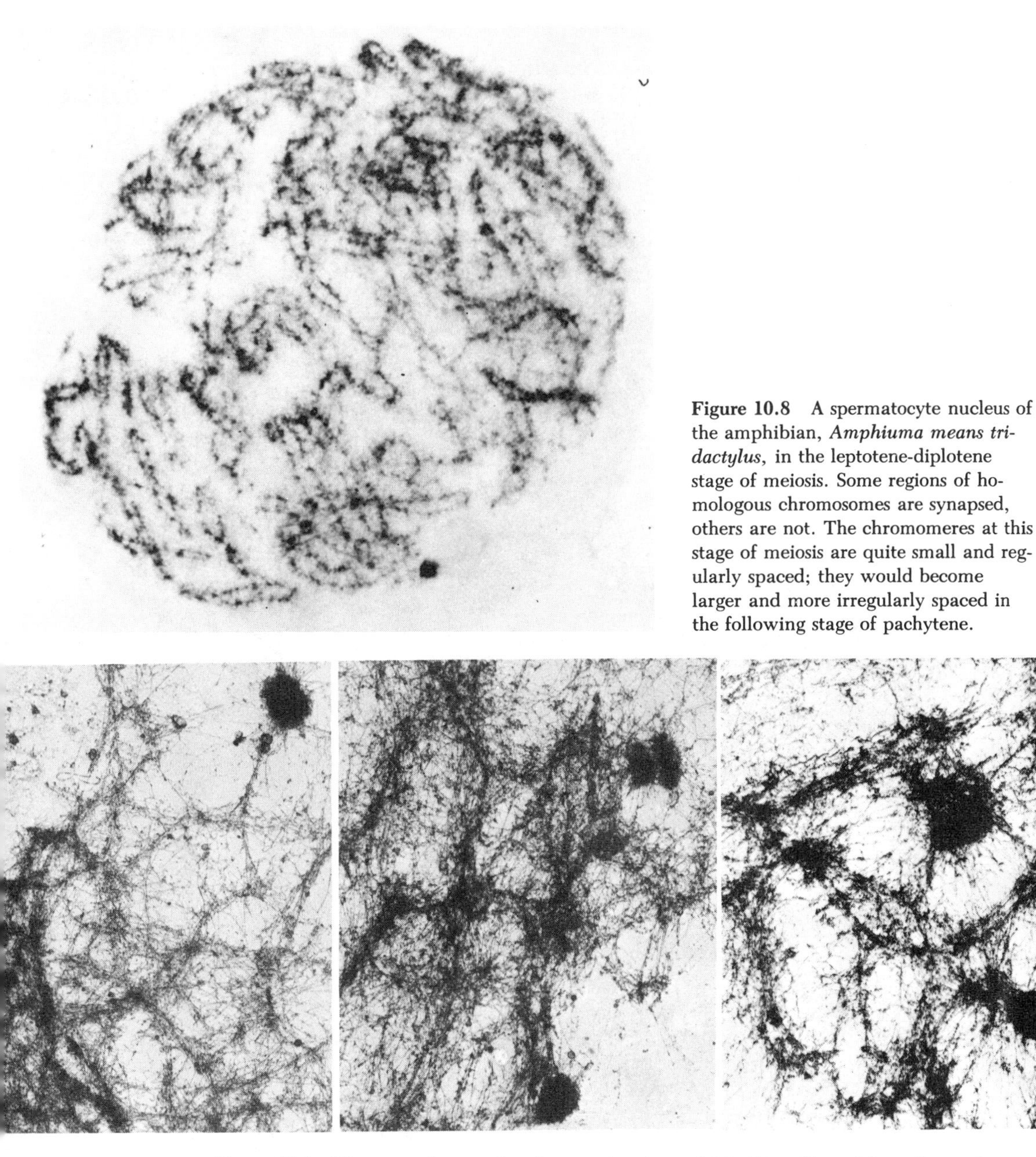

Figure 10.8 A spermatocyte nucleus of the amphibian, *Amphiuma means tridactylus*, in the leptotene-diplotene stage of meiosis. Some regions of homologous chromosomes are synapsed, others are not. The chromomeres at this stage of meiosis are quite small and regularly spaced; they would become larger and more irregularly spaced in the following stage of pachytene.

Figure 10.9 Electron micrographs of spermatocyte nuclei in the milkweed bug, *Oncopeltus*, prepared by being first floated onto a water surface, lifted off, and then dried by the critical point methods, a technique that dehydrates the specimen without serious distortion. Left, leptotene stage; middle, zygotene stage; and right, pachytene stage. The denser clumps of chromatin are sex chromosomes, which condense earlier than the autosomes. (Courtesy of Dr. S. Wolfe.)

stages where they are larger, fewer in number, and more readily identified. Chromomeres are regions of chromatin that have been compacted through localized contraction, most probably by association with the proteins of the chromosome, and in this state the DNA contained within them is thought to be metabolically inactive and nontranscribing. At one time it was believed that the chromomeres were visible manifestations of single genes, but since there are only about 2000 chromomeres among the 24 chromosomes of the garden lily, and far fewer in the woodrush *Luzula* or tomato (Figures 10.10 and 10.11), and since the pattern of the chromomeres changes with the stages of meiosis, it no longer seems reasonable to equate meiotic chromomeres with single genes.

Movement of the chromosomes initiates the *zygotene* stage, and this movement results from an attracting force of unknown nature that brings together

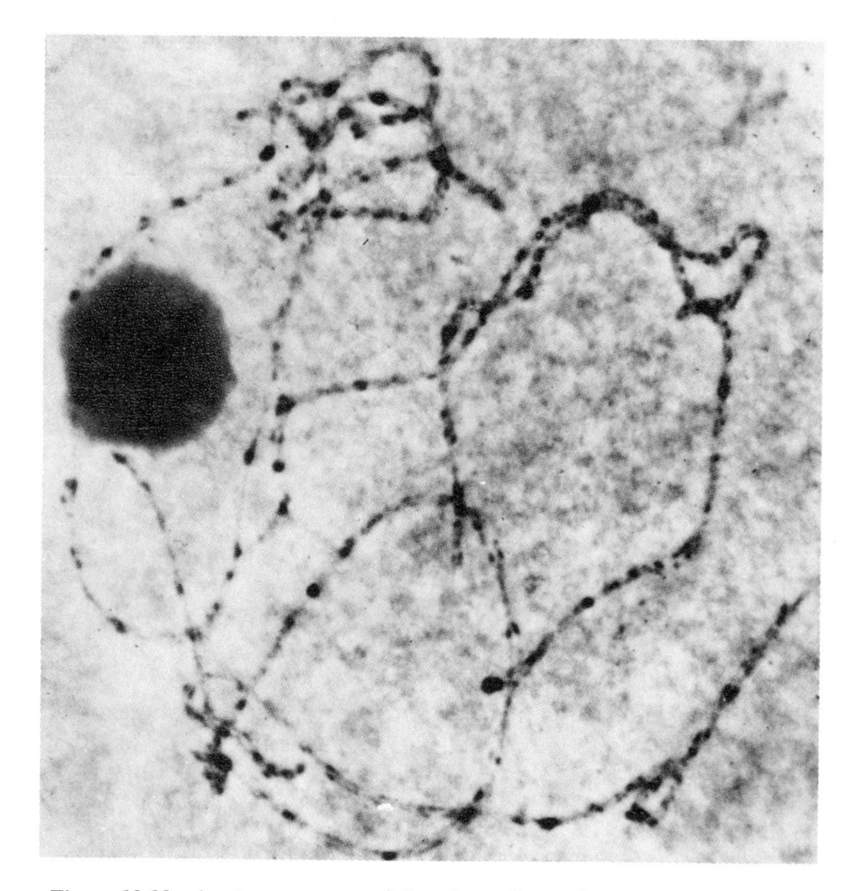

Figure 10.10 A microsporocyte of *Luzula* in the pachytene stage of meiosis, showing the chromomeres irregularly spaced along the length of the paired homologues. It can be seen that the pairing is chromomere-by-chromomere in a highly accurate fashion. The dark body at the left is the nucleolus.

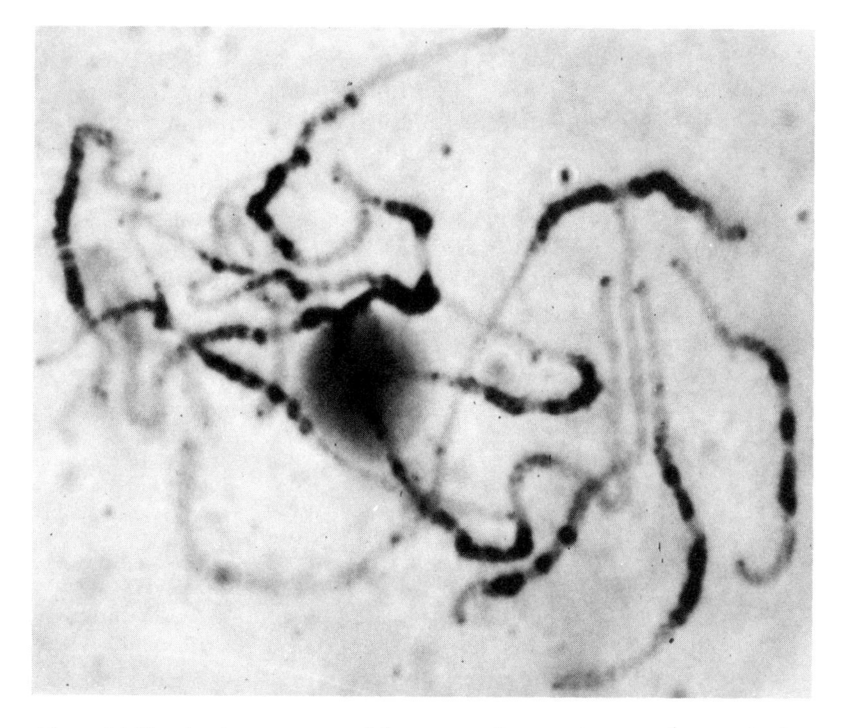

Figure 10.11 A microsporocyte of the tomato, *Lycopersicum esculentum*, in the pachytene stage of meiosis, and showing the large chromomeres concentrated around the centromeres of the paired homologues, with smaller chromomeres located at the ends of the chromosomes.

homologous chromosomes. The pairing of homologues, known as *synapsis*, begins at one or more points along the length of the chromosomes and then proceeds, much as a zipper would, to unite the homologues along their entire length. This is an exact, not a random, process, for the chromomeres in one homologue synapse exactly with their counterparts in the other (Figure 10.12). When synapsis is complete, the nucleus will appear as if only the haploid number of chromosomes is present. Each, however, is a pair of homologous chromosomes, and these are now referred to as *bivalents*.

At the level of resolution of the electron microscope, the synapsing homologues form a *synaptinemal complex* (Figure 10.13). The paired chromosomes seem to lie on either side of the complex, which consists of two denser outer boundaries and a less dense, ladderlike central core. There are various speculations as to the meaning of this structure, but as yet no certain information is known except that it seems to be correlated with synapsis followed by exchange of parts between homologues.

Zygonema is the period of active synapsis. The next, or *pachytene*, stage is distinguishable by the fact that, in some species, the paired chromosomes of each

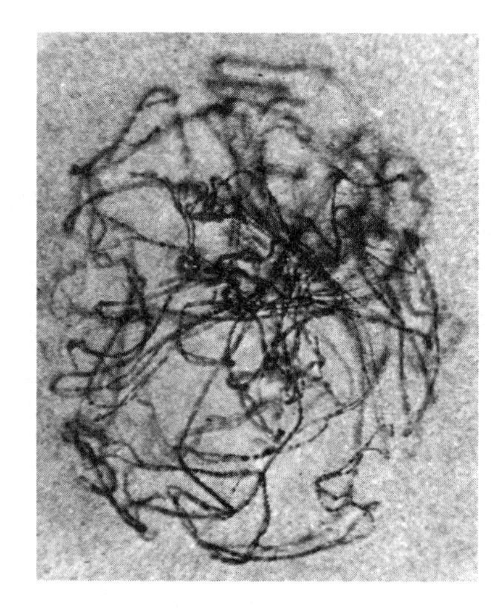

Figure 10.12 Zygotene stage of meiosis in a microsporocyte of the lily, ***Lilium regale***. In the lower right-hand corner, both paired and unpaired regions of homologues can be seen.

bivalent are easily seen (Figure 10.10), and since the chromosomes have continued to shorten and thicken by coiling, they are more readily identified one from the other. In other species the homologues are so tightly paired as to be indistinguishable one from the other (Figure 10.14). The chromomeres and the attachment of the nucleolus to a particular chromosome in maize may be visible with high magnification (Figures 10.15 and 10.16).

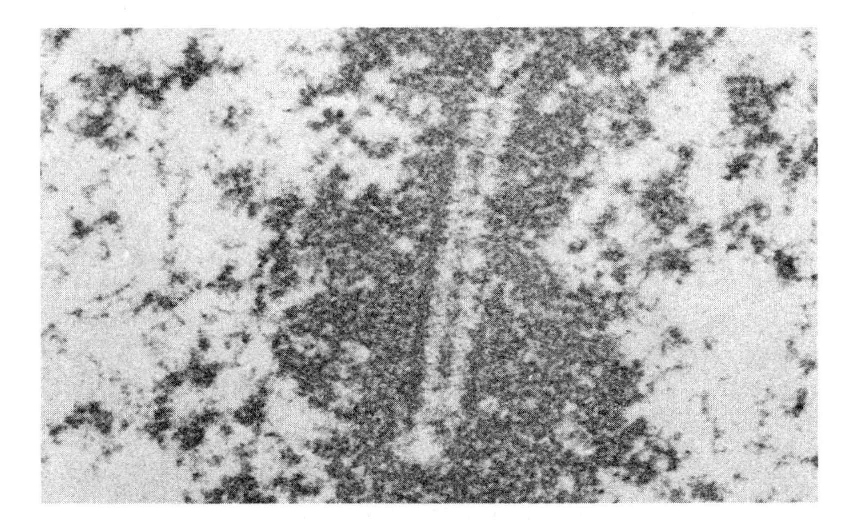

Figure 10.13 Electron micrography of a synaptinemal complex in a microsporocyte of *Tradescantia paludosa*. The dark masses are the chromatin of paired homologues, but an interpretation of the light and dark areas of the complex is not entirely clear at the present time even though synaptinemal complexes are obviously indicative of pairing relationships.

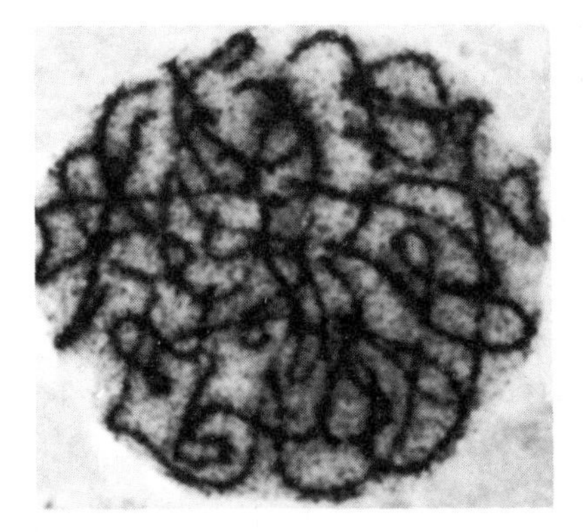

Figure 10.14 Pachytene stage in a spermatocyte of a salamander. The chromomeres are quite regularly spaced along the length of the paired homologues, and pairing seems to be complete for all of the chromosomes. (Courtesy of Dr. J. Kezer.)

The pachytene stage ends when the forces of attraction between homologues lapse and the homologous chromosomes separate from each other (Figure 10.17). This is the *diplotene* stage, and, as Figure 10.18 indicates, each chromosome now consists of two clearly visible chromatids. Each bivalent, therefore, is composed of four chromatids. Longitudinal replication of each chromosome took place prior to this stage, but it did not become obviously evident until the attraction between homologues ceased (Figure 10.19).

Separation of the homologues, however, is not complete. At one or more

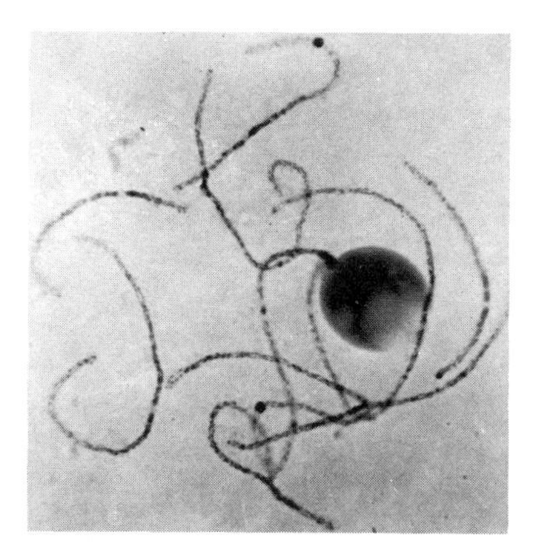

Figure 10.15 Pachytene stage in a microsporocyte of maize, with chromosome 6, in a paired condition, attached to the nucleolus by its nucleolar organizer. Maize chromosomes in meiosis are often characterized by prominent chromomeres called knobs, two of which are readily visible at the top and bottom of the figure, while a larger knob attaches chromosome 6 to the nucleolus.

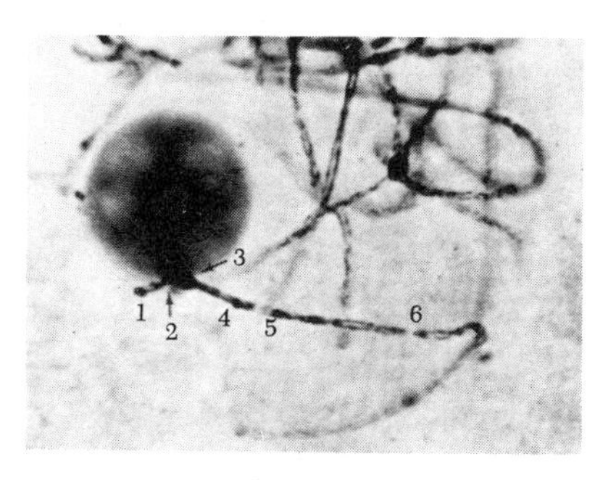

Figure 10.16 Chromosome 6 of maize magnified somewhat greater than that seen in Figure 10.15. The nucleolar organizer is indicated by the figure 3, while other recognizable regions of the paired homologues are also labeled with numbers.

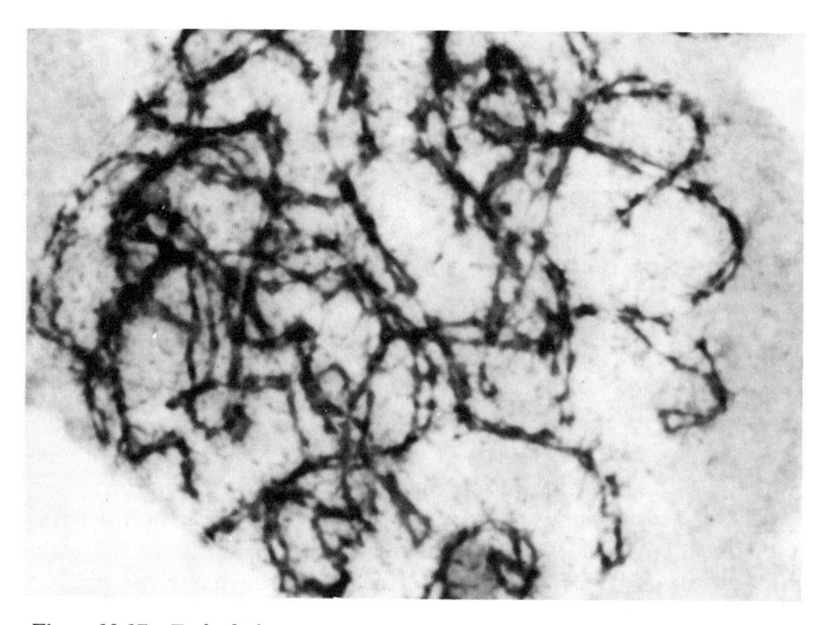

Figure 10.17 Early diplotene stage in a spermatocyte of a salamander. The paired homologues are beginning to fall apart as the synaptic attraction between them lapses. (Courtesy of Dr. J. Kezer.)

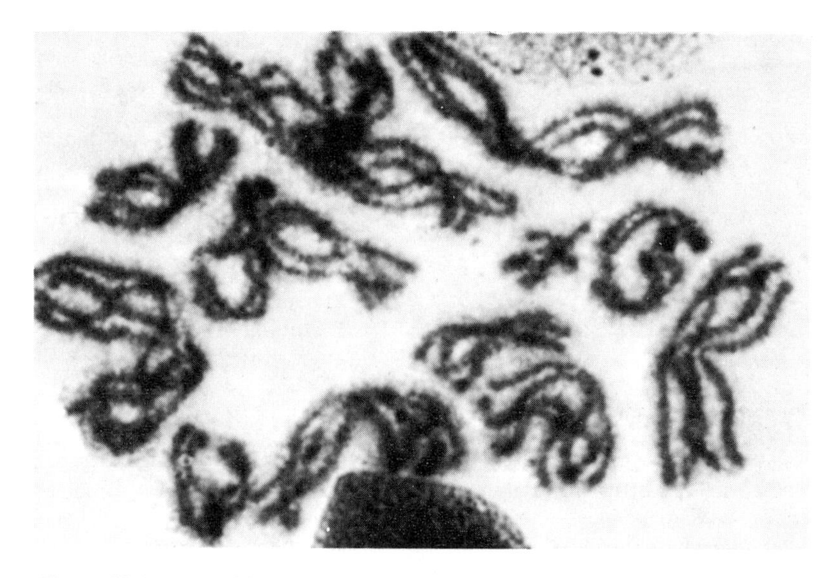

Figure 10.18 A middiplotene state in a salamander, with the position of chiasmata visible in a few of the homologous pairs. The four chromatids in each bivalent are readily apparent. (Courtesy of Dr. J. Kezer.)

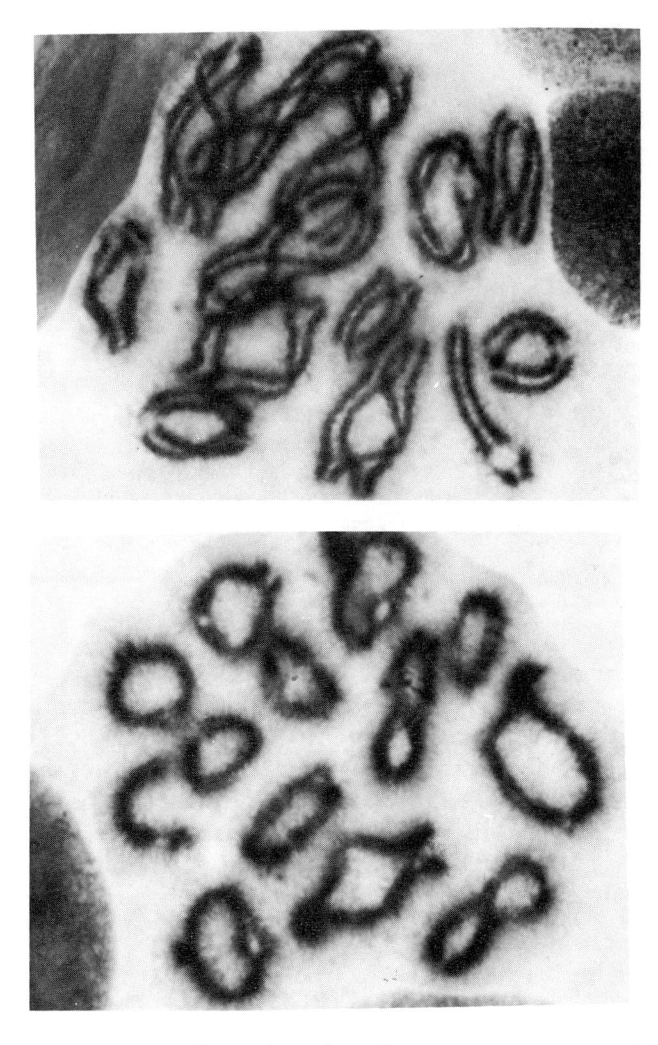

Figure 10.19 Somewhat later stages of diplotene than indicated in Figures 10.17 and 10.18. The bottom figure shows the characteristic fuzziness of the paired chromosomes at this stage, a fuzziness due to loops of chromatin extending beyond the main contracted portion of the chromosomes. (Courtesy of Dr. J. Kezer.)

points along their length, contact is retained by means of *chiasmata* (singular, *chiasma*). Each chiasma results from an exchange of chromatin between the chromatids of two homologues (we shall discuss later in this chapter the significance of this phenomenon as it relates to heredity), but the relationship of the chromatids to a chiasma is clearly indicated in Figure 10.20. To produce such an exchange of chromatin between two chromatids, the double helices of both chromatids must be broken and then reunited to effect the exchange. The repair of such broken ends requires the formation of new DNA; through the use of [^{3}H]TdR such repair has been detected.

When only one chiasma has formed, the bivalent in the diplotene stage appears as a cross. If two are formed, the bivalent is generally ring-shaped; if three or more form, the homologues develop a series of loops (Figures 10.19 and 10.20). In

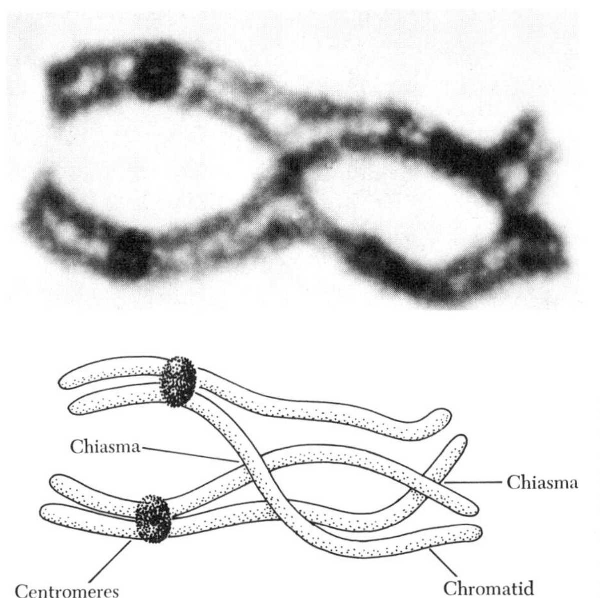

Figure 10.20 A single bivalent at the diplotene stage, which illustrates the position of the two chiasmata formed previously, and the position of the centromeres. The accompanying diagram traces all four chromatids throughout the length of the bivalent. (Courtesy of Dr. J. Kezer.)

different cells, the number and approximate positions of the chiasmata vary, even for the same bivalent, but as a rule long chromosomes have more chiasmata than short ones, although even the shortest seem able to form at least one chiasma.

The next prophase stage is called *diakinesis*, but the distinction is not sharp between it and the diplotene stage. During diakinesis the nucleolus becomes detached from its special bivalent and disappears, and the bivalents become considerably more contracted (Figure 10.21). Also, as contraction proceeds (Figure 10.19), the chiasmata tend to lose their original position and move toward the ends of the chromosomes.

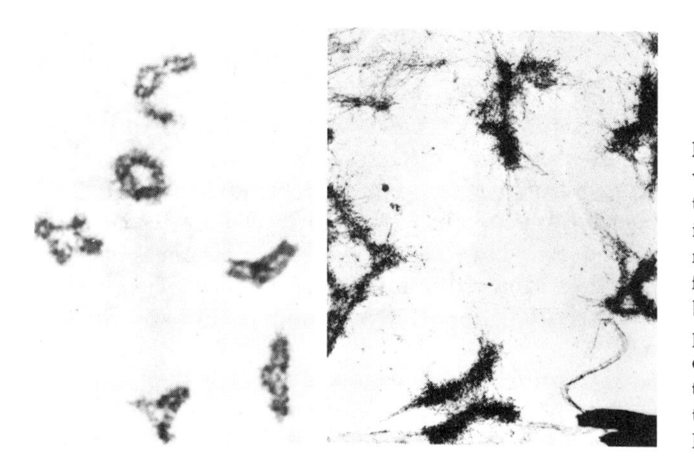

Figure 10.21 Diakinesis in the milkweed bug, *Oncopeltus*. The figure at the left was prepared by the usual fixing and staining methods for light microscopy; that on the right was floated onto a water surface before being lifted off and dried by the critical point method, a technique that dehydrates the specimen without serious distortion. It was then photographed in the electron microscope. (Courtesy of Dr. S. Wolfe.)

We have mentioned that the chromosomes shorten as they progress from the leptotene stage onward through prophase. This is accomplished by the development of a series of coils in each chromatid, which gradually decrease in number as their diameters increase. The process is no different from the shortening of chromosomes in mitosis; the coils here, however, are more easily observed, particularly when the cells have been pretreated with ammonia vapors or dilute cyanide solution before staining. Figure 10.22 illustrates the coils as they appear in the spiderwort, *Tradescantia*.

The breakdown of the nuclear membranes and the appearance of the spindle terminate prophase and initiate the *first metaphase of meiosis* (Figures 10.6 and 10.7). The bivalents then orient themselves on the spindle, but instead of all centromeres being on the equatorial plate, as in mitosis, each bivalent is so located that its centromeres lie on either side of, and equidistant from, the plate (Figures 10.23 and 10.24). This seems to be a position of equilibrium.

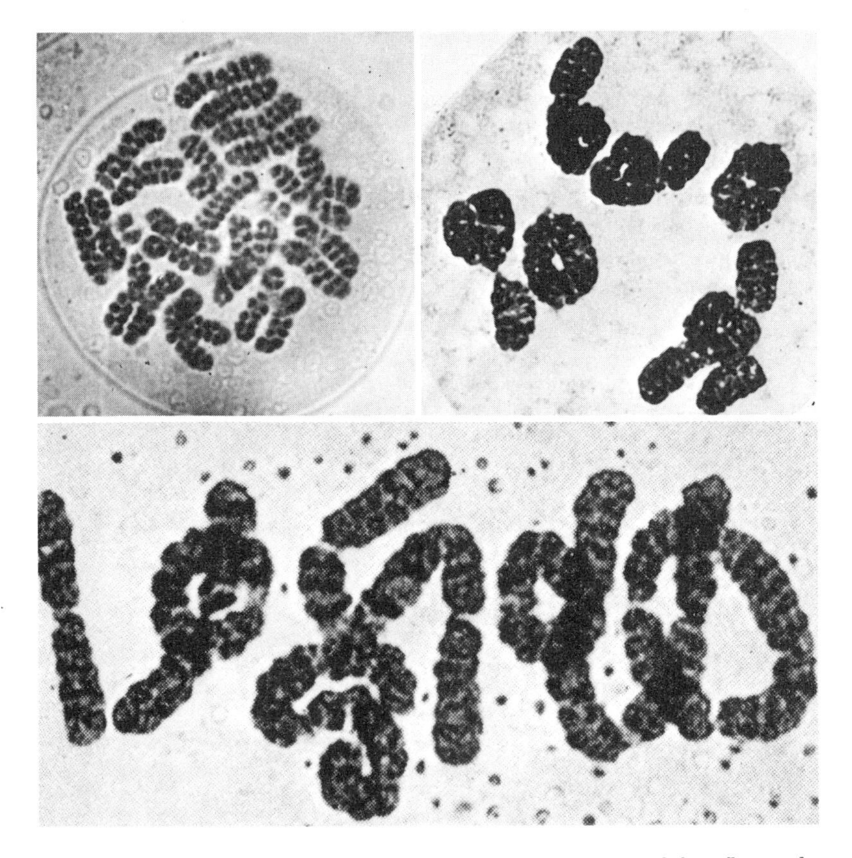

Figure 10.22 Coiling of the chromosomes of *Tradescantia*. Top left: a flattened anaphase I of a diploid form with the coils clearly evident in each chromatid; top right: a triploid form at metaphase I; bottom: a tetraploid form at metaphase I.

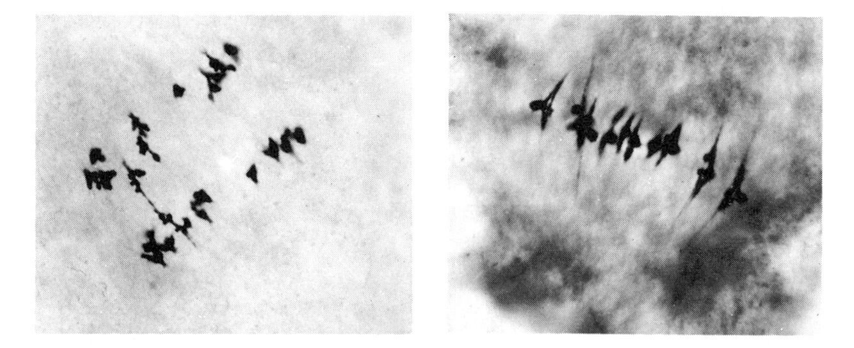

Figure 10.23 Metaphase I (left) and anaphase I (right) in maize. The stretching of the chromosomes to the poles can be seen in the left figure; the right one indicates that at times the homologues may have difficulty separating from each other.

The *first anaphase* of meiosis begins when the homologous centromeres of each bivalent start to separate toward opposite poles of the spindle (Figures 10.23, 10.24, and 10.25). As the homologous chromosomes are pulled toward the poles by their centromeres, any remaining chiasmata slip off and the homologues are completely separated from each other. During anaphase I, then, we can see that the haploid number of chromosomes, each chromosome still consisting of two chromatids, is being moved toward each pole. The two chromatids of each chromosome are held together only at the centromere, which, although structurally double, behaves as a single functional unit. When the chromosomes reach the

Figure 10.24 Anaphase I in a salamander. (Courtesy of Dr. J. Kezer.)

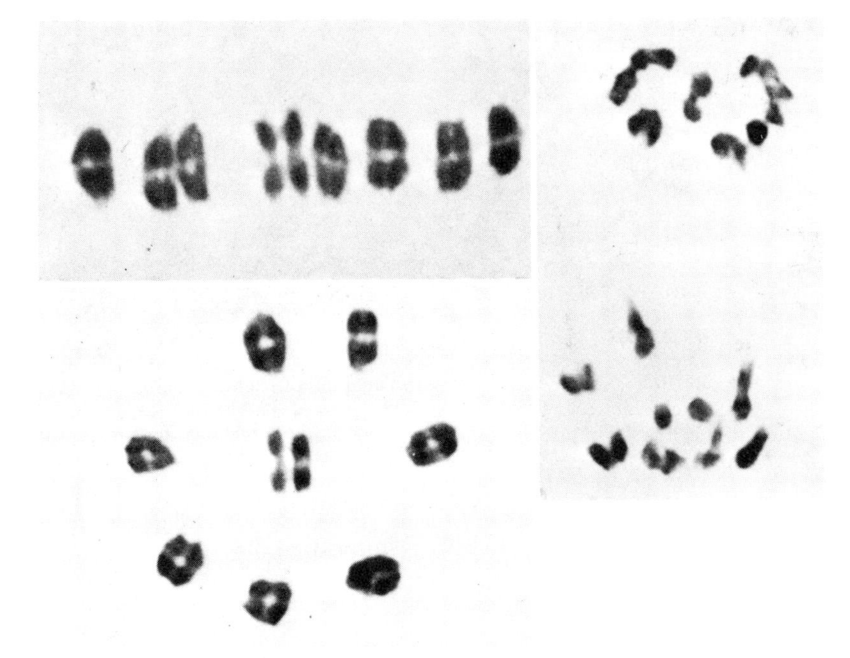

Figure 10.25 Metaphase I and anaphase I in the milkweed bug, *Oncopeltus*. Top left, a side view of metaphase I, with the sex chromosomes only loosely paired with each other; bottom left, a polar view of metaphase I, with the sex chromosomes occupying the center of the metaphase plate, and with the X chromosome to the right and the lighter Y chromosome to its left; right, anaphase I. (Courtesy of Dr. S. Wolfe.)

poles, two nuclei form, the chromosomes uncoil and the meiotic cell may be bisected by a membranous partition. This is the *first telophase* of meiosis (Figure 10.6).

After an interphase that, depending on the species involved, may be short or long—or even absent altogether—the chromosomes in each of the two haploid cells enter the *second meiotic division* (Figures 10.6 and 10.7). If an interphase is absent, the chromosomes pass directly from the first telophase to the *second prophase* without any great change in appearance (Figure 10.26). If an interphase is present, a nuclear envelope forms in telophase, the chromosomes uncoil, and a somewhat more prolonged second prophase is found. But whatever the case, the chromosomes reaching the *second metaphase* (Figure 10.27) are essentially unchanged from what they were in the previous anaphase; that is, *no chromosomal replication occurs during interphase*, and the centromere of each chromosome remains functionally undivided. A spindle forms in each of the two cells, and at the *second anaphase*, the centromeres separate and the chromosomes move to the poles (Figure 10.28). The nuclei are reorganized during the *second telophase*, giving four haploid nuclei that become segregated into individual cells by segmentation of the cytoplasm.

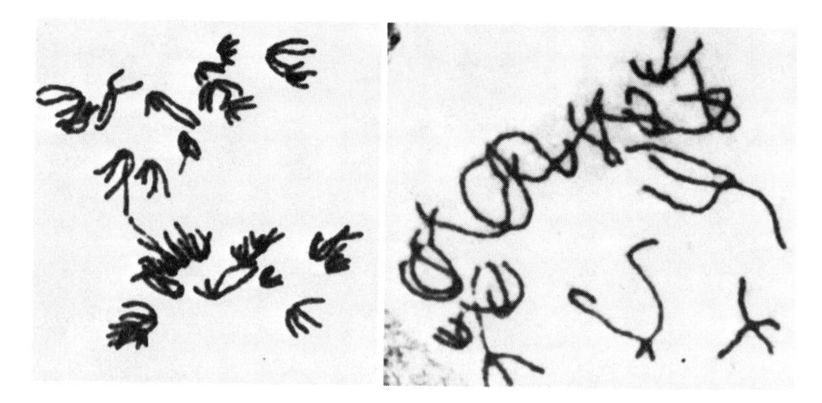

Figure 10.26 Anaphase I and prophase II in a salamander to show that there is relatively little change in the character of the chromosomes during the interval between the two meiotic stages. (Courtesy of Dr. J. Kezer.)

Looking back over the events of meiosis, we find that the chromosomes remained unchanged in longitudinal structure from the diplotene stage to the end of the second meiotic division. The replication of each chromosome occurred during the premeiotic interphase, but this was followed by two divisions; in the first, homologous centromeres and hence the homologues separated from each other at an anaphase I to reduce chromosome number, an event made possible because synapsis joined them and chiasmata held them together until metaphase I; in the second, sister centromeres and hence the two chromatids of each chromosome separated.

In the animal kingdom, meiosis leads to the formation of sexual gametes, the egg and sperm usually being the only cells carrying a haploid complement of chromosomes. In the plant kingdom, however, meiosis can occur at various times during the life cycle, and the haploid products may be sexual gametes or asexual spores, depending on the particular group of plants being studied. We shall con-

Figure 10.27 Anaphase I in a salamander to indicate that the chromatids are held together only at the centromere regions. These chromosomes are somewhat more compacted than those in Figure 10.26. (Courtesy of Dr. J. Kezer.)

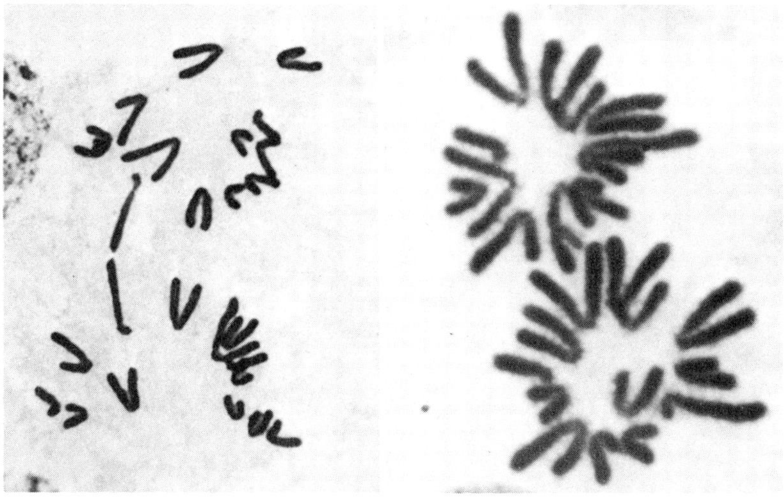

Figure 10.28 Side and polar views of anaphase II in a salamander. The chromatids are now separating, and each nucleus will round up and become, in this instance, the nucleus of a sperm. Notice that the distinctiveness of the chromosomes can be seen in that they are not of the same size and their centromeres are not similarly located. (Courtesy of Dr. J. Kezer.)

sider here only the formation of the products of vertebrate meiosis, that is, the egg and sperm, and meiosis as it occurs in the flowering plant *Zea mays*.

Meiosis in the Vertebrate Animal

The primordial germ cells of the human embryo make their appearance approximately 20 days after fertilization; they migrate from their origin in the wall of the yolk sac to the developing gonads during the fifth week of development. Once located in the female gonad, these cells become the source of the female germ cells. They divide rapidly to form clusters of *oogonia* near the outer wall of the ovary, and each cluster is transformed into a layer of flat epithelial, or nurse, cells surrounding a central cell, which, at the end of the third month of development, becomes the *primary oocyte*. The cluster is known as the *primordial follicle* (Figure 10.29). The primary oocyte enters meiosis as soon as it is formed, and by the seventh month of prenatal development, all of the oogonia have stopped dividing, and the oocytes have reached the *dictyotene* stage, which follows pachynema but which differs from the typical diplotene stage in that the chromatin is quite diffuse in appearance (Figure 10.30). The oocytes remain in this state until sexual maturity.

By birth, therefore, a human female will presumably have formed all of the oocytes she will have, and they will have progressed well into the prophase of meiosis. There is some uncertainty about this, however, for the number of oocytes

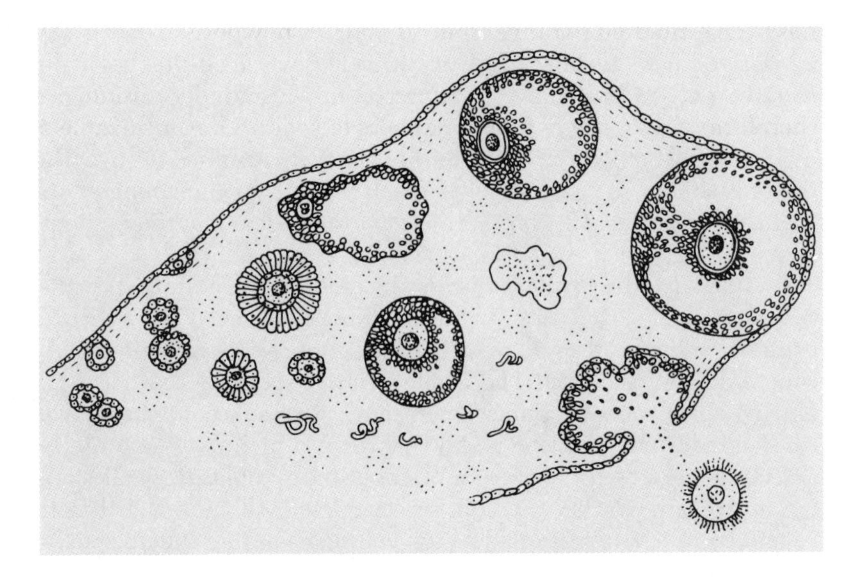

Figure 10.29 Diagrammatic view of a section of a mammalian oocyte, showing the progressive development of the oocytes as they arise from the germinal epithelium at the left, form a layer of nurse cells around them, increase in size, sink into the interior of the ovarian tissue, and finally escape to the outside by rupture of the wall of the Graafian follicle.

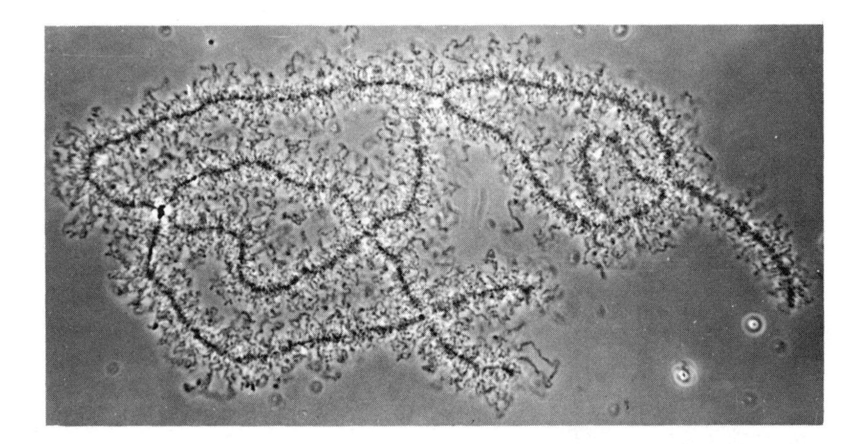

Figure 10.30 A meiotic bivalent of the newt, *Triturus*, consisting of two homologues held together at four points by chiasmata, and in the diffuse diplotene stage. The fuzziness results from the projection of loops of chromatin away from the linear body of the chromosomes, and these loops are in active states of synthesis, somewhat comparable to that depicted in Figure 4.28.

has been estimated to range from 40,000 in a newborn to over 400,000 in a woman 22 years of age. During each ovarian cycle, several oocytes begin development, but usually only one achieves maturity; the remainder disintegrate. As a rule, therefore, only a single functional oocyte produces a fertilizable egg during each ovarian cycle, and if the childbearing years are assumed to cover the period of 12 to 50, only 400 or so of the oocytes reach maturity. The youngest of them will have spent no less than 12 years in meiosis; the oldest may have spent as many as 50 years.

Initially, the *primary oocytes* lie close to the germinal epithelium, but later they increase in size and sink into the interior of the ovary where they become surrounded by *follicle cells,* which probably have both a protective and nutritive function (Figure 10.29). The whole structure is now known as a *Graafian follicle.* During this process of enlargement and encapsulation, the oocyte is building up reserve food material, the yolk. This food, which may be protein or fat in mammals, is generally distributed throughout the cytoplasm as yolk spheres or granules. In the frog, however, the yolk so completely fills the cell that the cytoplasm is restricted to a small fraction of the cell surrounding the nucleus; the well-known yolk in the hen's egg is also enormous compared to the amount of cytoplasm.

Eventually, the Graafian follicle ruptures and the egg (Figure 10.29), or *ovum,* is released from the ovary and passes into the *oviduct,* or Fallopian tube, where it can be fertilized by a sperm. By this time, however, meiosis has been resumed and has reached metaphase of the second meiotic division. Meiosis will be completed only if the egg is fertilized, the sperm acting as initiating agent. Only a *single* functional cell results, however. The other three cells, or *polar bodies,* are cast off and will degenerate, *but the process has effectively reduced the chromosome number without depriving the egg of the cytoplasm and yolk the embryo will need when it begins to develop.*

The first meiotic division in the primary oocyte takes place close to the cell membrane, and the outermost nucleus, together with a small amount of cytoplasm, is pinched off as a polar body (Figure 10.31). The second meiotic divi-

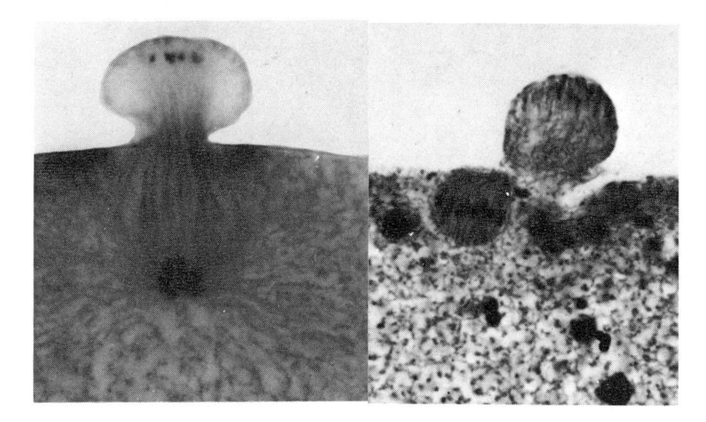

Figure 10.31 Polar body formation in the eggs of the whitefish, *Coregonus.* Left, anaphase of the first meiotic division, with the first polar body being pinched off. Right, metaphase of the second meiotic division, which will lead to the pinching off of a second polar body. The first polar body may or may not divide again.

sion results in the pinching-off of a second polar body; the first polar body, meanwhile, has also undergone a second meiotic division, thus giving a total of three polar bodies. The haploid nucleus remaining in the egg is now known as the *female pronucleus*. It sinks into the center of the cytoplasm and is ready for union with a similar haploid nucleus brought in by the sperm during fertilization.

The primitive germ cells of the human male enter the developing gonad during the fifth week of development. They become incorporated into the *sex cords*, which at first are solid structures but which after birth develop a lumen and become the *seminiferous tubules*. These make up about 90 percent of the bulk of the testes.

The germinal epithelium contains *spermatogonia*, cells that continue to increase their number by mitotic division throughout the sexual life of the male (see Figure 9.5, Chapter 9). These derivative cells mature into *primary spermatocytes*, which undergo a first meiotic division to produce *secondary spermatocytes;* the latter pass through a second meiotic division, giving four cells called *spermatids*. These become motile sperm by a remarkable transformation of the entire cell. The human male differs from the female in that the production of primary spermatocytes, and hence viable sperm, is initiated at the beginning of sexual maturity and continues until old age.

The mature sperm consists essentially of a head and a tail. The head is a highly compacted nucleus, capped by a structure known as the *acrosome* (Figure 10.32). It is derived from the Golgi materials of the spermatid and apparently functions as a device for penetrating the egg during fertilization. Just behind the compacted nucleus is the *middle piece*, formed by an aggregation of the mitochon-

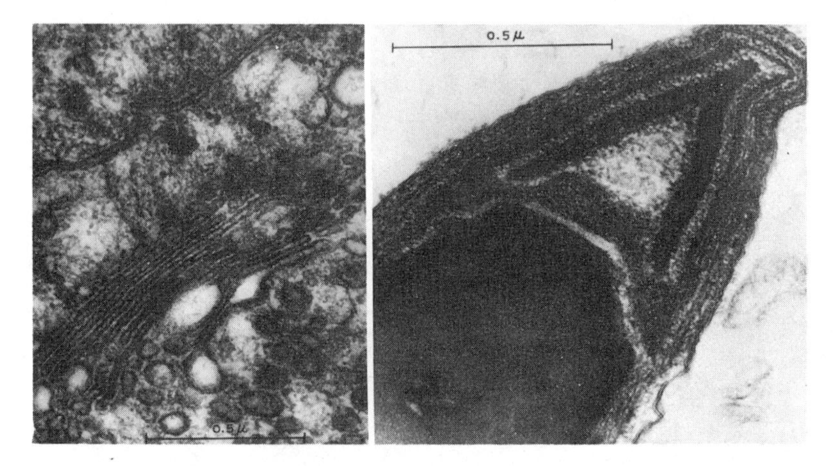

Figure 10.32 Electron micrographs of the Golgi complex (left) and the acrosome (right) of the sperm head of the house cricket, *Acheta domestica*. The cone-shaped acrosome results from the transformation of the Golgi complex during the development of the spermatozoan from the spermatid.

dria. It develops as a sheath around the filament, or tail, and provides the tail with energy for locomotion. The filament, in turn, has developed as the result of a tremendous growth from one of the centrioles; the other centriole remains just beneath the nucleus and at the time of fertilization enters the egg along with the male nucleus. Virtually no cytoplasm except particulate structures is used to form the mature sperm. Each spermatid, therefore, has been transformed from a rather undifferentiated cell into a highly specialized cell capable of reaching the egg under its own power and of penetrating it once it has made contact.

The egg and sperm differ not only in shape; they are vastly different in size. The egg is one of the largest cells of the human body, exceeded in volume only by some neurons. It is just barely visible to the naked eye, having a diameter of about 0.1 mm, and a volume of 2,000,000 μm^3. The sperm, on the other hand, is one of the smallest cells, having a volume of only 30 μm^3. The fact that their genetic contribution to the next generation is equal, once again points to the nucleus, the chromosomes, and to DNA as the crucial elements of inheritance.

The mature egg and sperm must unite within a limited period, for neither has an indefinite life span. The critical period may be a few minutes or it may be spread over several hours or days. In mammals, fertilization can occur as the egg leaves the ovary and passes down the oviduct on its way to the *uterus*. Insects, however, mate only once, and sperm are stored in the female and used throughout the entire egg-laying period; in the honeybee, for example, this period may last a year or more. It is now possible to store mammalian sperm for an indefinite period by freezing them, and by means of *artificial insemination* the sperm of a single sire may be used to fertilize the eggs of many females. This practice has been widely used in animal breeding programs, thus passing on the superior qualities of one sire to many offspring, and it has been successfully carried out in humans when, for one reason or another, normal conception fails.

The essential process of fertilization is the union of male and female pronuclei, but the sperm also functions as an activating agent. That is, nature has ensured against the egg's beginning its embryonic development in an unfertilized state; if it did, haploid embryos would result, and even if such embryos were viable and developed into sexually mature adults, the process of meiosis would be hopelessly complicated. Unfertilized eggs of mammals and other related vertebrates can be induced to initiate development by various artificial means, but this rarely occurs naturally.

Fertilization is also a specific process in that the sperm of one species will not, as a rule, fertilize the egg of another species. It now appears that several chemicals are present to ensure proper fertilization and to prevent the penetration of foreign sperm. The egg produces a protein substance called fertilizin, which reacts with an antifertilizin on the surface of the sperm; fertilizin may act to attract sperm of its own kind, but once the two substances interact, the sperm becomes firmly attached to the egg membrane and is then drawn into the interior of the egg. Other sperm are barred from entry by the changes that then take place in the *vitelline membrane* of the egg, an outer coating found on most eggs.

Only the nucleus and one centriole of the sperm enter the egg. The nucleus fuses with the female pronucleus, the centriole divides and begins formation of the first division spindle. In summary, therefore, the entry of the sperm into an egg contributes (1) a stimulus to development; (2) a set of haploid chromosomes, which is the paternal hereditary contribution to the newly formed zygote; and (3) a centriole, which is involved in the machinery of cell division.

Meiosis in a Flowering Plant

Meiosis in a flowering plant such as maize, or Indian corn, gives rise to asexual *spores* instead of gametes. These spores give rise to male and female *gametophytes*, which represent a haploid phase in the plant life cycle and which produce gametes by mitosis (Figure 10.33). A comparison with the life cycles of vertebrates can be gained from Figure 10.2.

The flowers of maize are unisexual although borne on the same plant. The male flowers are grouped into an inflorescence or tassel, and meiosis takes place in the sporogenous tissue of the anthers. The *pollen mother cells*, or *microsporocytes*,

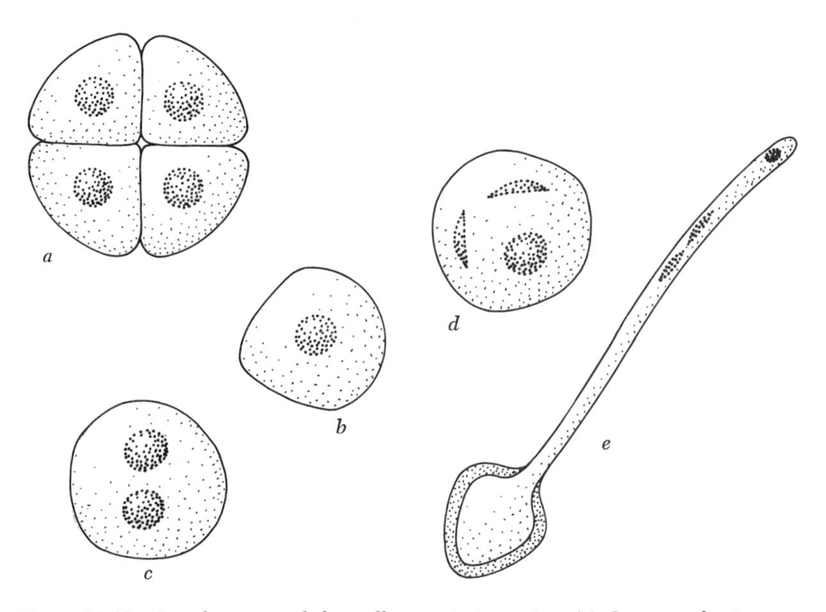

Figure 10.33 Development of the pollen grain in maize. (*a*) Quartet of microspores resulting from the two meiotic divisions of the microsporocyte. (*b*) A single microspore. (*c*) A binucleate microspore with tube and generative nuclei resulting from the first microspore division. (*d*) Mature pollen grain with the tube nucleus and the two sperm, which resulted from a mitotic division of the generative nucleus (the pollen grain is shed in this state). (*e*) A germinating pollen grain with the tube nucleus at the end of the pollen tube and the two sperm that will be carried to the embryo sac for purposes of fertilization.

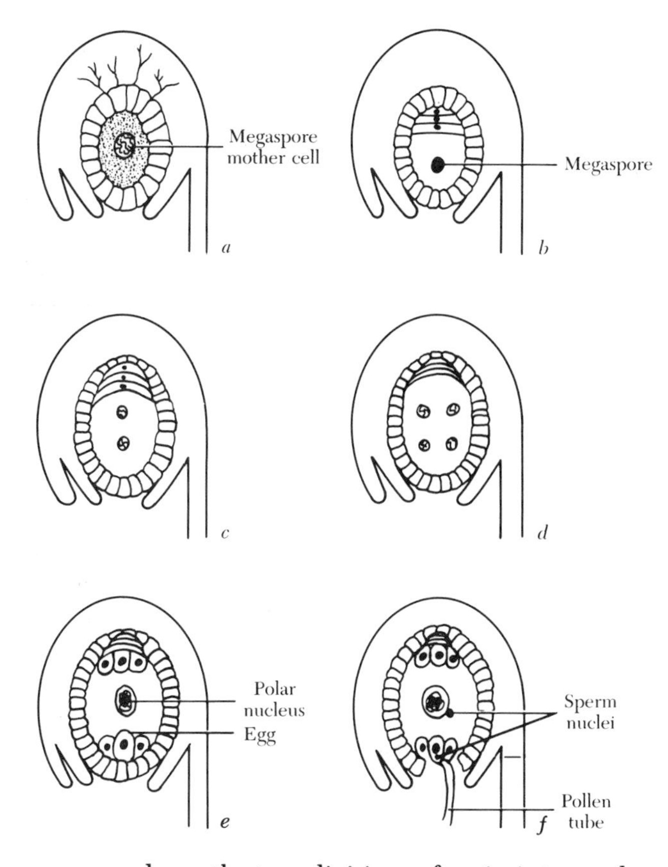

Figure 10.34 Formation of the embryo sac of maize within the ovule. (*a*) The megaspore mother cell, which will undergo meiosis to produce a linear tetrad of cells (*b*), three of which will disintegrate while the remaining one, the megaspore, enlarges to form the embryo sac. (*c*), (*d*), and (*e*). Successive divisions of the haploid megaspore produces eight nuclei, which in (*e*) arrange themselves as follows: three antipodals are at the innermost side of the embryo sac; three nuclei arrange themselves at the opposite end and consist of two synergids and one egg; the remaining two nuclei unite in the center of the embryo sac to form the fusion nucleus, now diploid. (*f*) Fertilization occurs by one of the sperm (solid bodies) uniting with the egg to form the zygote, while the other fuses with the fusion nucleus to form a triploid endosperm nucleus. As the ovule develops, the endosperm nucleus will divide repeatedly to form a nutritive tissue upon which the developing zygote will feed, particularly during the process of germination.

undergo the two divisions of meiosis to produce four haploid *microspores* (Figure 10.7). The single nucleus of each microspore divides mitotically again to produce two nuclei within its wall: a *generative nucleus* and a *tube nucleus*. The former will divide mitotically again to produce two sperm nuclei, after which the microspore matures and is shed as a three-nucleate *pollen grain* (Figure 10.33).

The female inflorescence, or cob, is made up of individual flowers, each consisting of an ovary and a long silk that is both style and stigma. The ovary bears a single ovule having the structure shown in Figure 10.34. The *megasporocyte* is the equivalent of the animal oocyte, but instead of producing an egg directly through meiosis, it forms a *linear tetrad* of four haploid *megaspores*, three of which abort. The remaining one develops into a *megagametophyte*, and three mitotic divisions convert the single nucleus into eight nuclei. At the end opposite the *micropyle*, three of the nuclei form a group called the *antipodals;* three remain at the micropylar end of the ovule, with one egg being flanked by two *synergids;* and the remaining two nuclei fuse in the center of the megagametophyte to produce the diploid *polar nucleus*. This is the state of the ovule when ready to be fertilized.

The mature pollen grain lands on the silk and germinates by producing a pollen tube, which carries the two sperm to the micropylar end of the ovule. Both sperm enter. One will fertilize the egg nucleus to give rise to a *diploid zygote;* this will eventually form a new plant or sporophyte. The other sperm fuses with the polar nucleus to form a *triploid endosperm nucleus;* this, by repeated mitotic divisions and wall formations, will give rise to the endosperm, a rich nutritive tissue that nourishes the growing embryo in its early stages. Flowering plants, therefore, have a double fertilization process, and the seed is a mosaic of tissues: the diploid integuments, which are maternal in origin and which became the hardened seed coats; the diploid zygote and the resultant embryo coming from the fused egg and sperm; and the triploid endosperm, consisting of two maternal nuclei and one male nucleus. Only the embryo, once it reaches sexual maturity, contributes to the next generation.

CHROMOSOME THEORY OF INHERITANCE

Meiosis has been discussed as a logical and necessary part of the life cycle of a sexually reproducing organism, and it is the antithesis of fertilization as regards the number of chromosomes (Figure 10.2). DNA also has been discussed as the molecular basis of inheritance. It is the crucial portion of the chromosome that maintains its linear integrity, and genes are segments of DNA in a chromosome. It follows, then, that the behavior of genes and chromosomes should parallel each other in a most exact manner, and that the phenomena of mitosis, meiosis, and fertilization can be interpreted in genetic as well as cytological or molecular terms. The merging of the genetical and cytological aspects has come to be known as the *chromosome theory of inheritance;* the molecular phenomena simply confirm and extend the theory.

Mendelian Laws of Inheritance

Gregor Mendel, the Austrian monk and botanist, knew nothing of chromosomes or of haploid and diploid stages of the life cycle when he published, in 1865, the results of his work on inheritance. However, in a series of carefully planned, executed, and interpreted experiments using the garden pea, he was able to elucidate the basic principles of inheritance in both plants and animals. Figure 10.35 serves to illustrate the type of experiment done by Mendel that allowed him to work out his first law of inheritance, the law of *segregation.* He started with *parental* (P) plants that when self-pollinated were true-breeding for one or another of a pair of contrasting traits, for example, tall vs. dwarf, round vs. wrinkled seed, and so on. When cross-pollinations were made between the tall and dwarf parents, all of the seed produced developed into tall plants in the next generation (*F_1*, first filial). The F_1 tall plants behaved differently from the parental talls, however; when they were self-pollinated, their offspring, the *F_2* generation, included both tall and dwarf plants. Mendel carried out the critical step of counting how many of each kind of offspring were produced, and noted that the ratio of tall to dwarf was close to 3:1.

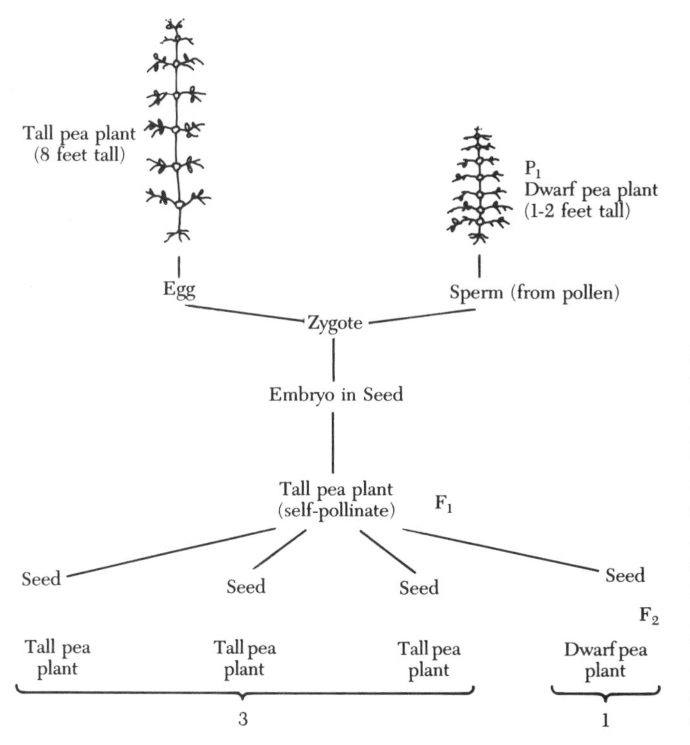

Figure 10.35 A diagrammatic representation of Mendel's experiment in which he crossed two distinct strains of peas—one tall, the other dwarf. These two strains would breed true to each other when self-pollinated. When crossed with each other, the resultant F_1 generation consisted of tall plants only; when these plants were self-pollinated, the seeds developed into an F_2 generation containing both tall and dwarf plants in a ratio of 3:1. The actual numbers of individuals in the F_2 generation were 787 tall and 277 dwarf.

The same kind of result was obtained when he followed the inheritance of other pairs of contrasting characters, indicating that the pattern observed was a general one, and not restricted to inheritance of plant height.

Mendel explained his results by assuming that for each type of trait (for example, plant height), each parent passed on to the offspring some *factor* that contributed to the expression of the trait; whichever two factors were inherited would determine exactly which trait would be expressed. Also, since F_1 plants were all tall, despite containing a factor for tallness and one for dwarfness, the factor for tall is *dominant* to dwarf, which is *recessive* and not expressed in the presence of tall. If an individual contained two factors that were identical (for example, the parental plants), that individual would be true-breeding for the trait; another individual, however, containing two different factors, one from each parent (for example, the F_1 plants) would not breed true, since it could pass on, via each of its gametes, either of the two alternative forms of the determinant; that is, the factors for tallness and dwarfness *segregate* from one another prior to the production of gametes. The different kinds of gamete, which are produced in equal frequencies, can come together in different combinations to produce the F_2 individuals in the ratios mentioned.

Figure 10.36 also interprets these events, and does so in terms of behavior of a pair of homologous chromosomes during meiosis and fertilization. We do so because, as we now know, the factors which determine the traits, the *genes*, are

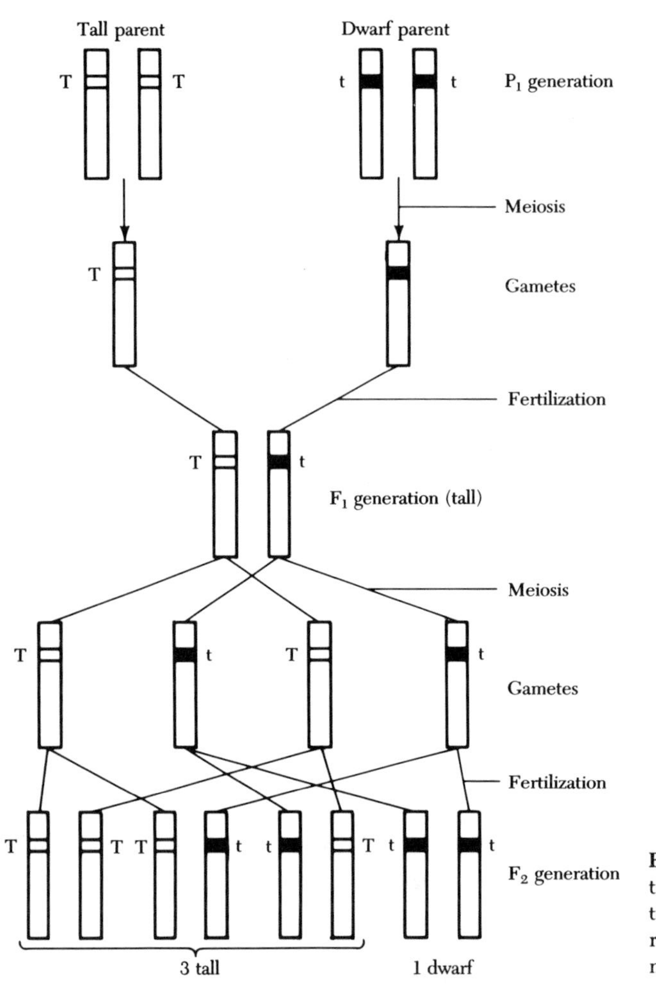

Figure 10.36 Diagrammatic representation of Mendel's first law of segregation, with the dominant (*T*) and recessive (*t*) alleles on particular chromosomes.

carried on the chromosomes, and the two alternate forms of each gene, the alleles, are carried on homologous chromosomes. Mendel's law was a brilliant piece of abstract reasoning from carefully kept quantitative data; our present knowledge of chromosomes permits us to place the same information on a physical and molecular basis.

Mendel's second law, of the *independent assortment of factors*, is illustrated in Figure 10.37. If it is assumed that each pair of genes is located on a different pair of homologues, and if it is assumed further that the pairs of homologues segregate independently of each other, then the behavior of the genes and the chromosomes is exactly parallel. This situation is made more complicated by considering what happens when four pairs of homologues segregate (Figure 10.38). If the bivalents are randomly oriented on the metaphase I spindle of meiosis, each gamete should contain four haploid chromosomes, with 16 different possible combinations of

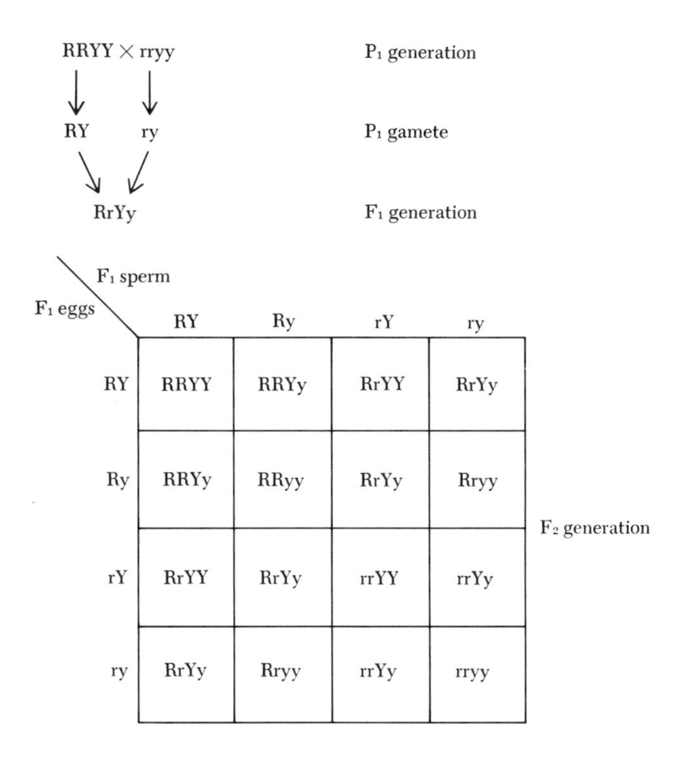

Figure 10.37 Diagrammatic representation of a Mendelian cross involving two independent traits or characters. Round (R) and Yellow (Y) are dominant to wrinkled (r) and green (y), and all are seed characters. The F_2 ratio is theoretically 9 *R-Y-*, 3 *R-yy*, 3*rrY-*, 1 *rryy*, and the actual numbers found by Mendel for this particular cross were, respectively, 315, 101, 108, 32.

maternal and paternal chromosomes. The number of gametic combinations possible can be readily obtained by calculating the value of 2^n, where n equals the number of pairs of chromosomes. In the human, having 23 pairs, the number of possible gametic chromosome combinations is 2^{23}, or 8,388,608. The chance of any single human egg or sperm containing only paternal or maternal chromosomes of the previous generation is, therefore, small indeed. Furthermore, since the same number of gametic combinations is also true for the other sex, the possibility of siblings other than identical twins being genetically identical is remote.

Fertilization, then, is the means whereby, through the union of the nuclei of egg and sperm, the genetic contributions of each parent are combined within a single cell. Mitosis will ensure that each cell of the body has the same genetic constitution. Meiosis, with synapsis bringing together homologous chromosomes, provides for the segregation of genes to the individual gametes. We may then ask what proof do we have that a particular gene is on a particular chromosome, and that

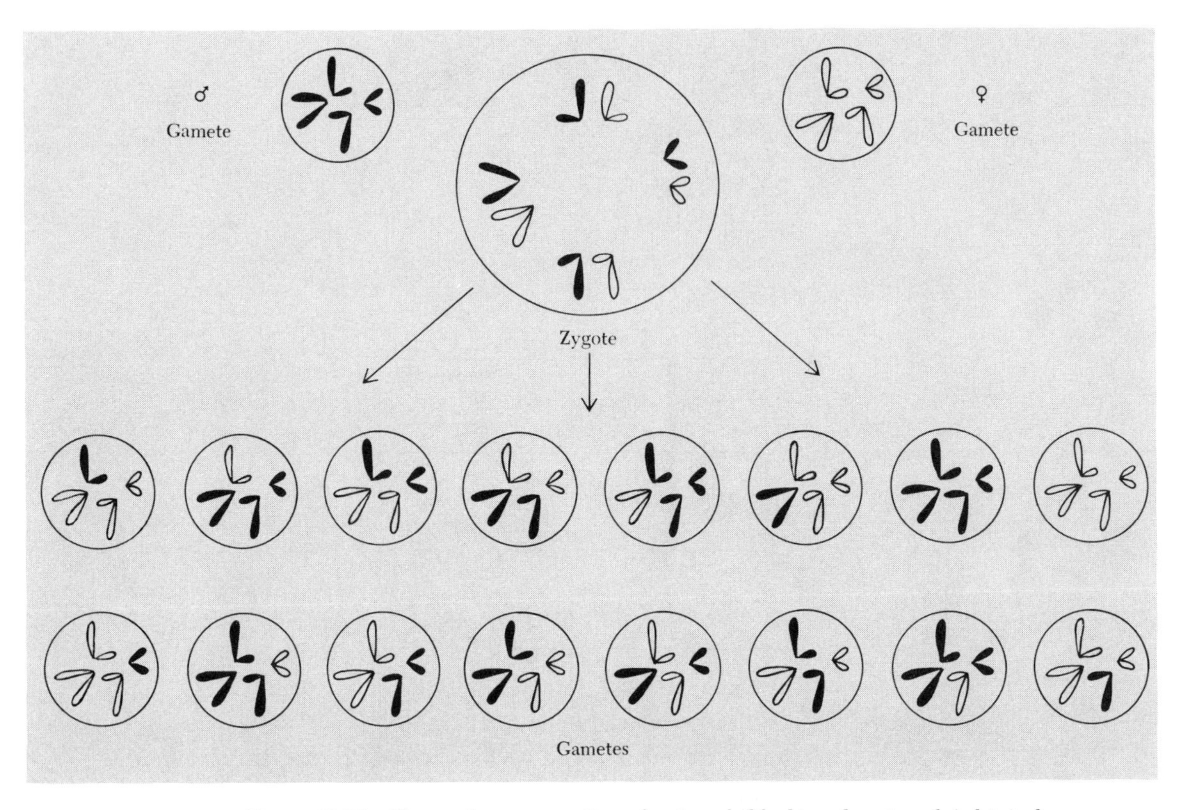

Figure 10.38 The random segregation of paternal (black) and maternal (white) chromosomes during meiosis to give 16 different gametic combinations when four pairs of homologues are involved.

pairs of homologues segregate independently from each other? The answers, coming long before DNA was known to be the crucial genetic substance, involved the relation of a particular chromosome with sex determination.

The X chromosome, so named like an algebraic expression because its value or function was once unknown, is associated with sex determination, and in *Drosophila melanogaster*, the fruit fly, as in human beings, females are XX and males XY. It was also suspected, around 1910, that the first mutation known in this organism, *white eye (w)*, was located on the X chromosome. As Figures 10.39 and 10.40 indicate, the inheritance pattern depended upon whether the white-eyed parent was male or female. The inheritance of *w* parallels exactly that of the X chromosome, provided it is assumed that *w* is recessive to the dominant red-eyed variant (*W*).

Occasionally, however, in crosses involving white-eyed females and red-eyed males, an exceptional white-eyed female or an exceptional red-eyed male appears

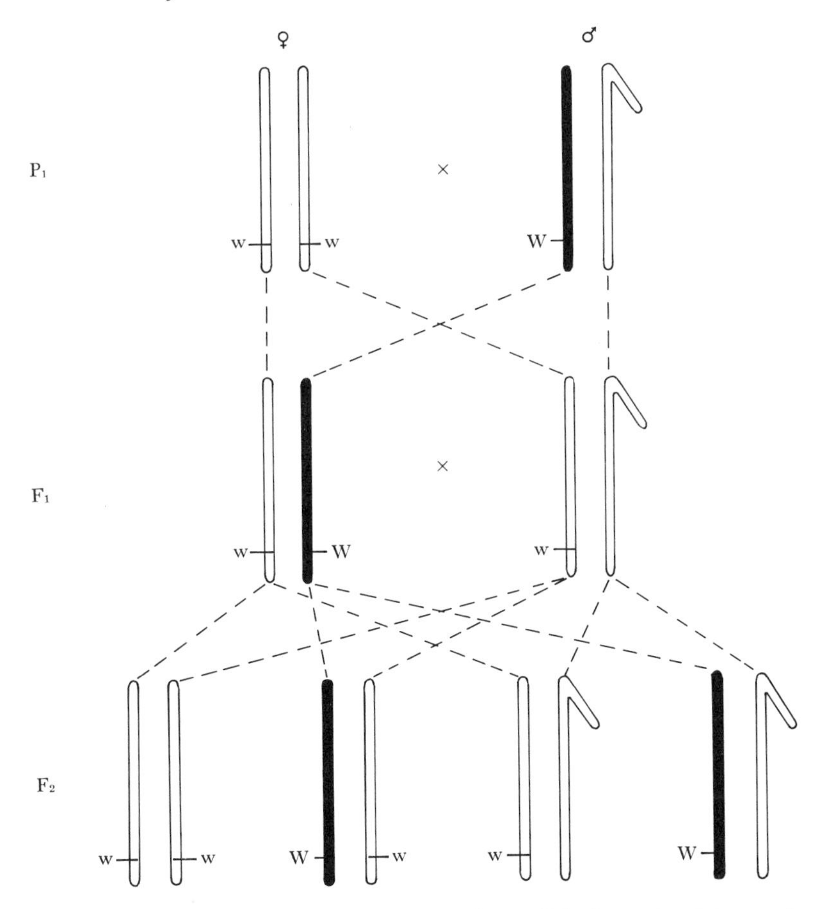

Figure 10.39 Diagrammatic representation of a sex-linked cross involving a red-eyed male (W) and a white-eyed (w) female. Such a cross yields white-eyed males and red-eyed females in the F_1 generation, and an F_2 generation of 1 w female to 1 W female to 1 w male to 1 W male. The Y (hooked) chromosome is devoid of genes.

in a culture (Figure 10.41). This would seem contrary to Mendelian segregation, unless it is assumed that an accident occurred in meiosis in the female such that either both X chromosomes or neither entered the egg nucleus. The exceptional females should, therefore, have two X chromosomes and a Y chromosome, the exceptional male one X chromosome and no Y chromosomes. Cytological examination of these flies showed this to be the case, plus the fact that the exceptional males were always sterile due to the absence of the Y chromosome. The exception, therefore, proved the rule in that the w gene is on the X chromosome and hence is sex-linked, with the deviant behavior of the genes due to the deviant behavior of

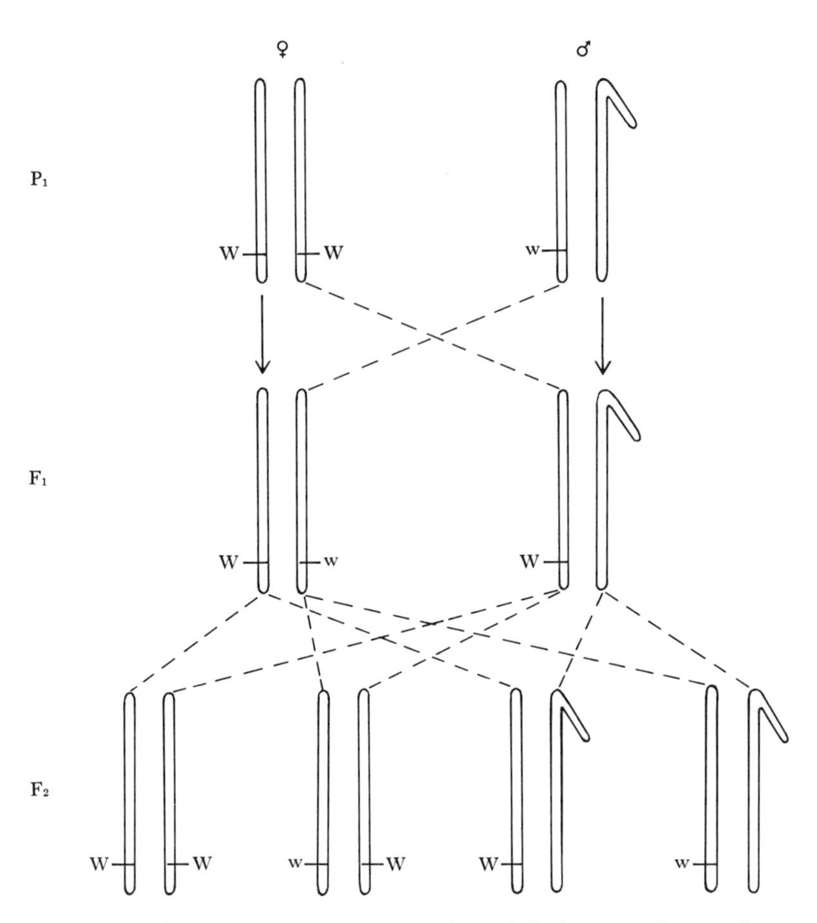

Figure 10.40 Diagrammatic representation of a sex-linked cross involving a white-eyed (*w*) male and a red-eyed (*W*) female. All F_1 individuals will be red-eyed; all F_2 females will be red-eyed; half of the males will be red-eyed; the other half white-eyed.

the chromosomes. This phenomenon of meiotic aberrancy was termed *nondisjunction,* and it is the cause in humans of many of the discovered aberrant sex-determining chromosomal situations.

Some organisms such as grasshoppers lack the Y chromosome, and the males, therefore, are XO instead of XY. Also, in some individuals the homologous chromosomes are morphologically different and distinguishable from each other. Taking advantage of these two aspects (Figure 10.42), it is possible to show that the single X chromosome goes to one pole as often with one homologue as it does with the other, supporting, through chromosomal observation, Mendel's contention that genes segregate independently of each other.

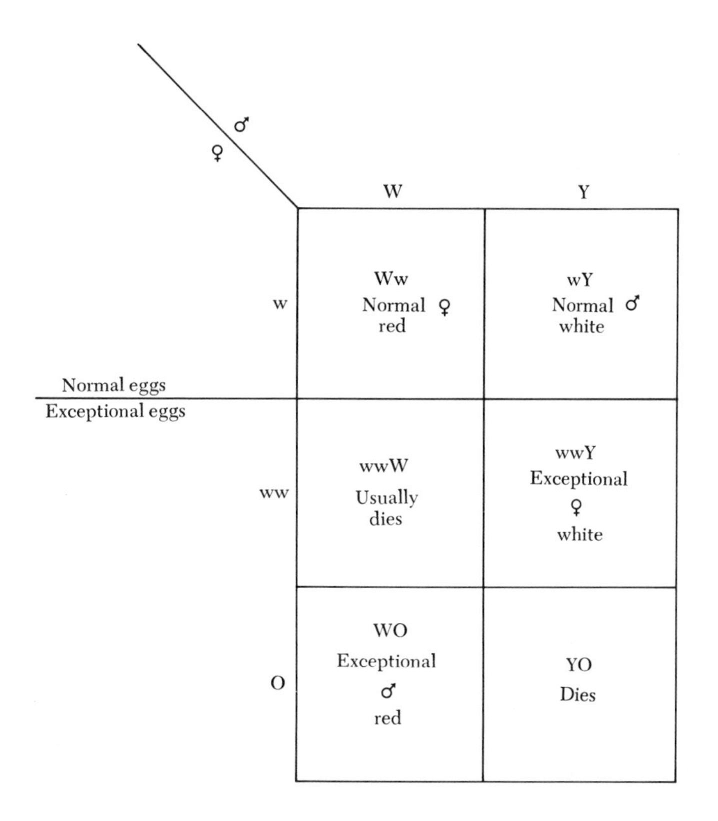

Figure 10.41 Diagrammatic representation of C. B. Bridges' proof that a particular gene is on a particular chromosome. The exceptional eggs—those with two X chromosomes or those with none—arise as a result of nondisjunction, that is, in anaphase I of meiosis the two X chromosomes failed to disjoin, and both went to one pole and none to the other. The exceptional white-eyed females would always possess a Y chromosome in addition to their two X chromosomes; the exceptional red-eyed males would be sterile because they lacked Y chromosomes. In *Drosophila*, the organism used in this instance, the Y chromosome only governs patterns of fertility or sterility; it possesses no other genetic function.

CROSSING OVER AND CHIASMA FORMATION

The number of haploid chromosomes in an organism is limited; four in *Drosophila melanogaster*, 23 in humans, 10 in maize. But as the recognized number of mutant genes increased, it became evident that some of the genes must occupy sites on the same chromosome. Each gene, tested individually as in Figure 10.35, would yield

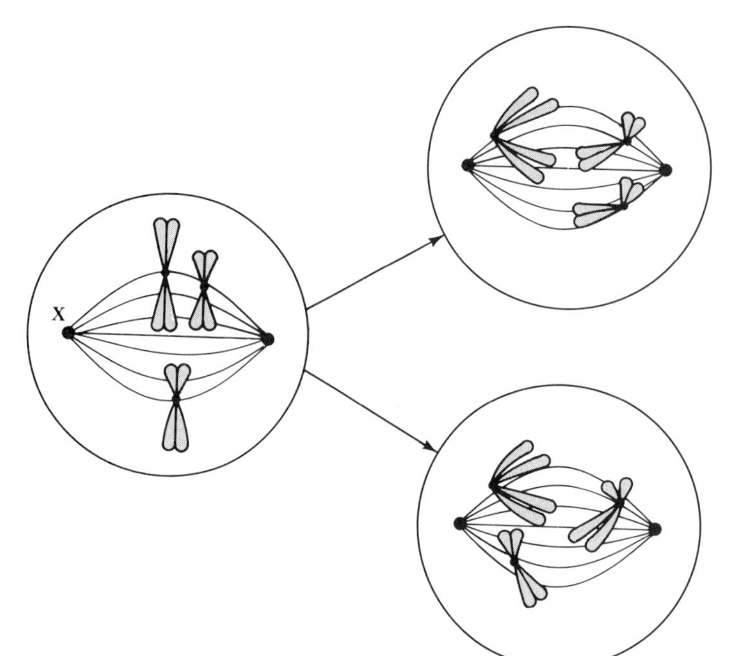

Figure 10.42 Grasshoppers generally have an XO sex-determining mechanism in males, and this is sometimes found with a pair of heteromorphic homologues, that is, two homologues that differ sufficiently in morphology to be distinguishable one from the other. In the above diagram, the single X chromosome in the spermatocytes moves to one pole or the other, and it is a matter of random chance as to which of the homologues it segregates with or from. By such observations, Eleanor Carothers showed that the independent segregation of genes is paralleled by an independent segregation of chromosomes.

F_2 ratios in conformity with Mendelian expectations, but when studied in pairs or in groups of three, the F_2 ratios would often depart significantly from the expected. When retested, the same discrepancies would appear. These findings tended to delay acceptance of the chromosome theory of inheritance until the phenomenon of *linkage* was discovered and understood. Mendelian law may hold, therefore, for the segregation of genes individually, but it is not universally applicable under all circumstances.

Figure 10.43 illustrates the kind of data that might be obtained from a cross when the genes are linked. Notice that the procedures differ from that in a typical Mendelian cross in that rather than obtaining F_2 ratios by the selfing or crossing of F_1 individuals, the F_1s are bred to double recessive individuals by what is termed a *testcross*. The purpose behind this procedure is to test for gametic ratios produced by the F_1s, and by the combining of all F_1 gametes with those from the double recessive, the gametic ratios will be revealed directly. The data reveal that all possible combinations of genes are recovered among the testcross progeny, but it is also evident that the parental types $A\,B$ and $a\,b$ are far more numerous than the new recombinants, $A\,b$ and $a\,B$. The degree of linkage, or the percent of recombination, is determined by adding all new types and dividing by the total number: $(149 + 151)/3000$. This gives 0.10 or 10.0 percent, a figure which is a genetic measure of distance between the two genes on the same chromosome. It would have made no difference in quantitative results if the parental combinations had been $A\,b$ and $a\,B$, indicating that the frequency of recombination is a function of chromosomal distance between genes and not of particular genetic combinations.

AABB × aabb — Parental cross

↓

AaBb × aabb — Testcross

Testcross progeny	Observed frequencies	Expected frequencies
AaBb	1,358	750
Aabb	149	750
aaBb	151	750
aabb	1,342	750
Totals	3,000	3,000

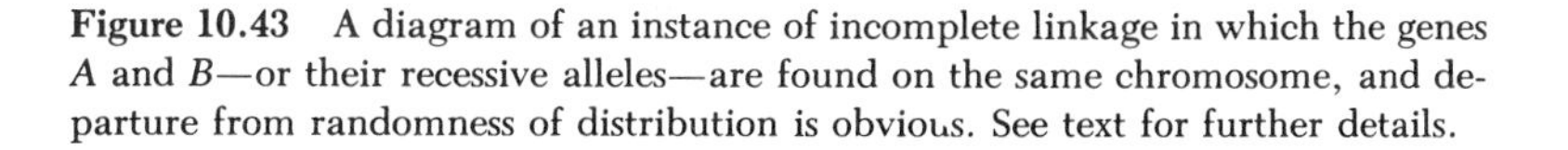

Figure 10.43 A diagram of an instance of incomplete linkage in which the genes *A* and *B*—or their recessive alleles—are found on the same chromosome, and departure from randomness of distribution is obvious. See text for further details.

What has been described is an instance of *incomplete linkage;* the genes are linked, but not absolutely, being recombined by a process of crossing over. The chromosome, then, which is a molecule of DNA extending from one end to the other, is a series of linked genes. If genes can be recombined by crossing over, homologous chromosomes must be able to exchange chromatin, with the frequency of chromatin exchange correlated in a quantitative way with the frequency of crossing over. The chiasmata visible in diplotene bivalents (Figures 10.19 and 10.20) are physical evidence that crossing over has taken place in the previous stages of synapsis.

It will be noticed in Figure 10.20 that the chiasmata are formed between chromatids of the two homologous chromosomes. Two pieces of evidence provide proof that crossing over (and chiasma formation) involves an actual exchange of chromatin between homologous chromosomes, and that it takes place after the chromosome has replicated itself and is longitudinally double. Chromosome 9 in maize has two forms: with and without a large terminal chromomere. This chromosome also carried the genes *C* or *c* (colored or uncolored seeds) and *Wx* or *wx* (starch or waxy endosperm). When the cross is made as indicated in Figure 10.44, with both genes and chromosomes in a heterozygous state, any testcross individual showing both recessive genes should show the chromomere on one homologue, and those showing both dominant genes should be without the terminal chromomere. No exceptions to these expectations were found. Crossing over between these two genes was, therefore, always correlated with an exchange of chromatin (Figure 10.45). Data from *Neurospora*, the pink bread mold, provided proof that crossing over takes place after, and not before, replication of the chromosomes (Figure 10.46).

The events taking place in meiosis—synapsis, chiasma formation, segregation of homologous centromeres and chromosomes at anaphase I and of sister cen-

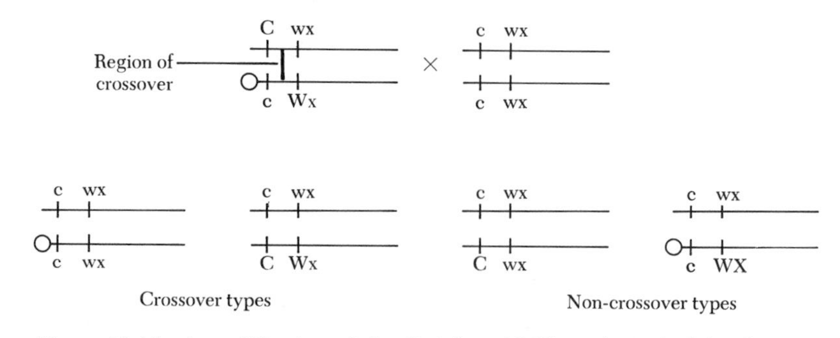

Figure 10.44 A modification of the Creighton-McClintock method for demonstrating that an exchange of genes, when both are on the same chromosome, is accompanied by an exchange of chromatin. The test plant on the left has both genes and chromosomes in heterozygous form, and an exchange of genes would put both recessive genes on the knobbed chromosome and the dominant genes on an unknobbed chromosome. The distance between *C* and the knob is too short to allow crossing over to occur at the same time that it takes place between *C* and *Wx*.

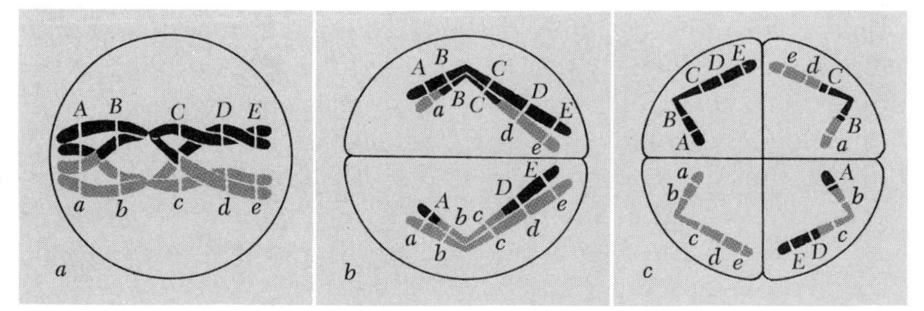

Figure 10.45 The genetic consequences of crossing over. (*a*) A bivalent, consisting of paternal (black) and a maternal (color) homologue, has formed, and crossing over has taken place between genes *A* and *B*, and *C* and *D*. (*b*) At anaphase the two chromatids in each segregating chromosome are no longer alike genetically. (*c*) The chromatids are now separated, and two of them have a different genetic composition, while the other two remain as before.

tromeres and chromatids at anaphase II—ensure that the haploid gametes or spores resulting from meiosis will have a variable combination of genes. Since these cells will contribute directly or indirectly through fertilization to the next generation, the individuals of that generation must exhibit a comparable genetic variation. It is this inherited variability upon which natural selection acts to bring about the evolution of organisms. Sexual reproduction, with its complementary phenomena of meiosis and fertilization, is a means not only for the production of new individuals but of new individuals that vary among themselves. In this sense, meiosis differs greatly from mitosis, which, in its production of similarly endowed cells, is a conservative process of reproduction.

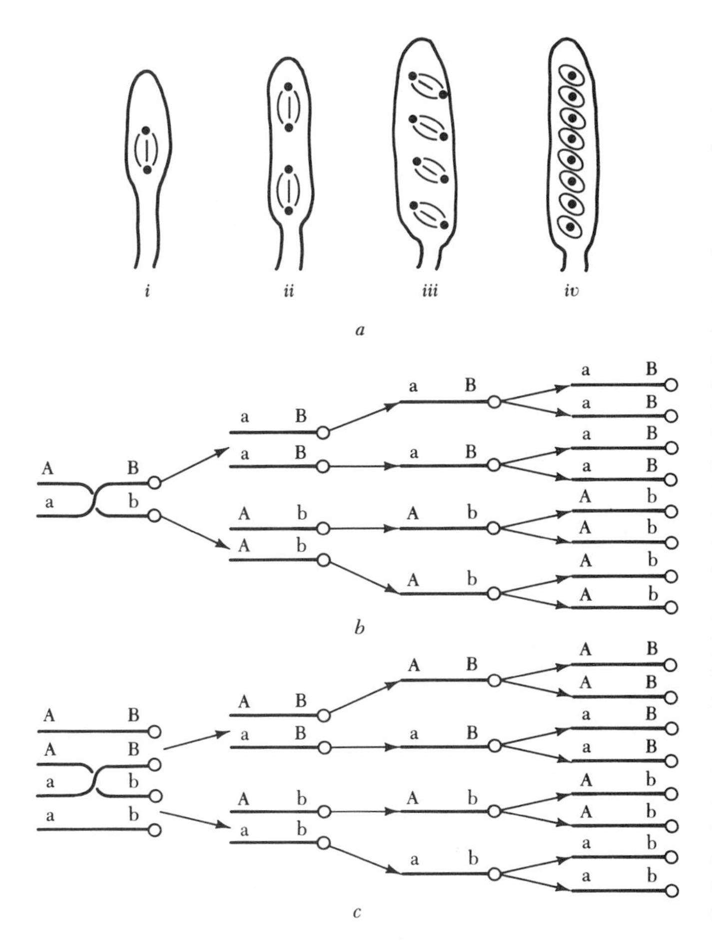

Figure 10.46 Meiotic divisions and segregation in *Neurospora.* (*a*) The two meiotic divisions followed by a mitotic division gives eight ascospores in each ascus. These can be isolated in serial order and their genotypes readily determined. (*b*) The sequence of events and the genotypes and serial order of the eight ascospores that would result if crossing over took place in the two-strand stage and before replication of the chromosomes; that is, if a crossover took place, all eight ascospores would show new combinations of genes from the original A B and a b combinations of the parental strains. (*c*) The sequence of events and the genotypes and serial order of the eight ascospores that would result if crossing over took place in the four-strand stage and after replication had occurred. Since a crossover between two genes would produce two crossover and two noncrossover chromatids, the map distance between two genes cannot exceed 50 map units when based on such a test. The circumstances in (*c*) are typical of such experimental testings, while that in (*b*) is not observed, leading to the conclusion that crossing over takes place in the four-strand stage.

The Mechanism of Crossing Over

The fact that crossing over involves the exchange of chromatin means that the helices of DNA somehow must be broken, with the broken ends rejoined in such a manner as to bring about recombination. The process of synapsis brings homologous chromosomes into close register with each other and makes crossing over possible, but from that point on the events are unclear. However, since all events taking place in the cell are chemical events mediated by enzymes, the most reasonable assumption is that crossing over is similarly determined. At least the requisite enzymes are known to exist, and to be present in meiotic cells. The events believed to take place are indicated in Figure 10.47, with only the two chromatids involved in crossing over being considered.

Enzymatically induced "nicks," caused by an enzyme, an *endonuclease*, in single strands of each double helix permit the double helices to open up through the breaking of hydrogen bonds, following which a rejoining takes place. The rejoin-

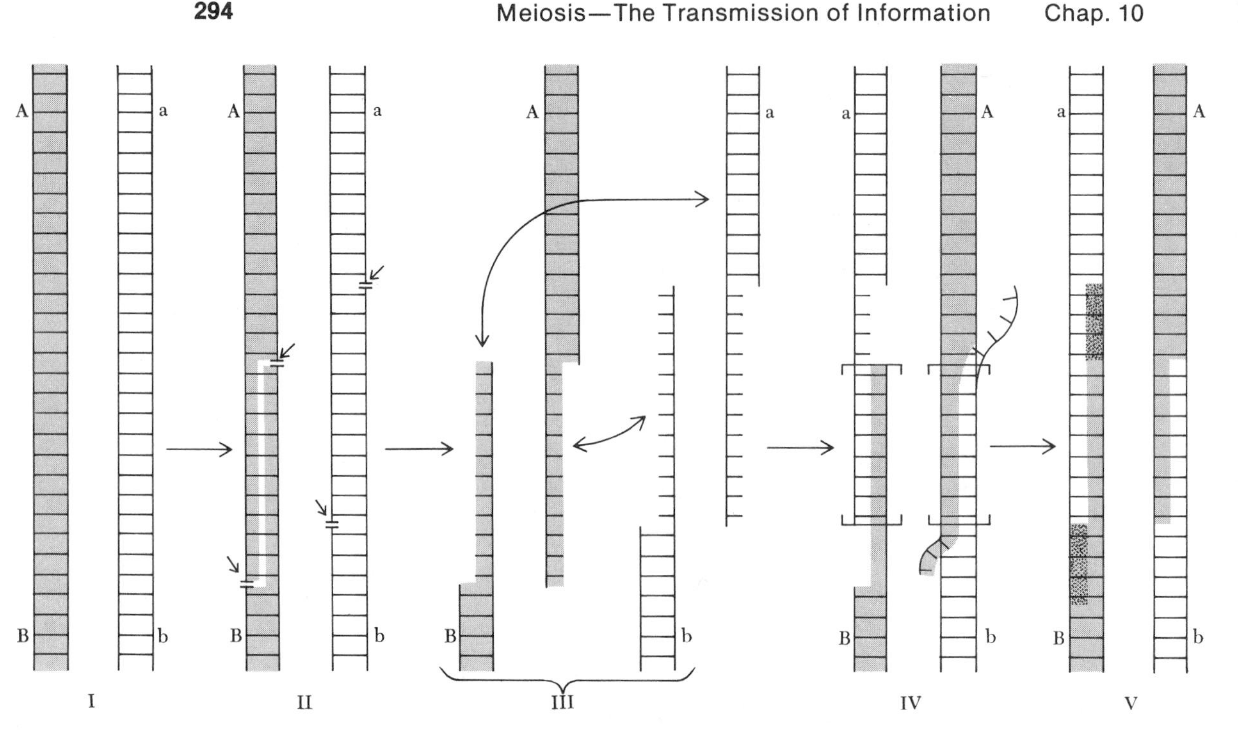

Figure 10.47 A mechanism of crossing over that is consistent with what is known of the process. (I) The two nonsister, but homologous, chromatids that will undergo crossing over between the two linked genes (the other two uninvolved chromatids are not included in this illustration). (II) The enzymatically induced "nicks" in the sugar-phosphate backbones of the polynucleotide strands of the two double helices. (III) The unraveling of the strands, which results from the breakage of hydrogen bonding and which precedes the rejoining to form new genetic combinations. (IV) The rejoined chromatids in a complementary fashion (those areas within the brackets), with gaps remaining to be filled and loose and excess polynucleotide strands remaining to be excised. (V) The finished result of crossing over, with the gaps filled (the darkest areas), the loose ends excised and the breaks in the sugar-phosphate backbone repaired. This is an enzymatically determined process of DNA breakdown and repair, which probably occurs in all cells (see Figure 1.16) but which, in meiotic cells and at the time of synapsis, leads to crossing over.

ing is exactly complementary, but this obviously leaves gaps in the double helices, as well as free, unpaired strands. The unpaired regions can be removed by another enzyme, an *exonuclease*, the gap can be filled by the action of a DNA polymerase I, an enzyme that can add nucleotides in a complementary way to a growing strand, and, finally, a polynucleotide ligase can close the gap by uniting the growing strand to the old one. The actual number of new nucleotides need not be great, and the small amount of DNA replication detected in zygotene-pachytene of meiotic cells could be related to these phenomena.

There is very little direct experimental evidence to prove that the indicated enzymes are actually involved in crossing over, and there is no evidence to indicate that synapsis sets up the enzymatic circumstances for crossing over. Rather, it would appear that nicks are frequently produced in double helices and in all kinds of cells, with the cells having a set of repair enzymes to keep the DNA sufficiently intact to perform properly. This is suggested by the observation that mitotic cells show reciprocal exchanges between the two chromatids of a single chromosome, indicating that breaks and rejoinings took place during or following the S period (see Figure 1.16). Synapsis, leading to the formation of a synaptinemal complex, then may be viewed as a means of bringing homologous chromosomes close enough so that the nicks in their respective chromatids may interact with each other, thereby affecting crossing over by increasing the frequency of chromatid exchange and ensuring its reciprocal nature. But since synapsis is characteristic primarily of meiotic cells, it is in these cells that crossover events take place normally, even though it is possible that the number of nicks are the same in both mitotic and meiotic cells.

MAPPING HUMAN CHROMOSOMES

There is an obvious desire and need to understand, to the extent possible, the organization and function of the human genotype. We are, of course, the products of our genes and the environment in which those genes are expressed, and our understanding of ourselves, in all of our varied manifestations of uniqueness, normal and abnormal, is dependent on knowledge of our genes: their number, organization, and function. Here we are concerned with their allocation to particular chromosomes.

It has been estimated that there are about 50,000 genes making up the human genotype. These are distributed among the 23 haploid chromosomes, with their allelic counterparts in diploid cells in the other haploid set. A single chromosome must, therefore, contain many genes, all linked to each other, loosely or tightly, and with the number of genes generally commensurate with the size of the chromosome. How then, can a particular gene be assigned to a particular chromosome, and how close or distant can that gene be to others on the same chromosome?

The classical method of mapping genes is described on page 291 (Figure 10.43). Two genes thought to be linked can be followed in a double hybrid cross, with the F_1 offspring being backcrossed to the double recessive. If the genes are linked, the parental combinations are recovered more frequently than are the new recombinants, with the frequency of the new recombinants providing a genetic measure of the distance between the two genes (1 percent recombination = 1 map unit). If the genes are not linked, the parental and recombinant types will be found with equal frequency.

Such a technique can be used to map the chromosomes of maize or *Drosophila*, but is wholly impractical with human subjects. Human family members are too small in number, critical genetic information on all family members is often difficult to obtain, the generation time is too long, and controlled human breeding is not socially acceptable. To be sure, family histories can be followed to determine the inheritance of variant forms of suspected genes, and they can at times be correlated with particular chromosomes, but uncertainty is often encountered. Much of the difficulty in mapping human chromosomes has, however, been overcome by the development of several recent techniques, with the result that well over several hundred traits, most of them of a biochemical nature, but an increasing number related to disease syndromes, including cancer, have been assigned to genes located in particular chromosomes. The linkage relations of genes on the same chromosome can thus be established, but the order of the genes and their proximity to each other must be determined by other means.

One of the most effective of these new techniques involves the fusion of human cells in culture with those of other mammalian species. Mouse cells are most frequently used because the mouse genome is sufficiently different from that of humans to be readily recognized, and a great number of mutant forms are available. Human fibrocytes from connective tissue and mouse cells from a similar source will fuse spontaneously at a slow rate in culture, but the frequency can be considerably enhanced by the presence in the medium of inactivated Sendai virus (SV40), or by the addition of appropriate amounts of polyethylene glycol. The technique and rationale are described in Figure 10.48. Once the cells have fused, they are grown in a selective medium which favors the fused cells, and inhibits or eliminates the unfused parental lines.

The procedure offers two critical advantages. One is that the mouse genome is retained intact while the human chromosomes are gradually and randomly eliminated with the exception of that chromosome (in addition to others that may be also retained) possessing the gene necessary for the viability of the hybrid line. Pure clones of hybrid cells can thus be established. A second advantage is that the mouse genome and the added human chromosomes are functionally active in that their genes are expressed at the same time. Certain requirements must be met, however, if the technique is to be effective in permitting the assignment of specific genes to specific chromosomes: (1) genic expression must be detectable at the cellular level; biochemical expression is therefore required, excluding, for example, not only morphological expressions but also a study of red blood cell antigens if only cultured fibrocytes are being assayed; and (2) the human chromosomes involved must be distinguishable from each other, a requirement readily met through the use of staining procedures which reveal the distinctive banding of mammalian chromosomes (see Figure 4.37).

The original hybrid cell would have contained 86 chromosomes, 46 human chromosomes, and 40 mouse ones. As the human chromosomes are eliminated in the successive cell divisions, the selectiveness of the medium will determine which of the hybrid cells will survive. The retention of a particular human chromosome together with the mouse genome indicates that the gene in question is a part of that chromosome. As a result of such testing, the gene coding for thymidine kinase (TK) has been assigned to chromosome 17; that for hexosamine A, the absence of which

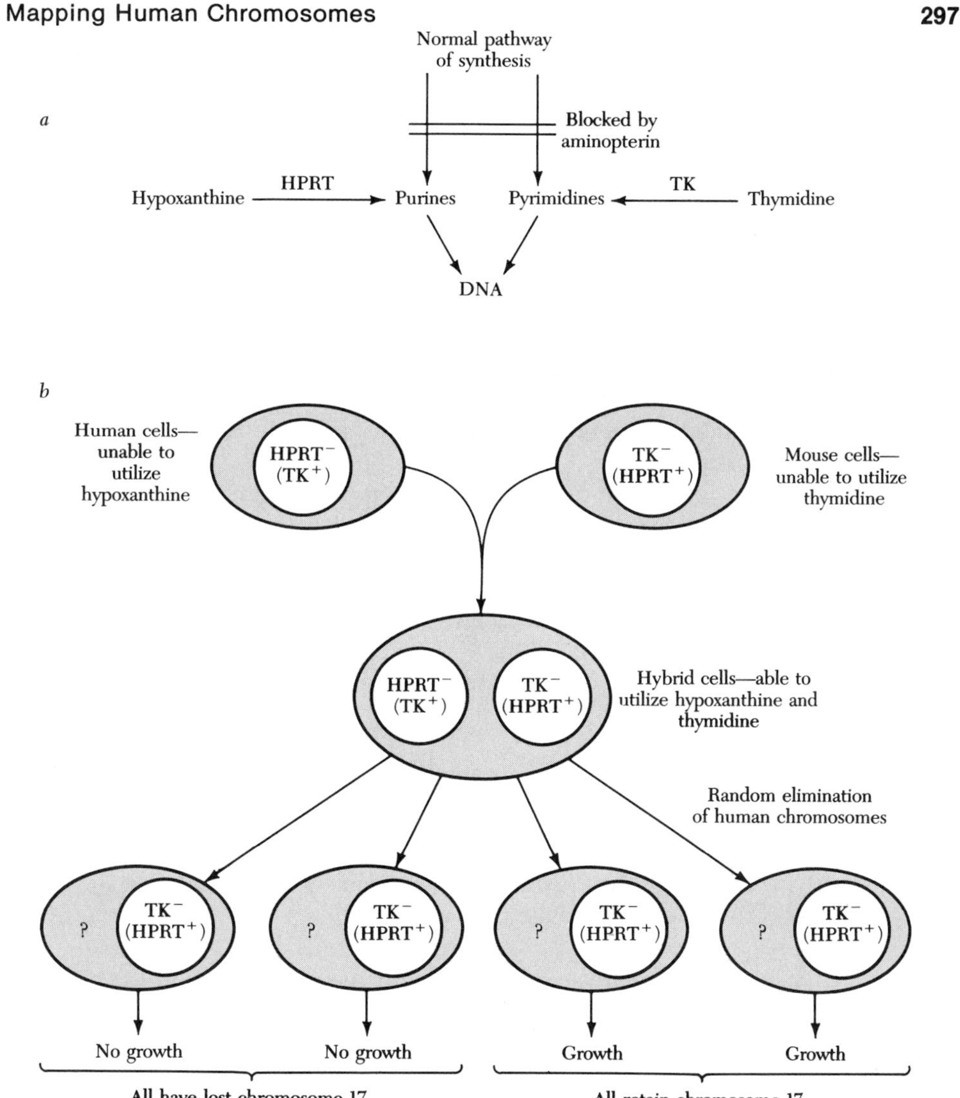

Figure 10.48 Method used to assign a particular gene (TK) to a particular human chromosome. (*a*) If the normal pathway of synthesis of purines and pyrimidines is blocked by the drug aminopterin, cells can convert exogenously supplied hypoxanthine and thymidine to purines and pyrimidines, but only if they contain the enzymes hypoxanthine phosphoribosyl transferase (HPRT) and thymidine kinase (TK), respectively. (*b*) Neither HPRT-deficient human cells nor TK-deficient mouse cells will grow on medium containing aminopterin, hypoxanthine, and thymidine. Hybrid cells resulting from fusion of the human and mouse cells, with their complementary nuclei, continue to grow, since the mouse nucleus provides the capacity to make HPRT and the human nucleus the capacity to make TK. As chromosomes are lost at random from the human complement, some hybrid cells are unable to grow, since they have lost the chromosome carrying the TK gene. By identifying which chromosomes are retained by the various surviving colonies, it is possible to assign the TK locus to chromosome number 17.

causes Tay-Sachs disease, to chromosome 15; and that responsible for galactose-1-phosphate uridyl transferase, whose absence is associated with galactosemia, to chromosome 3. The induction of interferon, a protein which inhibits virus replication, requires the presence of normal genes on chromosomes 2 and 5, as well as a gene on chromosome 21 which codes for a protein whose presence is necessary for interferon induction.

If two separate clones of hybrid cells, expressing two different biochemical traits, for example, two recognizably different enzymes, possess the same human chromosome(s) along with the mouse genome, it can then be assumed that the two genes responsible are linked on the same chromosome. Another technique must then be used to determine their location with respect to each other. This is best done through the use of translocations, and Figure 10.49 illustrates how this may be accomplished.

Many proteins, however, have no known variant forms, but the genes coding for them can be assigned through the use of RNAs. If the appropriate RNAs—mRNAs, tRNAs, or rRNAs—can be properly labeled with tritiated uridine for subsequent identification and then isolated, they can then be reannealed back onto the chromosome site from which they were transcribed. By this means, for exam-

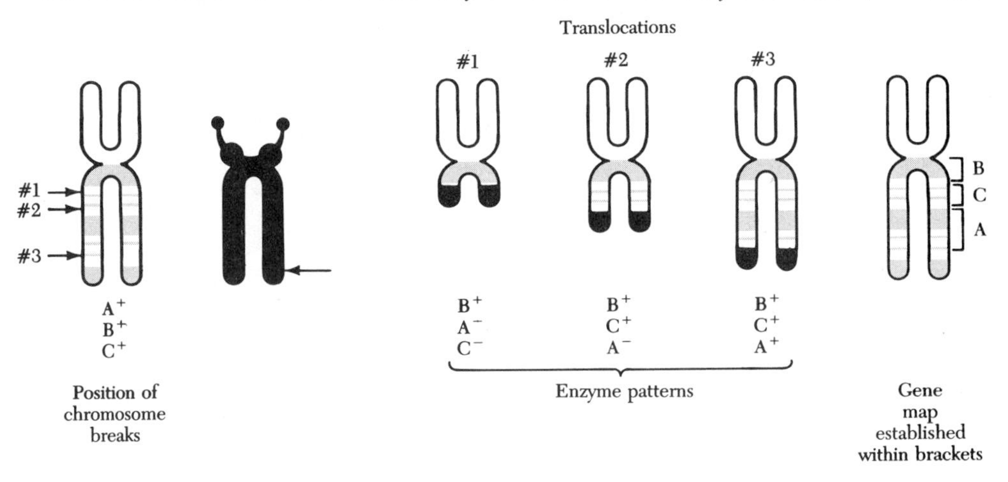

Figure 10.49 The use of translocations to allocate specific genes to specific sites on a given chromosome. Assume that the genes A, B, and C, coding for three different enzymes, are known to be on the long arm of a given human chromosome; this chromosome has an identifiable banding pattern, but the order of the genes is not known. These cultured cells are treated with some chromosome-breaking agent (x rays, ultraviolet light, etc.), and screened for the translocation of some portion of the long arm to some other chromosome in the genome. The translocation-bearing human cells are then fused with one of three strains of mouse cells, each strain being deficient for one of the three enzymes. Fused cells which can produce one or more of the enzymes can be identified by their ability to grow on particular selective media. A correlation of the banding pattern of the translocated piece of the long arm with the presence or absence of the enzymes under consideration can then permit the establishment of the order of the genes between the centromere and the end of the chromosome.

ple, the DNA that codes for the 18S and 28S rRNAs was found to be located in the short arms of the satellited chromosomes 13, 14, 15, 21, and 22, while the 5S rRNA locus was identified as being in the long arm of chromosome 1. Human chromosome maps are thus becoming more and more detailed, helping to put genetic counseling on firmer ground. It must be remembered, however, that the structural genes constitute only a small portion of the entire human genome, and many sections of human DNA may well be devoid of identifiable markers.

BIBLIOGRAPHY

BAKER, B. S., et al. 1976. The genetic control of meiosis. *Annu. Rev. Genet. 10:* 53–134.

CATCHESIDE, D. G. 1978. *The Genetics of Recombination.* University Park Press, Baltimore.

D'EUSTACHIO, P., and RUDDLE, F. H. 1983. Somatic cell genetics and gene families. *Science 220:* 919–924.

DICKINSON, H. G., and HESLOP-HARRISON, J. 1977. Ribosomes, membranes and organelles during meiosis in angiosperms. *Phil. Trans. R. Soc. Lond. B 277:* 327–342.

MCKUSICK, V. A. 1971. The mapping of human chromosomes. *Sci. Am. 227:* 104–113.

MCKUSICK, V. A., and RUDDLE, F. H. 1977. The status of the gene map of the human chromosomes. *Science 196:* 390–405.

PEACOCK, W. J., and BROCK, R. D. (eds.). 1968. *Replication and Recombination of Genetic Material.* Australian Academy of Science, Canberra.

RADDING, C. M. 1973. Molecular mechanisms of genetic recombination. *Annu. Rev. Genet. 7:* 87–112.

RILEY, R., et. al. (eds.). 1977. A discussion on the meiotic process. *Phil. Trans. R. Soc. Lond. B 277:* 183–376.

WESTERGAARD, M., and VON WETTSTEIN, D. 1972. The synaptinemal complex. *Annu. Rev. Genet. 6:* 71–110.

Note: The student might well consult any of the half dozen or more recent textbooks of genetics; most carry good accounts of meiosis in all its aspects.

11

The Cell in Development

In previous chapters, we have mentioned that multicellular organisms "develop" from a fertilized egg into a plant or animal of adult proportions. Each of us knows in a general way what is meant by development: it is a continuous and gradual process of change that takes time to be fully realized; it is generally accompanied by an increase in size and weight; it involves the appearance of new features and new functions; and it eventually slows down when mature dimensions are reached. Humans, for example, develop from the fertilized egg stage through embryonic and prenatal life, childhood, adolescence, sexual maturity, physical maturity, middle age, old age, and death. Development is, of course, one of the most prominent features in the early life of an organism, but the formation of new blood cells, gametes, and wound tissue, which may take place up to death at an advanced age, are also aspects of development. So too are those processes we associate with aging; for example, excess formation of collagen in the extracellular spaces and the calcification of joints. These, it would appear, are normal processes of development continuing beyond the point of a functional and developmental optimum. The terms we have used, however, are only broadly descriptive. They tell us very little about the mechanism of development as a biological phenomenon.

Development of a single-celled zygote into a multicellular organism, whether it be a human being or an oak tree, involves the processes of *growth* and *differentiation*. Growth can be simply defined as an increase in mass, and it results from assimilation of matter. It can involve an increase in cell size and an increase in cell number, the original cell taking from its environment the raw materials it needs and converting them into more substance and more cells like itself. Let us consider the human egg. It weighs about 1×10^{-6} g (grams), and the sperm, at fertilization, adds to it only another 5×10^{-9} g. At birth, however, a child will weigh around 7 pounds, or 3200 g, which is an increase of about 1 billion times during the 9-month prenatal period. Another 20-fold increase in mass occurs between birth and the achievement of full size of an average adult.

Increase in mass, however, is not sufficient to account for the particular *form* of an organism. Development of form requires differential growth; that is, different rates of growth and different rates and patterns of cell reproduction are involved in the determination of form, resulting in the majesty of a giant redwood or the "fearful symmetry" of Blake's tiger. In other words, some parts of the body grow at a faster or slower rate than others, and in development some features come into existence earlier than others. Figure 11.1 shows how the growth rate in the human being alters the relative proportions of bodily parts to one another. The head and neck increase in size rapidly during the early period of gestation, the arms grow faster at an earlier stage than do the legs, whereas the trunk progresses at a more or less steady rate until maturity.

Developmental growth, therefore, is not just the enlargement and multiplication of cells, it is a complicated pattern, with different centers of growth being active at different times and with different rates of development. These centers are coordinated to produce an unfolding of form, and it is form—as well as function, of course—that distinguishes humans from other animals, one human being from another, and an orchid from a lily.

Development also involves differentiation, the acquisition (or loss) of specific structural properties by different cells, such that these cells become specialized in different ways to carry out the various activities associated with living things. A generalized cell, therefore, is gradually transformed by a process of successive changes into a specialized one, and diversity is thereby introduced into a function-

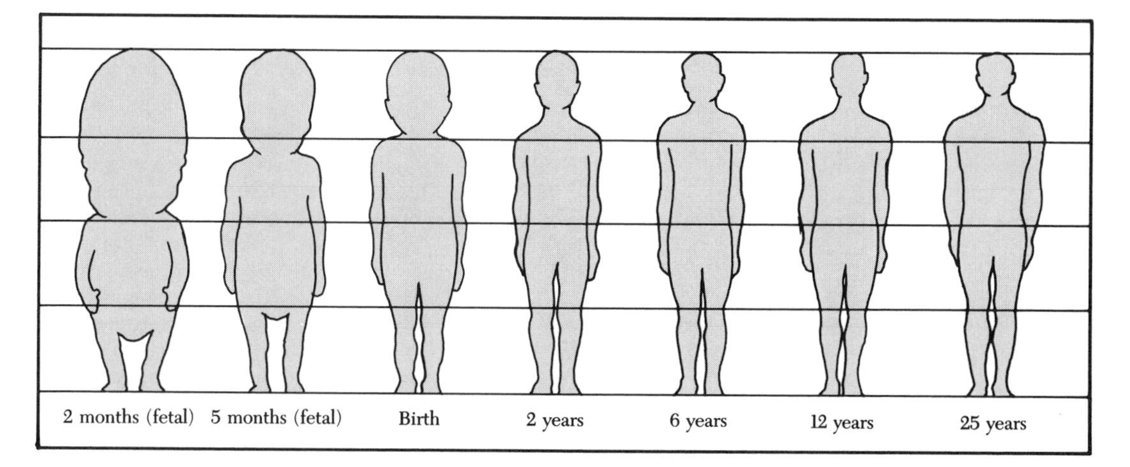

Figure 11.1 Changes in the form and proportion of the human body during fetal and postnatal life.

ing organism (Figure 11.2). In the human, for example, growing cells are transformed into the myriad of different cells that makes up the human body (Figure 11.3): cells of nervous, muscular, digestive, excretory, circulatory, and respiratory systems.

Differentiation, therefore, is a process of directed change. It is a phenomenon that has no counterpart in the nonliving world, and any information we have about it has been derived from observations of living systems. This process is creative in the sense that life is creative, for out of the general features common to all cells arise structures and functions that are peculiar to specialized cells. Differentiation, therefore, is to development what mutation is to biological inheritance and what imagination is to human endeavor; it provides variety of form, function, and behavior without at any time destroying the unity of an organism as an individual.

Both growth and differentiation occur during the life cycle of unicellular organisms as well as in multicellular organisms. However, an additional factor is

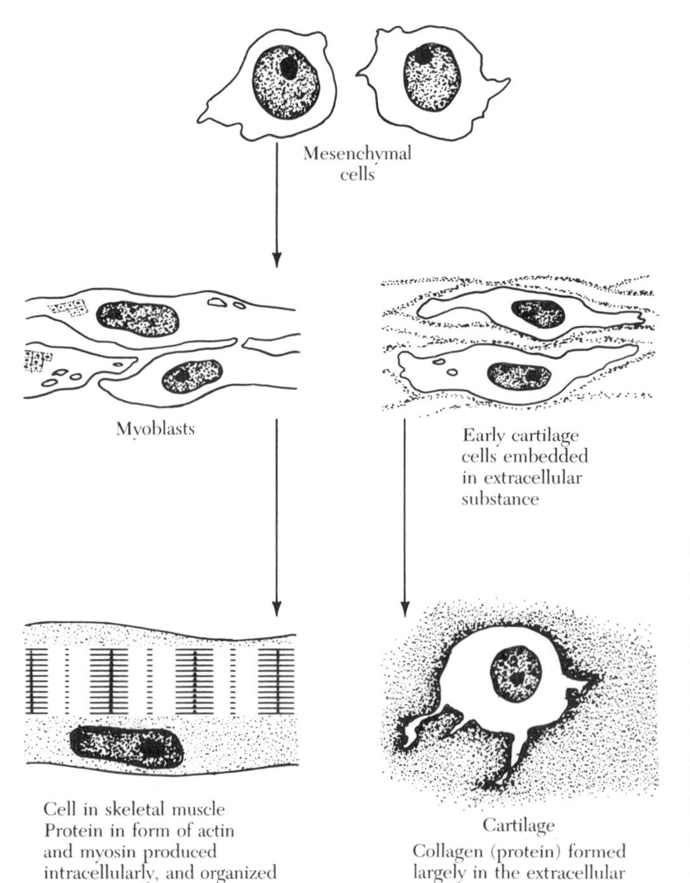

Figure 11.2 Differentiation of generalized mesoderm cells (mesenchyme) into two kinds of specialized cells, muscle and cartilage. Both these cells are similar in that they produce substantial amounts of protein, but in muscle cells the proteins (actin and myosin) are retained internally for contractile purposes, while in cartilage cells the protein is deposited as collagen outside the cell, where it plays a supportive role. The ultimate shape of the cells also changes as a result of differentiation.

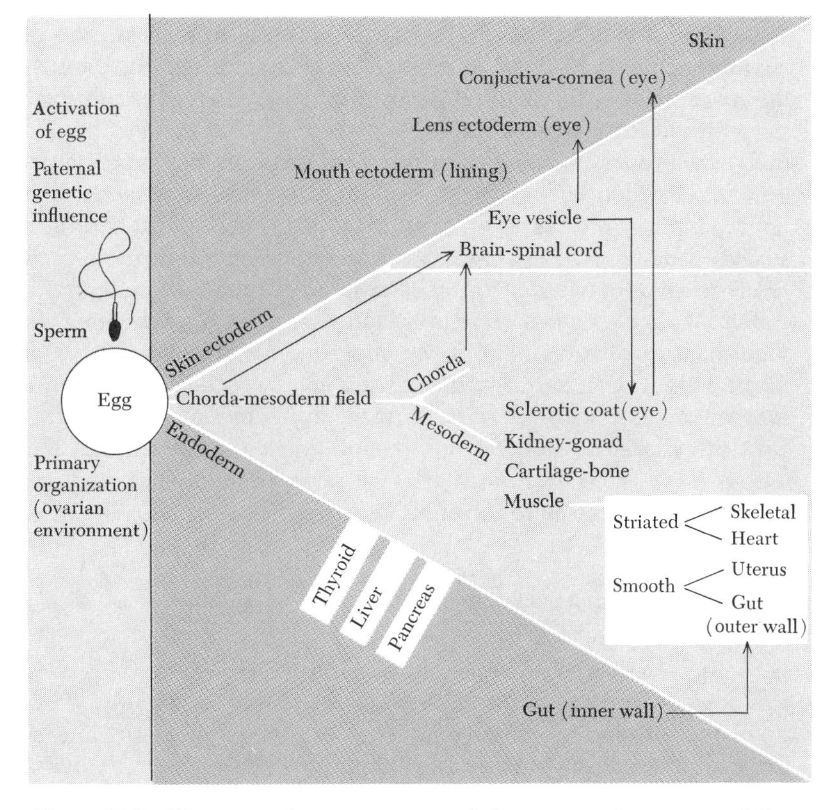

Figure 11.3 Diagrammatic representation of the pattern of progressive differentiation from unfertilized egg to mature tissues in a vertebrate. The three major tissue layers (ectoderm, mesoderm, and endoderm) originate early and progressively give rise to the cells of the major organs. Dashed lines indicate an influence of one tissue on another during the course of development. Note that the eye has a double origin from both ectoderm and mesoderm.

involved in development in multicellular types. For such development to proceed in an organized manner (which indeed it usually does), growth and differentiation of one cell must be coordinated with growth and differentiation of other cells. Otherwise, chaos would ensue. Such coordination requires some sort of communication between different cells in the developing organism. Development, therefore, depends on *integrated* growth and differentiation, and it is this integration that permits harmonious development and maintenance of unity of the whole organism.

The problems of development are many and complex, and we cannot deal with all its aspects. However, the cell is the basic building block of life, and since it is the cell that undergoes developmental changes, we must consider the cellular basis of development. Let us begin by examining the nature of the changes that cells undergo as they differentiate and become specialized.

What a cell is, and what it can do, are consequences of the chemical reactions that take place in that cell and that have taken place in the past. As we have seen in Chapter 3, the reactions that occur in a cell are determined by which enzymes are present and functionally active in the cell. Thus we can think of differentiation of cells as being essentially a change in cellular proteins, in particular a change in the enzyme complement of the cell. However, we also have seen that because of the exact replication and segregation of DNA during cell reproduction, all of the cells in a multicellular organism contain the same type and amount of nuclear information. Since the nucleus is the control center of the cell, all cells should have the same potential and be capable of manufacturing all of the proteins of the body. The obvious fact, however, is that they do not do so. Red blood cells produce hemoglobin, nerve cells do not; certain cells of the pancreas produce insulin, others do not; the mesophyll cells of a plant leaf contain the enzymes involved in photosynthesis, root cells in general do not. Since particular sets of proteins are determined by particular sets of genes, we must ask if differentiation is the result of particular sets of genes being expressed in different cells.

DIFFERENTIAL GENE EXPRESSION

It can be demonstrated in several ways that differentiation is not a consequence of loss of genetic information. Regeneration of whole organisms from small parts of an original individual is commonly encountered in plants and in lower animals. For example, a stem cutting can produce functional roots. Thus the differentiated cells of the stem have not lost the information necessary for the functioning of a root cell, since this information is expressed in their descendants. A more dramatic demonstration that differentiated cells retain the information required for all of the activities of the whole organism comes from experiments in which segments of carrot root, containing live, functional cells, are explanted and the cells induced to divide. The resulting disorganized clump of cells, or *callus* tissue, can then become organized in such a way that root and shoot meristems are formed and an intact, fertile carrot plant is produced (Figure 11.4). Such experiments tell us that once a differentiated cell is freed of the physical and chemical constraints imposed upon it in the intact organism, its full potential may be expressed.

Nuclear transplantation experiments carried out on certain amphibians also show that differentiation need not be a consequence of loss or addition of genetic information. Nuclei from intestinal epithelial cells of *Xenopus* tadpoles can be transferred by careful micromanipulation into eggs from which the original nucleus previously has been removed. These eggs can then give rise to tadpoles that ultimately develop into adult toads, showing that the nuclei of the epithelial cells retain the full potential of the nucleus of the zygote (Figure 11.5).

Heterokaryons, which are hybrid cells formed by fusion of cells of different types, also provide evidence for the reversible nature of differentiation. Chicken erythrocytes, unlike those of mammals, do not lose their nuclei when mature;

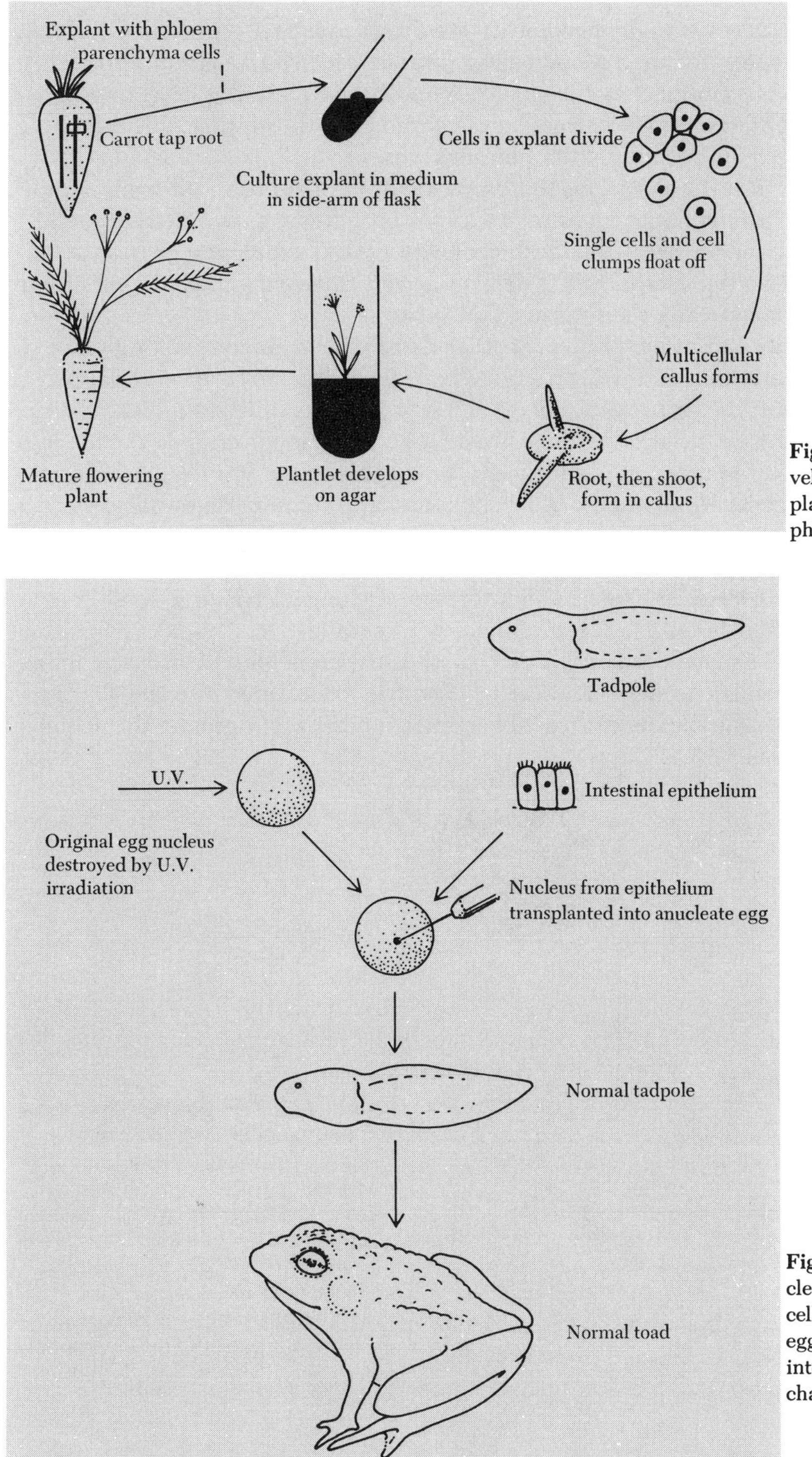

Figure 11.4 Diagram of development of whole carrot plant from cells derived from phloem parenchyma.

Figure 11.5 Diagram of nuclear transplant from tadpole cell into anucleate egg. The egg will develop normally into a mature toad, with the characteristics of the donor.

however, the nuclei eventually become inactive and no longer synthesize any DNA or RNA. When such chick red blood cells are fused with HeLa cells (a human cell line derived from a tumor), their nuclei respond by synthesizing DNA and RNA; chick-specific proteins, which normally are not made by the mature, differentiated erythrocyte, are then formed by the heterokaryon.

Such cases of differentiation, therefore, do not appear to result from permanent changes in the genetic potential of a cell (although, as we shall see, some developmental changes may), but rather of differential expression of that potential. How the potential is expressed is determined by the environment in which the cell, and more specifically the nucleus, finds itself.

We must also point out, however, that many of the changes taking place in cells as they differentiate may be difficult, if not impossible, to reverse. Indeed, it is generally true that the more specialized a cell type, the more difficult it is to change the pathway of differentiation, either directly or by transplantation of the nucleus (Figure 11.6). Differentiation in multicellular organisms is a progressive affair requiring several generations of cells before the final and stable differentiated state is achieved. At some point in the sequence, cells become committed to a course of action from which they cannot readily be diverted. C. H. Waddington, a British embryologist, has expressed the idea of commitment by his diagram of a developmental landscape (Figure 11.7). He visualizes a generalized cell as a ball rolling downhill toward its final destiny, a destiny that depends on which of the many valleys the ball rolls through. The farther the cell penetrates into the developmental landscape, the greater the loss of general properties, the greater the acquisi-

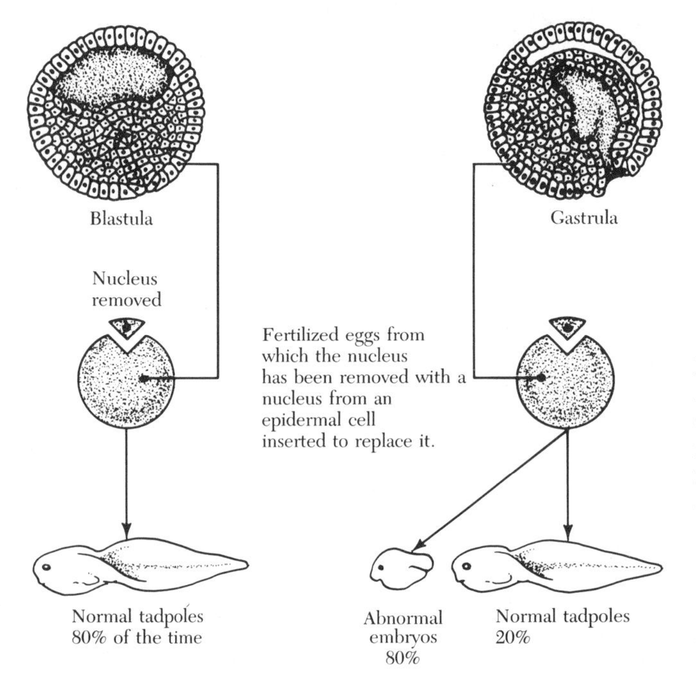

Figure 11.6 The two kinds of epidermal cells differ in their capacity to complete development when transplanted into anucleated eggs. About 40 percent of the eggs survive such treatment, but nuclei from cells of the blastula are far more likely to promote normal development as compared to those from the gastrula. The nuclei, therefore, are varied in their developmental potentiality as a result of differentiation.

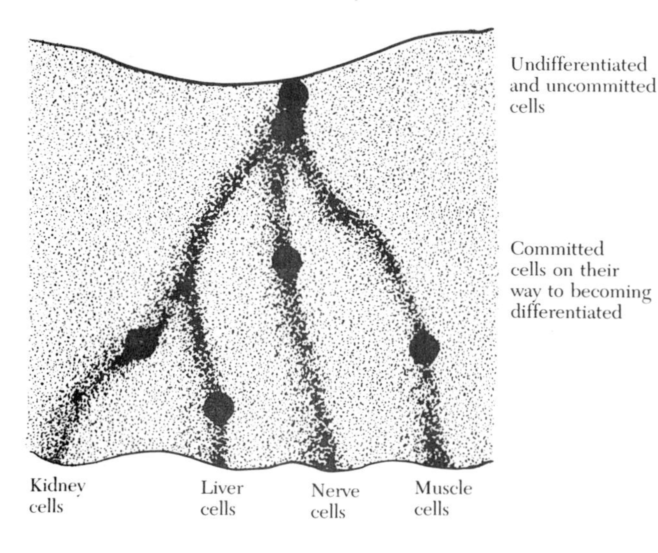

Figure 11.7 The uncommitted cells, depending on their location in an embryo, achieve their final destination as differentiated cells by moving along specific pathways of change.

tion of special features, and the less likely is it able to return to an undifferentiated state.

Let us express this in more specific terms. The embryologist can approach the problem by cutting certain cells out of an embryo and transplanting them to other embryos (Figure 11.8). If a group of young, undifferentiated cells are transplanted to the future head region of another embryo, the transplanted cells become part of the head region; if they are transplanted to the back, they will become part of the back musculature; if to the posterior part, they become part of the tail. But if the embryologist transplants "committed" cells from an older embryo in the same way then instead of becoming an integral part of the region to which they are transplanted, they tend rather to retain their own identity and even to modify the surrounding cells. This is well illustrated by an experiment done in the chick embryo. If the leg bud, which has no resemblance to a mature leg in any way, is removed from a young chick embryo and is transplanted to the body cavity of another embryo, the cells in the bud live, continue to increase in number, and eventually form therein a very well-developed leg with bones and muscles (Figure 11.9). Yet the bud at the time of transplantation had no obvious bone or muscle cells. In terms of Waddington's landscape, however, the cells had already entered a "valley" leading to leg formation, a valley down which they continued to roll and from which they could not escape; that is, they had become committed. Indeed, the limb bud cells can be separated from each other and then allowed to reassociate. Following reassociation, the tissue is still able to develop into a normal limb.

Experiments carried out on ferns also show that tissues become partially autonomous early in development. If very young leaf primordia are removed from the apex of *Osmunda* and cultured on a synthetic medium, they develop into shoots and ultimately give rise to whole plants. Older leaf primordia, however, develop

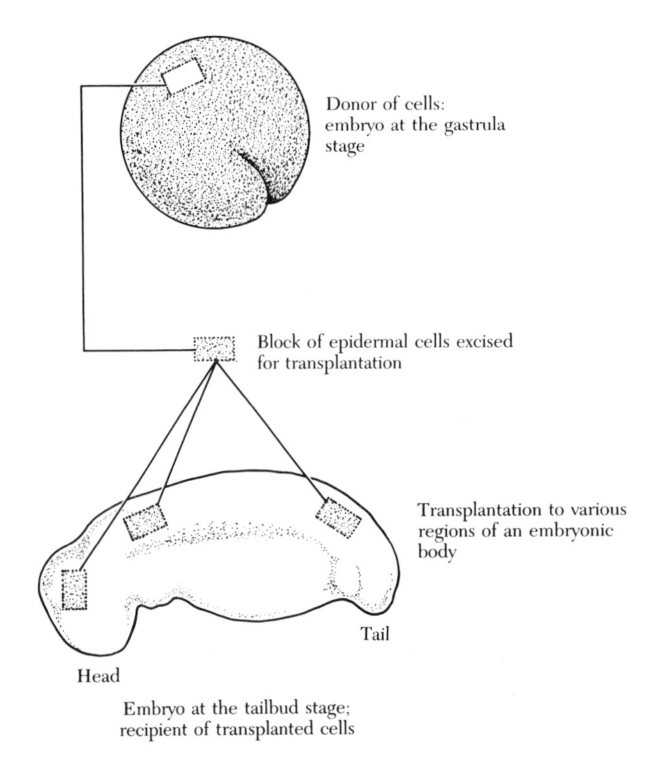

Figure 11.8 Epidermal cells from a gastrula are as yet uncommitted, and if transplanted to an embryo in the tail-bud stage, can become differentiated and integrated. Those transplanted to the head region can become part of the neural plate; if to the mid-body region, they can be differentiated into glands, muscle, or cartilage; if to the tail region, they can become part of the tail, kidney, or notochord. Final differentiation takes place, therefore, after transplantation.

into leaves, as they would have had they not been explanted. At some point, therefore, the cells acquire the capacity to develop as a leaf independently of the rest of the plant.

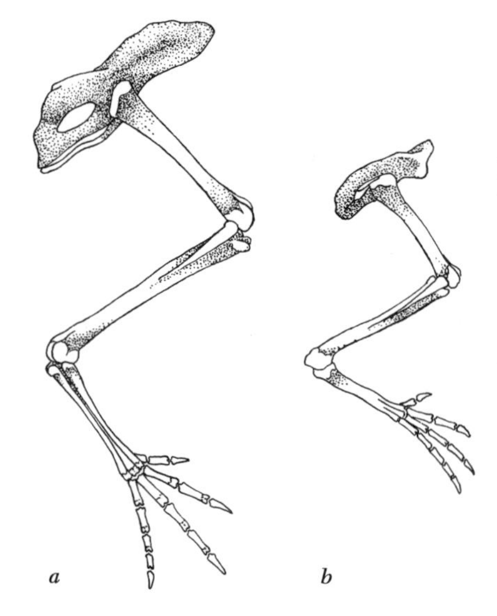

Figure 11.9 An example of a structure that develops after transplantation in a reasonably normal fashion. (*a*) Normal leg bones of a chick 18 days after incubation. (*b*) A slightly smaller but reasonably complete set of leg bones that developed after the hind limb bud was transplanted to the body cavity. At the time of transplantation, the limb bud showed no evidence of bone or muscle, but the cells had already been "committed" to leg formatin, a process of differentiation that continued even though the limb bud had been removed to a foreign location.

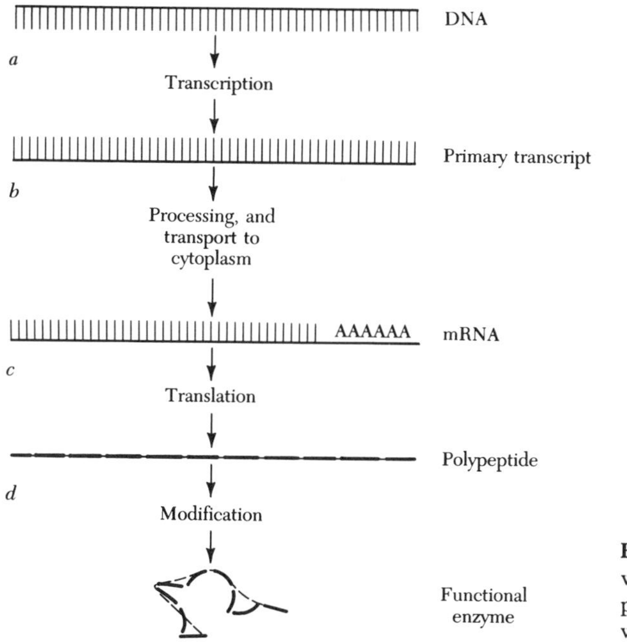

Figure 11.10 Information flow pathway involved in expression of genetic potential, indicating levels, *a–d*, at which control might be exercised.

Thus some developmental changes may be transient in nature, and fairly easy to reverse, while others result in a more "determined" state, one that is relatively stable and can persist over several generations of cells. Whatever may be the molecular mechanisms that selectively control expression of potential, some of them must be heritable and themselves perpetuated within cell lineages. The pathway of development of a given cell, therefore, depends not only on its genetic constitution, which supplies its potential, but also on its past history and present physical and chemical environment, which together determine those features of the potential that will be realized at any specific time.

When we say that differentiation is a consequence of differential gene expression, we are saying that it results from differential utilization of genetic information. As we have seen in Chapter 4, the pathway of information flow is from the DNA of the chromosome to the messenger RNA, and hence through the enzymes that are synthesized in the cytoplasm and that catalyze the reactions of the cell. The problem of differentiation, therefore, is how the flow of information is controlled in such a way that different enzyme complements can be established in different cells or in the same cell at different times during its lifetime. It now appears unlikely that there is a single solution to this problem, but rather that different control mechanisms operate in the cell and at several different levels.

Among the levels of the cellular information flow pathway at which control can be examined are those shown in Figure 11.10. The first is that of transcription of the DNA code into the base sequence of the RNA. Subsequent events leading to the translation of that base sequence into the amino acid sequence of a polypeptide chain can also be subject to regulation at several points, including processing of

RNA, transport to the ribosome, and initiation of translation on the ribosome. Post-translational modification of the protein itself can also play a part in determining the phenotypic characteristics of the cell. We shall now consider some of these levels and the possible mechanisms by which control may be exerted.

DIFFERENTIAL TRANSCRIPTION

Many studies indicate that selective expression of the information encoded in the genes of a cell is achieved by activation or repression of transcription of these genes; in other words, by selective synthesis of the primary gene product, the RNA that contains the information to be carried to the ribosomes.

Prokaryotic Cells: The Operon

One example of differential transcription and how it is regulated is the *lac operon* of *E. coli.* The operon hypothesis, developed by Francois Jacob and Jacques Monod, postulates the existence of both regulatory and structural genes, the latter being those that code for particular polypeptides while the former govern or regulate the activity of the structural genes.

Glucose is a normal energy and carbon source for *E. coli*, and the enzymes involved in its utilization are present in the cell at all times—they are *constitutive* enzymes. If, however, glucose is replaced by lactose, a sugar not normally used by *E. coli* if glucose is available, the cell can utilize it only after a lag period (Figure 11.11). Lactose utilization requires several enzymes which are not otherwise present in the cell, appearing only after lactose is presented. Lactose, therefore, is an inducer, and the enzymes are *inducible* rather than constitutive.

How expression of the genes for the lactose-metabolizing enzymes is regulated was worked out by Jacob and Monod in an elegant series of genetic analyses. The

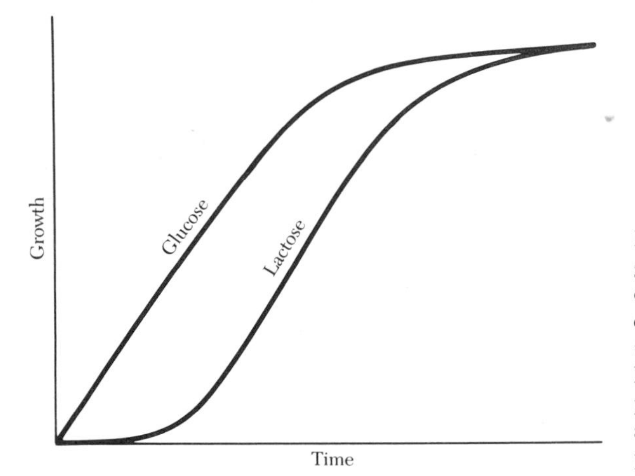

Figure 11.11 The growth of cells on glucose and lactose. The enzymes handling glucose are normal, or constitutive, elements of the cells, and glucose can be utilized without delay. The enzymes utilizing lactose, however, are not normally present and must be induced and synthesized before that sugar can serve the purposes of growth.

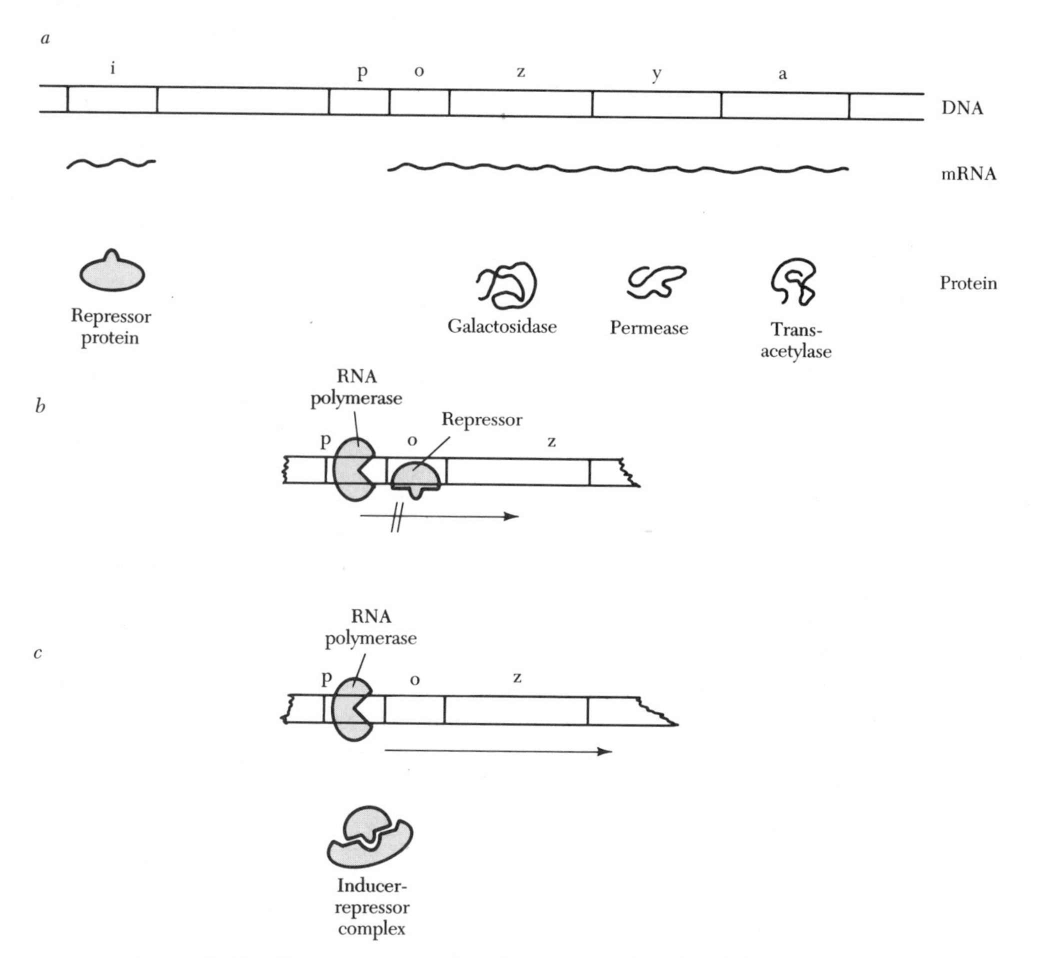

Figure 11.12 Negative control of the *lac* operon in *E. coli*. (*a*) Genetic structure of operon: the "i" gene codes for a repressor protein which, in the absence of lactose, (*b*), binds to the operator region, "o," and prevents RNA polymerase from initiating transcription from the promoter region. (*c*) When lactose, the inducer, is present it binds to the repressor and prevents it from binding to the operator. The RNA polymerase is then able to transcribe the structural genes z, y, and a.

structure of the *lac* operon, and how it is thought to work, are shown in Figure 11.12. Three contiguous structural genes, *z*, *y*, and *a*, code for the three necessary enzymes. In the absence of lactose, however, a *repressor* protein, coded for by the regulatory *i* gene, binds to a specific *operator* region of the DNA adjacent to the *z* gene. When the repressor is bound to the operator base sequences, RNA polymerase is unable to associate with the promotor and initiate transcription. The repressor protein also has a binding site for lactose (actually for allolactose, a related metabolic derivative); when lactose is present in the cell it binds to the

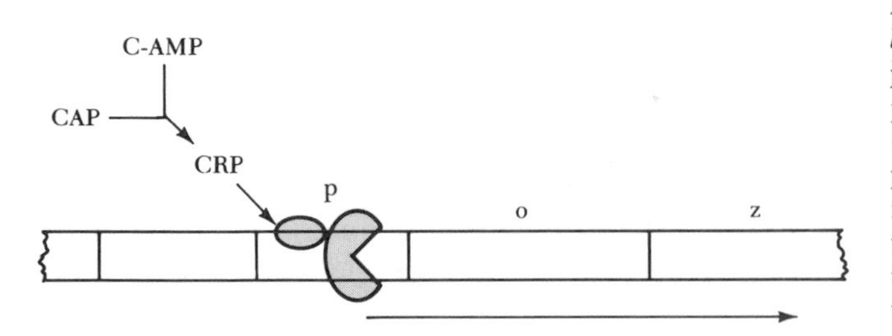

Figure 11.13 Positive control of the *lac* operon. The proper binding of RNA polymerase to the promoter requires the presence of an activator protein, CRP (cyclic AMP-receptor protein), at the promoter region. High levels of glucose reduce cAMP levels to the point where the activator protein no longer binds to the promoter. Thus transcription is prevented even though lactose is present.

repressor protein and removes it from the operator region. The RNA polymerase is then able to initiate transcription, and it transcribes the three structural genes as one unit, forming a polygenic (or polycistronic) mRNA. Thus the three genes are *coordinately* expressed in the presence of lactose and coordinately repressed in its absence.

The system we have just described represents negative control, in that the association between the protein and the DNA prevents transcription. The *lac* operon, however, is also subject to a positive type of control (Figure 11.13). If the sugar glucose is also present in the cell, lactose can no longer induce the *lac* operon, even though it removes the repressor protein from the operator. The reason for this is that another regulatory protein, an *activator,* must also bind to a specific nucleotide sequence in the DNA in order for the RNA polymerase to transcribe effectively. However, this activator can only bind to the DNA when it is complexed with an *effector* molecule, cyclic AMP (cAMP). In the presence of glucose the concentration of cAMP in the cell is low; only when the glucose level decreases does the concentration of cAMP reach a level sufficiently high to form a complex with the activator protein, allowing it to bind to the DNA and thereby permit expression of the *lac* operon genes.

The *lac* operon represents a control system in which transcription of specific genes can be switched on and off in a coordinated manner, involving interactions between proteins and DNA sequences and resulting in the capacity of the cell to cope efficiently with its carbohydrate needs. Similar regulatory mechanisms, involving both repressor and activator proteins, are widespread in prokaryotes. Although regulation in eukaryotic cells is likely to be much more complex than in prokaryotes, some features of eukaryotic control mechanisms may operate in ways similar to those of prokaryotes.

Eukaryotic Cells

There is certainly good evidence that eukaryotic genes can be switched on and off at the level of transcription. Much of this evidence is based on one or more of various techniques involving nucleic acid hybridization. The rationale behind these techniques is that the base sequence of an RNA molecule is complementary to the base sequence of the DNA strand from which it was transcribed; thus under appropriate conditions the template and the transcript are able to form a double-stranded, hybrid DNA/RNA molecule.

One can calculate how much of the total DNA of the nucleus is transcribed into mRNA in any cell type by determining how much of it will hybridize with the RNA from that cell type. In general fewer than 10 percent of the DNA sequences are recognized by the mRNA (Figure 11.14). Furthermore, when mRNA populations from two different cell types are compared it can be shown that while some of the sequences are common to both, many of them are different (Figure 11.14). However, since the mRNA found in the cytoplasm represents only a relatively small fraction of the heterogeneous nuclear RNA transcripts (Chapter 4), these experiments do not conclusively demonstrate transcriptional control. In order to do so, sequences in the primary transcripts must be compared. This can be done by using the enzyme reverse transcriptase to make DNA polynucleotide sequences that are complementary to mRNA. These cDNA sequences, after being cloned in

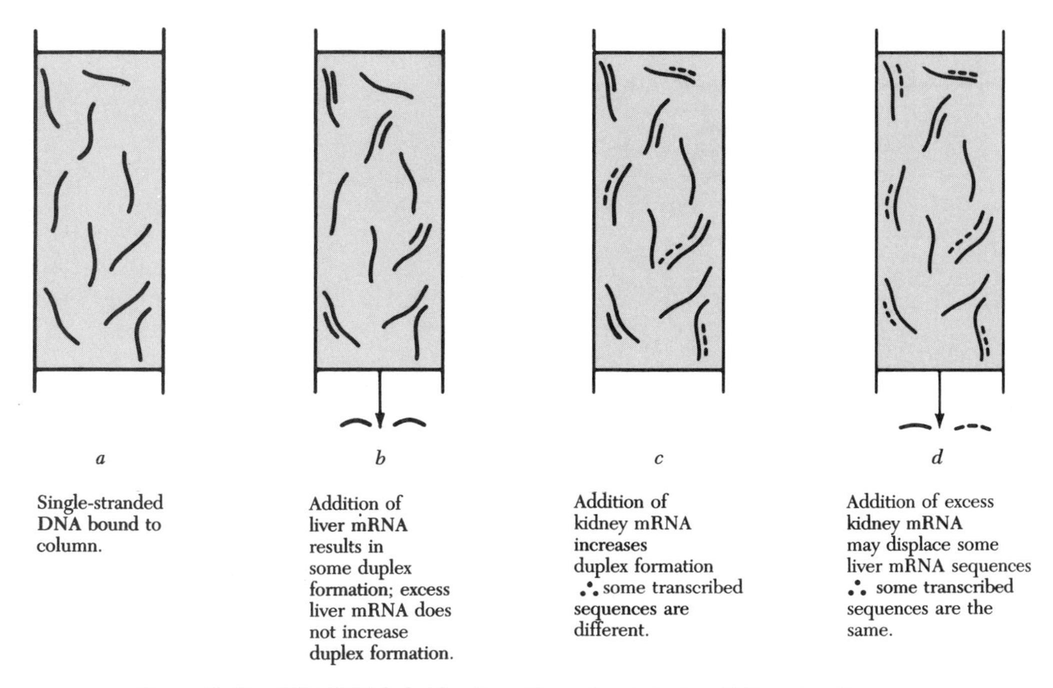

Figure 11.14 DNA/RNA hybridization. The rationale is that RNA nucleotide sequences are complementary to the DNA sequences from which they are transcribed, and therefore will hybridize by complementary base pairing with these DNA sequences to form duplex molecules. The two strands of the DNA double helices are separated by raising the temperature or the pH, and immobilized on a filter or column so that they cannot reanneal. If radioactive RNA is added and the temperature or pH lowered appropriately, any complementary sequences will form duplexes. RNA that fails to find a complementary sequence is washed out, and the amount of radioactivity left is a measure of how much hybridization has occurred.

plasmids to provide large amounts of material, can then be used to "probe" for the presence of complementary RNA sequences, that is, sequences identical to those of the original mRNA. When cDNAs complementary to mRNAs from liver cells are used as probes of nuclear RNA from liver or brain cells, it can be shown that while some sequences are expressed in both types of tissue, there are others that are not detected in the brain cells; that is, there are genes that are transcribed in liver cells but not in brain cells (Figure 11.15).

Differential transcription of specific identifiable genes can also be demonstrated in this way. For example, the activation of synthesis of the immunoglobulin J chain, a polypeptide required for the assembly and secretion of the immunoglobulin protein, represents a specific step in the differentiation of a class of white blood cells, the lymphocytes. By using cDNA copied from, and complementary to, J chain mRNA, it is possible to show that whereas these sequences

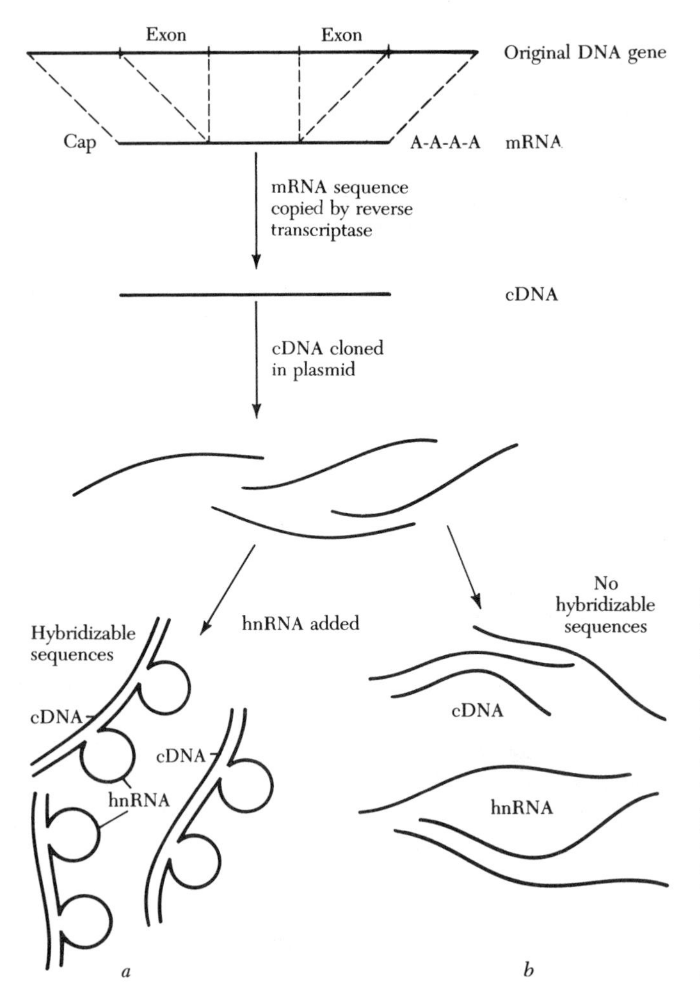

Figure 11.15 Use of cDNA to probe for specific mRNA sequences present among hnRNA from different cell types. Specific mRNA sequences from one cell type are used to make complementary DNA sequences. The cloned cDNA sequences are then used to detect, by hybrid duplex formation, the presence of complementary sequences in hn RNA. As expected, such sequences are present in hnRNA from the cell type in which the mRNA was formed (*a*); however, the hn RNA from a different cell type (*b*) does not include the sequences.

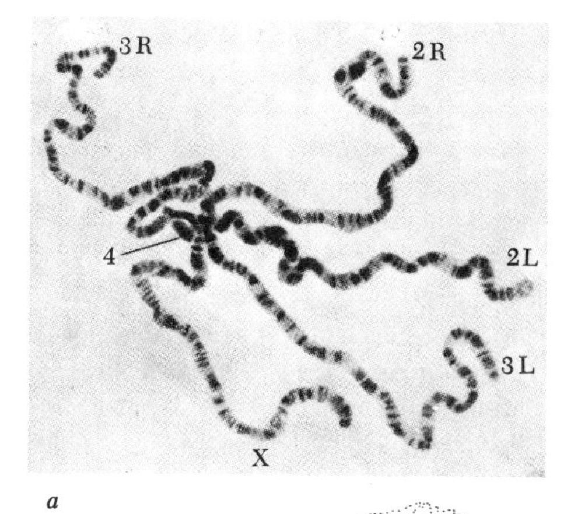

a

Figure 11.16 Salivary gland chromosomes of *Drosphilia melanogaster*. (*a*) A smear preparation from the salivary gland of a female, showing the X chromosome, the arms of the two autosomes (2L, 2R, 3L, and 3R), and the small chromosome 4. The diploid number of chromosomes is present, but the homologues are in intimate synapsis and are united by their centric heterochromatin into a chromocenter. (*b*) Enlarged drawing of chromosome 4, showing the banded structure; the diffuse chromocenter is at the left, and the two homologues are intimately paired. (*c*) Metaphase chromosomes from a ganglion cell, with an arrow pointing to the chromosomes 4 and with a scale to indicate differences in size between the two types of chromosomes.

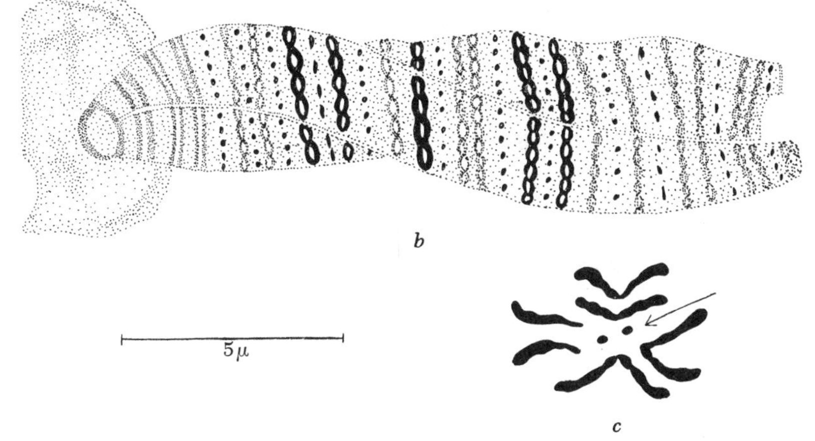

b

5μ

c

are transcribed in cell lines which synthesize the J polypeptide and secrete immunoglobulin, no such RNA transcripts are detected in cells which do not. Thus the differentiation step reflected by synthesis of this particular protein appears to involve the switching on of transcription of the specific coding gene.

Differential gene activity also can be observed directly at the chromosomal level in certain cells of the larvae of Dipteran insects. Cells of the salivary glands in particular contain very large chromosomes formed by successive duplications of the chromosome material without intervening segregation. These giant, polytene chromosomes display particular banding patterns (Figure 11.16), which are characteristic of the species, and the arrangement of bands is believed to reflect the linear sequence of genes along the chromosome. Tissues other than salivary glands also contain giant chromosomes, and the arrangement and sequence of bands is the same in all tissues.

At some, but not all, of the bands the chromosome material is less condensed, and "puffs" are seen (Figure 11.17); these can be shown to contain RNA. If autoradiographs are prepared following administration of radioactive precursors

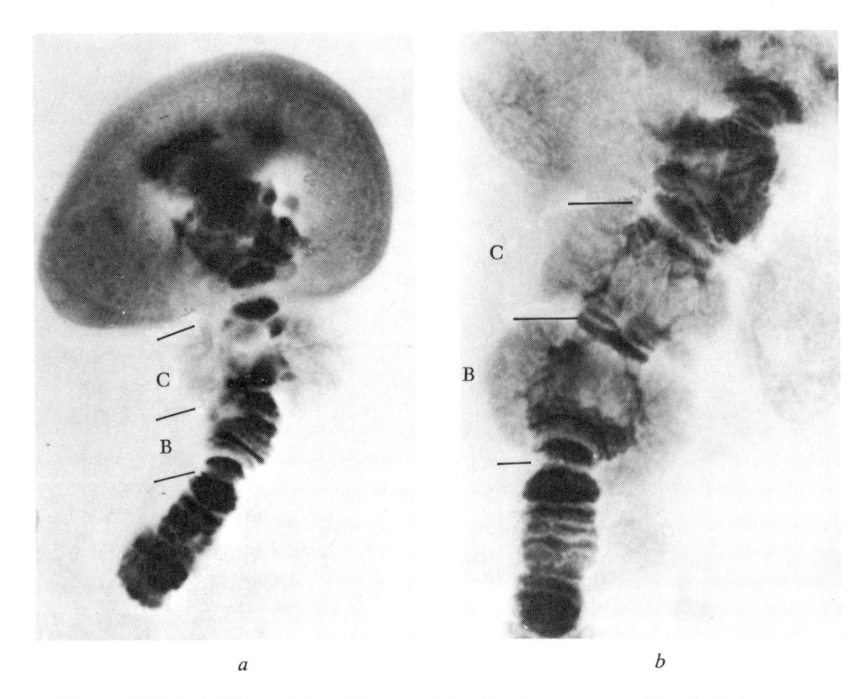

Figure 11.17 Differential puffing activity in chromosome IV of *Chironomus thummi.* (*a*) A puff at region C is present in late pre-pupal larvae. (*b*) In larvae treated with juvenile hormone a puff is induced at region B. The appearance of this puff coincides with the synthesis of RNA in this region. The upper portion of each chromosome is associated with the nucleolus. (From Laufer, H., and T. K. H. Holt, 1970. *Journal of Experimental Zoology* **173:** 341–351. Copyright, Wistar Institute Press, Philadelphia. With the permission of the authors and publisher.)

of RNA, the puffed regions are labeled, indicating that the puffs are sites of RNA synthesis and represent regions of gene activity. While the banding pattern is constant within a species, the puff patterns change during development. Administration of the hormone ecdysone, which induces molting in insects, results in the formation of developmentally specific puffs occurring in a regular time sequence. The hormone acts, therefore, by initiating a sequential series of gene activations that lead to subsequent developmental changes; how this is achieved by the hormone remains a mystery. Furthermore, different tissues show different patterns of puffing, indicating that some puffs are tissue-specific. An example of such differential gene activity can be seen in the salivary glands of the larval stage of the midge, *Chironomus.* In a few specialized cells of the gland, but not in others, a certain type of protein is formed. A specific puff on a particular chromosome is apparent only in the cells in which granules of the protein are present; this puff is not present on the chromosomes of the other salivary gland cells, nor on the chromosomes of a related species that does not form granules at all. Genetic tests have shown that the gene (or genes) for granule formation is (are) localized within the region of the chromosome from which the puffs arise. These giant chromosomes, therefore, allow us to see that different genes are active during different stages of development and in different tissues; furthermore, we can correlate the appearance of a particular protein in a cell type with the synthesis of RNA at a specific region of the chromosome, that is, with the activity of a particular gene.

Another type of chromosomal differentiation is shown in the behavior of the sex chromosomes of mammals. For example, humans possess an XX/XY sex-determining mechanism, with females being XX and males XY. Early in the prenatal life of the female embryo, one of the X chromosomes, in all cells except those destined for the germ line, becomes transformed into a heterochromatic, inactive body. This chromosome continues to replicate prior to each cell division but no longer synthesizes RNA. Direct confirmation that the genes on the heterochromatic X chromosome are not expressed comes from experiments in which cultured cells taken from females who are heterozygous for sex-linked genes are used to form clones. One such gene codes for the enzyme glucose-6-phosphate dehydrogenase (G-6-PDH). Two alternative allelic forms of this gene code for enzymes with slightly different properties, and heterozygous females produce both types of enzyme. However, if single cells from such a female are cloned, each clone produces only one of the two types of molecule. This, together with similar results for other genes of the X chromosome, shows not only that one of the two X chromosomes in a female cell is inactive, but also that once the inactivation has occurred, all of the descendants of that chromosome remain inactive. Clearly, in this case the capacity for repression of particular genes is stable and is passed on from cell to cell during development.

Regulation of Gene Transcription

Changes in chromatin structure. The complexity of chromatin structure, involving as it does interactions between DNA and both histone and nonhistone proteins, would suggest that the mechanisms by which transcription of the DNA might be regulated are themselves complex, and possibly superimposed on one another. Certainly, there are recognizable differences that can be detected between actively transcribing and inactive chromatin that may be related to such regulatory mechanisms.

The behavior of the X chromosome in female mammals provides a clue as to one level of gene inactivation at the chromosomal level. Chromatin can exist in two states: heterochromatin, which, as in the inactive X, is tightly condensed, and euchromatin, which is more dispersed. Various lines of evidence suggest that genes in heterochromatic regions of the chromosomes are inactive; furthermore, heterochromatin can be physically separated from euchromatin and shown to be relatively inactive compared to the euchromatin in supporting RNA synthesis. It appears likely, therefore, that certain regions of the chromosome, and perhaps

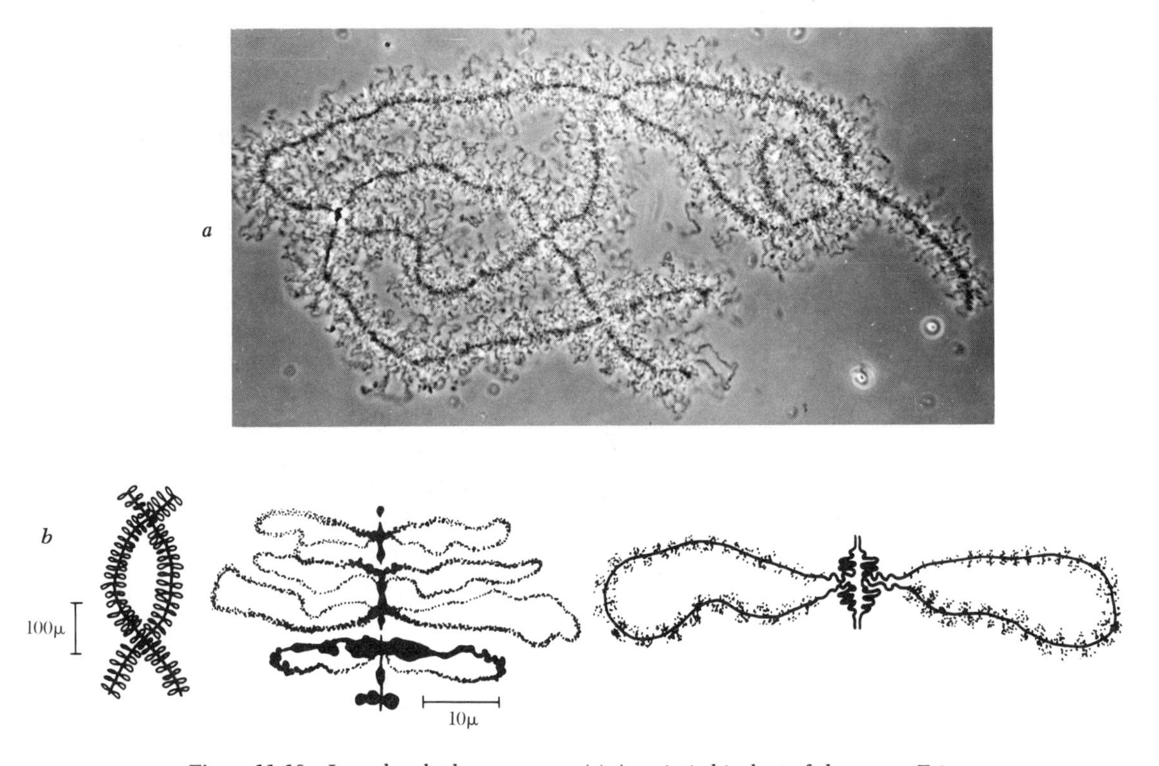

Figure 11.18 Lampbrush chromosomes. (*a*) A meiotic bivalent of the newt, *Triturus*, consisting of two homologues held together at four points by chiasmata, and in the diffuse diplotene stage. The fuzziness results from the projection of loops of chromatin away from the linear body of the chromosomes, and these loops are in active states of synthesis, somewhat comparable to that depicted in Figure 4.28. (Courtesy of Dr. J. Gall.) (*b*) An interpretation of the loops of a lampbrush chromosome. Top right, the chromomeres are of varying sizes and shapes, as are the loops that spin out from them. Below, an interpretation, with RNA synthesis proceeding as the loop spins out from the chromomere at the top and is then compacted back into the chromomere at the bottom; each loop is a portion of a single chromatid (courtesy of Dr. J. Gall).

even specific genes, can undergo heterochromatization and thus be rendered inactive, or inaccessible to the transcription machinery. Similar reductions in the degree of chromatin compaction can be observed directly in the actively transcribing regions of polytene and lampbrush chromosomes (Figure 11.18), where the chromatin uncoils to form extended lateral loops. Unfortunately, we do not yet know the molecular basis for such differential decondensation of the chromosomal materials.

Modulation of the physical configuration of the DNA double helix, or of its packaging in nucleosomes, also appears to accompany transcription. For example, DNA of actively transcribing genes is more sensitive to digestion by certain DNAase enzymes, suggesting that the conformation of DNA in the nucleosomes of active chromatin is somehow different from that of inactive chromatin. Certainly, both initiation and continued transcription of DNA by RNA polymerase would be expected to require changes in the double-helical conformation in order to accommodate both the polymerase and the growing RNA transcript.

Chromosomal proteins. The major non-DNA component of chromatin is, of course, represented by the histone proteins of the nucleosomes. While the ability of DNA to support RNA synthesis increases if the histones are removed, it is unlikely that they are specific repressors since their composition and arrangement are remarkably uniform. However, histones can be selectively methylated, acetylated, and phosphorylated, and such changes may facilitate localized transcription of the DNA.

Other chromosomal proteins are much more heterogeneous than the histones and show some tissue specificity. These do not all appear to be necessary for the maintenance of chromatin structure, and it is likely that at least some of them play a regulatory role. One fraction of nonhistone chromosomal proteins, named because of this electrophoretic mobility, are the HMG (high mobility group) proteins. Two of these, called HMGs 14 and 17, are present in nucleosomes of active but not of inactive chromatin, where they appear to interact with the DNA and to alter its conformation to one that is potentially more available for transcription.

Specificity of Transcription

While it is clear from the above that the specific physical conformation of DNA and/or the nature of the associations between DNA and other components of chromatin may cooperatively provide necessary conditions for transcription of genes, they appear to reflect general regulatory mechanisms; none of these changes suggest how specific genes might be selected for transcription.

As we have already discussed (Chapter 4), most eukaryotic genes have specific promotor sequences which are necessary for accurate initiation of transcription; only when "initiation factors" bind to the promotor sites can the RNA polymerase initiate transcription at the proper place in the DNA. Similar proteins which bind to specific sequences in the DNA may be responsible for recognition of specific genes by the transcription apparatus and, therefore, activate transcription in a selective fashion. For example, hormones such as progesterone are able to stimulate synthesis of specific proteins in target cells. The response is mediated by certain receptor molecules in the cytoplasm which are able to bind to the hormones. The receptor–hormone complex then moves into the nucleus, and after becoming associated with the chromatin, stimulates synthesis of the RNA molecules which code for the called-for proteins. It has been shown that associated with the steroid-regulated genes, such as the ovalbumin gene, are specific DNA sequences which are recognized by and which bind the steroid receptor; these binding sequences are absent from DNA of genes which are not activated by steroid hormones, such as the actin and globin genes.

Thus not only is the physical state of the chromatin important in determining whether genes can be transcribed, but also there are interactions between activator proteins and DNA sequences which allow specific genes to be expressed.

Nucleocytoplasmic Interactions

Whatever the mechanisms might be that regulate transcription of genes at the chromosomal level, they must be influenced at least in part by the cytoplasm, the milieu in which the nucleus functions. An example of what appears to be cytoplasmic determination of nuclear activity is seen during development of the pollen grain in higher plants. Following meiosis, each haploid spore undergoes a mitotic division, forming a binucleate pollen grain. One of these nuclei becomes very condensed and elongate, and is completely inactive in RNA synthesis. The sister nucleus, on the other hand, maintains a normal appearance, synthesizes RNA, and supports all further development of the germinating pollen grain (Figure 11.19). Prior to the mitotic division, however, a high degree of polarity is established in the immature pollen grain, such that the two telophase nuclei end up in very different types of cytoplasm (Figure 11.20) and subsequently behave very differently.

An elegant demonstration of the importance of the cytoplasm in determining nuclear activity comes from experiments in which differentiated muscle cells of the mouse are fused with human nonmuscle cells. Although the nuclei in these heterokaryons do not themselves fuse, but remain separate, human muscle proteins, normally not found in the nonmuscle cells, are synthesized by the heterokaryons soon after cell fusion occurs. Clearly, the muscle cytoplasm is able to influence the human nuclei in such a way that expression of muscle-specific genes is evoked.

Cytoplasmic influence over nuclear expression is also important during early embryonic development. For example, in *Xenopus* successive synchronous cleavage divisions of the egg, forming smaller and smaller cells, result in formation of the blastula. During this period very little, if any, RNA of any kind is synthesized by the nuclei. (The RNA required by these cells has already been synthesized during prophase of meiosis in the oocyte and is present in the egg at the time of fertilization.) At about the twelfth cleavage cycle, however, RNA synthesis is suddenly switched on. Experiments in which division is delayed in one half of the embryo, or in which extra DNA is injected into fertilized eggs, have shown that the onset of transcription occurs once the ratio of cytoplasmic to nuclear material, specifically DNA, reaches a sufficiently low level. Thus some cytoplasmic component appears to keep the nuclei in a repressed state until it reaches some critically low level, when transcription can resume. If nuclei from cells of older embryos are transplanted into *Xenopus* eggs, they undergo morphological changes, RNA synthesis stops, and the nuclei synthesize DNA and undergo cleavage divisions. Later in development, the patterns of RNA synthesis characteristic of normal em-

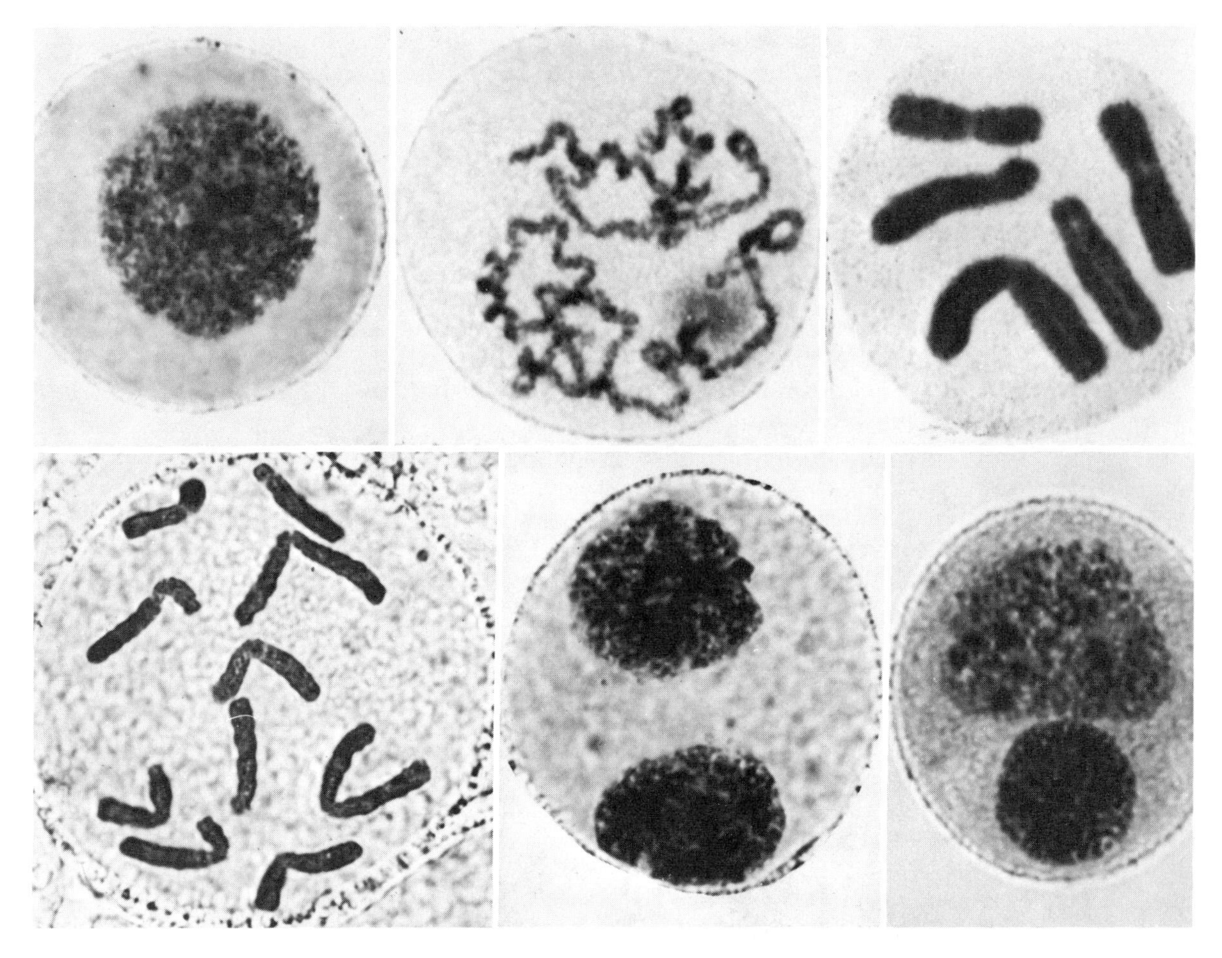

Figure 11.19 Mitosis in the haploid microspore of *Trillium erectum*. Following telophase the vegetative nucleus (top) and the generative nucleus (bottom) begin differentiation.

bryogenesis are reestablished. Clearly, it is the cytoplasm that determines the patterns of transcription in nuclei during early development.

From the examples above, we can see that while the direction of flow of genetic information is from nucleus to cytoplasm, information from the cytoplasm must somehow elicit responses from the nucleus. It is likely that proteins which can migrate into the nucleus are involved in this; in the case of the nuclear transplants into amphibian eggs, it has been shown that proteins from the cytoplasm do become localized in cleavage nuclei. The steroid receptor proteins that we have already discussed represent a situation where movement of a protein from the cytoplasm into the nucleus leads to expression of particular genes.

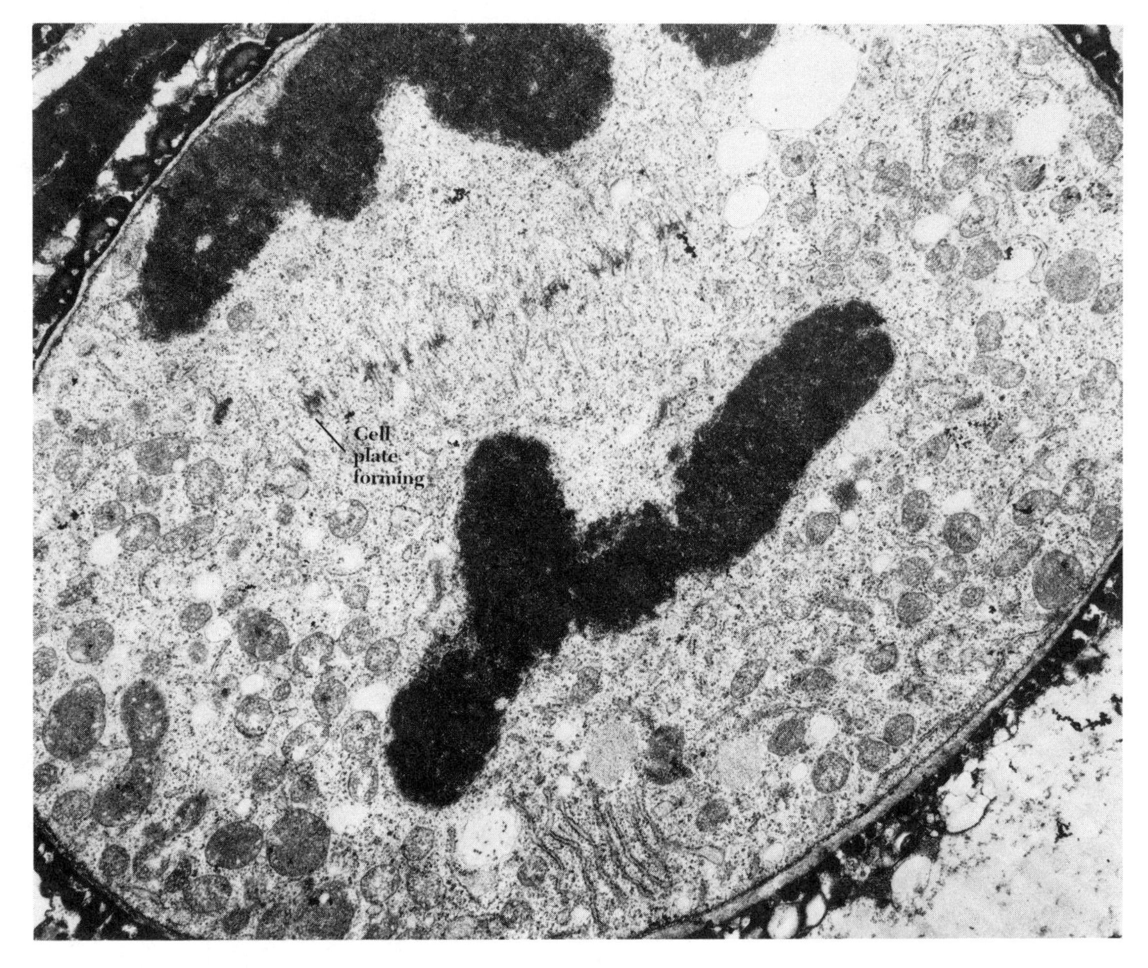

Figure 11.20 Electron micrograph of pollen grain (microspore) mitosis in *Tradescantia*. Note the unequal distribution of cytoplasmic components.

Post-Transcriptional Regulation

Although there is ample evidence that gene expression can be regulated at the level of production of primary transcripts, we also know that the steps between transcription and translation of mRNA into protein are complex; it is not surprising, therefore, that these steps also appear to be subject to some forms of selective control that influence which proteins are ultimately produced by the cell. We shall now look at some examples that demonstrate such regulation.

Processing of primary transcript. As we saw in Chapter 4, the coding sequences of the RNA molecules transcribed from DNA are interrupted by noncoding intron sequences. The intron sequences are subsequently removed, and the coding sequences spliced together. It has now been shown that in several genes, alternative patterns of splicing of coding sequences can generate different mRNAs (Figure 11.21). How the selection of which coding sequences are to be included and which excluded is made is not clear; however, it does appear to be developmentally regulated, since different mRNAs (and hence different polypeptides) are produced in different tissues and at different stages of development.

Other modifications of the primary transcript, such as the addition of the "cap" at the 5′ end of the initiation site and of the poly-A tail at the 3′ end, take place within the nucleus. Although the significance of these processes is not understood, they may play a role in regulating which messages reach the ribosome and which are to be translated. An increasingly important question, therefore, is how the cell knows which gene transcripts to select for conversion to functional mRNA.

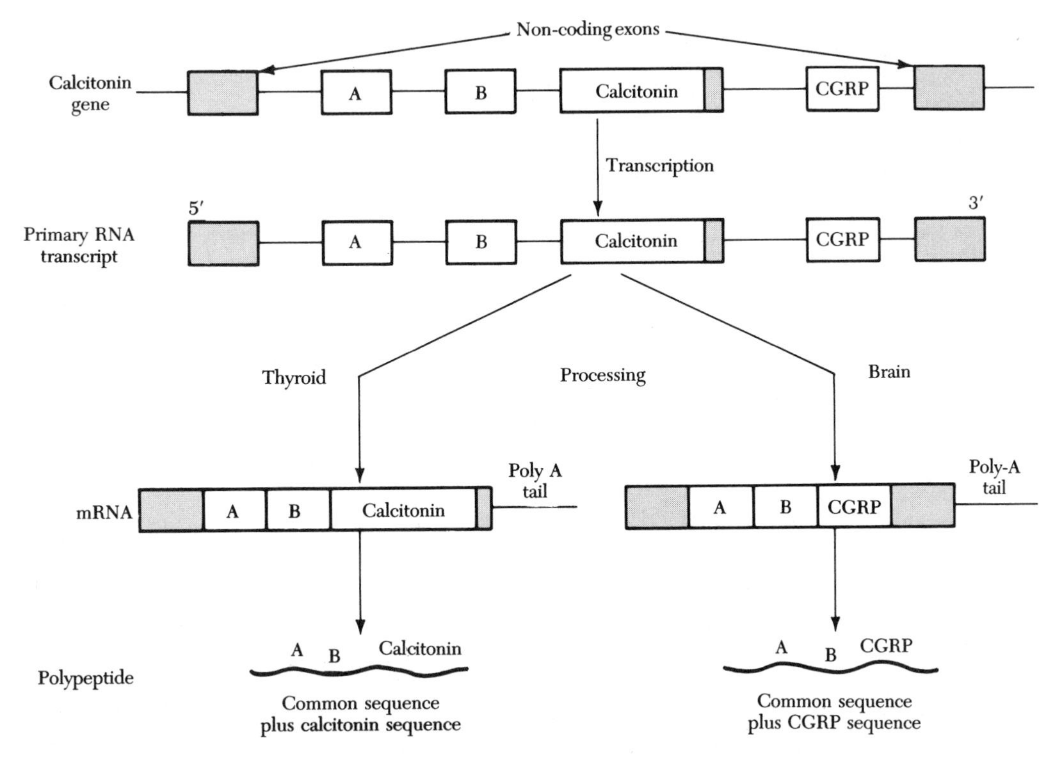

Figure 11.21 Tissue-specific processing of calcitonin gene product. The calcitonin gene consists of 5′ and 3′ noncoding exons, two common coding exons A and B, an exon coding for the calcitonin amino acid sequence and an exon coding for the amino acid sequence of the CGRP peptide. In thyroid cells the primary transcript is processed in such a way that the B exon is spliced to the calcitonin exon and the CGRP exon is not included in the mRNA; in brain cells the B exon is spliced to the CGRP exon, and the calcitonin exon is lost from the mRNA. Thus the polypeptide translation products, which are themselves further processed (not shown), are different in the two cell types.

Differential translation of mRNA. Once the mRNA molecules have been formed, they must associate with ribosomes in order to be translated into proteins. However, we know that not all mRNA in the cell need be translated, and that selective recruitment of specific mRNAs can determine which proteins are to be formed by the cell. This is clearly seen during early embryogenesis; as we have already mentioned, mature oocytes contain all of the mRNA sequences required for early development, although they are not translated. Only after fertilization are proteins made from these templates. Furthermore, although there are no changes in the mRNAs present during early stages of development in at least one type of embryo, that of the surf clam, different mRNAs become associated with polyribosomes at different stages, resulting in sequential changes in the proteins that are synthesized.

Such selective translation of mRNA is not confined to early embryo development, having been demonstrated in several other types of cell. Although this selecting mechanism is also not known, cytoplasmic extracts are able to confer specificity of translation of mRNAs in *in vitro* protein-synthesizing systems. It is likely that proteins with which mRNAs can be associated in the cytoplasm play a role in regulation of gene expression at this level.

Changes in the Genetic Material

Although we pointed out earlier that in general differentiation does not appear to involve permanent changes in the genetic potential of the cell, but rather in the expression of that potential, there are some cases where indeed developmental steps are accomplished as a consequence of changes in either the amount or the type of nuclear DNA.

Gene amplification. Regulation of the rate of synthesis of a specific gene product, ribosomal RNA, can be exercised in one of the above ways. During the extended prophase of immature oocytes of *Xenopus*, ribosomal RNA synthesis proceeds at a rapid rate. By the time the oocytes mature, rRNA synthesis can no longer be detected. The high rates of rRNA synthesis during this stage in development of the oocyte are achieved in part by the production during pachytene of large numbers of copies of the genes that code for rRNA. This differential replication of a specific part of the genome is followed by release of the DNA into a large number of nucleoli, each of which is then capable of making the rRNA. These amplified genes can be isolated from nucleoli and photographed (see Figure 4.28, Chapter 4).

This mechanism of ensuring the availability of sufficient copies of a gene to meet the large demands of the cell for the specific gene product are not restricted to the rRNA genes. We have already discussed the extensive polytenization undergone by cells of *Drosophila* salivary glands; many other differentiated cell types, not only in *Drosophila* but also in other organisms, have polytene chromosomes. However, additional amplification of genes controlling specific developmental steps can also occur. In nuclei of the *Drosophila* follicle cells which produce the eggshell, or chorion, proteins, but not in nuclei of other tissues, the clusters of genes coding for the chorion proteins are selectively replicated several times; this results in localized amplification of these genes at a time when large amounts of mRNA are required to support the synthesis of large amounts of the eggshell proteins.

Immunoglobulin genes. Another type of genetic change which is related to a developmental switch and which influences gene expression is known from the immune system of mammals. Immunology is a large area of study in its own right, and one which is beyond the scope of this book; however, we can discuss one aspect of the immune response that is related to our consideration of regulation of gene expression. This is the role played by gene rearrangements in the differentiation of lymphocytes, the white blood cells that participate in the immune system.

The immune system is a defense mechanism of the body that allows it to combat infection by destroying invading cells and their harmful products. Part of the response depends on the production of antibodies by lymphocytes; these antibodies, which are able to recognize and inactivate the foreign substances, or *antigens*, are produced in large quantities in response to the presence of the antigen in the body. Millions of different antibodies are capable of being produced by mammals, enabling them to cope with the myriad of antigens that might be encountered. Each type of antibody is produced by a particular class of lymphocytes, all members of which are descended from an original cell which somehow had been selected originally to be able to make that particular antibody. Thus the bloodstream contains millions of "clones" of lymphocytes, each clone being responsible for dealing with a specific antigen.

The basis of at least some of this diversity can be understood if we examine the structure of antibodies which are made up of molecules called *immunoglobulins* (Igs). Each class of immunoglobulins consists of two identical heavy (H) and two identical light (L) polypeptide chains, attached to a carbohydrate moiety (Figure 11.22). By analyzing the amino acid sequences of different heavy and light chains, it can be shown that each chain contains a constant (C) sequence and a variable (V) sequence. The antigen-binding sites are within the V regions, and the antigen-binding specificity is determined by the amino acid sequences of the V regions.

Let us now look at the genetic basis of immunoglobulin diversity, so that we can understand how antibody production is controlled. Each particular type of immunoglobulin chain is coded for by a C gene, which codes for the constant sequence, plus one of several hundred different V genes, each coding for a different V sequence (Figure 11.23). In embryonic cells, which do not synthesize immunoglobulins, the cluster of different V genes is situated far from the C coding region. At some point during lymphocyte differentiation, however, the genes are rearranged in such a way that in any one lymphocyte a particular V coding segment is transferred to a site adjacent to the C region and the intervening DNA se-

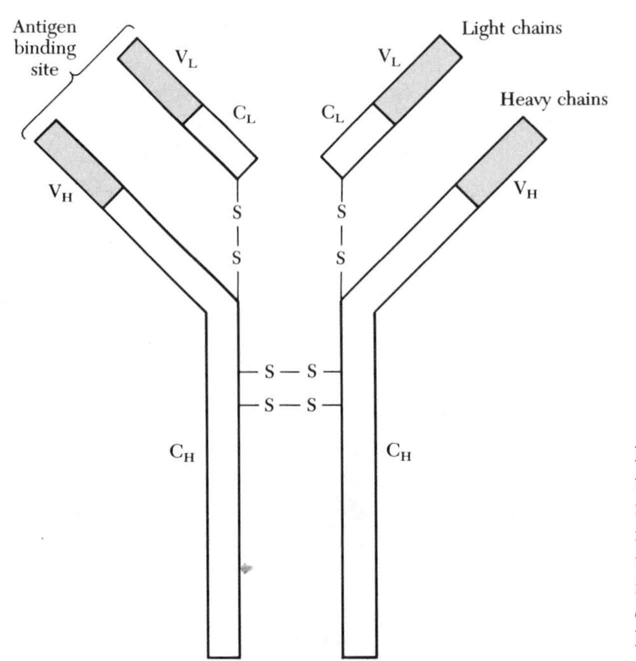

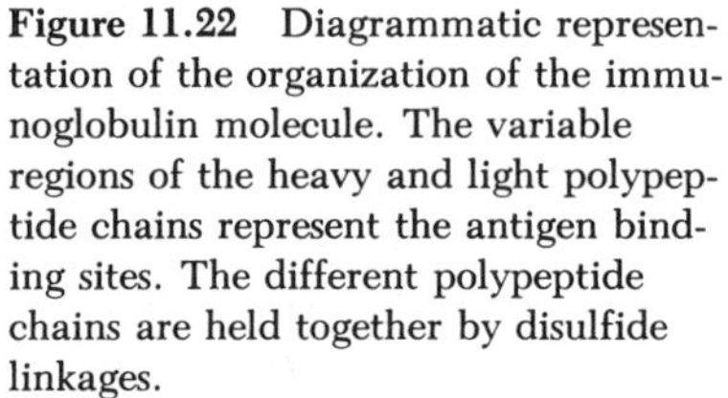

Figure 11.22 Diagrammatic representation of the organization of the immunoglobulin molecule. The variable regions of the heavy and light polypeptide chains represent the antigen binding sites. The different polypeptide chains are held together by disulfide linkages.

quences are lost (Figure 11.23). Once the two coding regions, V and C, are brought into conjunction they can be coordinately transcribed into an RNA sequence that will code for the particular type of polypeptide chain (Figure 11.23) to be synthesized by the lymphocyte and by its descendants.

We must point out that the account above is an oversimplified one, and that

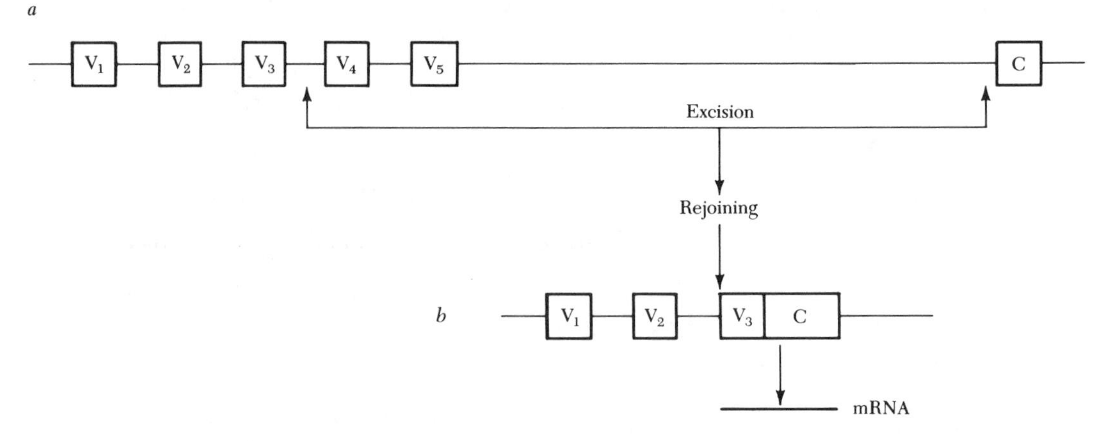

Figure 11.23 Rearrangement of immunoglobulin genes. (*a*) In embryonic cells a series of V genes is situated at some distance from the C gene. Excision and rejoining of DNA segments results in a particular V gene being brought into contact with the C gene in the mature lymphocyte (*b*). The mRNA transcript produced then codes for a polypeptide with a specific V region.

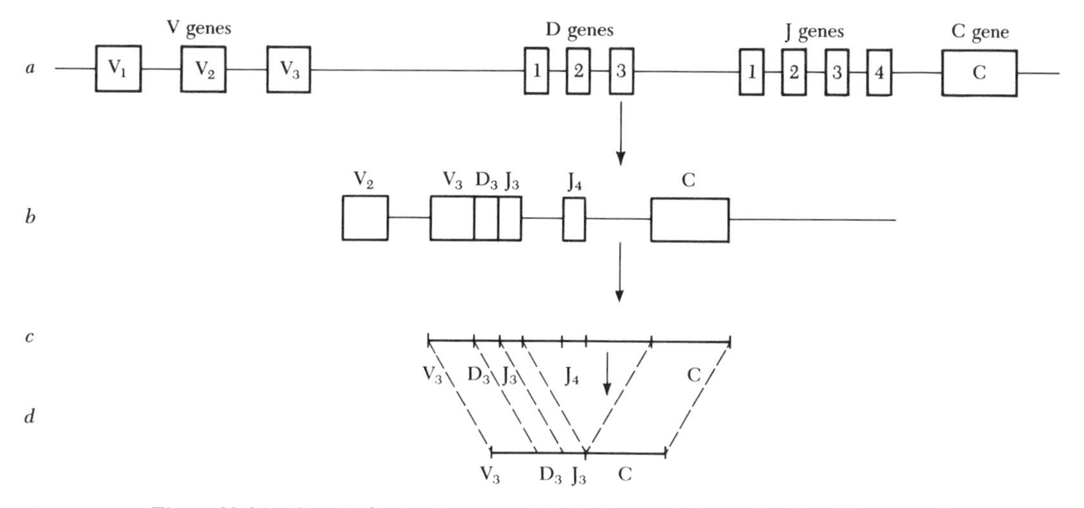

Figure 11.24 Genetic basis of immunoglobulin heavy chain production. Three sets of genes, V, D, and J, code for different parts of the variable segment of the heavy chain. Rearrangement of the original sequences (*a*) results in a particular combination in the lymphocyte DNA (*b*). The primary transcript (*c*) is then processed by removal of an intron containing the unwanted J sequence, and the resulting mRNA (*d*) is translated to produce a heavy chain with a particular variable region.

the situation is in fact much more complex (see, for example, Figure 11.24). However, it does indicate (a) that lymphocyte differentiation is accompanied by somatic recombination of specific DNA sequences and loss of others, leading to activation of expression of these sequences, and (b) that the selection of different V sequences for transposition results in the synthesis of different immunoglobulin chains.

Yeast mating-type genes. Although the type of rearrangement described above involves permanent loss of some DNA sequences, and hence is irreversible, another type of genetic change which is reversible has been recognized in cells of the yeast, *Saccharomyces cerevisiae*. Sexual reproduction in yeast involves fusion of two cells of opposite mating types, designated α and a. The mating type of a cell is determined by whichever of two alleles, α or a, representing different DNA base sequences, is present at a particular mating type locus, MAT (Figure 11.25). However, cells of one mating type are able to switch to the opposite type by the substitution of one allele for the other at the MAT locus. Each cell, in addition to the particular coding sequence present and expressed at the MAT locus, has copies of both the α and a sequences which are normally not expressed from their sites elsewhere in the chromosomes (Figure 11.25). The switch in mating type occurs when the DNA at the active MAT locus is replaced by a copy of the DNA from the "silent" locus of the opposite type; this newly inserted DNA sequence can then be expressed, presumably as a consequence of being transferred to its new position and subject to the controls exercised over the MAT region. Since the

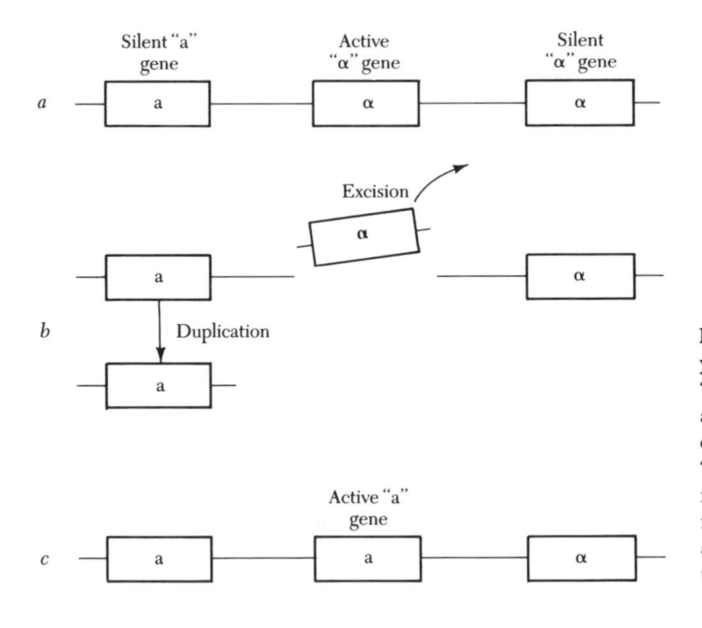

Figure 11.25 Mating-type switch in yeast. (*a*) Gene arrangement in type "α" cells: An "α" sequence is present at the active locus. (*b*) The "α" sequence is removed, and a copy of the "a" sequence present at the silent locus is insertėd in its place (*c*). The cell is now of mating type "a." The silent loci are kept in the repressed state by products of other regulatory genes.

"silent" copies of the α and a alleles are always present, the potential for mating-type changes is always retained.

ONCOGENES AND CANCER

Cancer is a disease which represents development gone awry, in that cancer cells assume an autonomy which allows them to escape from normal control and to proliferate and spread throughout the body, with often disastrous consequences. It is impossible here to deal with this topic in a fashion that would do justice to its importance; the transition from a normal cell to a cancer cell represents a complex series of changes, few of which are properly understood. However, we shall discuss one aspect of current cancer research, the results of which suggest that tumorigenesis is at least in part a consequence of the alteration of expression of specific genes which play a role in the activity of normal cells.

Oncogenes, or tumor-inducing genes, were first identified in retroviruses, which contain RNA rather than DNA as their genetic material. After infection by the virus, the RNA base sequence is copied into a complementary DNA sequence, which is then inserted into the DNA of the host cell nucleus. Following infection and integration of viral DNA, the host cells become transformed to the malignant state. If the DNA from transformed cells is then transferred to other, nontumorous cells, it can in turn induce transformation of the latter. The DNA sequences responsible for the transformation are termed viral *oncogenes*.

Subsequently, oncogene sequences were shown to be present in the nuclear

DNA of uninfected cells; some of these are homologous to the viral oncogenes, while others are quite unrelated. The onset of the tumorous state in uninfected cells appears to result from some sort of modification of the oncogenes; oncogene-containing DNA from tumor cells can, in turn, transform nontumor cells, although the same oncogenes from nontumor cells do not transform. Thus the transforming oncogenes in cancer cells have been activated in some way.

The manner in which the oncogenes in cancer cells are activated and the cells transformed is not yet understood, although several possibilities have been presented. One transforming cellular oncogene is known to differ from its non-transforming counterpart by a single base substitution, resulting in the production of an altered protein. Higher levels of the RNAs and proteins coded for by other oncogenes have been detected in transformed cells, suggesting that increased expression of what are otherwise normal genes may be in part responsible for the neoplastic state. Yet other transforming oncogenes may have other sequences within, or close to, the coding sequences, possibly resulting in changes in how the oncogenes are regulated. Of interest in this respect is that many types of cancer in humans can be correlated with chromosome translocations and deletions; the breakage points in these chromosome aberrations have been shown to be at, or very close to, the site where the oncogene sequences reside, suggesting that the positional changes may activate the oncogene sequences by altering the spatial relationships between them and possible regulatory sequences.

However the oncogene sequences might be activated to induce a tumorous state, it appears that some oncogene products are similar, if not identical, to proteins that are involved in regulation of cell proliferation during normal development. Mitogenic growth factors are substances that can induce certain cells to divide if these cells carry specific membrane receptors that can be recognized by the growth factors. Two such mitogens are platelet-derived growth factor (PDGF) and epidermal growth factor (EGF), both of which play specific roles in the normal processes of cell division in the body. One viral oncogene codes for a protein that is structurally very similar to PDGF, and that can affect cells with PDGF receptors; a different oncogene codes for an EGF receptor, and presumably can transform otherwise nonsusceptible cells by allowing them to recognize and respond to the mitogen. Yet other oncogenes may code for proteins which stimulate synthesis or release of natural growth factors, or which act on the receptor-activated pathway leading to cell proliferation, suggesting again that in at least some cases the neoplastic state results from untimely expression of genes controlling normal developmental processes.

INTERCELLULAR COMMUNICATION

As we pointed out earlier in this chapter, development in multicellular organisms is a coordinated process; while we have discussed how specific pieces of information may be expressed in individual cells, it is important to remember that the activities of different cells must be integrated and controlled. We further indicated that the pathway of development undergone by a cell must depend in large measure on its environment; recognition of that environment, therefore, is an important factor in specifying how a cell will behave.

One type of intercellular communication is mediated by substances called *hormones*, chemicals that are produced in one part of the organism and that exercise their effect on cells in a different part. Cells that respond to a hormone must possess receptors with which the hormone can interact, plus the ability to respond to that interaction in a way that results in specific biochemical changes in the cell. The part of the cell first encountered by a hormone, or indeed any other intercellular signal, is the cell surface. As we discussed in Chapter 5, specific receptor molecules are present at the cell membrane, and these are able to recognize and respond to the hormonal signal. Some hormones may exercise their effects without entering the cell at all; rather the interaction with the cell surface receptors causes changes to occur inside the cell. For example, increased synthesis of the compound cyclic AMP results from activation by hormones of an enzyme located at the inner surface of the cell membrane. The cyclic AMP is then able to modify other proteins in the cell with a resulting change in cellular activities. Other hormones, such as the lipid-soluble steroid hormones described earlier, enter the cell and bind with cytoplasmic receptors. In this case, the response of the cell is increased transcription of specific genes, followed by changes in the proteins synthesized.

Extracellular influences other than hormones can modify various cellular activities, although again it is the cell surface which may first respond to these influences. Intercellular contact plays an important role in cellular recognition and aggregation, which appear to depend on the specific chemical nature of the glycocalyx. Such recognition, which is likely to mediate some aspects of embryonic development, certainly can affect cell growth and reproduction. When cultured cells come into contact with one another, their activity changes dramatically; they no longer move around on the surface of the culture medium, and they cease proliferation. Cancer cells, which often have altered glycoproteins and glycolipids at their surfaces, not only have reduced recognition capabilities but also fail to be subject to contact inhibition.

Our discussion of what are only selected aspects of cell differentiation should illustrate that the problem is a complex one, and that cell activity can be modulated at many different levels. However, the goal of explaining differentiation in molecular terms is being approached, as is an understanding of how the organism accomplishes the spatial and temporal division of labor necessary for it to carry out its various activities and to cope with its environment. And at a vital, practical level, our ability to control cancer, which results from the escape of cells from regulative control, rests on our comprehension of the functional cell and the processes of development as rigidly governed systems of chemical checks and balances.

ADDITIONAL READING

AMARA, S. G., JONAS, V., ROSENFELD, M. G., ONG, E. S., and EVANS, R. M. 1982. Alternative RNA processing in calcitonin gene expression generates m-RNAs encoding different polypeptide products. *Nature 298:* 240–244.

BISHOP, J. M. 1982. Oncogenes. *Sci. Am. 246*(3): 81–92.

BLAU, H. M., CHIU, C.-P., and WEBSTER, C. 1983. Cytoplasmic activation of human nuclear genes in stable heterokaryons. *Cell 32:* 1171–1180.

BROWN, D. D. 1981. Gene expression in eukaryotes. *Science 211:* 667–674.

DARNELL, J. E. 1982. Variety in the level of gene control in eukaryotic cells. *Nature 297:* 365–371.

DERMAN, E., KRAUTER, K., WALLING, L., WEINBERGER, C., RAY, M., and DARNELL, J. E., JR. 1981. Transcriptional control in the production of liver-specific mRNAs. *Cell 23:* 731–739.

GRAHAM, C. F., and WAREING, P. F. (eds.). 1976. *The Developmental Biology of Plants and Animals.* W. B. Saunders Company, Philadelphia.

GURDON, J. B. 1968. Transplanted nuclei and cell differentiation. *Sci. Am. 219*(6): 24–35.

JACOB, F., and MONOD, J. 1961. Genetic regulatory mechanisms in the synthesis of proteins. *J. Mol. Biol. 3:* 318–356.

LEDER, P. 1982. The genetics of antibody diversity. *Sci. Am. 246*(5): 102–115.

LEWIN, B. 1980. *Gene Expression*, Vol. 2: *Eukaryotic Chromosomes*, 2nd ed. John Wiley & Sons, Inc., New York.

MARX, J. L. 1981. Antibodies: getting their genes together. *Science 212:* 1015–1017.

MATHER, E. L., ALT, F. W., BOTHWELL, A. L. M., BALTIMORE, D., and KOSHLAND, M. E. 1981. Expression of J chain RNA in cell lines representing different stages of B lymphocyte differentiation. *Cell 23:* 369–378.

ROSENTHAL, E. T., HUNT, J., and RUDERMAN, J. V. 1980. Selective translation of m-RNA controls the pattern of protein synthesis during early development of the surf clam, *Spisula solidissima. Cell 20:* 487–494.

WEISBROD, S. 1982. Active chromatin. *Nature 297:* 289–295.

12

Cells through Time

The point has been stressed that the cell is the most elementary unit of organization through which life is expressed. Stated somewhat differently, it can be said that life manifests itself when matter achieves a special state of organization, one that evolved from a noncellular condition to a degree of complexity represented by cells. Admittedly, we do not know with any degree of certainty just how life began, or how cells were first formed, but there is a good deal of evidence to suggest that a variety of chemical and physical processes, taking place on earth or its surrounding atmosphere, led to the appearance of simple organic (carbon-containing) compounds. These compounds interacted with each other to give more and more complex chemical groupings and structures until that special state of organization made its appearance, and life as we know it expressed itself.

Life, therefore, and hence cells, have a history, one that the fossil record suggests began about 3.5 billion years ago. The cells we can examine today, even the simplest of which are highly complex and intricately organized, are products, consequently, of a long process of time and change. We can dismantle a cell, and by piecemeal examination learn a great deal about its structures and how these are related to functions, but it is quite a different thing to carry out the opposite task—to visualize how simple components came into being and ultimately organized themselves into a life-producing state of arrangement. Much is known, and a good deal of experimental evidence is available, but two major gaps (as well as many minor ones) exist in our knowledge of the evolution of cells: the origin of the prokaryotic cell from noncellular materials, which is the more elusive problem; and the origin of the eukaryotic cell from prokaryotic ancestors, the subject of much speculation.

PREBIOTIC CONDITIONS

The solar system and its planets, including the earth, were formed about 4.5×10^9 years ago from a diffuse dust cloud condensed through gravitational force. At the time when life originated, it seems very likely that the temperature and mineral composition of the earth were much as they are today, with the remains of exploded stars, comets, and meteorites contributing to the variety of elements and surface substances. The early atmosphere of the earth was, in all likelihood, mildly reducing, with no free molecular oxygen (O_2) present. Nitrogen was probably present as N_2 with ammonia (NH_3) in small amounts. Carbon could have been present as methane (CH_4), carbon monoxide (CO), carbon dioxide (CO_2), or a mixture of all three. Water seems to have covered a portion of the earth within a short time, derived possibly from volcanic activity. These molecules would react only very slowly with each other if left undisturbed, but energy in the form of electrical discharges (lightning) and ultraviolet light from the sun could have activated these simple molecules and caused them to form more complex ones in the atmosphere or in water. Such molecules as those indicated in Table 12.1, which have been identified as existing in outer space, were almost certainly important intermediates to the larger macromolecules. Among these intermediates, hydrogen cyanide, formaldehyde, more complex aldehydes, and the acetylenes are thought to have been particularly important as precursor molecules, especially when dissolved in water and concentrated by the evaporation of lakes or tidepools, or by being adsorbed onto clays, which served as activating sites.

This seems like an inauspicious group of molecules out of which to derive the proteins, nucleic acids, and carbohydrates so commonly associated with life, but during the 1920s the Russian Oparin and the Englishman Haldane put forth the proposition that life arose spontaneously from these meager beginnings. This re-

TABLE 12.1 A PARTIAL LIST * OF MOLECULES FOUND IN INTERSTELLAR SPACE THAT COULD HAVE PLAYED A ROLE AS PREBIOTIC PRECURSORS OF AMINO ACIDS, SUGARS, AND NUCLEIC ACIDS.

Molecule	Formula	Molecule	Formula
Cyanogen radical	CN^-	Carbon monosulfide	CS
Hydroxyl radical	OH^-	Formamide	$HC(NH_2)O$
Ammonia	NH_3	Silicon monoxide	SiO
Water	H_2O	Carbonyl sulfide	OCS
Formaldehyde	$H_2C=O$	Acetonitrile	$CH_3C\equiv N$
Carbon monoxide	CO	Isocyanic acid	$HN=C=O$
Hydrogen	H_2	Methyl acetylene	$CH_3C\equiv CH$
Hydrogen cyanide	HCN	Acetaldehyde	CH_3CHO
Cyanoacetylene	$HC=C\text{–}N$	Thioformaldehyde	$H_2C=S$
Methyl alcohol	CH_3OH	Hydrogen sulfide	H_2S
Formic acid	HCOOH	Methylene imine	$H_2C=NH$

* From G. H. Herbig, *American Scientist 62:* 200 (1970).

mained no more than an interesting proposal until the 1950s, when a number of investigators demonstrated that an electrical discharge passed through an enclosed mixture of ammonia, methane, hydrogen, and water vapor led to the formation of a number of amino acids, plus other molecules of biological interest. Ultraviolet light and very high temperatures, possibly from volcanoes or hot springs, also serve as sources of energy for bringing about similar molecular changes. With carbon dioxide, carbon monoxide, and nitrogen added to the mixture above, nearly all of the amino acids known from living organisms, some that were unknown in biological systems, and some simple polypeptides (proteins) were also found. Seventeen of the 20 amino acids that occur in proteins have been obtained from experimental situations simulating the presumed prebiotic conditions, sometimes with yields of up to five percent. Sugars can be similarly formed from formaldehyde (H_2CO), and hydrogen cyanide (HCN) and related compounds are sources of the bases entering into the nucleic acids. Adenine, in fact, is formed in high yields, and perhaps it is not surprising that adenine would become a molecular species in centrally important structures such as DNA, RNA, and ATP.

The fact that amino acids can be formed under prebiotic conditions suggests that they should be found where life does not now exist but where the conditions are appropriate. As Table 12.2 indicates, amino acids are present in meteorites and in samples from the moon, in roughly the same kinds and proportions as in comparable simulated laboratory experiments. This suggests, of course, that the laboratory experiments are on the right track, and are not far removed from the terrestrial conditions just preceding the beginning of life. This also suggests that we exist in an orderly universe: the chemistry of 4.5×10^9 years ago on earth, on the moon today, on a recent meteorite from the asteroid belt, or in a planned laboratory experiment are strictly comparable to each other. The same physical laws operate independent of time and place.

With 1×10^9 years intervening between the formation of the earth and the first positive appearance of life, much could have happened. There were no organisms to use up these compounds as they formed, and with evaporation occurring in some areas, the molecules could have reached relatively high concentrations. Eigen and his colleagues point out that if all of the carbon in the fossil fuels, carbonate rocks, and living material were distributed in the oceans of today, a carbon solution "as concentrated as a strong bouillon" would result. It has often been argued that life began in the molecular soups of a warm tidepool, but it is also possible that it occurred in a cold, concentrated mixture in which the degradation of newly formed molecules was taking place at a slower rate than their spontaneous formation. Under these conditions, larger and larger molecules such as simple proteins and nucleic acids could have had a more favorable opportunity for coming into being, particularly in the presence of activating clays.

An additional possibility is that life originated elsewhere in the universe—on an asteroid, some other planet whose fragments entered our solar system, or in interstellar space—and that the earth was inoculated with primitive cells at the time of collision or when passing through enriched space. This suggestion was advanced

TABLE 12.2 PROTEINACEOUS AMINO ACIDS FROM A VARIETY OF ABIOTIC SOURCES, COMPARED WITH ANIMAL SOURCES *

Terrestrial lava	Moon samples	Murchison meteorite	Laboratory samples from CH_4, N_2, H_2O, and traces of NH_3	By sparking: from CH_4, NH_3, H_2O, and H_2	Animal sources
Glycine	Glycine	Glycine	Alanine	Glycine	Aspartic acid
Alanine	Alanine	Alanine	Glycine	Alanine	Glutamic acid
Glutamic acid	Glutamic acid	Glutamic acid	Aspartic acid	Glutamic acid	Glycine
Aspartic acid	Aspartic acid	Aspartic acid	Valine	Aspartic acid	Serine
Serine	Serine	Valine	Leucine		Alanine
Threonine	Threonine	Proline	Glutamic acid		Leucine
Isoleucine			Serine		Isoleucine
Leucine			Isoleucine		Proline
Valine			Proline		Valine
			Threonine		Lysine
					Threonine
					Arginine
					Cystine
					Phenylalanine
					Tyrosine
					Histidine
					Tryptophan
					Methionine

* Laval, lunar and animal data from S. W. Fox, 1975; other data from S. L. Miller, and L. E. Orgel, 1974 (see Bibliography). A large number of amino acids of nonbiological significance, and other organic molecules, were also present in the laboratory samples. The meteorite and laboratory-derived amino acids are listed in order of prevalence of formation.

by the Swedish chemist Arrhenius more than half a century ago, and continues to be given credence today by some reputable scientists, but if correct it does not dispose of the problem of the origin of life, but merely pushes it back in time to some earlier period.

It is a far cry, certainly, from molecules in solutions to organized cells. One can, indeed, ask if there is a predisposition, due to chemical affinities or structures of these molecules, to assemble into cell-like forms. Sidney Fox and others have argued that this is so to a degree; that is, there is stereochemical information in these prebiotic protein molecules that caused them to assemble from amino acids in preferred sequences without the guidance of nucleic acid codes. He has demonstrated in the laboratory that these molecules possess the ability to aggregate upon contact with water, producing "proteinoid" droplets, which possess a membranelike outer surface and a fluid interior, and which can enlarge and divide. In addition, these proteinoids exhibit enzymatic and active transport activities of a primitive but qualitatively distinct character, just as do artificially prepared membranes composed of proteins and phospholipids. These discoveries would seem to point to an important step in our understanding of the evolution of the cell from inanimate beginnings, but unless the sequence of amino acids in these prebiotic proteins was preferentially ordered because of stereochemical relations, the proteins and the proteinoids would be merely statistical examples of what was possible, and the structural or enzymatic properties would be unpredictable, uncoordinated, and not likely to manipulate a steady-state flow of energy through the system. However, laboratory studies have shown that although any amino acid in solution can spontaneously link up with any other one with a loss of water (Figure 12.1), the process does exhibit a degree of specificity. In the presence of activated clay, which acts as a catalyst, proteins of 20 to 40 amino acids long are possible. The intriguing fact is that the sequences of amino acids in these spontaneously formed proteins are statistically similar to those found in today's organisms. The significance of this fact is that it is the sequence of amino acids in a protein that determines both its three-dimensional shape through folding and its active site if it functions as an enzyme. The same geometric property also permits the proteins to assemble into higher orders of structure. This can be readily demonstrated with artificial membranes or with the tobacco mosaic virus. This virus can be decomposed into protein and RNA molecules. If separated from each other, the proteins alone will reassemble spontaneously to form rodlike particles. If the RNA is added to the solution, the intact virus will be reconstituted and will be infective. A comparable situation occurs with ribosomes: if decomposed into their proteins and rRNA, spontaneous reassembly will occur under the proper conditions. There is, therefore, not only a selectivity in noncatalyzed sequencing of amino acids into primitive proteins, but also a selectivity in the spontaneous union of an amino acid with a nucleotide. It is this coupling that is central to the beginning of a genetic code, that is, the coupling of an amino acid to a specific tRNA.

The moon, meteorite, and laboratory data all suggest that given time, which was abundantly available, a primitive kind of life, represented by something like the Fox proteinoids, could have arisen, or been introduced, many times and in

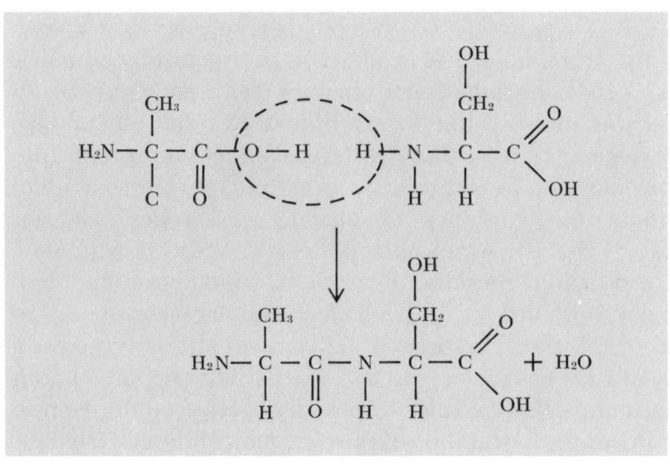

Figure 12.1 Linkage of the amino acids alanine and serine to form the dipeptide alanylserine with loss of water.

many locales on earth, the only impetus being the ability of molecules to react selectively with each other. The subsequent development of primitive catalytic systems could enhance the survival of these proteinoids by improving the flow of energy. Such a catalytic function may have been based on the ability of metallic ions to act as catalyst. For example, the capacity of the ferric ion, Fe^{3+}, to catalyze the decomposition of H_2O_2 to H_2O and O_2 is increased manyfold when the ferric ion is incorporated into an organic molecule such as a heme, and even more so when incorporated into a protein to form the enzyme catalase. It is also now known that artificial catalysts of a small and nonproteinaceous nature can be synthesized possessing the active sites and selectivity of reaction of certain known enzymes. In some instances large rates of acceleration of specific reactions have been achieved by such geometric control. Early catalysis necessary to carry out the simpler processes of life need not, therefore, have been based on larger protein structures; the latter could have come into existence at later stages of evolution. Improved catalysis would favor survival of those proteinoids possessing it, and they could interact more successfully with their environment than their statistically less favored neighbors.

Proteinoids, however, cannot be the first step toward the evolution of a cellular state. There is no known chemical way that the informational character of the nucleic acids can be specified by, or transferred from, the amino acid sequence of a protein. Assuming that the initial system must have been of a macromolecular nature in order to provide the complexity necessary for a minimal informational code, two additional features would seem to be required. Following the arguments of Eigen and his colleagues, the first would be the possession of an "organizing principle" that, for purposes of replication and on a selective basis, could extract the "right" molecules from the primordial soup, and avoid those that were "wrong." That is, the spontaneously formed molecules were of many kinds, but only a select few could have become incorporated as subunits into the macromolecular structures of a living system, however primitive. Also, the initial system, by some mechanism, would have been able to extract energy from other compounds around it in solution, to store this energy in a usable form, and then to deliver it wherever it was needed. Undoubtedly this was a trial-and-error period of unguided experimentation, with errors predominating, and leading to a great many short-lived systems incapable of going beyond initial formation. Selection would favor any system characterized by stability and the possibilities of self-replication since variations can persist only if copied.

It can reasonably be assumed that the energy problem was solved by making use of some kind of available phosphate molecules until such time as a primitive fermentative, and eventually a photosynthetic, system became established. But when both "functional and instructional properties are required" of a macromolecule, RNA is the molecule of choice, now and very probably when life began. The backbone structure of ribose and phosphate linkages not only provides stability, but the conformation of the molecule lends itself to secondary folding which is a deterrent to hydrolysis in aqueous solution. The projecting bases, through hydrogen bonding, lead to complementary pairing, and thus to self-replication. RNA, of whatever size, can therefore be viewed as a source of instruction as well as template for the production of similar or complementary likenesses.

One can imagine a number of similar, but sequentially different, RNA "species" competing for nucleotides in their neighborhood, with stability, rate of self-replication, and a relatively error-free replicative ability determining survival. G-C base pairs would probably predominate since their bonding is considerably stronger than that of the A-U base pairs, and it has been estimated that in an enzyme-free environment the less-than-perfect replication would set a length limit of about 100 nucleotides for such primitive RNA molecules. The nearest approach to such a molecule in present-day organisms is that of the transfer RNAs (see Figure 4.26), and the thought that they might have played such an early role is given credence by the fact that, evolutionarily, the tRNAs are extraordinarily conservative. This suggestion is made more credible by the recent observation that certain RNAs have an enzymelike action in that they can, by excision and splicing, create their own finished form, ready for use. In addition, they have the capacity to carry an amino acid at the -C-C-A end when activated. The nature of the initial coding system based on the nucleic acids was, therefore, DNA ⟵ RNA ⟶ protein, and not the DNA ⟶ RNA ⟶ protein sequence of present-day organisms. Only when the double-stranded DNA came into existence, and the RNAs acquired the capacity to translate the encoded information of DNA into enzymatic proteins, could the size of the coding units be extended beyond the 100-nucleotide limits; that is, it would be the prevalence of replicative errors that would limit the size of any reliable coding system, and it is the DNA polymerases that have an error-correcting capability. How DNA came into being from a primitive RNA state is problematical (some viruses can do this, but only with the requisite enzymes), but a possible accidental fit between spontaneously formed proteins and RNAs could have set the stage for the initiation of translation.

It might be well, at this point, to state that the discussion above is speculative, and that we do not know how the organic world came into being. If life is monophyletic, it goes back to one point in time which marks the transition from abiotic synthesis to the start of all organic evolution (Figure 12.2). The many common features of all cells would suggest that life originated only once (monophylesis), but it is also possible that there were many beginnings (polyphylesis), with only one surviving to give rise to present-day forms.

The presumed nucleic acid sequence that was assembled abiotically can be thought of as *the primordial gene*, capable of acting as a template for copying, and possessing a phenotype subject to natural selection. The phenotype, consisting of the stability and copying capability of this molecule, must have been determined by relatively short sequences of nucleotides; only these would have a chance of persisting since longer sequences would have been eliminated as a result of errors of replication. As the replicative process itself became more and more error-free, longer sequences could come into existence with improved probabilities of persistence. K. C. Atwood, for example, has estimated that all genes, however large or

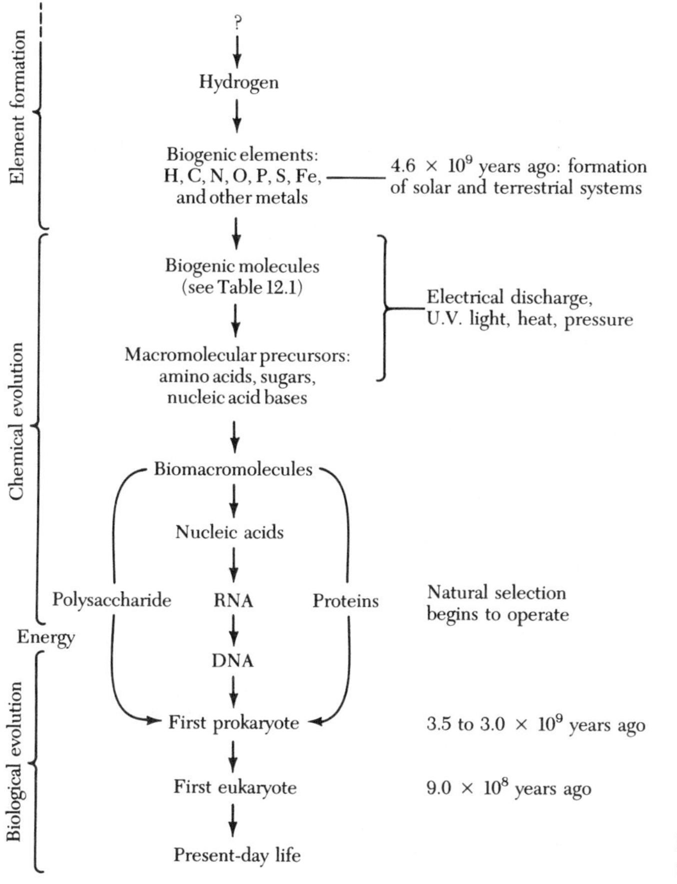

Figure 12.2 Time sequence of evolution from hydrogen to present-day life.

small, are descendants of nucleotide sequences of 60 or fewer. A similar argument can be advanced for the translational products as well; that is, the active sites of enzymes constitute a small portion of the total molecule, and the additional amino acids would have been acquired as they improved upon the basic operation of the active site. Genes, therefore, and enzymes as well are not randomly arrayed sequences of smaller units; they have a long history, and the genetic diversity of the organisms of the world is but a negligible fraction of that which is potentially possible.

PRIMITIVE CELLS

The difference between a precellular chemical system and the simplest prokaryote is enormous. This can be best appreciated by pointing out that among the mechanisms needed to bridge the gap are the following: a compartmentalized arrangement which, by means of membranes, would separate the cell from its environment; a system for the precise sequencing of amino acids in the proteins for structural and enzymatic purposes; an equally precise system for the formation and assembly of the several kinds of nucleic acids; a genetic code that sets the pattern for all cellular activity through organized structure and behavior; and one or more means for extracting energy from the environment in a systematic and sustained manner.

It is presumed that the first organisms, when once formed, drew their sustenance for growth and reproduction from the prebiotically produced organic molecules in their immediate neighborhood. How long they could have depended upon such a supply, with little or no expense of cellular energy for synthesis, is open to question, but it can be assumed that as the demand for these molecules began to exceed the supply, and survival became precarious, more and more emphasis had to be placed on the development of mechanisms of synthesis as well as mechanisms to ensure a reliable flow of energy through the cell. Such metabolic transactions, at least as understood today, require two features: enzymes and other molecules that act as intermediaries in energy-exchange processes. The fortuitous origin of prebiotically formed proteins with enzymatic properties could serve as a start for promoting and maintaining crucial chemical reactions; any circumstances, intra- or extracellular, that would favor their formation and continuance would be evolutionarily advantageous. As to energy-exchange systems, it is no coincidence that adenine is not only a part of the informational system of all cells—that is, in DNA and the several RNAs—but is also a part of some of the most basic energy-related molecules: the mono-, di-, and tri-phosphonucleotides (AMP, ADP, and ATP), cyclic AMP, nicotinamide adenine dinucleotide (NAD), nicotinamide adenine dinucleotide phosphate (NADP), and flavin adenine dinucleotide (FAD). Adenine is very readily formed under prebiotic conditions if hydrogen cyanide is present, so its central role in energy exchanges must have been established very early in the history of life. ATP is, of course, the universal source of energy for so many basic reactions; cyclic AMP is involved in a host of cellular events; and the other adenine-containing molecules are intimately related to the activity of enzymes.

It is difficult to determine which of the energy-exchange systems are primitive; probably, the simplest systems are those which formed earliest. Indeed, it is equally difficult to determine which, among present-day prokaryotic species, are most primitive and consequently closest in function to the earliest formed cells. Practically every known biochemical pathway exists in one or another species of the prokaryotes. But in view of the reducing character of the precellular atmosphere, it is probable that the first organisms that would have been recognized as cellular were prokaryotic anaerobes, with their catalytic systems based upon primitive iron- and sulfur-containing enzymes such as hydrogenase and ferredoxin (Figure 12.3). Of the three principal types of energy-yielding metabolism—fermentation, respiration, and photosynthesis—fermentation is the simplest, as well as the most inefficient, and it is widespread among prokaryotes. The fermentation of glucose—presumably a prevalent prebiotic molecule—to lactic

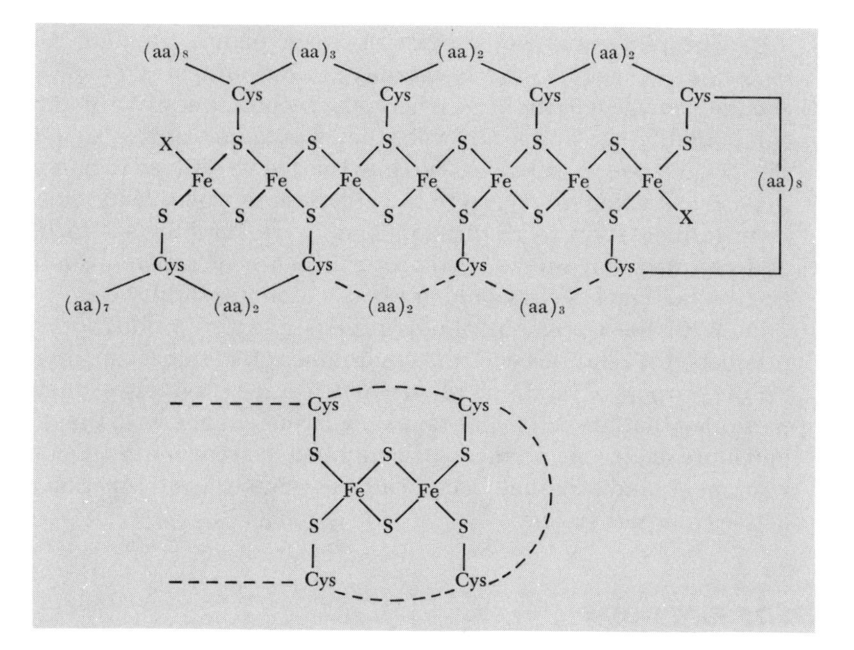

Figure 12.3 The structure of bacterial ferredoxin (above) and the proposed structure of hydrogenase and green plant ferredoxin (below). The former is a small molecule of about 5,000 molecular weight, built around 4 to 7 iron atoms held in place by the sulfhydryl groups of cysteine. The latter is a larger molecule, about 13,000 molecular weight, but with only 2 iron atoms linked together by 2 sulfur atoms and 4 cysteines; the amino acids of this iron-sulfur protein extend off to the left. (Above: after H. T. Yost, *Cellular Physiology*. Englewood Cliffs, N.J.: Prentice-Hall, Inc., 1972.)

acid or to ethyl alcohol and carbon dioxide is not an efficient source of energy as compared to the breakdown of glucose to carbon dioxide and water through oxidative phosphorylation. In fermentation, only two net moles of ATP are realized per mole of glucose, whereas 36 moles of ATP are obtained via the oxidative route. These facts, among others, would argue for the primitiveness of fermentation.

As prebiotically formed molecules in the environment were being depleted, other sources of energy and means of energy exchange had to be developed. Anaerobic photosynthesis, trapping solar energy for conversion into chemical energy, was probably an early development. This would require concentrations of chlorophyll and related light-trapping pigments; enzymes such as hydrogenase and ferredoxin, which could function anaerobically; and either internal membranes formed by extensions of the cell membrane, on which the photosynthetic apparatus was layered, or chromatophores, the photosynthetic particles of some bacteria. No oxygen would be involved, organic compounds other than carbon dioxide would serve as carbon sources, water would not be oxidized, and the only biosynthetic product would be the ATP needed for cellular fuel.

The photosynthetic process of green plants, yielding ATP, glucose, and molecular oxygen, would eventually evolve, but at a much later time. As O_2 became prevalent in the atmosphere, the accumulation of prebiotically synthesized compounds could not occur, since they would be oxidized as soon as formed, and since O_2 is a poison, means would have had to be evolved to handle it. The increasing concentration of oxygen in the atmosphere would lead some species to evolve from fermentation to respiration—many prokaryotes are facultative anaerobes and can function in the presence or absence of oxygen—and a more efficient energy-yielding system would gradually become established.

With the advent of photosynthesis and the production of photosynthetic products, the relations with the environment had to become more selective if only for the purpose of retaining these newly formed products within the cell. It can be presumed that the cell membrane underwent change with time, becoming a more and more selectively permeable barrier between the inner cell contents and the environment, and eventually acquiring the structure and function appropriate to an active transport system.

ORIGIN OF THE CODE

In considering the origin and evolution of the informational system of cells, we need first to be reminded again that life makes use only of L-amino acids (left-handed) and D-sugars (right-handed) (Figure 12.4). This well may be an accident of history, for double-stranded nucleic acids, capable of replicating, theoretically can contain either D-ribose or L-ribose but not both. Similarly, spontaneous group-

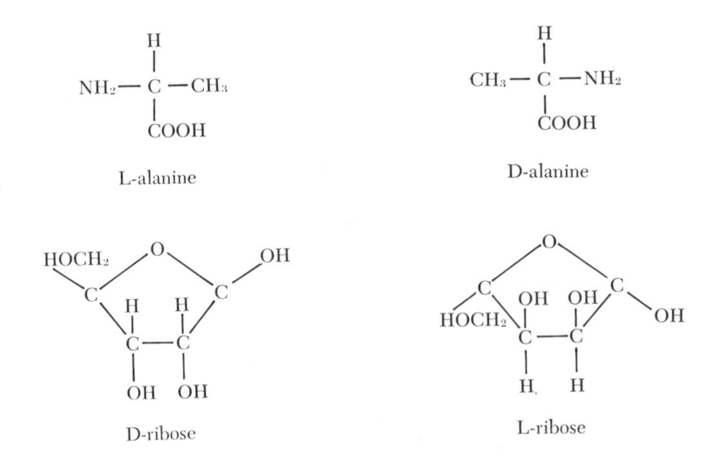

Figure 12.4 Right-(D-) and left-(L-) handed amino acids and sugars. The amino acids are mirror images of each other, but if rotated 180° cannot be superimposed upon each other.

ings of amino acids into short stretches of protein usually contain only D or L forms but not both. In addition, certain D-amino acids do not fit well into certain secondary forms of proteins such as helices, but there is no overwhelming and compelling argument as to why L-amino acids became the exclusive choice of living forms. It may be simply that L-forms competed better and won out.

There is very general agreement that the evolutionary beginning of an informational system started with the tRNAs. It is this molecule that is central in the coding process, for it couples in an exact manner with a particular amino acid, and brings that amino acid into position and in an activated state so that it can react with other amino acids to form proteins. The anticode, therefore, would seem to have preceded the code in evolution.

Was the code established initially on the basis of a trinucleotide, or triplet, code sequence? There is no compelling argument suggesting that the code evolved in stepwise fashion from a singlet to a doublet to a triplet state. A shift from one to the other would have resulted in informational chaos until sorted out, and continued survival would scarcely have permitted such a traumatic luxury. There is, on the other hand, evidence to suggest that it might have started initially as a doublet code, and then at a later stage became triplet in nature. As Miller and Orgel point out, a glance at the code as we now know it will indicate why this might be so (see Table 4.4). If we let *xy* stand for any pair of the four nucleotides, certain facts emerge:

1. *xy*U and *xy*C *always* code for the same amino acid.
2. *xy*A and *xy*G *usually* code for the same amino acid, but when they do not, one or both of the codes are message terminators (UAA, UAG, and UGA), or an initiator (AUG).
3. In eight cases, *xy*U, *xy*C, *xy*A, and *xy*G all code for the same amino acid.
4. Except for leucine, serine, and arginine, which are coded for by six triplets each, all codes for the same amino acid start with the same pair of bases. It seems most unlikely that the informational content of a code, so strongly dependent on the first two letters of the triplet, could have evolved in stepwise fashion from a singlet code, but it does seem probable that the third letter of the code, responsible for a lesser degree of selectivity, might have been a later addition.

Other pieces of information are extractable from Table 4.4. When U is the second letter, all of the amino acids are hydrophobic, that is, insoluble in water; and when a purine, A or G, is the second letter, the amino acids coded for are always charged instead of being neutral. The codes, consequently, are not a haphazard array of cryptic devices; rather, the regularities argue for some kind of precise chemical specificities, the nature of which resides in unknown fashion in the structure of the tRNAs. It is at this point that knowledge of the early evolution of the informational system ceases. The origin of the code is not known, nor is it yet known how the chromosome, the mRNAs, the rRNAs, and the ribosomes came into existence and were fitted into the scheme as parts of the transcriptional and translational machinery of the cell.

Eigen and his colleagues, looking at the coding system somewhat differently, argue that the code and its reading direction was initially fixed by the triplet code RNY, with R being a purine nucleotide (G or A), N signifying any one of the four nucleotides, and Y being a pyrimidine (C or U). In looking for "remembrances of ancestral sequences" among the tRNAs and the genes of a number of primitive as well as higher organisms, the RNY triplet tends to appear more commonly than would be expected on the basis of randomness. There is the further suggestion made that the strength of the G-C base pairing would favor the formation, and perhaps, at the early stages of life, limit the number of codes to four: these would be GGC coding for the amino acid glycine, GCC for alanine, GAC for aspartic acid, and GUC for valine. It is precisely these four amino acids that are most commonly found in experiments which are thought to simulate prebiotic conditions (Table 12.2). In addition, there is a strong interaction of the carboxylate groups of such amino acids as aspartic acid and the G nucleotides of coded information, setting the stage possibly for a primitive "enzymatically" fostered kind of replication and translation.

ORIGIN OF THE EUKARYOTIC CELL

Evolution tells us that all living things share a common ancestry, and if what has just been stated about primitive cells is reasonably correct, that ancestral form was a prokaryotic cell. In this state, the prokaryote underwent a good deal of change to produce a variety of species differing in their energy-yielding metabolic processes. The first were probably anaerobic, deriving energy from chemical substances formed earlier by abiotic processes, and, as these chemicals in the environment became depleted, perfecting fermentative pathways that provided sources of carbon and nitrogen as well as energy. Some forms were anaerobic photosynthesizers, capable of forming chlorophyll for light absorption, deriving their hydrogen to reduce CO_2 from hydrogen sulfide and gaseous hydrogen rather than water, and producing ATP as a by-product, but no oxygen and no carbohydrates.

A further evolution of the photosynthetic mechanism produced a system in which water was split to yield the hydrogen that entered into the carbohydrates being formed; oxygen was released into the atmosphere as a waste product. Oxygen, however, is a metabolic poison to an obligate anaerobic species, with the result that as the O_2 in the environment increased in concentration, the anaerobic bacteria retreated into oxygen-free locations, while aerobic forms evolved and made their appearance. These were able to adapt to the presence of molecular oxygen and, in fact, to change their energy-yielding processes from a fermentative to a respiratory form. These changes must have taken millions and millions of years, and among the aerobes that evolved were the blue-green algae, the first photosynthesizing aerobes.

If we think of the first primitive cells arising about 3.5×10^9 years ago, blue-green algal fossils (stromatolites) from the Precambrian are at least 2.5×10^9 years old, and possibly even older. Fossils judged to be of eukaryotic origin are no older than 0.9×10^9 years ago. We must search, therefore, for eukaryotic origins from among the prokaryotes that preceded them in time. All eukaryotes are aerobic, so it is from among this group of prokaryotes that the search must be focused. Several hypotheses account for the origin of the eukaryotic cell, but before examining these speculations, let us consider again how these two types of cells differ from each other.

1. Prokaryotes are unicellular or filamentous forms, and the cells are small, not exceeding 10 μm in diameter. Eukaryotes have large cells generally, only a few kinds being under 10 μm in diameter. They include not only unicellular and filamentous forms, but also all of the varied two- and three-dimensional forms of the plant and animal world. The latter result from the division of cells in several planes, with the cells adhering to each other after division. However, such prokaryotic cells in groups do not communicate with each other, as do similarly grouped eukaryotic cells.
2. The vast majority of eukaryotes are aerobic, although some can function as facultative anaerobes; the prokaryotes are enormously varied metabolically. Prokaryotic cells, nevertheless, are structurally simple; the eukaryotic ones are extensively partitioned into double-membraned organelles—nuclei, mitochondria, and plastids—and single-membraned cytoplasmic systems—endoplasmic reticulum, Golgi apparatus, vacuoles, lysosomes, and so on. The variety of eukaryotic membranes permits the separation of various metabolic pathways, while the membrane-associated pathways in the prokaryotes must be confined to the single cell membrane or to the membranes concerned with photosynthesis.
3. Both prokaryotic and eukaryotic cells have motile systems based on cilia or flagella, but these differ chemically and structurally. Those of the flagellated bacteria are made of the protein flagellin and are constructed as solid single strands with no discernable internal structure. Eukaryotic cilia and flagella are based on a 9 + 2 microtubular system, connected to a 9 + 0 basal body (Figure 12.5), with tubulin the basic protein component. Microtubules made of tubulin have been found in several species of prokaryote, but they play no significant role other than that of movement.
4. Multiplication of cells is by binary fission in the prokaryotes (Figure 12.6). The naked DNA of the nuclear area is believed to be attached to a mesosome, a structure in the cell membrane that seems to divide and pull apart as the DNA is replicating. The actual details, however, remain uncertain. Cell division in the eukaryotes is by mitosis with the spindle being composed of microtubules, and there is no similarity between the two kinds of divisional processes even though the end results are the same.

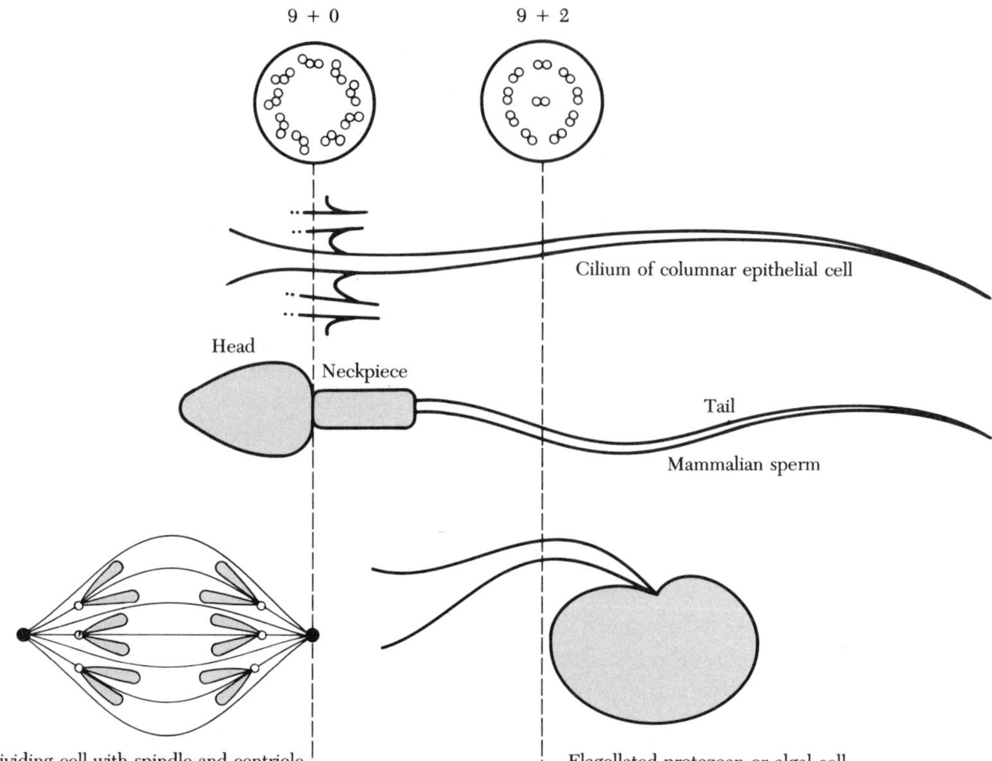

Figure 12.5 Arrangement of microtubular elements found in eukaryotic organelles concerned with movement. Left: the 9 + 0 arrangement characteristic of centrioles, basal bodies, or kinetosomes, at the proximal ends of flagella and cilia, and of sperm just below the attachment of the middle piece to the head. Right: the 9 + 2 arrangement of the cross-sections of the main body of flagella, cilia, and the tails of sperm.

5. Inheritance in eukaryotes is generally biparental where sexual reproduction is concerned, although more recent evidence indicates that the gain or loss of coding and noncoding DNA can take place independently of the meiotic process, and that mobile units of DNA can be shifted about in the genome, or introduced from external sources. In some bacteria transfer of genetic information on a piecemeal basis from one cell to another can occur, but not on a reciprocal basis; it is always a one-way transfer. The genomes of both prokaryotes and eukaryotes are, therefore, far more labile than was once thought to be the case.

   All prokaryotes exist only as haploids, whereas most eukaryotes show an alternation of haploid and diploid states, with fertilization and meiosis making sexual reproduction a workable system. Also, no prokaryote is known to possess more than a single naked chromosome per genome, whereas the eukaryotes have acquired a multichromosomal state with the

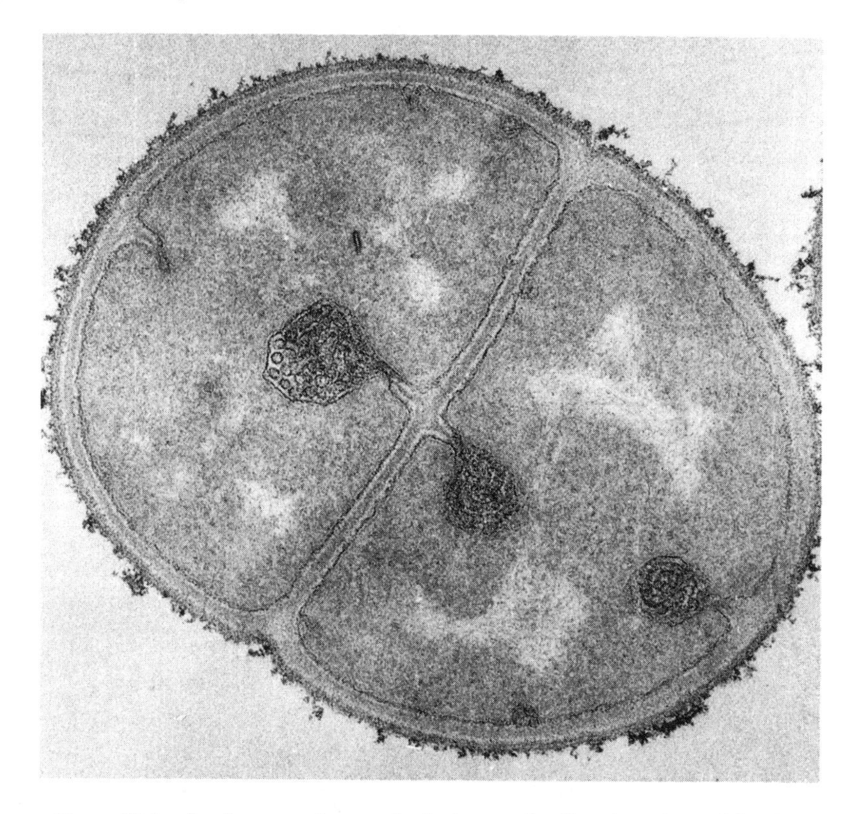

Figure 12.6 An electron micrograph of a bacterial cell in the process of dividing. The rounded structures are mesosomes, formed by a folding in of the plasma membrane. The function of the mesosome is not known for certain, but the different functions suggested include: a localized center of respiratory activity; a means for wall formation in dividing cells; and a means for the separation of nuclei after replication has taken place. In support of the latter suggestion is the observation, made in some bacteria, that the bacterial chromosome is attached to the mesosome. (Courtesy of Dr. Stanley C. Holt.)

DNA associated very generally with histones to form microscopically conspicuous nucleoprotein structures.

It is obvious, therefore, that the eukaryotic cell architecturally is far more complex than its prokaryotic counterpart. It contains several protein-synthesizing units while its prokaryotic ancestor contains but a single such unit. In attempting to understand the origin of the eukaryotic cell, the question, consequently, is whether the prokaryotic ancestral cell, by the gradual origin, development and differentiation of internal organelles, evolved into a eukaryotic status, or whether the structural complexity was acquired by a different evolutionary route. Three of the hypotheses that have been advanced are illustrated in Figures 12.7, 12.9, and 12.11.

The cell symbiosis theory, most recently advanced and elaborated upon by Margulis, holds that the eukaryotic cell is a composite structure, made up of several kinds of cells living in symbiotic relations with each other and within a common cell membrane (Figure 12.7). The concept is not new, but it can be supported to-

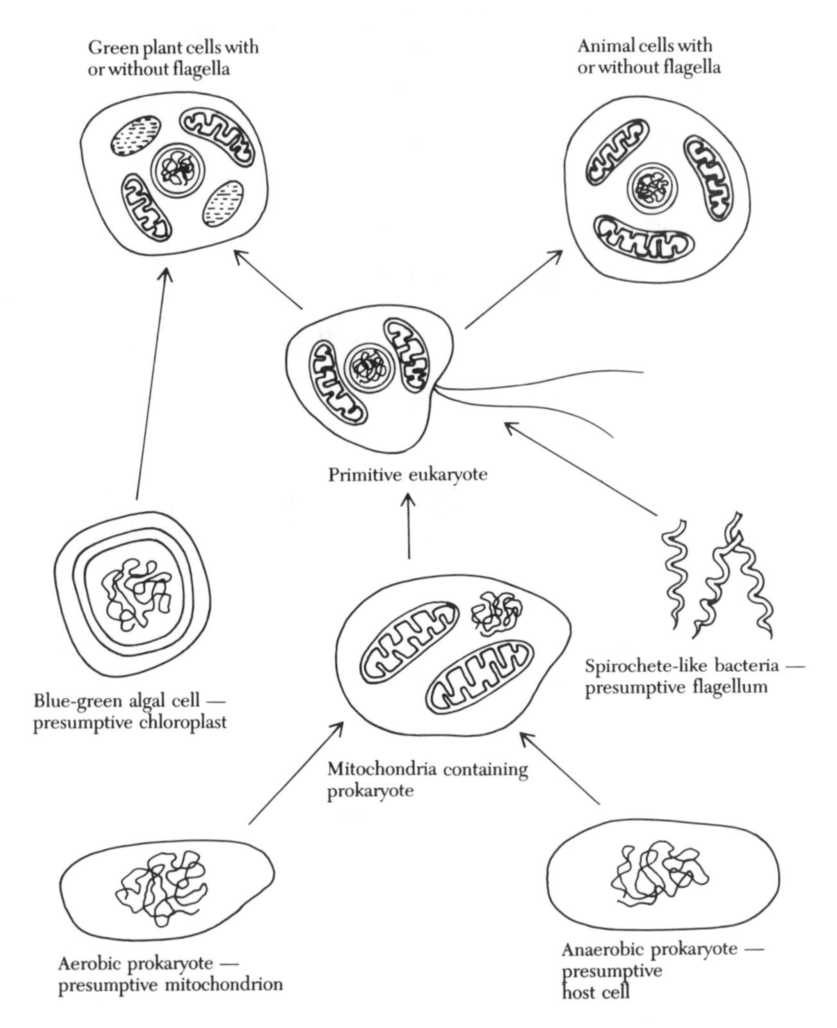

Figure 12.7 The origin of the several kinds of eukaryotic cells according to the symbiosis theory of Margulis. The host cell is presumed to be an anaerobic, prokaryotic bacterium. Aerobic bacteria are assumed to be the precursors of the eukaryotic mitochondrion, and a blue-green algal cell the forerunner of the chloroplast. The origin of the double-membraned nucleus is obscure, but it is probably the DNA of the host cell that became enclosed. Once the nucleus is formed, the endoplasmic reticulum and the Golgi complex could be derived from the outer nuclear membrane, with the lysosomes, vacuoles, and so on derived from the Golgi complex (see Figures 12.8 and 12.10).

day with greater conviction because of a more extensive knowledge of comparative biochemistry and ultrastructure.

Symbiosis can be defined as the cooperative existence of two or more organisms for mutual benefit. Symbiosis exists today at many levels of biological organization, but it is those of a cellular, and particularly of an intracellular, nature that are of interest here. The green alga *Chlorella* can be incorporated into the cytoplasm of the slipper animalicule, *Paramecium bursaria*, where it continues to photosynthesize and to provide nutrients to the host cell, even in a subminimal medium. The blue-green alga *Cyanocyta* has been similarly found in symbiotic relations with the protozoan *Cyanophora*, while certain snails have the ability to extract chloroplasts from plant cells and incorporate them intracellularly, where they continue to function photosynthetically. In these instances, the relationships seem not to be mandatory for the host, but may well be for the invading plant.

The validity of the symbiotic theory has been given added credence by the discovery of what seems to be an instance of genuine symbiosis established in the laboratory, although selection cannot be totally ruled out. K. W. Jeon found a culture of *Amoeba proteus* which became infected with a strain of bacteria, and which caused a significant drop in growth rate of the amoeba. Eventually, however, the amoebae recovered their rate of growth and their health, but it turned out that not only were the bacteria still present, but the amoebae had become dependent on them as symbionts. Transplanted nuclei from the amoebae could not function in the absence of the bacteria.

The cell symbiosis theory holds, therefore, that the particulate organelles of the eukaryotic cell had independent origins as prokaryotic cells. Thus, the plastids of the green plants were thought to be derived from a symbiotic algal cell capable of aerobic photosynthesis; this kind of cell was also ancestral to our present-day blue-green algae. The mitochondria of the eukaryotic cell were similarly derived, but in this instance from an aerobic bacterium. The location of respiratory enzymes in the cell membranes of bacteria and the membranes of mitochondria reinforce this point of view. The host cell is presumed to be an anaerobic, nonphotosynthesizing cell, which, because of the necessity of adapting to atmospheric oxygen, acquired a symbiont capable of managing oxygen through respiratory pathways. The heat- and acid-resistant *Thermoplasmas*, which possess histones to protect their DNA, are thought to be a presumptive host cell.

How is a symbiont, acquired as described above, to be distinguished from an organelle that evolved intracellularly and progressively over long periods of time? Are there criteria that point to the symbiotic acquisition rather than to the evolutionary developments of these organelles? The arguments are compelling in favor of the symbiotic origin.

First, a newly acquired symbiont of cellular nature would possess its own hereditary system, although time might well have brought about alterations in hereditary function and dependence of both host and invader. Mitochondria and plastids fulfill these criteria; they are of the right size, and they possess DNA, which is usually circular in form, mRNAs that are complementary to the DNA, rRNAs, tRNAs, and the enzymes necessary for these to function. Their tRNAs are different in certain nucleotide sequences from those found in the cytoplasm of the host cell, the rRNAs of both mitochondria and chloroplasts are very similar in sequence to prokaryotic rRNA, and their ribosomes are smaller, being much more like those of bacteria and blue-green algae than the larger ones of either animal or higher plant species. In addition, protein synthesis in mitochondria is inhibited at the ribosomal sites by the antibiotic chloramphenicol, but is insensitive to cycloheximide. A similar situation occurs in bacteria and blue-green algae, whereas protein synthesis in the cytoplasm of the eukaryotic cell is inhibited by cycloheximide and unaffected by chloramphenicol. There are striking parallels, therefore, in the structure and physiology of eukaryotic organelles and of prokaryotic cells, just as there are striking differences in resistance to antibiotics between the organelles and the nucleocytoplasmic portion of a eukaryotic cell. However, the interrupted structure of some mitochondrial genes into exons and introns is more a eukaryotic than a prokaryotic feature, and raises a question of origin not as yet answered.

The hereditary system of present-day organelles is, however, insufficient in complexity and diversity of function to fulfill all of the needs of the organelle. This well might be expected, since the continued evolution of the symbiotic relation might lead to a loss or reduction of independent synthetic and informational capabilities, particularly if these were duplicated in the host cell. For example, in yeast, parts of two mitochondrial enzymes, ATPase and cytochrome oxidase, are synthesized on the mitochondrial ribosomes, other parts on cytoplasmic ribosomes. In green plants, the chloroplast enzyme, ribulose-1,5-diphosphate carboxylase, consists of two subunits, one unit with a molecular weight of 14,000, the other about 50,000. Synthesis of the smaller subunit occurs on cytoplasmic ribosomes and is preferentially inhibited by cycloheximide, that of the larger on plastid ribosomes, and is preferentially inhibited by chloramphenicol. It would be expected, of course, that the ribosomes of an invading symbiont and those of the host cell would differ in their protein structure; a long evolutionary period, prior to the establishment of a symbiotic relation, would almost ensure such a difference. This is strikingly borne out, for example, in the flagellated, photosynthesizing unicell, *Chlamydomonas*. The chloroplast ribosomes contain a total of 48 different proteins in the two subunits; the larger cytoplasmic ribosomes contain a total of 65 proteins. Based on electrophoretic mobility, no more than four proteins from the two types of ribosomes show any degree of similarity, and even these may show differences when examined by more discriminatory methods.

A second argument in favor of symbiotic origin concerns the fact that the organelles exhibit a pattern of division very similar to that of prokaryotes but quite dissimilar to that of eukaryotes. Mitochondria, chloroplasts, and prokaryotes increase their numbers by binary fission, eukaryotes by mitosis with its structured spindle apparatus. Whether mitochondria and chloroplasts have their DNA attached to their inner membranes is not entirely clear, but individual plastids and mitochondria can possess a number of copies of its genomic DNA, so that division of an organelle and its hereditary content need not be done with the same precision as the eukaryotic nuclear genome. It is known that when cells contain only a single plastid, as in *Spirogyra*, or a single mitochondrion, as in *Microsterias*, division of the cell is accompanied by division of the organelle, but the great majority of eukaryotic animals and plants have sufficient numbers of these two kinds of organelles to ensure that each daughter cell receives a share at division.

The presence of DNA in organelles indicates that they are capable of mutation and that, with patterns of variation possible, patterns of inheritance of organellar traits should be discernible that depend upon modes of transmission from one generation to the next. A number of species of green plants exhibit a kind of heredity known as maternal inheritance, the trait in question being transmitted only through the maternal side, that is, through the egg but not through the pollen. In barley, for example, a large number of chlorophyll or chloroplast mutants have been identified. The great majority of these are inherited in a Mendelian fashion, and the mutant genes are, therefore, believed to be in the nuclear genome. Some chlorophyll mutants, however, exhibit maternal inheritance, and these have been traced to mutant chloroplasts. Since chloroplasts are only rarely transmitted through the male pollen grain, the inheritance must of necessity be maternal through the embryo sac. Similarly, uniparental inheritance of mutant mitochondria in yeast has been identified with mutations occurring in mitochondrial DNA.

The theory of the symbiotic origin of the mitochondria and plastids rests, therefore, on a substantial body of circumstantial evidence. On less firm ground is the concept, also a part of the symbiosis theory, that flagella, cilia, centrioles, and the mitotic spindle also owe their origins to symbiotic organisms that attached themselves initially to the outside of the host cell, later to become incorporated into the internal organization of the cell. According to Margulis, these include certain spirochetes inhabiting the guts of termites as well as other motile bacteria, which, upon external attachment, provided the host cell with mobility, an obvious adaptive mechanism if the food supply were in any way depleted in the neighborhood. A number of observations support this view. In the first place, the kind of symbiosis envisioned exists today. The protozoan, *Myxotricha paradoxa*, an intestinal parasite of termites, has a limited mobility of its own, but thousands of attached and motile bacteria, embedded in the thick outer pellicle, provide motion through symbiosis. In addition, other bacteria, internally located or externally attached, are also symbiotic, but their precise roles remain unknown at present.

Such attachments have been proposed as a first step toward the acquisition of eukaryotic motility and, eventually, replacement of fission with mitotic cell division. At the point of attachment of a flagellum or a cilium to a eukaryotic cell, there is a structure, a basal body, comparable to that of a centriole—an arrangement of microtubules in a 9 + 0 pattern (Figure 12.5). Since it is known that centrioles can function as basal bodies, it is presumed that they are homologous structures. In addition, both centrioles and basal bodies function as microtubule-orienting centers, the basal bodies being concerned with the orientation of microtubules along the length of the flagella and cilia, the centrioles adjacent to the outer edge of the nucleus with the formation of spindle fibers that extend from pole to pole. At a somewhat later time in evolution, the centromeres of the chromosomes also became involved in microtubule-orienting processes, interacting with the poles to form half-spindle fibers stretching from the chromosome at metaphase to one or the other of the two poles. Microtubules are, therefore, involved with movement, whether this is external to the cell by flagellar or ciliary activity, or internal and expressed as metaphase orientation or anaphase movement of chromosomes. This concept makes the assumption, therefore, that in the course of evolution the nucleic acid of attached spirochetes responsible for the coding of tubulin became transferred to, and incorporated into, the genome of the host cell, where it now codes for eukaryotic tubulin. This assumption is supported by the observation that all tests so far conducted indicate that the tubulin of spirochetes and that of eukaryotes are immunologically and electrophoretically identical.

It is possible that the nucleus, like the mitochondria and plastids in being enclosed by a double membrane, is also the evolutionary remnant of another invading symbiont. In the course of time this symbiont would have lost most of its cytoplasm, and with that remaining becoming nucleoplasm. If so, the host cell must have evolved in the opposite direction, retaining its cytoplasm but losing its hereditary features. A far more likely possibility, however, is that the prokaryotic genome, together with the mesosome that attaches it to the cell membrane, became enveloped by a phagocytotic inward extension of the cell membrane. Such an origin would account for the origin of both the inner and outer portions of the double nuclear membrane as well as the observation that many eukaryotic chromosomes seem to be attached to the inner nuclear membrane for a good part of the cell cycle. The archaebacteria, a more primitive group than the eubacteria represented by *E. coli*, have been suggested as a source of the eukaryotic genome, since Thermoplasma, as a representative type, has histone associated with its DNA, and the ribosomal RNAs and proteins are more eukaryotic than typically prokaryotic.

Given the double nature of the nuclear membrane, and the attachment of ribosomes to the outer membrane, it is not too difficult to envision the origin of the endoplasmic reticulum (Figure 12.8) and, by further elaboration, the Golgi apparatus and the several kinds of lysosomes and lysosomic-like vesicles.

Figure 12.9 illustrates a second hypothesis to account for the origin of the eukaryotic cell. This is an evolutionary model in which the several organelles are produced initially by invaginations of the cell membrane. The ancestral cell is assumed to be an aerobic and multigenomic prokaryote, with each separate genome attached by a mesosome to the cell membrane. All of the invaginated bodies would be presumed to be similar initially and possibly possessing respiratory and photosynthesizing capabilities. With time, a division of labor would occur: the nucleus would lose its respiratory and photosynthetic aspects, these becoming emphasized respectively in the mitochondria and plastids; surface area in the mitochondria and plastids would be greatly increased by additional invaginations of the inner membranes to form, respectively, cristae and thylakoids; the

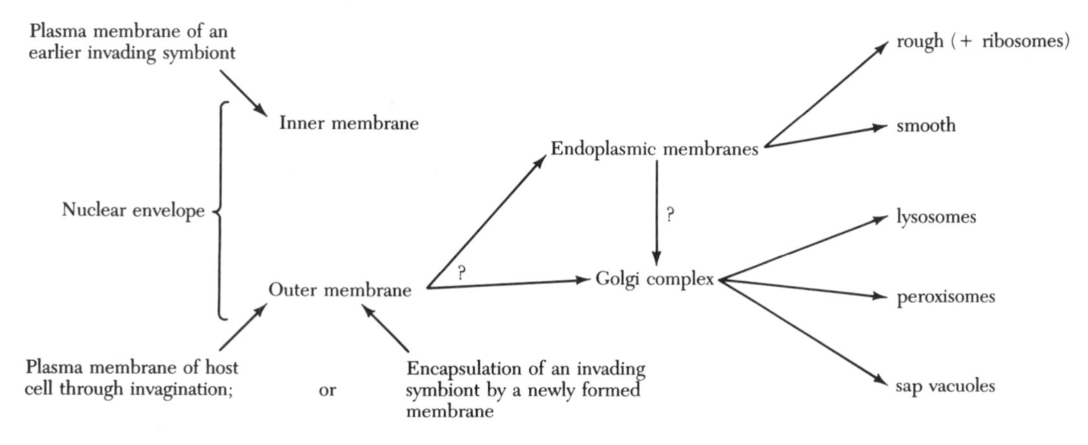

Figure 12.8 Possible origins and subsequent derivations of the inner membranes of the eukaryotic cell.

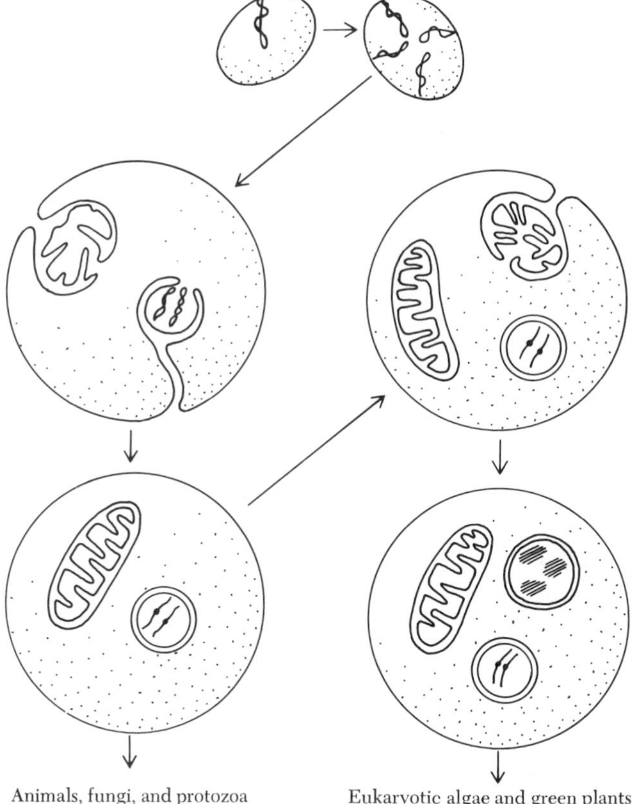

Figure 12.9 The origin of the eukaryotic cell by invagination of the cell membrane to form double-membraned organelles and nucleus. It is assumed by Uzzell and Spolsky (1974) that the invaginations did not occur simultaneously and that the nuclear and mitochondrial invaginations occurred first and led in the direction of the animal, fungal, and protozoan groups, while the photosynthetic invaginations occurred later and led in the direction of the eukaryotic algal and higher plant groups (see Figure 12.10). The ancestral form is assumed to be an aerobic bacterium that had acquired duplicate genomes prior to invagination, a feature seen in many present-day bacterial cells.

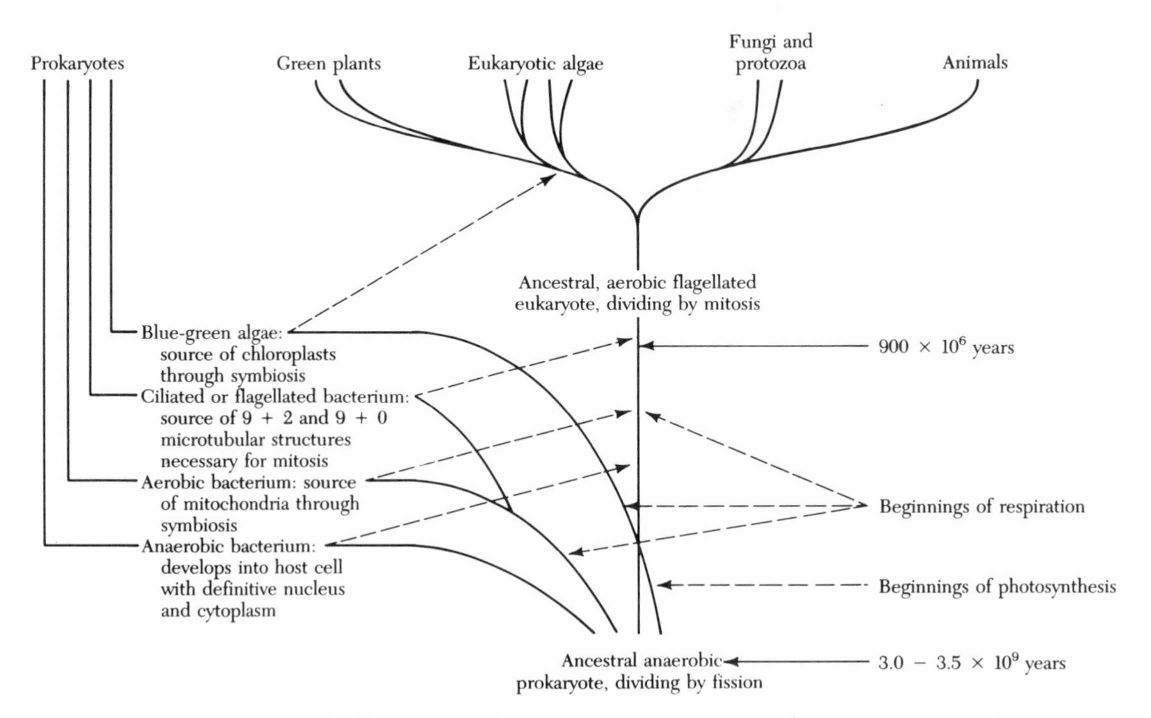

Figure 12.10 An interpretation of evolutionary trends according to symbiosis hypothesis of Margulis (left) and the invagination hypothesis of Uzzell and Spolsky (right).

organellar genomes would lose many of the genic functions duplicated in the nuclear genome; and the nuclear genome would become much more complex as the functional and structural complexity and diversity of species increased. Figure 12.10 provides a family tree of ancestry, one based on the invagination hypothesis and the other on the symbiosis hypothesis, and with acquisitions or losses of function indicated along the evolutionary route.

The invagination hypothesis, therefore, proposes a unicellular rather than a multicellular origin for the eukaryotic cell. The features that the organelles share with the prokaryotes—single, small, circular, and naked genomes; small ribosomes; chloramphenicol sensitivity; cycloheximide resistance; and so on—are regarded as retained primitive characteristics rather than ones acquired through symbiosis. A more positive feature of the invagination hypothesis is that it provides a plausible explanation for the origin of the double membranes of nucleus, mitochondria, and plastids. In addition, where intracellular membranes do exist in the prokaryotes, as in photosynthesizing bacteria, these are formed by invagination of the cell membrane. The larger size of the eukaryotic cell could have been an adjustment to maintain a proper genomic/cytoplasmic ratio. In the final analysis,

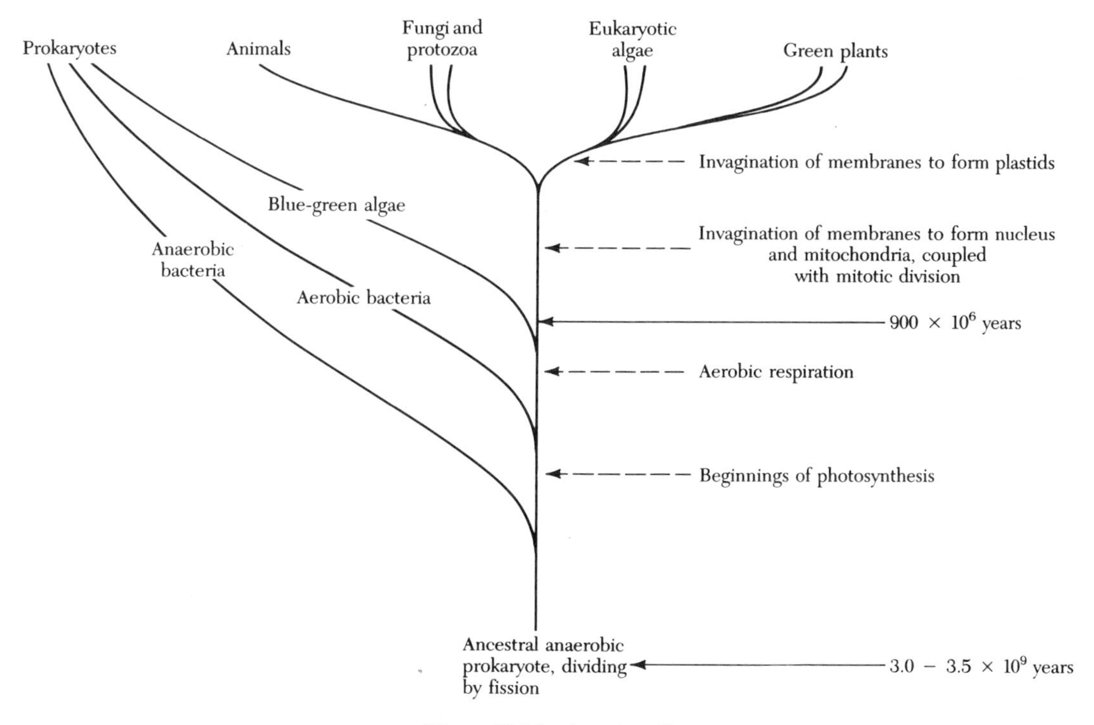

Figure 12.10 *(continued)*

however, there is no single piece of critical evidence that permits one to choose between the symbiotic and the invagination hypotheses, although the former is clearly the more favored point of view.

A third hypothesis accounting for the origin of eukaryotic cells suggests that they arose by *cluster cloning* of the elements in the genome (Figure 12.11). A basic assumption would be either that an ancestral prokaryote possessing multiple chromosomal copies segregated these separate genomes into independent vesicles, or that a single genome would break up in parts, with these becoming isolated into vesicles of different functions. The multigenomic assumption is made plausible by the fact that the organellar and nuclear coding mechanisms for protein formation are basically similar, with control of organellar function coming eventually to be shared by both organelle and nucleus. It is, in fact, the shared nuclear-organellar control of respiratory and photosynthetic activities that is advanced as support for this kind of intracellular evolution, but since gene transfer, gains, and losses can occur among prokaryotes, and in unidirectional ways, it should also be a possible avenue of exchange between invading symbionts as well as between cloned and membraned segments of an original genome.

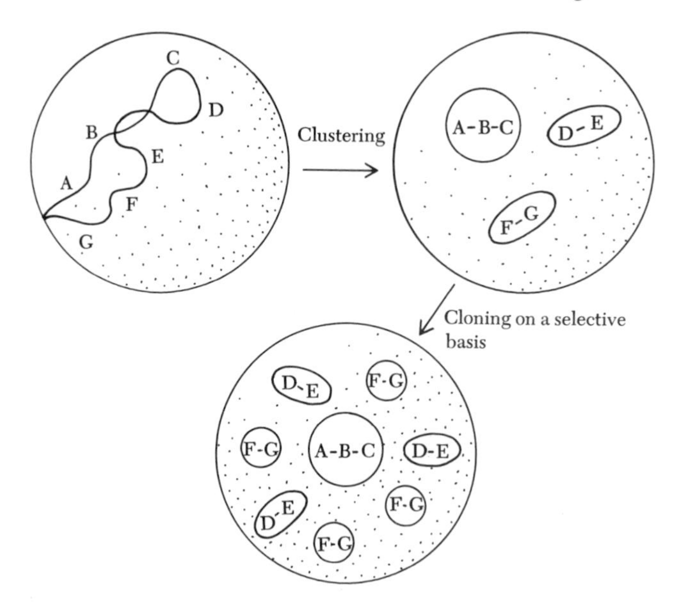

Figure 12.11 Possible origin of the eukaryotic cell through a system of cluster-cloning, beginning with a single genome within a prokaryotic cell. It is assumed by Bogorad (1975) that genes having special functions would become clustered within the genome, that is, those having to do with mitochondrial functions would become clustered in one part of the genome to form the D-E segment, those having to do with plastid functions becoming clustered in another part of the genome to form the F-G segment, leaving the A-B-C segment to govern the remaining functions of the cell. The genome would then become fragmented, with each special segment remaining intact to become enclosed within a double membrane, this leading to the formation of the nucleus, mitochondrion, and plastid of the newly created eukaryotic cell. Further cloning of the mitochondrial and plastid systems would give multiple copies within a single cell. The microtubular system, in all of its manifestations, could be presumed to arise at a later time through mutational events. It is also assumed that the original prokaryotic cell giving rise to the above system was an aerobic, photosynthetic organism. In the diagram, the A-B-C system is nuclear, the D-E system mitochondrial, and the F-G system related to the plastids.

ORIGIN OF MITOSIS

Nuclear membranes, microtubules, centrioles, and mitosis are uniquely eukaryotic phenomena, and they are somehow related to each other in the process of cell division. The behavior of present-day eukaryotes during division makes this relation obvious, but questions of origin and of evolutionary integration of parts remain unanswered. It is clear, however, that the enclosure of the genome within a nuclear membrane, the partitioning of the genome into chromosomal units and the isolation of the nucleus in the cytoplasm, required the parallel development of a mechanism that would distribute the replicated chromosome(s) in a regular quantitative and qualitative manner.

A good many variations of the mitotic theme exist among the lesser known algal groups, the dinoflagellates and the ciliated protists. As these become better known cytologically, it well may be that the origins of mitosis can be ascertained with greater certainty than we now possess, but there is general agreement that a "closed" mitosis is primitive. This is a process in which the nuclear membranes never break down during the cell cycle, the chromosomes are continuously attached either to the nuclear membrane or to a spindlelike structure outside of the membrane, and separation of the replicated chromosomes occurs when the nucleus elongates and pulls apart as a result of elongation of the spindle. Such a spindle consists of microtubular fibers extending from pole to pole (centriole to centriole?) and following the outer contours of the nucleus. Later modifications led to complete or partial breakdown of the nuclear membrane in prophase and the origin of half spindle fibers extending from the poles to the centromeres. The breakdown of the nuclear membrane before metaphase and the absence of centrioles in many species indicate that the aggregation and behavior of microtubules alone can bring about the required separation of chromatids or chromosomes, but whether the microtubules are primitive or more recently derived structures is an actively debated subject.

The evolutionary significance of mitosis is undisputed, however. In addition to its significance as a means for maintaining a proper nucleocytoplasmic ratio, the precision with which chromosomes are segregated at anaphase guarantees a steady succession of genetically similar cells, a necessary condition upon which multicellularity is dependent, even though the process of cell differentiation may be accompanied by a parallel differentiation of the nuclear genome. Nuclear differentiation, however, occurs only in somatic cells and not in the germ line.

Meiosis is a variant of mitosis made possible by diploidy, but according to Margulis it may have been selected because it rid diploid cells of the excess chromatin acquired by cell fusion. This is pure speculation at this time, but in any event it is the chromosomal process basic to an alternation of generations in the plant kingdom, to sexual reproduction and the formation of eggs and sperm in all of the eukaryotes, to the whole gamut of patterns of biparental inheritance, Mendelian and otherwise, and to the recombination of genes that promoted diversity among individuals and made rapid evolutionary change possible. There can be no doubt that the introduction of mitosis and meiosis in eukaryotic organisms was an evolutionary innovation of far-reaching consequences.

CHANGES IN EUKARYOTIC CELLS

Eukaryotic cells originated less than a billion years ago, and the appearance of a vast array of different forms of life in the early Cambrian period would suggest that this newly arisen cell possessed enormous potential for evolutionary change from the time of its inception. Perhaps this is not surprising for two reasons. In the first instance, each eukaryotic cell had at least two genomes, nuclear and mitochondrial, either or both of which could undergo heritable change through mutation, while plant cells had an additional genome in the plastids that could do likewise. The genomes of plastids and mitochondria are small in size and gene content, being comparable in these respects to bacterial genomes, but they are present in multiple copies; for example, the chloroplast of *Euglena*, a single-celled green alga, has many copies of its genome, thus increasing mutational possibilities to the same extent. The nuclear genome of eukaryotes is, on average, many times greater in size than that of the prokaryotes, but this well may be a more recently derived feature. If it is assumed that the eukaryotic cell of a billion years ago contained nuclear, mitochondrial, and plastid genomes, all of a similar size and genetically self-sufficient—a not unreasonable assumption on either the symbiotic or invaginational hypothesis of eukaryotic origin—one can then inquire as to the changes that have led to present-day cells and their behavior. All of the genomes have retained the same basic mechanisms of transcription and translation, and with the triplet code universally applicable, so one can assume this to be a system relatively intolerant of major mutational change. In other respects, the genomes of mitochondria and plastids have probably undergone less change than has that of the nucleus. Neither organelle has a self-sufficient genome, which suggests that over long periods a loss of genetic function has occurred, the loss being compensated for by nuclear genes that were either present from the beginning, obtained by transfer from the organelles, or developed anew by mutational process. In any event, the mitochondria and plastids lost whatever independence they once might have had. Mitochondria, on the other hand, have retained an aerobic metabolic system that is basically similar throughout the eukaryotic world. The enzymes involved may be of mitochondrial or of nuclear origin, but the mitochondria remain the basic converters of chemical energy into useable form wherever they are found.

The several kinds of algae—red, brown, golden, yellow-green, and green—as well as all of the variety of higher green plants would suggest that the chloroplasts, the site of color as well as of photosynthesis, have undergone considerable changes. This is so only in part. The process of photosynthesis has undergone little alteration in essential biochemical steps, and chlorophyll *a* and phycobilin are the basic photosynthetic pigments in all forms of algal species. However, each kind of alga has additional secondary pigments, giving each form its characteristic color and, because the secondary pigments are also light absorbing, shifting the peak of photosynthetic efficiency to different wavelengths. The chloroplasts are also structurally characteristic. In all except the green algae (*Chlorophyta*), where the thylakoids are stacked in grana as in higher plants, the thylakoids run from one end of the plastid to the other (Figure 12.12). In the red algae (*Rhodophyta*), the thylakoids are single; in the *Cryptophyta*, in pairs; in the *Chrysophyta* (golden), in groups of three; and in the *Phaeophyta* (brown), in groups of three or four. It is quite probable on the basis of secondary pigment, thylakoid character, and lack of flagellated cells that the red algae are most

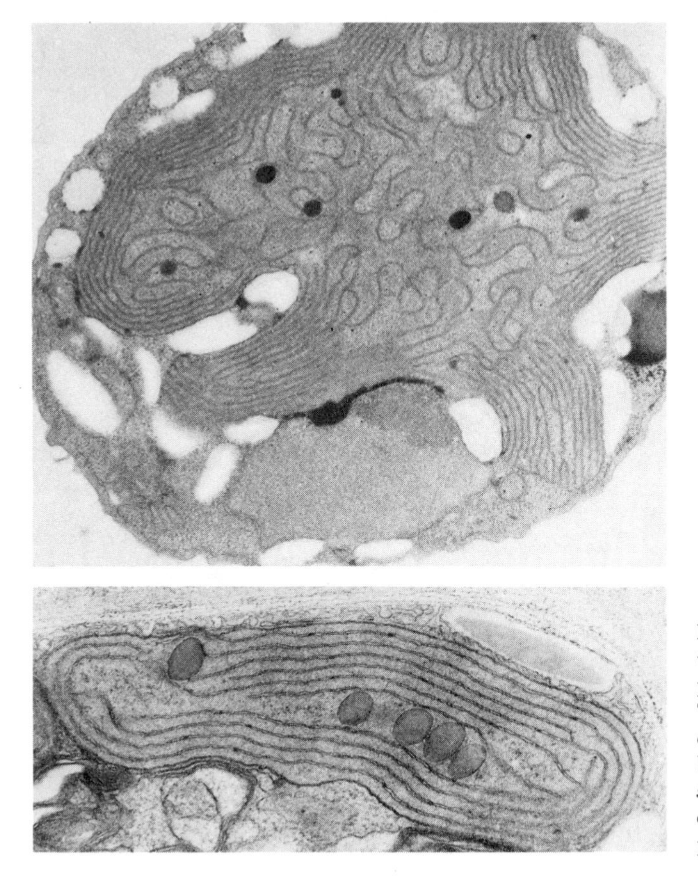

Figure 12.12 Chloroplast structure in two species of red algae in which the photosynthetic membranes exist singly. Top: *Porphyridium*, a single-celled species containing but a single chloroplast; bottom: a chloroplast from *Rhodophysema*, a multicellular species with each cell containing more than one chloroplast.

primitive and most similar to the blue-greens, the supposed source of symbiotic plastids.

Just as the eukaryotic cell has its major biochemical systems compartmentalized into organelles and other membrane systems, so has it introduced comparable changes in its hereditary system. One of the most striking differences between prokaryotes and eukaryotes is the shift from single, naked DNA molecules to multiple, histone-clad chromosomes. These chromosomes vary in number, size, and genetic content, and are equipped with a centromere, which can interact with the spindle and the poles in the process of division.

There is little discernible evolutionary trend that relates to chromosome number. No eukaryotic organism is known that has a haploid number of one. A member of the plant family Compositae has a haploid number of two (*Haplopappus*, a tiny annual species), while an ancient fern, *Ophioglossum*, has a diploid number of over 1000. Tables 12.3 and 12.4 list a number of common species of organisms and their haploid number, and Table 12.5 lists the chromosome numbers in the order Primates, to which humans belong. The most primitive

TABLE 12.3 HAPLOID CHROMOSOME NUMBERS IN SOME COMMON EUKARYOTIC SPECIES OF PLANTS (SEE ALSO TABLE 12.6).

Species	Number	Species	Number
Fungi		*Lycopsida*	
Aspergillus nidulans (black bread mold)	8	*Lycopodium lucidulum* (club moss)	132
Neurospora crassa (pink bread mold)	7	*Selaginella kraussiana* (club moss)	10
Saccharomyces cerevisiae (yeast)	15	*Sphenopsida*	
Saccharomyces pombe (yeast)	2	*Equisetum arvense* (horsetail)	108
Penicillium notatum	5	*Pteropsida*	
Fomes annosus (bracket fungus)	7	*Onoclea sensibilis* (sensitive fern)	37
Ustilago maydis (smut fungus)	2	*Ophioglossum petiolatum* (ancient fern)	ca. 520
Algae		*Osmunda regalis* (royal fern)	22
Amphidinium carteri (dinoflagellate)	25	*Pteris aquilina* (bracken)	58
Spirogyra weberi (green alga)	2	*Gymnospermae*	
Oedogonium cardiacum (green alga)	19	*Juniperus virginiana* (red cedar)	11
Triploceras verticillatum (desmid)	ca. 15	*Larix laricina* (larch)	12
Chlamydomonas reinhardi (protozoa)	15	*Pinus* sp.	12
Chara braunii (green alga)	14	*Tsuga canadensis* (hemlock)	12
Bryophyta		*Angiospermae*	
Marchantia polymorpha (liverwort)	9	*Allium cepa* (onion)	8
Sphagnum sp. (moss)	19	*Brassica oleracea* (cabbage)	9
Anthocerus husnotii (liverwort)	6	*Raphanus sativa* (radish)	9
Funaria hygrometrica (moss)	14, 28	*Zea mays* (maize or Indian corn)	10
Psilopsida		*Lilium canadensis* (common lily)	12
Psilotum nudum	104	*Lycopersicum esculentum* (tomato)	12
Tmesipteris sp.	208	*Triticum diccocum* (macaroni wheat)	14
		Triticum aestivum (bread wheat)	21

members of the groups have the highest number, and the most specialized the lowest. Humans are about in the middle of the Primate groups.

Plants and animals differ to the extent that *polyploidy* has played an evolutionary role; this is when the basic genome is represented more than twice in somatic cells. For example, the grass family, Graminae, has a basic haploid chromosome number of 7, but more than 75 percent of all species exhibit some degree of polyploidy, a rather rare circumstance among animals. Among higher plants as a whole, about 35 percent are polyploid, but polyploidy is not evenly distributed throughout the plant families.

Chromosome size differs as widely as does chromosome number among the eukaryotes. Among the animals, the Orthoptera (grasshoppers, mantids, and their allies) and the Amphibia possess the largest chromosomes, while among the plants the monocots have generally larger ones than the dicots, with the lily and grass families particularly noteworthy.

The amount of DNA per nucleus is, as one would expect, correlated with chromosome number and size (Table 12.6), and these relations are generally but not absolutely correlated with evolutionary advancement. Evolution has led to greater and greater structural, functional, and regulatory complexity, and it is obvious that this must be determined or governed by an increasing number of genes. In attaining this complexity, accompanied by varying degrees of specialization,

TABLE 12.4 HAPLOID CHROMOSOME NUMBERS OF SOME COMMON SPECIES OF ANIMALS, WITH THE PROTOZOA INCLUDED FOR CONVENIENCE (SEE TABLE 12.5 FOR CHROMOSOME NUMBERS AMONG THE PRIMATES).

Species	Number
Protozoa	
Amoeba proteus	ca. 250
Barbulanympha ufalula (flagellate)	26
Trichonympha sp. (flagellates)	24
Invertebrata	
Strongylocentratus purpuratus (sea urchin)	18
Drosophila virilis (fruit fly)	6
Nemobius fasciatus (ground cricket)	8
Astacus trowbridgei (crayfish)	188
Tityus bahiensis (scorpion)	2
Phigalia pedaria (geometrid moth)	112
Papilio sp. (swallowtails)	30
Pisces	
Torpedo ocellata (shark)	ca. 20
Lepidosiren paradoxa (S. African lungfish)	19
Petromyzon marinus (lamprey eel)	84
Salmo satar (Atlantic salmon)	30
Salmo trutta (brown trout)	40
Amphibia	
Ambystoma tigrinum (salamander)	14
Amphiuma means (Congo eel)	12
Triturus cristatus (newt)	12
Bufo bufo (common toad)	11
Hyla arborea (tree frog)	12
Rana catesbiana (bullfrog)	13
Reptilia	
Alligator mississippiensis	16
Chelonia sp. (turtles)	28
Natrix sp. (water snakes)	20
Aves	
Columbia livia (pigeon)	40
Sirinus canarius (canary)	ca. 40
Numida meleagris (guinea hen)	33
Mammalia	
Macropus rufus (kangaroo)	10
Trichosurus vulpecula (opossum)	10
Bos taurus (cattle)	30
Canis familiaris (dog)	39
Equus caballos (horse)	32
Felis catus (cat)	19
Ovis aries (sheep)	27
Rattus rattus (rat)	21
Sus scrofa (pig)	20

there has been a parallel specialization of chromatin. Some chromosomes have become specialized in being concerned exclusively, or in part, with sex determination, and there are many kinds of chromosomally based sex-determining mechanisms. Specialization is also seen in that chromatin can divide into two kinds: euchromatin containing the genes that govern readily recognizable traits, and heterochromatin, which may be inactive genetically, regulatory in some as yet undetermined way, or concerned with their formation of rRNA in the nucleolar-organizing region, a subdivision of specialization within heterochromatin.

The amount of DNA in some cells cannot, however, be explained totally by an increase in the number of structural genes, that is, those genes coding for proteins of an enzymatic or structural nature (Table 12.6). For example, the lungfish, *Amphiuma*, and *Necturus* among the Amphibia, and *Vicia faba* among the legumes, have many times greater amounts of DNA per nucleus than their very close relatives. Part of this may be due to gene redundancy, that is, genes of particular kinds may be represented in the genome by more than one copy. Structural genes are generally represented by nonredundant genes, but those for the several kinds of rRNAs, tRNAs and histones have been shown to be highly redundant, sometimes with thousands of copies of the same gene present in a single

TABLE 12.5 HAPLOID CHROMOSOME NUMBERS OF SOME SPECIES OF PRIMATES, INCLUDING HUMANS.

Prosimii		Anthropoidae	
Tupaiidae (tree shrews)		*Cebidae* (New World monkeys)	
Tupaia glis	30	*Lagothrix ubericola* (wooly monkey)	31
T. chinensis	31	*Ateles* sp. (spider monkeys)	17
T. minor	33	*Alouatta seniculus* (red howler)	22
T. montana	34	*A. caraya* (black howler)	26
T. palowensis	26	*Cebus capucinus* (organ-grinder monkey)	27
Lemuridae (lemurs)		*Saimiri sciureus* (squirrel monkey)	22
Lemur macaco	22	*Callimico jacchus* (marmoset)	23
L. catta	28	*Cercopithecidae* (Old World monkeys)	
L. mongoz	30	*Macaca mulatta* (Rhesus macaque)	21
L. variegatus	23	*Papio sphinx* (mandrill)	21
Lorisidae (loris and bushbaby)		*P. hamadryas* (baboon)	21
Nycticebus coucang (slow loris)	25	*Presbytis entellus* (common langur)	22
Loris tardigradus (slender loris)	31	*Colobus polykomos* (colobus monkey)	22
Galago senegalensis (lesser bushbaby)	19	*Nasulis larvatus* (proboscis monkey)	24
G. cassicaudatus (thick-tailed bushbaby)	31	*Cercopithecus aethiops* (green vervet monkey)	30
Tarsioidae (tarsiers)		*C. mitis* (guenon)	36
Tarsius bancanus	40	*Hominoidae* (great apes)	
T. syrichta	40	*Hylobates lar* (gibbon)	22
		Symphalangus syndactylus (siamang)	25
		Pongo pigmaeus (orangutan)	24
		Gorilla gorilla	24
		Pan troglodytes (chimpanzee)	24
		Hominidae	
		Homo sapiens (humans)	23

chromosome. How this has come about is not fully understood, but it is an obvious device for maintaining a rich supply of RNAs and proteins for biosynthesis purposes. In addition to these redundant genes, there is also in some species DNA whose function is not known but whose presence can account for a substantial amount of the DNA of each cell. In the fly, *Drosophila virilis*, for example, a "simple sequence" DNA has been recognized; each block consists of seven nucleotides and is represented 1 million or more times in the same genome. Similar sequences have been found in other species as well, much if not all of it located in centric heterochromatin. Other redundant sequences of DNA consist of families of similar sequences, each family of many units differing from other families, and all scattered throughout the genome rather than being concentrated in particular chromosomes. The function of these families is not known, although a regulatory function has been suggested.

The eukaryotic genome is a complex one, and not all of the evolutionary innovations that have occurred are understood. But the evolutionary potential residing in the eukaryotic cell can be dramatized by pointing out that it took a far longer period of time for the eukaryotic cell to evolve from a prokaryotic ancestor(s) (about 2 billion years) than for the earliest eukaryotic cell to evolve into

TABLE 12.6 HAPLOID CHROMOSOME NUMBERS AND DNA VALUES PER HAPLOID CELL AND PER CHROMOSOME (GIVEN IN TERMS OF NUMBERS OF NUCLEOTIDES) IN A GROUP OF REPRESENTATIVE ORGANISMS, WITH A RANGE OF FIGURES OF MITOCHONDRIA, CHLOROPLASTS, AND VIRUSES INCLUDED FOR COMPARATIVE PURPOSES.

Organism	Chromosome number	DNA/cell	DNA/ chromosome
Mitochondria	1	$5.2 \times 10^4 - 1.0 \times 10^6$	Same
Chloroplasts	1	$1.8 \times 10^5 - 2.8 \times 10^7$	Same
Tobacco necrosis satellite virus (RNA)	1	1.3×10^3	Same
DNA poxviruses	1	5.3×10^4	Same
Bacillus subtilis (bacterium)	1	1.3×10^7	Same
Escherichia coli (colon bacterium)	1	$9.0 \times 10^6 - 3.4 \times 10^7$	Same
Hemophilus influenzae (influenza bacterium)	1	4.0×10^7	Same
Anacystis nidulans (blue-green alga)	?	1.2×10^7	?
Oscillatoria linosa (blue-green alga)	?	1.6×10^{10}	?
Spirogyra setiformis (green alga)	4	7.0×10^9	1.8×10^9
Euglena gracilis (green flagellate)	45	5.8×10^9	1.3×10^8
Aspergillus nidulans (black mold)	8	8.8×10^7	1.1×10^7
Neurospora crassa (pink bread mold)	7	8.6×10^7	1.2×10^7
Osmunda cinnamomea (cinnamon fern)	22	9.6×10^{10}	4.4×10^9
Pinus strobus (white pine)	12	8.4×10^{10}	7.0×10^9
Lilium longiflorum (Easter lily)	12	1.1×10^{11}	9.2×10^9
Chrysanthemum sp.	18	1.0×10^{10}	5.6×10^8
Drosophila melanogaster (fruit fly)	4	1.7×10^8	4.2×10^7
Gryllus domesticus (cricket)	11	1.1×10^{10}	1.0×10^9
Esox lucias (pike)	9	1.7×10^9	1.9×10^8
Protopterus sp. (lungfish)	17	1.0×10^{11}	1.4×10^{10}
Rana pipiens (leopard frog)	13	1.4×10^{10}	1.1×10^9
Necturus maculosus (mudpuppy)	12	1.7×10^{11}	1.4×10^{10}
Boa constrictor	18	3.5×10^9	1.9×10^8
Gallus domesticus (chicken)	39	2.3×10^9	5.7×10^7
Mus musculus (mouse)	20	6.5×10^9	3.2×10^8
Homo sapiens (humans)	23	6.0×10^9	2.6×10^8

humans (about 900 million years). Prokaryotic species show very little differentiation of cells, the heterocysts of the blue-green algae being one of the few kinds of differentiated cells. Multicellularity of an organized sort is also absent. The blue-green algae exhibit a pseudomulticellularity in the form of filaments or flat sheets of cells, but the cells are held together by their gelatinous sheaths and possess no true intercellular connections through plasmodesmata or pits. Evolutionary change in the prokaryotic system took the form of varied metabolic pathways, whereas the eukaryotic system retained, for the most part, a common aerobic metabolism but showed great potential in exploiting aspects of form and size. This was made possible by the acquisition of multicellularity, intercellular communication, and the potential for cellular differentiation as a developmental phenomenon, features which in turn required the evolution of genomic control systems. We are just beginning to realize the intricacies and complexities of this kind of cellular behavior.

BIBLIOGRAPHY

BOGORAD, L. 1975. Evolution of organelles and eukaryotic genomes. *Science 188:* 891–898.

CALVIN, M. 1975. Chemical evolution. *Am. Sci. 63:* 169–177.

COHEN, S. 1973. Mitochondria and chloroplasts revisited. *Am. Sci. 61:* 437–443.

EHRENSVARD, G. 1962. *Life: Origin and Development.* University of Chicago Press, Chicago.

EIGEN, M., GARDINER, W., SCHUSTER, P., and WINKLER-OSWATITSCH, R. 1981. The origin of genetic information. *Sci. Am. 244:* 88–118.

FOX, S. W. 1975. Looking forward to the present. *BioSystems 6:* 165–175.

FOX, S. W., and DOSE, K. 1972. *Molecular Evolution and the Origin of Life.* W. H. Freeman & Co., San Francisco.

GIBBS, M. (ed.) 1971. *The Structure and Function of Chloroplasts.* Springer-Verlag, New York.

KIRK, J. T. O., and TILNEY-BASSETT, R. A. E. 1967. *The Plastids.* W. H. Freeman & Co., San Francisco.

LIMA-DE-FARIA, A. (ed.) 1969. *Handbook of Molecular Cytology.* North Holland Publishing Co., Amsterdam.

MARGULIS, L. 1970. *Origin of Eukaryotic Cells.* Yale University Press, New Haven.

———. 1971. Symbiosis and evolution. *Sci. Am. 225:* 48–57.

———. 1974. On the evolutionary origin and possible mechanism of colchicine-sensitive mitotic movements. *BioSystems 6:* 16–36.

MILLER, S. L., and ORGEL, L. E. 1974. *The Origins of Life on Earth.* Prentice-Hall, Inc., Englewood Cliffs, N.J.

NASS, S. 1970. The significance of the structural and functional similarities of bacteria and mitochondria. *International Review of Cytology 28:* 55.

PICKETT-HEAPS, J. D. 1969. The evolution of the mitotic apparatus: an attempt at comparative ultrastructural cytology in dividing cells. *Cytobios 1:* 257.

———. 1974*a.* The evolution of mitosis and the eukaryotic condition. *BioSystems 6:* 37–48.

———. 1974*b*. *Structure, Reproduction and Evolution in Some Green Algae.* Sinauer Associates, Stamford, Conn.

PONNAMPERUMA, C. (ed.). 1972. *Exobiology.* North Holland Publishing Company, Amsterdam.

RAFF, R. A., and MAHLER, H. R. 1972. The non-symbiotic origin of mitochondria. *Science 177:* 575–582.

RAVEN, P. H. 1970. A multiple origin for plastids and mitochondria. *Science 169:* 641.

SAGER, R. 1972. *Cytoplasmic Genes and Organelles.* Academic Press, New York.

SLOBODKIN, L. B. 1964. The strategy of evolution. *Am. Sci. 52:* 342–353.

UZZELL, T., and SPOLSKY, C. 1974. Mitochondria and plastids as endosymbionts: a revival of special creation? *Am. Sci. 62:* 334–343.

Appendix A

TEXTBOOKS AND MONOGRAPHS OF GENERAL CYTOLOGICAL INTEREST

ALBERTS, B., BRAY, D., LEWIS, J., RAFF, M., ROBERTS, K. and WATSON, J. D. 1983. *Molecular Biology of the Cell.* Garland Publishing, New York.

AVERS, C. J. 1981. *Cell Biology,* (2nd ed.). Van Nostrand, Co., New York.

BECKER, W. M. 1977. *Energy and the Living Cell.* J. B. Lippincott, Co., Philadelphia.

DEROBERTIS, E. D. P. and DEROBERTIS, E. M. F., JR. 1980. *Cell and Molecular Biology.* Saunders, Philadelphia.

FAWCETT, D. W. 1981. *The Cell,* (2nd ed.). W. B. Saunders Co., Philadelphia.

GALL, J. G., PORTER, K. R. and SIEKEWITZ, P. (eds.) 1981. Discovery in cell biology, *J. Cell Biology, 91,* No. 3, pt 2.

GUNNING, G. E. S. and STEER, M. W. 1975. *Ultrastructure and the Biology of Plant Cells.* Arnold, London.

HALL, J. L., FLOWERS, T. J. and ROBERTS, R. M. 1982. *Plant cell structure and metabolism.* Longman, New York.

KARP, G. 1979. *Cell Biology.* McGraw-Hill, New York.

LEDBETTER, M. C. and PORTER, K. R. 1970. *An Introduction to the Fine Structure of Plant Cells.* Springer-Verlag, New York.

LEHNINGER, A. L. 1982. *Principles of Biochemistry.* Worth, New York.

PRESCOTT, D. M. and GOLDSTEIN, L. *Cell Biology—A Comprehensive Treatise.* Academic Press, New York.

STRYER, L. 1981. *Biochemistry,* (2nd ed.). W. H. Freeman, San Francisco.

STUMPF, P. K. and CONN, E. E. (eds.) *The Biochemistry of plants—a comprehensive treatise.* Vol. 1. The Plant Cell, ed. Tolbert, N. W. Academic Press, New York.

WATSON, J. D. 1976. *The Molecular Biology of the Gene,* 3rd ed. W. A. Benjamin, Menlo Park, California.

WOLFE, S. L. 1981. *Biology of the Cell,* (2nd ed.). Wadsworth Publishing Co., Inc., Belmont, California.

APPENDIX B

REVIEWS CONTAINING ARTICLES OF CYTOLOGICAL INTEREST

Annual Review of Biochemistry. Ann. Rev. Inc., Palo Alto, California.
Annual Review of Genetics. Ann. Rev. Inc., Palo Alto, California.
Annual Review of Physiology. Ann. Rev. Inc., Palo Alto, California.
Annual Review of Plant Physiology. Ann. Rev. Inc., Palo Alto, California.
Cold Spring Harbor Symposia on Quantitative Biology, Cold Spring Harbor Laboratory, New York.
- Vol. 31. 1966. *The Genetic Code*
- Vol. 33. 1968. *Replication of DNA in Microorganisms*
- Vol. 34. 1969. *The Mechanism of Protein Synthesis*
- Vol. 35. 1970. *Transcription of Genetic Material*
- Vol. 38. 1973. *Chromosome Structure and Function*
- Vol. 42. 1977. *Chromatin*
- Vol. 42. 1978. *DNA: replication and recombination*
- Vol. 46. 1981. *Organization of the Cytoplasm*
- Vol. 47. 1982. *DNA structures*

International Review of Cytology, Academic Press, New York.

Appendix C

JOURNALS CONTAINING ARTICLES OF CYTOLOGICAL INTEREST

American Scientist
Biology of the Cell
BioScience
Cell
Cell Motility
Cellular and Molecular Biology
Chromosoma
Developmental Biology
European Journal of Cell Biology
Experimental Cell Research
Journal of Cell Biology
Journal of Cell Science
Journal of Cellular Physiology
Journal of Histochemistry and Cytochemistry
Journal of Membrane Biology
Journal of Molecular Biology
Journal of Ultrastructure Research
Molecular and Cellular Biochemistry
Nature
Proceedings of the National Academy of Science
Protoplasma
Science
Scientific American

INDEX